Range Development and Improvements

Second Edition

John F. Vallentine
Professor of Range Science
Brigham Young University

Brigham Young University Press
Provo, Utah

Library of Congress Cataloging in Publication Data

Vallentine, John F
    Range development and improvements.

    Includes bibliographies and index.
    1. Range management. 2. Pastures. I. Title.
SB199.V32   1979   633.2'02   79-26676
ISBN 0-8425-1708-1

International Standard Book Number: 0-8425-1708-1

Brigham Young University Press, Provo, Utah 84602
© 1971 by Brigham Young University Press. All rights reserved
Second Edition 1980
Printed in the United States of America

80   2.5Mc   41398

Range Development and Improvements

Second Edition

Preface

Many aspects of range development and improvement have been extensively researched in the United States and Canada since World War II. This has resulted in a rather voluminous literature on this phase of range science. But the data has remained widely scattered in various journals, bulletins, handbooks, and research reports. Although individual chapters in a limited number of textbooks have dealt with range improvements and closely related subjects, no known effort has previously been made to bring this information together under one cover. *Range Development and Improvements* has as its goal the survey of this scientific field, the selection of principles and results, and organizing this information into a useful and readily accessible form.

In developing this textbook and reference manual, attempts have been made to include both principles and practices. Principles have been emphasized since these are the basis of solving the varied problems found on rangeland and of achieving high levels of productivity. Rigid formulas lack sufficient flexibility for uniform application. However, stepwise procedures recommended in certain instances have been included to exemplify the adaptation of principles to applied range improvement.

Basic principles provided in this manual should allow its use over broad geographical areas. However, regional aspects over the United States and adjacent areas in Canada and Mexico have been emphasized. Local aspects of range improvement have frequently been referred to as a means of making geographical association.

It is hoped this book will prove useful not only as a textbook but also as a reference manual for ranchers, range technicians, public land administrators, agribusiness personnel, educators, and students.

Range improvement is a rapidly advancing phase of range management. Researchers in the state agricultural experiment stations, the Forest and Range Experiment Stations, the SEA—Agricultural Research, and various private research foundations are continually adding to information available in this field.

The Society for Range Management through its 5500 membership has actively promoted the advancement of grazing land management including improvements. Additional professional organizations associated with weed science, ecology, agronomy, animal science, wildlife science, soil conservation, and farm and ranch management have also promoted certain aspects of range improvements.

A unique interagency committee active in the field of range-improvements is the Range Seeding Equipment Committee. This committee was organized to evaluate and develop better equipment and methods for range seeding, weed and brush control, and related range improvements. In cooperation with private and governmental equipment-development centers, many important contributions have been made on range-improvement equipment. Membership of this committee has come principally from the public land managing agencies but also from various universities, research organizations, and commercial enterprises.

The original idea for a textbook on range improvements was provided by Dr. L. A. Stoddart and Dr. C. Wayne Cook, former associates of the author in the Department of Range Science, Utah State University. Teaching courses in range improvements at Utah State University and Brigham Young University and associated extension and research work by the author in the Intermountain Region and in the Great Plains provided the impetus for developing this idea into an active plan. However, the basis of this text is the written reports of many people in range science and related fields.

Helpful comments and suggestions on the content and format of *Range Development and Improvements* were received from many reviewers. Many of the pictures and drawings used in the text were provided by various agencies, organizations, and individuals, and acknowledgment is made in the respective figure captions. Appreciation is expressed to Wayne Patton, a former range graduate student at Brigham Young University, for making many of the drawings not otherwise credited in the text. And finally, appreciation is expressed to members of the Brigham Young University Press for their initiative and extra efforts made in developing a quality publication from the original manuscript.

The Author

Contents

Chapter 9

Special range seeding and treatment techniques

Chapter 10
Range fertilization

Chapter 11
Rodent and insect control

Chapter 12
Range animal handling facilities

Chapter 1

Planning range improvements

The Role of Range Improvements

Range management can be defined as the art and science of planning and directing the use of rangelands to obtain optimum, sustained returns based on the objectives of land ownership and on the needs and desires of society [46]. Definitions of range management* have commonly given domestic livestock production and big game a high priority while recognizing other goods and services that can be provided by rangelands [2, 38]. Scientific range management stands on the premise that vegetation can be used perpetually for grazing while simultaneously providing society with high quality air and water, open space, and recreation [42].

Range improvements are special treatments, developments, and structures used to improve range forage resources or to facilitate their use by grazing animals. Range seeding, control of undesirable range plants, applying fertilizer, and pitting, furrowing, and water-spreading are direct means of developing and improving range forage resources. Range improvements such as stockwater developments, fences, and trails provide the means for more effective management of grazing and thus indirect improvement of the forage resources. They also enable more efficient utilization of the forage resource and thus increase animal production.

In planning range improvements and in realizing their full potential benefits, a working concept of range must be realistically broad in definition. *Range* is defined as uncultivated grasslands, shrublands,

---

*When viewed within the ecosystem framework, range management has been defined as *management of a renewable natural resource composed of one or more range ecosystems for the optimum, sustained yield of the optimum combination of goods and services* [23].

1

or forested lands with an herbaceous and/or shrubby understory, particularly those areas producing forage for grazing or browsing by domestic and wild animals [46]. Range includes lands with native vegetation cover, but also lands naturally or artificially revegetated with native or even adapted, introduced forage plant species not requiring periodic reestablishment, and subsequently managed like native range.

Rangelands are highly diversified and include meadows, grasslands, shrublands, and deserts. Ranges may be treeless or consist of understory plants beneath open forests. Logging and burning of dense forests provide rangeland of temporary to extended longevity by allowing the growth of herbs and shrubs.

In contrast with *range,* which implies permanence, *cultivated pasture* (which includes irrigated perennial pasture and agronomic rotation pasture, annual pasture, and crop aftermath pasture) is cropland used temporarily for the production of pasture [45]. Cultivated pasture has commonly been differentiated from rangeland pasture on the basis of receiving more intensive management—close control of grazing animals and frequent use of cultural treatments such as tillage, mowing, weed and brush control, irrigation, and fertilization. However, the intensification of management and improvement practices on rangelands in recent years have materially reduced these differences. This trend is expected to continue.

Rangelands actually grazed by livestock comprise 43.5% of the land area of the contiguous 48 states, according to the Forest-Range Environmental Study (FRES) report [40]. Another 19.5% is forest-range not grazed by livestock but having grazing potential. The study revealed that range provided 213 million animal unit months (*aums*) of livestock grazing in 1970.

The FRES report concluded that grazing capacity could be increased by 49% to 317 aums by extensive management systems and techniques, including fencing and water developments to obtain uniform livestock distribution and plant use and to maintain plant vigor. It was further concluded that grazing capacity could be increased an additional 117% to 566 aums annually through intensive range improvements such as reseeding, noxious plant control, fertilization, and mechanical range treatments, while still maintaining the environment and providing for multiple use.

Improvements in
Range Ecosystems

Range improvements must be based on ecological principles, particularly competition and succession. A first step in improving range

2

Table 1. Induced Progression by Manipulation of Factors in Range Ecosystems*

Factors	Manipulation
Controlling factors	
Climatic elements	Weather modification, *burning*
Geological materials (edaphic)	*Waterspreading,* land leveling, terracing, *fertilization,* ground water recharge, drainage
Available organisms (plants and animals)	*Species introduction,* elimination, and genetic improvement
Dependent factors	
Consumers	
Wild fauna	*Grazing management,* wildlife management, *insect and rodent control*
Livestock	*Grazing management* (through handling facilities), livestock management
Vegetation	*Plant control, revegetation,* hay management, plant disease control
Soil	*Mechanical treatments, nitrogen fertilization*
Decomposers and transformers	Direct manipulation probably not feasible
Microclimate	Shades, shelters, *mulch manipulation*

*Table adapted from Lewis [23]. Induced progression evaluated in terms of increasing ecosystem productivity from the standpoint of both quality and quantity. Manipulative practices emphasized in *Range Development and Improvements* are italicized.

forage resources is providing the desirable forage species with a competitive advantage for water, sunlight, and soil nutrients. The reduction of competition from undesirable plants through biological or herbicidal control is a means of inducing succession in the desired direction. But complete replacement of one plant community by another is often the basis of range seeding.

Range ecosystems comprise plant and animal communities along with the abiotic factors of soil, topography, water, temperature, air, and solar energy. The outputs of range ecosystems are many but include forage, fish and wildlife, livestock, water, air, recreation, landscape, and open space. Man is part of the ecosystem. But man is also a directing influence capable of manipulating the productivity of the ecosystem to his advantage. Range improvement is principally involved in manipulating factors leading to increased productivity from rangelands. (See Table 1.)

Lewis [23] has recommended the following as the basis for managing the range resource: "Induced progression must be our aim and the management equilibrium our goal." This suggests that the productivity and biological efficiency presently being obtained from range ecosystems can be greatly increased. It also suggests that range

3

improvement cannot be increased indefinitely because the controlling factors, which man either cannot or should not manipulate because of economic constraints, place ceilings on productivity obtainable from range ecosystems. The rate of induced progression is highly variable and is determined by (1) the kind of range ecosystem, (2) the extent of depletion, (3) climatic fluctuations, (4) the improvement plan put into operation, and (5) the efficiency of subsequent management [23].

Range improvements are frequently used to restore depleted ranges to higher levels of productivity. However, range improvements are not limited to restoration or rehabilitation of ranges in low condition. Fertilization, waterspreading, and herbicides are means of increasing productivity beyond pristine conditions. Where undesirable brush species are a major part of the climax vegetation, replacing this portion of the climax vegetation with desirable, productive forage species is a means of raising the forage productivity ceiling previously imposed by natural factors. However, some means of keeping the climax brush in control is a necessity.

Range Improvements Aid Management

Range improvements have many management implications. It is important that range improvements be made a part of the planning and directing of range use rather than being considered separately. Range improvements are probably best considered as special aids

FIG. 1. *Range developments and improvements must be part of the planning and directing of range use rather than considered separately.* (Soil Conservation Service photo of Cherry County, Nebraska)

available for achieving the objectives of range management. For example, only slight improvement can be expected on many big sagebrush cattle ranges, even after good grazing management for fifteen to thirty years, unless special brush-control practices are applied [29].

However, the full expected benefits from special range treatments, developments, and structures are realized only when accompanied by good grazing management. The ability of range improvements on ranches to pay back total costs depends on inherent soil capacities, climate, and price of livestock. Readjustments in grazing use such as season of use, stocking rates, distribution of grazing, or class or stage of production of livestock must often be made before range improvements can become profitable.

Range improvement plans must provide for plant vigor and rapid establishment of desirable plants following treatment if range condition is to be improved. This often requires from six months to a year of complete protection from grazing following plant control and two to three consecutive growing seasons following range seeding. Control of big game as well as livestock grazing is often required. On heavily grazed ranges where desirable grasses are found only under the protection of the brush, removal of the brush without adjustment of grazing will only further eliminate the desirable plants.

Evaluating the Desirability of Range Plants

Before the principles of competition and succession can be applied in the management of grazing lands and be accelerated beneficially by range improvements, resident plants as well as those capable of being introduced must be evaluated as to desirability. Changes in plant cover induced by range improvements must be determined in advance to be necessary and desirable. The following terms are commonly used in evaluating range plants:

Feed—any noninjurious, edible material having nutritive value for animals.

Forage—herbaceous feeds available for grazing by livestock or big game animals or harvested for feeding; also commonly used to include browse.

Herbage—nonwoody, flowering plants taken collectively; commonly used in the same sense as forage except that it may include material unacceptable or unavailable to animals.

Browse—leaf and twig growth of shrubs, woody vines, and trees available for animal consumption, usually based on current year's growth. Greater attention is now being given to useful shrubs in range improvement programs than previously.

Pasturage (or pasture)—plant materials harvested directly by grazing animals.

Forage value—a subjective evaluation of forage plants giving consideration to some or all of the following: palatability, length of palatable period, nutritive value, and productivity under grazing. (Commonly classified as good, fair, and poor.)

Noxious plant—a plant which is undesirable in light of planned land use or which is unwholesome to rangelands or range animals. (Not to be confused with plant species arbitrarily defined by state law as noxious weeds because they are especially undesirable, troublesome, and difficult to control—generally referred to as *primary noxious* weeds.)

Weed (agronomic use)—any plant growing out of place.

Brush—shrubs or small trees considered undesirable from the standpoint of planned use of the area.

Poisonous plant—a plant which injures animal health when taken internally because of a toxic agent present.

Proper classification of desirability or undesirability must consider why a plant species is desirable, how desirable, when desirable, where desirable, and for what is it desirable. Although tall larkspur (*Delphinium* spp.)* is highly poisonous to cattle, it provides palatable, nutritious forage for sheep and deer. Some woody plants classified as brush on cattle range may be desirable plants on deer range or sheep range. In the Intermountain Region big sagebrush (*Artemisia tridentata*) is considered undesirable on most mountain summer ranges for cattle, sheep, and deer. Although used to a considerable extent by sheep on fall and winter range, big sagebrush in large amounts on livestock range is generally considered undesirable because it greatly reduces the production of the better forage plants, prevents grazing of grasses growing underneath sagebrush, hampers movement of livestock, especially sheep, snags wool from fleeces, causes lambs to stray and become lost, and aids predators [29].

In Utah studies each 1 percent increase in sagebrush cover decreased crested wheatgrass (*Agropyron desertorum*) production by 6.6 pounds per acre, or each sagebrush plant per 50-foot line transect decreased crested wheatgrass production by 12.1 pounds per acre [8]. Big sagebrush control has increased carrying capacity for livestock by two to twenty-five times [29]. However, on deer winter range big sagebrush is considered one of the desirable forage species. On some sites, association with sagebrush increases forb production. But where sagebrush has usurped the site and excluded understory species, the

*Scientific names of all plant species referred to in the text are given in the "Common Name–Scientific Name Index of Plants" near the end of the book.

6

stand should be thinned to permit recovery of grasses and forbs for spring use by deer [32]. In Idaho sagebrush control reduced nesting of sage grouse on treated areas but had little effect on use of these areas by young broods [22].

Based on seeding studies with crested wheatgrass in southern Idaho, Hull and Klomp [20] concluded that all big sagebrush should be killed where not needed by wildlife or domestic livestock. They found that any remaining sagebrush suppressed grass growth while producing seed for reinvasion. The last remaining sagebrush reduced grass production the most. Killing the last 25 percent of the big sagebrush plants more than doubled the grass production over killing only the first 75 percent of the brush. When only 75 percent of the sagebrush plants were killed, the remaining plants rapidly increased in size and canopy cover.

Extensive areas of rangeland in southern Texas are occupied by dense stands of pricklypear (*Opuntia* spp.) and its net value is controversial (Figure 2). The disadvantages of pricklypear are (1) reduction of grass production, (2) low forage value and the high volume of pricklypear herbage required for animal maintenance, (3) development of "pear eaters" among cattle, (4) increased difficulty of managing livestock, and (5) protection for undesirable gophers and rats. However, uses of pricklypear have been listed as (1) emergency livestock feed, (2) food for some species of wildlife, (3) half-developed pads sometimes eaten as a human delicacy, (4) fruit used to make jelly, candy, or syrup, and (5) decorations and landscaping. It can also (6) prevent soil erosion to some extent and (7) protect grasses when range is in poor condition and overstocked [18].

FIG. 2. *Pricklypear being eaten by cattle as emergency feed near Zapata, Texas, after the spines have been burned off with a backpack burner. Note that range is in very poor condition, has been heavily grazed, and reduced to a cover of low quality plants.* (Soil Conservation Service photo)

7

Many woody plants are important as part of the natural habitat in providing wildlife with food, nesting and roosting sites, dens, and cover. On ranches where deer and other wildlife provide recreation and income for ranchers, selective control is recommended to leave beneficial woody plants in proper amounts [34] (Figure 22, page 73). Selective control is generally the policy on public grazing lands. It is probable that beneficial uses will be found for many woody plants and forbs now considered undesirable. The introduction and genetic improvement of exotic species of grazing animals to make efficient use of brush species shows some promise.

On the Edwards Plateau of Texas, grazing studies were made with cattle, sheep, goats, and deer grazing in common on fenced range [44]. Where oak occurred in open stands that did not greatly restrict grazing, it was concluded that beneficial results were obtained by leaving live oak (*Quercus virginiana*), an evergreen species, and white shin oak (*Quercus sinuata* var. *breviloba*), a moderately palatable species. It was also concluded that Spanish oak (*Q. texana*), post oak (*Q. stellata*), and Ashe juniper (*Juniperus ashei*) were undesirable and should be selectively cleared. In a drought year during the study, deer died from apparent starvation even when large amounts of juniper were available for grazing.

Dense brush has sharply decreased understory herbage production in many areas. Rangeland in Arizona with no pinyon-juniper overstory has produced 600 pounds of herbage annually per acre but only 100 pounds when the overstory increased to 60 percent and 50 pounds when it increased to 80 percent [1]. Carrying capacity on grass-shrub range in southern Arizona with 25 mesquite (*Prosopis juliflora*) plants or less per acre was twenty-five cattle per section in 1922 but was reduced to seven animals per section by 1947 afer mesquite had increased on the same area to 150 plants per acre [35]. Thinning overstocked ponderosa pine forests in eastern Washington increased understory production from 75 pounds to 417 pounds per acre by eight years after treatment [24]. The increase of native forage included 51 percent grass, 37 percent forbs, and 12 percent shrubs.

On Missouri Ozark range, one treeless acre can produce as much forage as forty to sixty acres of hardwood forest (1,800 pounds versus 30 pounds of herbage per acre) [14]. It was concluded that when hardwood forests in the Missouri Ozarks are thinned to produce more forage, thinning well over 50 percent is necessary to get large increases in herbage production [15]. In the Southeast, 400 pine trees per acre decreased annual production of air dry grass herbage to 400 pounds per acre compared to 450 pounds with 225 pines and 850 pounds with 30 pines per acre [37]. However, thinning of trees to increase forage production should also be evaluated in terms of production of forest products.

Evaluating the desirability of range plants on a species basis alone may be inadequate. It should be recognized that within plant species generally considered undesirable are often found good subspecies and ecotypes that have unusually high palatability or productivity. Although big sagebrush and big rabbitbrush are generally unpalatable to cattle, certain ecotypes have been found to have unusually high palatability. Certain juniper trees have been observed to have been closely browsed by deer while other plants of the same species nearby were apparently avoided. Natural genetic variations within western wheatgrass (*Agropyron smithii*) may account for conflicting reports on its palatability.

Benefits from Range Improvements

Although there may be only one primary objective sought in a specific range improvement program, there are usually one or more secondary benefits of range improvement. Possible benefits from range improvements include:

1. *Increased quantity of forage.* The production, accessibility, and maintenance of an adequate supply of high quality forage is the basis for a successful ranch operation and the production of range livestock and big game (Figure 3). The balancing of seasonal grazing capacity, reducing pressure on overstocked ranges, or replacing grazing capacity lost in reductions in federal grazing permits are problems that increased forage production could solve. Plans for increasing forage production must consider seasonal use. For example, spring-fall carrying capacity frequently limits cattle and sheep production on Intermountain ranches [8]. But, for mule deer over much of the Intermountain Region, improvement of winter range is considered the most critical need [32].

2. *Increased quality of forage.* Providing forage of greater palatability, of higher nutritive content, or of longer green growth period is often desired. However, quality of forage for domestic livestock is important only as it adds to gains or saleable products from the range. The goal of game range restoration may be the development of a balance of browse for winter grazing and herbaceous plants for succulent forage for late winter and early spring use [32].

3. *Increased animal production.* This is often the primary goal and considers increased numbers of animals, greater number of

9

offspring weaned per breeding female, increased weaning weights of lambs and calves, increased size and condition of big game, increased fleece weights, or reduced death losses. The removal of undesirable brush may reduce wool pulled from fleeces.

4. *Facilitate handling of and caring for range animals; keep livestock more tame and docile; reduce number of breeding males required.* This can be accomplished by brush control, range fencing, corrals, stockwater developments and trails. Increasing hunter access to big game ranges through building access roads provides a means of controlling big game numbers. Brush control may reduce lamb losses from straying and predation.

FIG. 3. *Range development increased grazing capacity threefold on a bottomland site in Texas.* Top, *dense infestation of mesquite before treatment;* bottom, *after mesquite control, moderate stocking, and waterspreading.* (Photos by Texas A&M University Research Station, Spur)

5. *Control poisoning of livestock by poisonous plants.* This can be accomplished by removing poisonous plants, replacing existing vegetation with nonpoisonous species, or by providing alternative sources of palatable, nonpoisonous forage. Also, injury and associated diseases and parasites can be reduced by removing mechanically injurious species. On one grazing allotment on the Manti-La Sal National forest in Utah, tall larkspur (*Delphinium barbeyi*) killed annually about 8 percent of the permitted cattle from 1956 through 1959, representing a loss of $14,126 each year [36]. A survey of this large allotment in 1960 revealed that tall larkspur occurred in scattered patches totaling only 343 acres. On this basis, ranchers could afford to spend $41 per acre of larkspur annually to prevent poisoning.

6. *Reduced fire hazard.* Possibilities include prescribed burning of forest slash or other flammable material during low hazard periods, replacing species such as big sagebrush and cheatgrass (*Bromus tectorum*) with less flammable species, and constructing fire guards.

7. *Increased water yields on watershed by replacing woody species with herbaceous plants.* Studies in Colorado revealed that water is yielded to stream flow on the average only after aspen sites receive 14 inches but by grass types after receiving only 8.3 inches [41]. Replacement of chaparral on the deeper upland soils in California [4] and on canyon-bottom brush woodland by grass [17] increased water yields. Brush or phreatophyte control often activates springs or increases spring flow [5].

8. *Replacement control*—control of insects and small animals by replacing their host plants, often weeds and ephemeral plants, with desirable forage species. An example is the replacement of certain weeds which host the beet leafhopper, which in turn is a carrier of curly top disease of tomatoes [31]. Another example is the replacement of bulbous and tuberous plants, which encourage high pocket gopher populations, with fibrous-root grasses.

9. *Control erosion by stabilizing erosive soils.* On low potential or frail sites, soil stabilization may justify restoration with only secondary consideration given to forage production.

10. *Reduce conflicts between multiple uses of range resources.* Access roads can permit better distribution of livestock as well as proper harvesting of big game by hunters. Reseeding denuded ranges can provide needed forage for livestock and clear water for fishing streams. Williamson and Currier [48] have indicated that applied landscape management enables natural beauty to

11

be retained and even enhanced while accomplishing the basic objectives of mechanical brush control programs (Figure 4).

The visual landscape is a basic resource of rangelands, and the application of landscape management concepts and principles to the visual aspects of range improvements should be considered, particularly on public lands but on private lands also. The U.S. Forest Service [43] has suggested three basic concepts in evaluating visual range improvement and management practices:

(1) Landscape has an identifiable character, regardless of the size or segment being viewed.
(2) Landscapes with visual variety are more desirable and appealing than monotonous landscapes.
(3) Deviations from a characteristic landscape vary in degree of contrast and can usually be designed to achieve visually acceptable variety.

The Forest Service report [43] concluded that range structures and vegetation manipulation can add visually acceptable variety and may provide the only variety in an otherwise monotonous landscape. It was pointed out that ranching is a part of our cultural heritage, and that range structures (such as fences and corrals), water developments, and livestock can enhance the visual environment. However, it was recommended that brush piles not needed for wildlife cover be burned or removed, or at least shaped to complement the landscape. The visual impact of root wads and other large debris should be minimized.

FIG. 4. *Landscape management applied to brush control in New Mexico to enhance natural beauty while increasing forage production.* (U.S. Forest Service photo)

On the more productive portions of the California chaparral, the native brushlands can be converted to a grassland cover that is fairly stable under continuous maintenance [4]. In addition to increasing water yields, other multiple use advantages in converting selected areas of chaparral to grassland have been (1) more manageable units for effective fire control by breaking up expansive brush areas, (2) new habitat for wildlife, (3) greater access for hunting and other forms of recreation, and (4) new range for domestic livestock [17]. Chaparral conversion to grass increased the frequency of soil slips by five times and area involved in soil slips by five times on steep slopes (those over 55 percent). However, it was concluded that areas with slopes under 40 percent were relatively safe for conversion [10].

Selecting Range Improvements

The type of range improvement must be carefully considered and be properly located and utilized to give maximum benefits. Guidelines to consider in selecting and locating range improvements include:

1. *Use only proven methods except on small-scale trial basis.* Undertake large-scale projects only where practical, effective, and economical procedures can be used. The high costs of some range improvements often discourage ranchers from doing an adequate or complete job, thus resulting in unsatisfactory results. Information on new range improvement techniques can be obtained from university, experiment station, extension, and other range scientists.

2. *Range improvements must be compatible with the goals of ownership.* These goals may be similar on private and public grazing lands or may differ rather widely.

3. *Availability of local or contract labor, needed equipment, and supervisory or consultative assistance needed.* On-the-ground assistance in planning range improvements can be provided ranchers by personnel of the SCS and extension services or professional range consultants.

4. *Evaluate what range improvements can be most effectively utilized in the herd or range management plan.* Determine the factors limiting animal production that can be solved by range improvements.

5. *Changes in management practices that will be required, and maintenance that will be needed* in order to obtain full benefits from the range improvement practice.

6. *Expected cost-benefit ratios.* Range improvements judged to offer the greatest returns on investments should be given priority. However, the risk factor of failure should be weighed against the potential returns on investment.

7. *Apply range improvements at appropriate stages of range deterioration.* Invading poisonous plants should be treated when they first appear since control is generally much less costly and more effective at that time. However, native undesirable plants not invading new sites can generally be economically controlled by fixed cost methods only after their density seriously reduces production of forage species. Fixed cost methods such as aerial spraying of herbicides and some newer mechanical control methods using large machines have nearly the same per acre costs regardless of the amount of tree cover [21]. However, treatment costs per acre of variable-cost methods such as individual tree burning, bulldozing, or spraying increase with increasing stand density.

8. *Amount and character of residual forage cover.* Is the existing forage plant stand adequate to satisfactorily respond to range treatments, or will seeding be additionally required?

9. *Locate range developments on areas of greatest potential for increasing range productivity* (Figure 5). Areas with shallow or infertile soil, low site potential, low rainfall, or steep topography often produce too little forage to justify expensive treatments. Soils that contain 1 percent soluble salts—particularly sodium—are considered unsuitable for big game restoration in Utah [32]. Avoid areas where range treatment may seriously increase the

FIG. 5. *Silty range site in Nebraska with high potential for forage production. The lack of perennial grasses and predominance of weeds will require intensive treatment before it can be made productive, however.*

risk of accelerated erosion or interfere with livestock or big game management.

10. *Plan livestock handling facilities that are practical and beneficial both to the rangeland and to the range animals.* Do not create new grazing distribution problems by spot treatments such as herbicide spraying, mowing, or fertilizing or improper location of livestock handling facilities.

After the type of range improvement most needed is decided upon, there are often several alternative methods or procedures from which to select. Additional stockwater on a particular range might be provided by developing existing springs or seeps or by constructing reservoirs, digging wells, or hauling water. Alternative methods of controlling big sagebrush include planned burning, spraying with herbicides, plowing or disking, anchor chaining, railing, pipe harrowing, or mowing, with each method having its own limitations and advantages [29].

The best combination of range improvements rather than the best single range improvement practice should generally be considered. Range seedings generally require prior control of perennial vegetation and often additional fencing, stockwater developments, fertiliza-

FIG. 6. *Chamise chaparral converted to grassland on a selected, productive site by controlled burning, artificial seeding, and herbicide application.* (U.S. Forest Service photo)

15

tion, and prevention of reinvasion of undesirable plants. The usual steps in converting brushland to grassland in California include (1) preparation of brush for burning by clearance of firebreaks and sometimes mashing with a bulldozer, (2) controlled burning, (3) reseeding with adapted grasses, and (4) application of herbicides to control regrowth [30] (Figure 6).

Selecting the area for location of any range improvement is particularly important. Considering Utah rangelands, it was estimated in 1955 that only about 5 percent of the area was topographically suited, had sufficient precipitation and good soil, and was sufficiently devoid of good natural forage to make complete seedbed preparation and artificial seeding economically justified [29]. However, with present day advancements in seeding methods, including only partial vegetation replacement, this figure could undoubtedly be considerably expanded. But in portions of the Northern Great Plains, precipitation, soils, and site characteristics allow a majority of the land to be artificially seeded. The improvement of intensively treated areas is often associated with the natural recovery of adjacent but untreated areas within the same pasture.

In converting chaparral in California to grassland, researchers have suggested the areas selected first be those with the best combination of soil depth and productivity, steepness of slope, suitability for grazing, feasibility of economical conversion, advisability of conversion, kind of brush and control methods needed, and location for fire control [4]. In considering tractor-use safety on the contour, drill use, grass establishment, and grazability, it was concluded that slope gradients of 40 percent or less should be considered and 30 percent or less on easily erodible soils.

Disking and seeding subalpine range on the Wasatch Plateau in Utah greatly improved usable forage for cattle but had no significant effect on infiltration or soil stability when compared with adjacent untreated range seven years after treatment [26]. However, this treatment did decrease organic matter and capillary porosity in the surface soil, increased soil bulk density, and decreased plant litter cover. Temporary removal of vegetational cover in a reseeding program on highly erosive areas requires that special precautions be taken against erosion during the interim period of establishment. Great diversity of site characteristics commonly occurs even locally, and this adds to the difficulty in selecting the best range improvement practices.

The complexity of brush and weed problems and control methods generally makes it imperative that control be conducted as part of a complete plan of range restoration and use. Plant control is often

16

complex because of the many species involved, often in mixtures but of different climax status and requiring different treatment. Control of one species may release another to become a major problem. Killing big sagebrush on Intermountain ranges by some mechanical methods releases rabbitbrush (*Chrysothamnus* spp.) and creates a more difficult problem since the latter is unpalatable to livestock, difficult to eradicate, and quick to invade deteriorated ranges [9]. Sites in Texas may be cleared of mesquite by root plowing, only to be rapidly invaded by pricklypear unless special precautions are taken [34].

Plant control methods are often only partially effective, and reinfestation frequently requires additional treatment. Many shrubs are good browse plants, particularly for big game, and provide cover for wildlife; but untreated brush areas are important sources of seed to reinfest treated areas [34].

Some ranges in relatively low condition for livestock may support many big game animals such as mule deer because they utilize large quantities of browse [11]. Many range improvements that aid in livestock production also aid in the production of big game. However, most range improvements have been planned on the basis of giving priority considerations to only one or two species of grazing animals. The impact of range improvements on all species of grazing animals should be considered. Not only do elk, deer, antelope, and other wildlife inhabit many private as well as public ranges, but they are rapidly becoming an important source of ranch income in many sections of the range country.

Economics of
Range Improvements

Private versus public sectors. Range improvements offer many opportunities for increasing ranch profits. In contrast to relatively low returns on total ranch investment in recent years, sound investments in range improvements are often capable of producing annual returns of 10 to 25 percent or even more. However, many types of range improvements are expensive. Before beginning such improvements, ranch owners and managers must carefully consider whether the proposed improvements will be profitable.

Range improvements offer many opportunities in the private sector for correcting seasonal imbalance in carrying capacity, meeting special forage needs on the ranch, or correcting problems associated with grazing management. Additional forage provided through range development will be worth a great deal more if made available during a season of short feed supply rather than in a season of plentiful feed.

Ranches must be large enough in land and livestock to be physically and economically efficient and provide a satisfactory income for the owner or owner-manager. Even when the percentage of calf or lamb crops and weaning weights is high and annual unit costs of carrying a ewe or cow are acceptable, inefficient use of labor, machinery, and managerial ability is characteristic of small ranches. The rancher is then faced with the decision of whether to expand externally through land purchase or internally though intensification of production factors.

Additional rangeland is often difficult to acquire as well as being uneconomically priced. High prices for additional land with its associated overhead expenses such as taxes, fencing, and water development often make purchasing land an unattractive approach to increasing ranch size. Development and improvement of range and other forage lands already part of the ranch is generally a less costly way of increasing grazing capacity and thus effective ranch size. The value of added grazing capacity that improves economies of scale may be considerably higher than their base value alone.

Society has also expressed its interest in improving private range lands by providing cost-sharing programs to encourage the application of conservation practices on private lands. Cost-sharing assistance is available for many types of range improvements on private and Indian lands through the Agricultural Conservation Program administered by the Agricultural Stabilization and Conservation Service (ASCS) with technical assistance from the Soil Conservation Service (SCS). Also, cost-sharing assistance is available through programs administered directly by the SCS such as the Great Plains Program.

Since these government cost-sharing programs commonly cover from 30 to 60 percent of the initial cost of selected range improvements, the availability of approved practices and supporting funds should be considered as a means of reducing the private costs of range improvements. Reduced interest-rate loans are also made available by some state governments for improvement of privately owned or state-lease rangelands.

Range improvements on public lands generally benefit both the users and the general public. For this reason range users of Forest Service and Bureau of Land Management lands such as livestock graziers, state wildlife agencies, water users, and even sportsmen and recreationists often participate in the initial cost or maintenance of range improvements. However, since funds, equipment, and manpower needed for range improvements are limiting factors on public grazing lands, the cost-benefit ratio must be carefully considered here as well as on private lands.

Although range improvements may increase receipts from public grazing lands, other society benefits including extra-market returns generally receive first consideration. Such benefits are often difficult to evaluate in terms of dollars and cents, but are nevertheless real and include maintaining environmental quality, strengthening rural economies, and providing recreational opportunities [42]. Improved watershed conditions, reduced erosion, enhanced esthetics, and the improvement of big game and other wildlife habitat are other socio-economic benefits of range improvements. Range improvements often provide the principal means of coordinating livestock and big game grazing with the many other multiple uses of public range-lands.

Restrictions other than economic are commonly placed on range improvements on public range lands. Such restrictions may be legislative or executive in origin, or court ordered. The coordination of multiple uses or the designation of a dominant use other than grazing through administrative decision may restrict or prevent certain types of range improvements on public range lands. The requirement of environmental impact statements before proceeding may delay or greatly modify range improvement plans. More recently another concern—and a potentially serious obstacle to range improvements on public lands if not properly guided—has been the protection of threatened and endangered plant and animal species.

The Public Land Law Review Commission [33] has recommended that deteriorated or frail rangelands retained in federal ownership should be protected from further deterioration and rehabilitated where possible with the use of federal funds. However, the Commission recommended that investments made in higher quality lands for better forage conditions should be shared between the federal government and the users on the basis of identifiable benefits to each. It further recommended that the rancher be protected in his range improvement investments on public lands and be credited for his investment as he pays his grazing fees. A final recommendation was that range improvement and rehabilitation funds not be limited to a portion of the grazing fees collected, but that they be appropriated from general funds on the basis of need.

Determining economic returns. The economic benefits from range improvements should be carefully estimated before funds are expended. Expected rates of return, risk of failure, and availability and source of capital must all be considered. The final decision for making range improvements on private lands must eventually be based on a comparison of the added costs of the project with the expected added market returns.

19

The market value of an animal unit month of grazing at the prevailing rental rate is one index to value of increased forage production [28]. Several proposed range treatments can be compared on the basis of cost per unit increase in grazing capacity. For short term effects from fertilizer, the cumulative increase in forage production has been used as a direct measure of benefits [71]. However, McCormick and Workman [25] used hay replacement value to evaluate nitrogen fertilization of seeded spring range based on the finding that 25 to 30 pounds of nitrogen per acre hastened spring range readiness by 11 to 13 days. However, they noted that this approach did not consider the additional benefits of increased total grazing capacity or possibilities of carrying over advantages into the second year.

When range improvements are expected to permanently increase grazing capacity, one approach to figuring profitability has been to capitalize the present annual value of increased grazing capacity at the going interest rate [6]. Break-even production rate increases have also been used in decision making. For example, Dahl et al. [12] calculated that an 80 percent root kill of a minimum of 30 percent honey mesquite canopy cover was required to provide the necessary increases in perennial grass yield to pay the cost of spraying.

The quandry of whether all or only part of the increased herbage production resulting from range improvement should be calculated as increased grazing capacity is a definite problem in using this shortcut approach. Where the original stand of forage plants was basically left intact, but the range improvement practice increased forage production from this stand, Bayoumi and Smith [3] posed two contrasting approaches to evaluating the increased forage production:

(1) All of the increase is available for use, since it represents an additional increment over that otherwise available for grazing and for protection of the forage plants.
(2) The increase should be reduced by an appropriate percentage to maintain productivity of plants.

The former approach assumes that the amount of ungrazed herbage or unbrowsed current twig growth remaining is of primary importance. The latter approach assumes that protection requirements are a percentage function of herbage or twig growth. These workers assumed that the former is more nearly correct on normally productive ranges where restoring the vigor of plants was unnecessary. However, allowances for ungrazed herbage production necessary to promote or maintain forage plant vigor must be realistic, whether a partial or complete conversion of the vegetation on the site has been made or forage production of the resident stand increased.

Caton et al. [6] have pointed out that the time lapse between incurring costs and realizing benefits is an important economic factor to consider in evaluating range improvements. The time lapse often results in a direct cost in interest on the investment and may have substantial indirect costs in deferred income or reorganization of livestock operations while waiting for the treated range to be ready for use.

Multiple economic benefits often result from range improvement. For example, planting wheatgrasses on range in poor condition, in addition to increasing grazing capacity, may also increase per head livestock performance and reduce losses from poisonous plants. Two methods of estimating the profitability of a range improvement practice based on comparing the added costs of the practice with the added returns from the practice over time are as follows [28]:

(1) Discount* the added future returns to enable comparison with initial cost plus capitalized annual operation and maintenance costs.
(2) Amortize** the initial cost and add additional annual operation and maintenance costs to compare with the expected additional annual return.

The period of amortization or discounting should not exceed the expected life of the range improvement project. Even if the improvement has the potential of long-term benefits, this period should not normally be extended over a period of thirty years. If money to make the improvement is borrowed, the length of the loan will determine the period of amortization or discounting.

Partial budgeting. Partial budgeting and the use of internal rates of return is probably the most universally adapted method of determining the profitability of investing in range improvements. Providing the projected range improvement will not necessitate complete ranch reorganization or a substitution of a different earning enterprise for the present livestock enterprise, a partial budget can be used, and requires only that the additional costs and returns associated with the range improvement practice be included. Partial budgets are much easier to prepare than complete earning enterprise budgets and are sufficient for most range improvements.

The decision of whether or not to invest in range improvements is simplified when based on the internal rate of return [27]. The inter-

*Discounting is the process of determining the present value of a stream of future returns.

**Amortization is the process of paying initial costs plus subsequent interest costs over a repayment period, usually in equal annual (or sometimes biennial or monthly) installments.

nal rate of return is, in fact, the discount rate which brings the discounted future returns equal to the cost of obtaining the income stream. This approach circumvents the problem of having to select an arbitrary interest rate or discount rate. It allows the residual or additional annual income from the project to be expressed as a true rate of return when compared with the initial investment.

Table 2 shows the procedures and format used in partial budgeting of expected costs and returns from range improvements, using range seeding as an example. It also shows the steps for obtaining an internal rate of return through the use of Table 3. The decision of whether or not to make the investment is based on comparing the internal rate of return with current money lending rates or with the opportunity cost* of using owned money.

The costs of range improvements including range seeding can be broken down into (1) initial investment (fixed costs), including direct and indirect costs, (2) interest on initial investment during an amortization period (an indirect but variable cost), and (3) the annual using costs. The direct costs of seeding include some or all of the following: (1) removing unwanted vegetation, (2) seedbed preparation, (3) seed, (4) planting, (5) new fencing, (6) new stockwater developments, (7) pest control during establishment, and (8) overhead.

The indirect costs of seeding include (1) risk cost** of seeding failure, (2) nonuse during establishment, and (3) interest on direct costs during nonuse. Additional using costs of seeding may include (1) additional fence maintenance, (2) additional stockwater maintenance or water hauling, (3) renovation of seeded stands including weed control, fertilization, and insect and rodent control, (4) interest and taxes on investment in extra livestock, (5) additional livestock operating expenses, and (6) increase in land taxes (if any). To omit any of these costs from consideration will be to underestimate the cost of seeding and to overestimate the net returns from the seeding project.

Sources of cost and return data. Research can supply information relative to expected yield and animal gain advantages from range treatments. However, Wight [47] has pointed out that the economic feasibility of a range improvement practice, such as fertilization, cannot be determined by research alone, since it is subject to the fluctuations of prices in the livestock industry and needs of individ-

---

Opportunity cost is the return given up by not putting the factor of production to a different use. (For example, the owned money could have been put in an insured savings account or used to buy government bonds rather than invested in the range improvement.)

**Risk cost is the average percent of the initial investment in the range seeding that must be added for reseeding following seeding failures.

ual ranching situations. Caton et al. [6] recognized that a large task remained in determining the costs of range improvement but concluded that the most formidable task is the evaluation of benefits. They pointed out that nonmarket benefits, in particular, are difficult to evaluate.

The best source of price and cost data are current, local sources. Costs of range improvements can be obtained from conservation contractors, public land management agencies, SCS and Extension Service, ACP offices, SWCD supervisors, agribusiness representatives, or even neighboring ranchers. Current and projected livestock market information will be required to evaluate benefits in terms of increased livestock production.

Caton et al. [6] suggested that average costs of range improvements over a wide area have meaning only within limits. However, they concluded that costs for representative situations can be used as guides to probable costs, provided the resources, the operation sequence, and the cost accounting procedure are fully identified.

Cost figures quoted throughout this book have been provided primarily for comparative rather than absolute values. Although inflation during the last 30 years has raised the costs of most range improvements, tabular material on costs (standardized to a given year) is useful in making projections and comparisons.

Gray et al. [16] provided costs and returns from range improvements in New Mexico, Arizona, and Utah for the base year 1961. Duran and Kaiser [13] tabulated investment costs for 18 selected range improvement practices used in the FRES study. These values were determined from published and unpublished sources—including U.S. Forest Service field personnel—and standardized to 1970. Horvath et al. [19] made a survey to determine local costs of range improvements prevailing in early 1976 in the northern Rocky Mountain states and estimated the value of existing improvements.

Table 2. Partial Budgeting for Range Improvements

1. Select a range improvement project with possible income advantage to the ranch. Estimate lifetime expectancy of the project.

Seed 640 acres of poor condition range to Russian Wild rye. Twenty years life expectancy. Expected to increase grazing capacity from 8 to 4 acres / AUM. When grazed from April 20 to June 20, expected to increase ave. daily gains of yearlings from 1.6 to 2.0 lbs.

23

2. Estimate the initial investment that will be required, including direct costs and indirect costs.

(a) Plowing – 640a. @ $12.00 7680
(b) Drilling – 640 a. @ $ 2.⁰⁰ 1280
(c) Seed – 6 lb./acre @ $1.20¢ x 640 a. . . . 4608
(d) Fencing - 1 mile @ $ 1500 1500
(e) Spring Development 600
(f) Non-Use: 80 AUM's @ $ 7.00 x 3 years . . 1680
(g) Interest during non-use period
    ($17,348 @ 10% for 3 years) 5204
(h) Risk of failure (20% $5888) 1178
                              TOTAL 23,730

3. Determine the net additional annual income expected from the project as follows:
I. Plus factors affecting income
  (a) Additional annual returns

      1 - 62 head x 60 days x .4 lb. additional daily gain
        @ 75¢ per lb. $ 1115
      2 - 62 additional yearlings sold – 700 lb.
        @ 75¢ per lb. $ 32,550 $ 33,665

  (b) Reduced annual costs

      1 - Water hauling - 62 head
        @ $1.⁰⁰ per mo. x 2 mo. $124

                          124
  (c) Total plus factors $ 33,789
II. Minus factors affecting income
  (d) Additional annual costs (using costs)
    1 - 62 head x 580 lb. @ 75¢ per lb. $26,970
    2 - Fence maintenance   150
    3 - Spring Development   60
    4 - Veterinary - 62 head @ 1.00   62
    5 - Salt - 62 head x 8 lb. @ 2¢   10
    6 - Interest 580 lb x 62 head x
             75¢ per lb. x 3/12 year @ 11% 742
                         $27,994

    (e) Reduced annual returns

1 - Death loss - 62 head x 700 lb. x
75¢ per lb. x .5% death loss   163

    (f) Total minus factors   *$ 28,157*

    (g) Net additional annual income   *$ 5,508*

4. Determine the appropriate table factor for use with Table 3 as follows:

$$\text{Table factor} = \frac{\text{Initial investment}}{\text{Net additional annual return}}$$

$$F = \frac{23,730}{5,508} = 4.31$$

5. Using the table factor determined above and the estimated lifetime expectancy of the project, determine the appropriate internal rate of return from Table 3, Table of Factors for Use in Computing Internal Rate of Return. Enter from the left and read along the row to the entry closest to the table factor determined and note the percentage internal rate of return at the top of the column. Interpolate for a more precise percentage.

Internal rate of return is about 23%.

6. Compare the calculated percentage internal rate of return with the alternative uses of owned capital or with interest charges if the investment money is to be borrowed. Make the decision on initiating the proposed range improvement project on the basis of this comparison. If the internal rate of return exceeds the alternative uses of capital and/or the lending rate, the range improvement project under consideration would be concluded to be profitable.

With real estate loans at 10%, it is concluded that the range seeding project should be completed. A net profit of about 13% annually is calculated.

7. Describe any extra-market returns that should be considered in making the final decision. These include conservation, stabilization of the ranching industry, increased feed for game, aesthetics, etc.

Erosion prevalent on part of the area to be seeded would be controlled. Some Spring-Fall grazing by deer is expected.

Table 3. Table of Factors for Use in Computing Internal Rate of Return (Discounting)
(Present Value of $1 Received Annually for N Years)

Year (N)	1%	2%	4%	6%	8%	10%	12%	14%	15%	16%	18%
1	0.990	0.980	0.962	0.943	0.926	0.909	0.893	0.877	0.870	0.862	0.847
2	1.970	1.942	1.886	1.833	1.783	1.736	1.690	1.647	1.626	1.605	1.566
3	2.941	2.884	2.775	2.673	2.577	2.487	2.402	2.322	2.283	2.246	2.174
4	3.902	3.808	3.630	3.465	3.312	3.170	3.037	2.914	2.855	2.798	2.690
5	4.853	4.713	4.452	4.212	3.993	3.791	3.605	3.433	3.352	3.274	3.127
6	5.795	5.601	5.242	4.917	4.623	4.355	4.111	3.889	3.784	3.685	3.498
7	6.728	6.472	6.002	5.582	5.206	4.868	4.564	4.288	4.160	4.039	3.812
8	7.652	7.325	6.733	6.210	5.747	5.335	4.968	4.639	4.487	4.344	4.078
9	8.566	8.162	7.435	6.802	6.247	5.759	5.328	4.946	4.772	4.607	4.303
10	9.471	8.983	8.111	7.360	6.710	6.145	5.650	5.216	5.019	4.833	4.494
11	10.368	9.787	8.760	7.887	7.139	6.495	5.988	5.453	5.234	5.029	4.656
12	11.255	10.575	9.385	8.384	7.536	6.814	6.194	5.660	5.421	5.197	4.793
13	12.134	11.343	9.986	8.853	7.904	7.103	6.424	5.842	5.583	5.342	4.910
14	13.004	12.106	10.563	9.295	8.244	7.367	6.628	6.002	5.724	5.468	5.008
15	13.865	12.849	11.118	9.712	8.559	7.606	6.811	6.142	5.847	5.575	5.092
16	14.718	13.578	11.652	10.106	8.851	7.824	6.974	6.265	5.954	5.669	5.162
17	15.562	14.292	12.166	10.477	9.122	8.022	7.120	6.373	6.047	5.749	5.222
18	16.398	14.992	12.659	10.828	9.372	8.201	7.250	6.467	6.128	5.818	5.273
19	17.226	15.678	13.134	11.158	9.604	8.365	7.366	6.550	6.198	5.877	5.316
20	18.046	16.351	13.590	11.470	9.818	8.514	7.469	6.623	6.259	5.929	5.353
21	18.857	17.011	14.029	11.764	10.017	8.649	7.562	6.687	6.312	5.973	5.384
22	19.660	17.658	14.451	12.042	10.201	8.772	7.645	6.743	6.359	6.011	5.410
23	20.456	18.292	14.857	12.303	10.371	8.883	7.718	6.792	6.399	6.044	5.432
24	21.243	18.914	15.247	12.550	10.529	8.985	7.784	6.835	6.434	6.073	5.451
25	22.023	19.523	15.622	12.783	10.675	9.077	7.843	6.873	6.464	6.097	5.467
26	22.795	20.121	15.983	13.003	10.810	9.161	7.896	6.906	6.491	6.118	5.480
27	23.560	20.707	16.330	13.211	10.935	9.237	7.943	6.935	6.514	6.136	5.492
28	24.316	21.281	16.663	13.406	11.051	9.307	7.984	6.961	6.534	6.152	5.502
29	25.066	21.844	16.984	13.591	11.158	9.370	8.022	6.983	6.551	6.166	5.510
30	25.808	22.396	17.292	13.765	11.258	9.427	8.055	7.003	6.566	6.177	5.517
40	32.835	27.355	19.793	15.046	11.925	9.779	8.244	7.105	6.642	6.234	5.548
50	39.196	31.424	21.482	15.762	12.234	9.915	8.304	7.133	6.661	6.246	5.554

Table 3. (Continued)

Year (N)	20%	22%	24%	25%	26%	28%	30%	35%	40%	45%	50%
1	0.833	0.820	0.806	0.800	0.794	0.781	0.769	0.741	0.714	0.690	0.667
2	1.528	1.492	1.457	1.440	1.424	1.392	1.361	1.289	1.224	1.165	1.111
3	2.106	2.042	1.981	1.952	1.923	1.868	1.816	1.696	1.589	1.493	1.407
4	2.589	2.494	2.404	2.362	2.320	2.241	2.166	1.997	1.849	1.720	1.605
5	2.991	2.864	2.745	2.689	2.635	2.532	2.436	2.220	2.035	1.876	1.737
6	3.326	3.167	3.020	2.951	2.885	2.759	2.643	2.385	2.168	1.983	1.824
7	3.605	3.416	3.242	3.161	3.083	2.937	2.802	2.508	2.263	2.057	1.883
8	3.837	3.619	3.421	3.329	3.241	3.076	2.925	2.598	2.331	2.108	1.922
9	4.031	3.786	3.566	3.463	3.366	3.184	3.019	2.665	2.379	2.144	1.948
10	4.192	3.923	3.682	3.571	3.465	3.269	3.092	2.715	2.414	2.168	1.965
11	4.327	4.035	3.776	3.656	3.544	3.335	3.147	2.752	2.438	2.185	1.977
12	4.439	4.127	3.851	3.725	3.606	3.387	3.190	2.779	2.456	2.196	1.985
13	4.533	4.203	3.912	3.780	3.656	3.427	3.223	2.799	2.468	2.204	1.990
14	4.611	4.265	3.962	3.824	3.695	3.459	3.249	2.814	2.477	2.210	1.993
15	4.675	4.315	4.001	3.859	3.726	3.483	3.268	2.825	2.484	2.214	1.995
16	4.730	4.357	4.033	3.887	3.751	3.503	3.283	2.834	2.489	2.216	1.997
17	4.775	4.391	4.059	3.910	3.771	3.518	3.295	2.840	2.492	2.218	1.998
18	4.812	4.419	4.080	3.928	3.786	3.529	3.304	2.844	2.494	2.219	1.999
19	4.844	4.442	4.097	3.942	3.799	3.539	3.311	2.848	2.496	2.220	1.999
20	4.870	4.460	4.110	3.954	3.808	3.546	3.316	2.850	2.497	2.221	1.999
21	4.891	4.476	4.121	3.963	3.816	3.551	3.320	2.852	2.498	2.221	2.000
22	4.909	4.488	4.130	3.970	3.822	3.556	3.323	2.853	2.498	2.222	2.000
23	4.925	4.499	4.137	3.976	3.827	3.559	3.325	2.854	2.499	2.222	2.000
24	4.937	4.507	4.143	3.981	3.831	3.562	3.327	2.855	2.499	2.222	2.000
25	4.948	4.514	4.147	3.985	3.834	3.564	3.329	2.856	2.499	2.222	2.000
26	4.956	4.520	4.151	3.988	3.837	3.566	3.330	2.856	2.500	2.222	2.000
27	4.964	4.524	4.154	3.990	3.839	3.567	3.331	2.856	2.500	2.222	2.000
28	4.970	4.528	4.157	3.992	3.840	3.568	3.331	2.857	2.500	2.222	2.000
29	4.975	4.531	4.159	3.994	3.841	3.569	3.332	2.857	2.500	2.222	2.000
30	4.979	4.534	4.160	3.995	3.842	3.569	3.332	2.857	2.500	2.222	2.000
40	4.997	4.544	4.166	3.999	3.846	3.571	3.333	2.857	2.500	2.222	2.000
50	4.999	4.545	4.167	4.000	3.846	3.571	3.333	2.857	2.500	2.222	2.000

Table 4. Table of Factors for Amortizing Investments (Loans) into Equal Annual Payments ($1 Units)*

Years	Interest rates (percent per year)										
	5%	6%	7%	8%	9%	10%	11%	12%	13%	14%	15%
1	1.05000	1.06000	1.07000	1.08000	1.09000	1.10000	1.11000	1.12000	1.13000	1.14000	1.15000
2	0.53781	0.54544	0.55310	0.56077	0.56847	0.57620	0.58394	0.59170	0.59949	0.60729	0.61512
3	0.36721	0.37411	0.38106	0.38804	0.39506	0.40212	0.40922	0.41635	0.42353	0.43074	0.43798
4	0.28202	0.28860	0.29523	0.30193	0.30867	0.31548	0.32233	0.32924	0.33620	0.34321	0.35027
5	0.23098	0.23740	0.24390	0.25046	0.25710	0.26380	0.27058	0.27741	0.28432	0.29129	0.29832
6	0.19702	0.20337	0.20980	0.21632	0.22292	0.22961	0.23638	0.24323	0.25016	0.25716	0.26424
7	0.17282	0.17914	0.18556	0.19208	0.19870	0.20541	0.21222	0.21912	0.22612	0.23320	0.24037
8	0.15473	0.16104	0.16747	0.17402	0.18068	0.18745	0.19433	0.20131	0.20839	0.21558	0.22286
9	0.14070	0.14703	0.15349	0.16008	0.16680	0.17365	0.18061	0.18768	0.19487	0.20217	0.20958
10	0.12951	0.13587	0.14238	0.14903	0.15583	0.16275	0.16981	0.17699	0.18429	0.19172	0.19926
15	0.09635	0.10297	0.10980	0.11683	0.12406	0.13148	0.13907	0.14683	0.15475	0.16281	0.17102
20	0.08025	0.08719	0.09440	0.10186	0.10955	0.11746	0.12558	0.13388	0.14236	0.15099	0.15977
25	0.07096	0.07823	0.08582	0.09368	0.10181	0.11017	0.11875	0.12750	0.13643	0.14550	0.15470
30	0.06506	0.07265	0.08059	0.08883	0.09734	0.10608	0.11503	0.12415	0.13342	0.14281	0.15231
35	0.06108	0.06898	0.07724	0.08581	0.09464	0.10369	0.11293	0.12232	0.13183	0.14145	0.15114
40	0.05828	0.06647	0.07501	0.08387	0.09296	0.10226	0.11172	0.12131	0.13099	0.14075	0.15057
45	0.05627	0.06471	0.07350	0.08259	0.09191	0.10140	0.11102	0.12074	0.13054	0.14039	0.15028
50	0.05478	0.06345	0.07246	0.08175	0.09123	0.10086	0.11060	0.12042	0.13029	0.14021	0.15014

*Figures are for combined principal and interest payments payable at the end of each year.

Literature Cited

1. Arnold, Joseph F., Donald A. Jameson, and Elbert H. Reid. 1964. The pinyon-juniper type of Arizona: effects of grazing, fire, and tree control. USDA, Agric. Res. Serv. Production Res. Rpt. 84.
2. ASRM, Range Term Glossary Comm. 1964. A glossary of terms used in range management. American Society of Range Management, Portland, Oregon. (New address: 2760 West Fifth Ave., Denver, Colo.)
3. Bayoumi, Mohamed A., and Arthur D. Smith. 1976. Response of big game winter range vegetation to fertilization. J. Range Mgt. 29(1):44–48.
4. Bentley, Jay R. 1967. Conversion of chaparral areas to grassland. USDA Agric. Handbook 328.
5. Biswell, H. H. 1954. The brush control problem in California. J. Range Mgt. 7(2):57–62.
6. Caton, Douglas D., Chester O. McCorkle, and M. L. Upchurch. 1960. Economics of improvement of western grazing land. J. Range Mgt. 13(3):143–151.
7. Clary, Warren P. 1975. Range management and its ecological basis in the ponderosa pine type of Arizona: the status of our knowledge. USDA, For. Serv. Res. Paper RM-158.
8. Cook, C. Wayne. 1958. Sagebrush eradication and broadcast seeding. Utah Agric. Expt. Sta. Bul. 404.
9. Cook, C. Wayne, Paul D. Leonard, and Charles D. Bonham. 1965. Rabbitbrush competition and control on Utah rangelands. Utah Agric. Expt. Sta. Bul. 454.
10. Corbett, Edward S., and Raymond M. Rice. 1966. Soil slippage increased by brush conversion. USDA, For. Serv. Res. Note PSW-128.
11. Costley, R. J., P. F. Allan, Odell Julander, and D. I. Rasmussen. 1948. Wildlife, a resource of the range. *In Grass*, 1948 USDA Yearbook of Agric., pp. 243–248.
12. Dahl, B. E., R. E. Sosebee, J. P. Goen, and C. S. Brumley. 1978. Will mesquite control with 2,4,5-T enhance grass production? J. Range Mgt. 31(2):129–131.
13. Duran, Gilbert, and H. F. Kaiser. 1972. Range management practices: investment costs, 1970. USDA Agric. Handbook 435.
14. Ehrenreich, John H., and Robert F. Buttery. 1960. Increasing forage on Ozark wooded range. USDA, Central States Forest Sta. Tech. Paper 177.
15. Ehrenreich, J. H., and J. S. Crosby. 1960. Herbage production is related to hardwood crown cover. J. Forestry 58(7):564–565.
16. Gray, James R., Thomas M. Stubblefield, and N. Reith Roberts. 1965. Economic aspects of range improvements in the Southwest. New Mexico Agric. Expt. Sta. Bul. 498.
17. Hill, Lawrence W., and Raymond M. Rice. 1963. Converting from brush to grass increases water yield in southern California. J. Range Mgt. 16(6):300–305.
18. Hoffman, G. O., and R. A. Darrow, 1964 (Rev.). Pricklypear—good or bad? Texas Agric. Ext. Serv. Bul. 806.
19. Horvath, Joseph, Dennis Schweitzer, and Enoch Bell. 1978. Grazing on national forest system lands: costs of increasing capacity in the Northern Region. USDA, For. Serv. Res. Paper INT-215.
20. Hull, A. C., Jr., and G. J. Klomp. 1974. Yield of crested wheatgrass under four densities of big sagebrush in southern Idaho. USDA Tech. Bul. 1483.
21. Jameson, Donald A. 1971. Optimum stand selection for juniper control on southwestern woodland ranges. J. Range Mgt. 24(2):94–99.
22. Klebenow, Donald A. 1970. Sage grouse versus sagebrush control in Idaho. J. Range Mgt. 23(6):396–400.
23. Lewis, James K. 1969. Range management viewed in the ecosystem framework. Chapter VI. *In* the ecosystem concept in natural resource management. Academic Press, New York and London.
24. McConnell, Burt R., and Justin G. Smith. 1970. Response of understory vegetation to ponderosa pine thinning in eastern Washington. J. Range Mgt. 23(3):208–212.

25. McCormick, Paul W., and John P. Workman. 1975. Early range readiness with nitrogen fertilizer: an economic analysis. J. Range Mgt. 28(3):181–184.
26. Meeuwig, Richard O. 1965. Effects of seeding and grazing on infiltration capacity and soil stability of a subalpine range in central Utah. J. Range Mgt. 18(4):173–180.
27. Nielsen, Darwin B. 1967. Economics of range improvements. Utah Agric. Expt. Sta. Bul. 466.
28. Olson, Carl E., William A. Daley, and Charles C. McAfee. 1977. An economic evaluation of range resource improvement. Wyo. Agric. Expt. Sta. Bul. 650.
29. Pechanec, Joseph F., A. Perry Plummer, Joseph H. Robertson, and A. C. Hull, Jr. 1965. Sagebrush control on rangelands. USDA Agric. Handbook 277.
30. Perry, Chester A., Cyrus M. McKell, Joe R. Goodin, and Thomas M. Little. 1967. Chemical control of an old stand of chaparral to increase range productivity. J. Range Mgt. 20(3):166–169.
31. Piemeisel, Robert L. 1954. Replacement control: changes in vegetation in relation to control of pests and diseases. Bot. Rev. 20(1):1–32.
32. Plummer, A. Perry, Donald R. Christensen, and Stephen B. Monsen. 1969. Restoring big game range in Utah. Utah Div. of Fish & Game Pub. 68-3.
33. Public Land Law Review Commission. 1970. One third of the nation's land. U.S. Govt. Printing Office, Washington, D.C.
34. Rechenthin, C. A., H. M. Bell, R. J. Pederson, and D. B. Polk. 1964. Grassland restoration. II. Brush control. USDA, Soil Cons. Serv., Temple, Texas.
35. Reynolds, Hudson G., and S. Clark Martin. 1968 (Rev.). Managing grass-shrub cattle ranges in the Southwest. USDA Agric. Handbook 162.
36. Richman, Lavar M. 1961. Economics of controlling tall larkspur. Master's Thesis, Utah State University, Logan, Utah.
37. Smith, L. F., R. S. Campbell;, and Clyde L. Blount. 1958. Cattle grazing in longleaf pine forests of south Mississippi. USDA, Southern Forest Expt. Sta. Occasional Paper 162.
38. Stoddard, L. A. 1967. What is range management? J. Range Mgt. 20(5):304–307.
39. Stoddard, L. A., and Arthur D. Smith. 1955. Grassland improvement: the Intermountain Region. J. Agric. and Food Chem. 3(4):303–305.
40. USDA, Forest-Range Task Force. 1972. The nation's range resources—a forest-range environmental study. USDA, For. Serv. For. Resource Rep. 19.
41. USDA, Rocky Mountain Forest & Range Expt. Sta. 1959. Research on Black Mesa—a progress report. USDA, Rocky Mountain Forest and Range Expt. Sta. Paper 41.
42. USDA, Forest Service. 1970. Range ecosystem research—the challenge of change. USDA Agric. Info. Bul. 346.
43. USDA, For. Serv. 1977. National forest landscape management, volume 2, chapter 3, range. USDA Agric. Handbook 484.
44. Vallentine, John F. 1960. Live oak and shin oak as desirable plants on Edwards Plateau ranges. Ecology 41(3):545–548.
45. Vallentine, John F. 1978. More pasture or just range for rangemen? Rangeman's J. 5(2):37–38.
46. Vallentine, John F., and Phillip L. Sims. 1980. Range science—a guide to information sources. Gale Res. Co., Detroit, Mich.
47. Wight, J. Ross. 1976. Range fertilization in the Northern Great Plains. J. Range Mgt. 29(3):180–185.
48. Williamson, Robert M., and W. F. Currier. 1971. Applied landscape management in plant control. J. Range Mgt. 24(1):2–6.

Chapter 2

Noxious plant problems and plant control

Causes of Plant Invasions

One of the most pressing problems of range management in western United States is the invasion and increased density of noxious (undesirable) plants and the associated reduction in forage supply. These invasions, both external and internal, of both native and introduced plant species began in some cases as early as 1850 but have been particularly evident since 1900.

Recent estimates indicate 88.5 million acres, or 82 percent of Texas's once luxuriant grasslands, are infested with one or more worthless brush species and that 54 million are so densely covered with brush that little improvement can be expected without reduction of the brush [102]. Woody plants are considered a problem on at least 16 million acres in Arizona and on an equal acreage in New Mexico [88]. Such invasions are by no means restricted to the Southwest but are common throughout much of the United States on grassland and savannah sites. Klingman [53] has estimated that brush infests 320 million acres of range and pastureland in the United States and that brush removal can improve forage production on two-thirds of this infested area.

The replacement of desirable forage species by low-value plant species is causing much concern among ranchers and land administrators; and researchers continue to investigate the factors responsible for these changes. Natural succession can cause an increase of undesirable climax species; but outside factors associated with man's activities are generally contributive if not the major causal factors of most plant invasions. Intensive competition by vigorous perennial grasses is a major deterrent to noxious plant invasions; but reduced competition from desirable forage species is often more an indirect effect of other natural or introduced factors than being the direct cause of the problem.

Primary factors causing or contributing to the increase, spread, and invasion of noxious plants include:

1. *Grazing by domestic livestock.* Reduced density, production, and seed development or vegetative reproduction of the more desirable forage species can result from selective grazing practices such as overgrazing, improper grazing season, or rigid livestock numbers. However, this must be considered only a partial cause, since neither protection from grazing nor competition by grasses has been able to prevent the increase of shrubs in some cases [37]. Removal of fuel by grazing also tends to restrict fires.

2. *Reduction of fire.* Fire originally played a major role in keeping many shrubs in check and preventing their spread into grasslands. The reduced role of fire today in checking invasions or natural succession of shrubs has been attributed to expanded efforts in fire control, reduced Indian influence, and reduction of ground fuel necessary for a destructive fire.

3. *Seed transport by grazing animals.* Livestock readily disseminate barbed or awned seeds such as cocklebur (*Xanthium* spp.), houndstongue (*Cynoglossum officinale*), threeawns (*Aristida* spp.), and cheatgrass brome (*Bromus tectorum*) by mechanical attachment to hair and wool. Also, both domestic livestock and big game are unable to digest or kill all of the weed seeds that pass through their digestive tracts. One study indicated that most of the weed seeds were voided within a seventy-two-hour period after consumption by sheep [40], but in another study viable seeds of halogeton and medusahead remained in the digestive tracts of sheep for as long as nine days [58]. These undigested seeds remaining in the digestive tracts and later appearing in the droppings provide a means of spreading noxious plant seeds over great distances in short periods of time.

4. *Dissemination by small animals.* Small animals such as jackrabbits, kangaroo rats, mice, birds, and coyotes also play a similar role in spreading weed seeds through incomplete digestion, attachment to parts of the body, and planting seed caches.

5. *Climatic fluctuations.* Major shifts in climate such as temperature or rainfall do not account for the consistent and widespread increase of brush species. However, drought cycles are conducive to the invasion of undesirable plants. Prolonged or severe drought may reduce the herbaceous and other competing vegetation and open up the site to establishment of noxious species.

Subsequent precipitation sufficient to germinate noxious plant seeds and permit them to become well rooted may result in lasting changes in plant composition. Drought may be an important means of triggering noxious plant invasions on range previously weakened by grazing or other factors. Perennial vegetation can also be damaged by hurricanes in coastal areas.

6. *Cultivation and subsequent abandonment.* Removal of original cover in farming operations and later abandonment without artificial revegetation has opened up large acreages in marginal rainfall areas to invasion of noxious plants. Such areas have been widespread in western portions of the Great Plains area and in dryland farming areas of the Intermountain and Southwest.

7. *Local denudation.* Road and railroad right-of-ways, stock trails, industrial areas, mining locations, farmsteads, and other areas locally denuded of vegetation allow initial establishment of invading species from which subsequent spread can initiate. Plant diseases or insects may also partially or completely destroy established vegetation.

8. *Increase in commerce.* The increase in highway, railroad, and other forms of transportation makes it increasingly difficult to prevent transport and establishment of weed seeds in new areas. Weed seeds are readily spread in the transport of hay, seed, manure, and farming and earth moving equipment.

Successful techniques for preventing continued increase or invasion of undesirable species depend largely upon an understanding of factors contributing to these changes. Knowledge of the life cycle, ecological requirements, physiological capabilities, and means of spread are necessary in halting the advance of each noxious plant and in its removal where practical. Brief reviews follow for selected species considered especially troublesome on U.S. rangelands.

Individual Problem Plants

Mesquite (*Prosopis juliflora*). Three varieties of mesquite are commonly recognized [31]. Honey mesquite (*P. juliflora* var. *glandulosa*) occurs principally in eastern New Mexico, southern and western Texas, and western Oklahoma. Velvet mesquite (*P. juliflora* var. *velutina*) predominates in Arizona, Lower California, and Mexico. Western honey mesquite (*P. juliflora* var. *torreyana*) is found principally in southwestern New Mexico, southeastern and western Arizona, and California. All varieties vary from large single-trunk trees to small shrubs.

33

Mesquite often forms dense jungles of brush which sharply reduce range forage production and its accessibility to grazing animals. It is an aggressive competitor because (1) seed production is high and seeds remain viable in the soil for many years, (2) seeds germinate and seedlings are able to establish under a wide range of temperature and moisture conditions, (3) root systems extend both laterally and to considerable depths, and (4) dormant buds, located under the bark on the root crown two to twelve inches below the ground surface, readily sprout following injury or removal of the canopy [31, 37, 92, 111]. In Arizona 90 percent of mesquite seeds buried in the soil had sprouted or decayed after ten years, but the remaining 10 percent were still sound [62]. After twenty years in the ground 0.2 percent of the seeds still sprouted after mechanical scarification of the seed coat.

In 1964 mesquite was estimated to cover 56 million acres in Texas (52 percent of the state) and to have increased 1.25 million acres during the previous fifteen years [102]. It was estimated in 1959 to cover an additional 15 to 20 million acres in New Mexico and Arizona [31]. More than half of the infestation in Texas is in moderate to dense stands that seriously affect production of forage and livestock [31]. Mesquite now covers almost twice as much rangeland in the Southwest as it did in 1900 [61]. This native species was originally confined mostly to the valley bottoms and drainage courses with only a few scattered trees in the upland [74]. As mesquite invaded the desert grasslands and portions of the Southern Great Plains and thickened in stand, grasslands were converted to a shrub type.

Early spread of mesquite in Texas is attributed to the buffalo and later to Spanish horses and trail herds [31]. Seeds of mesquite are readily eaten by domestic livestock and native animals. Since the seed coats are nearly impervious to digestion, viable seeds are readily disseminated through the digestive systems as animals move about on the range (Figure 7) [37]. Animals regarded as effective seed disseminators by this means are cattle, horses, mules, goats, deer, peccaries, cottontails, jackrabbits, other rodents, and birds [2, 74].

In feeding studies with sheep, 68 percent of the mesquite seeds fed were destroyed beyond recognition by mastication and digestion [36]. However, 27.3 percent of the seeds fed were passed intact with only slightly reduced germination ability. Feeding sheep 2,4-D in the drinking water has been of little value in reducing the viability of the seeds ingested. The digestive system of cattle was also found ineffective in killing mesquite seeds [61]. A single cow chip has contained up to 1,500 or more mesquite seeds of which one-half to three-fourths were still viable. Although 95 percent of the intact seeds fed to sheep were eliminated by the fifth day, viable seeds have been re-covered from both cattle and sheep as late as eight days after feeding [36, 61].

Other factors also contribute to the spread of mesquite. The re-moval of fuel on the ground by grazing reduces the occurrence of fire and associated damage to mesquite plants [37, 31]. Heavy grazing pressures and drought reduce grass vigor and density and provide openings in the sod favorable to mesquite establishment [37, 11]. However, once established in an area, mesquite tends to thicken whether the range is grazed or not [74, 41]. Rodents such as the

FIG. 7. *Mesquite seeds are distributed in the droppings of cattle and other animals.* Top, *cattle readily graze mesquite pods;* bottom, *partially disintegrated cow chips containing up to 1,670 seeds per chip* [61]. (U.S. Forest Service photos)

Merriam kangaroo rat (*Dipodomys merriami*) have been found to store mesquite seeds in shallow caches under the ground [87, 78]. Many of these seeds are never retrieved but instead germinate and increase the infestation. As rodent populations expand, they tend to move their caches farther into grassland areas. Although rodents may kill considerable numbers of mesquite seedlings by grazing, this is more than offset by their seed distribution activities [77].

Three stages of mesquite invasion on sandy soil in New Mexico are recognized: (1) young plants hidden among the grasses, (2) older plants with sand blowouts around them, and, finally, (3) mesquite sand dunes (Figure 8) [41]. This deterioration has been associated with a reduction in carrying capacity from eighteen to three or less animal units per section and a condition uneconomical to reclaim at present. Cyclic rainfall patterns favor mesquite over herbaceous plants with shallower roots [11, 95]. Periods of below average moisture followed by drought-breaking rains intensify mesquite emergence and establishment at the expense of the grasses. Although record-breaking droughts have been observed to thin out mature mesquite stands, young or resprouted mesquite are not generally affected [17].

No one control treatment is considered capable of completely eliminating mesquite at present [95]. Repeated treatments, together with sound range management practices to develop and maintain a good grass cover, are necessary. Cable and Martin [15] found that bulldozing or burning alone only interrupted the invasion process and recommended that treatment should be both early and vigorous. Considering the potential for further mesquite invasion into adjacent grasslands, prevention should include (1) maintaining maximum development of grass stands, (2) minimizing seed dissemination, and (3) prescribed burning [37].

Chaparral. Chaparral refers to dense stands of shrubby plants dominated by broadleaf and narrowleaf, nondeciduous species, many of which vigorously sprout following removal of the aboveground parts (Figure 9). Chaparral in California occurs on about eleven million acres in the foothills and another nine million acres intermixed with woodland species at higher elevations [7]. Similar types of chaparral occur in Arizona. Chaparral vegetation is characteristic of many critical watersheds, particularly in California. Many such areas are steep and erosive, are located near and above densely populated residential areas, and are subject to torrential rains.

Chamise (*Adenostema fasciculatum*) predominates over much of the California chaparral. Associated species include sprouting and nonsprouting species of ceanothus (*Ceanothus* spp.), California scrub oak

(*Quercus dumosa*), sprouting species of manzanita (*Arctostaphylos* spp.), interior live oak (*Quercus wislizenii*), and others.

Several factors encourage the increase of chaparral in California [8]. Regenerative characteristics of these woody species include prolific seed production and ready establishment of seedlings, spread by rhizomes, and sprouting from root crowns following fire or cutting

FIG. 8. *Sandy soils on College Ranch near Las Cruces, New Mexico, subject to mesquite invasion.* Top, *black grama range in high condition with only scattered mesquite seedlings;* bottom, *similar range deteriorated to mesquite sand-dune stage.*

37

which kills only the tops. Fires followed by protection from fire allow brush seedlings to establish, particularly where the perennial herbaceous competition has been removed or weakened by grazing. Fires appear to increase the germination of the seeds of many chaparral species. Also, rains late in the growing season after grasses have dried favor brush growth and vigor. Light, shallow soils are generally invaded first, but chaparral species subsequently spread to the better soils.

Chaparral in Arizona is found at elevations of about 4,500 feet in the central and southeastern portions of the state. The principal shrubby species are shrub live oak (*Quercus turbinella*), pointleaf manzanita (*Arctostaphylos pungens*), Wright silktassel (*Garrya wrightii*), mountainmahogany (*Cercocarpus* spp.), sugar sumac (*Rhus ovata*), and

FIG. 9. *Dense chaparral on foothill range in California nearly prevents livestock and big game entry; woody plants cleared at bottom of photo.* (Towner Manufacturing Co.)

skunkbush sumac or squawbush (*Rhus trilobata*). Arizona chaparral commonly occurs in dense thickets and tends to invade former grasslands. Most of the shrubs resprout following top removal.

Although the sprouts and regrowth of many of the chaparral species provide fair to good forage, particularly for deer, dense and mature stands are relatively unproductive for domestic livestock, game, timber, recreation, or water yield [8]. The conversion of selected, potentially productive chaparral sites in California to grassland is recommended. The attributes of grasses on these sites include good soil cover, reduction of erosion, excellent forage for livestock, good supplement for deer browse, lower water use than brush, low fuel volume for easier fire control, making recreation resources more usable, and breaking up extensive brush fields for effective fire control [7].

Pricklypear and cholla (*Opuntia* spp.). Pricklypear (Figure 10) includes the several flat-jointed species of *Opuntia*, which are widespread in the Great Plains and portions of the Southwest and Southern Intermountain areas. The response of pricklypear to grazing has been controversial. In south Texas it is reported to increase on poor condition range and decrease when ranges improve and produce abundant grass [42]. On Colorado and Montana shortgrass range, grazing intensity was found not to affect the density of pricklypear [6, 45, 115]. It was concluded that changes in grazing intensity could not be depended upon as a management practice to reduce pricklypear. Removal of pricklypear, in Colorado studies, did not increase blue grama yield but did make almost 20 percent more of the forage produced available to cattle grazing [6].

Weather in the Great Plains was found to have the greatest influence on the abundance and vigor of pricklypear [45, 107]. Pricklypear spread in dry years and tended to recede in wet years. High rainfall apparently was associated with accelerated insect and disease damage of the pricklypear. Pricklypear also varied with soil type and condition.

Cholla, the several round-stemmed species of *Opuntia*, is found in the western part of the Southern Great Plains and west to Arizona. Cholla often invades grasslands, reduces carrying capacity, and makes livestock handling difficult because of its sharp and abundant spines [92]. However, in Arizona it was concluded that cholla was more objectionable in interfering with the grazing and handling of livestock than for decreases it caused in grass production [63]. The removal of walkingstick cholla (*Opuntia imbricata*) in light to dense stands in New Mexico increased blue grama yield only slightly [81]. The low apparent competition between the two species was attributed to different growth periods and rooting zones, since cholla grows earlier than blue grama and has very shallow roots. However, control of cholla makes handling of livestock easier and improves wool quality. Utilization of grass plants growing next to cholla is restricted, and livestock are reluctant to graze in dense cholla stands.

The density of both cholla and pricklypear plants is quite cyclic in nature, but both species groups are well adapted to and favored by drought conditions because of the ability of their shallow but wide-spreading root systems to absorb and hold water. Pricklypear reproduces by seeds, adventitious roots, and root sprouts [45], and the seeds are widely spread by rabbits. Both cattle and sheep mechanically spread pricklypear, and the joints readily root down when placed in contact with mineral soil [102]. That cactus is a wide-spread problem is indicated by the fact that *Opuntia* species occupy more than 35 million acres of range in Texas alone [102]. In Wyoming pricklypear now infests 1.5 million acres to the extent that control measures are necessary to permit restoration [106].

Big sagebrush (*Artemisia tridentata*). Sagebrush, primarily big sagebrush, occupies about ninety-six million acres of land mostly lying between the Rocky Mountains and the Sierra Nevada Mountains. Big sagebrush grows chiefly on low foothill ranges and adjacent valley slopes. It is also found in stands of mountain brush and adjacent to aspen in the subalpine zone and intermixed with blackbrush (*Coleogyne ramosissima*) in the southern desert shrub [83].

FIG. 10. *Dense stand of pricklypear in Jim Hogg County, Texas, reduces livestock access and grass production. Herbicide application has killed many of the pads; affected pads are more curled, whiter, and more dehydrated.* (Soil Conservation Service photo)

Over much of its range today big sagebrush is found in excessively dense stands. Big sagebrush is a natural component in grass-sagebrush communities but readily replaces the grasses under heavy livestock grazing pressures (Figure 11) [108]. The result has been lowered forage production, increased noxious weed and insect pest populations, soil erosion, and excess runoff. Heavy grazing has generally been considered the principal reason for thickening up of big sagebrush and its limited invasion into adjacent grasslands and into higher elevations [44, 79]; but the reduction of fire and occasional severe drought greatly favors sagebrush over perennial grasses. On sagebrush lands where the brush has been removed and introduced wheatgrasses seeded, sagebrush reinvasion is accelerated by the remaining sagebrush seed plants, heavy grazing, and drought periods followed by periods ideal for seedling germination and establishment [33].

In western Montana, factors credited with favoring the invasion of sagebrush were plowing and subsequent abandonment, rodent activity, road construction and abandonment, and competition-free sites allowing rapid seedling establishment [68]. The source of sagebrush seed permitting reestablishment following burning and other control practices has been a matter of concern. Seedlings following fires have been shown to come both from seed stored in the soil that remains viable after burning and from windblown seed in areas near unburned seed sources [69]. However, the residual seed was the greater source of big sagebrush seed.

Rabbitbrush (*Chrysothamnus* spp.). Big rabbitbrush (*Chrysothamnus nauseosus*) and little rabbitbrush (*Chrysothamnus viscidiflorus*) are small to medium shrubs that grow on open plains and foothills. They are

FIG. 11. *Big sagebrush occupies extensive areas in the Intermountain Region. This site near Scipio, Utah, shows reduced understory of herbaceous plants and a scattering of juniper.*

41

most abundant in the Great Basin, where they appear on a variety of sites from salt meadows to dry upland sites with sandy soils. Rabbitbrush spreads through removal of competing vegetation by close grazing, fire, and cultivation to occupy extensive areas of rangeland [66].

Removal of competition by perennial grasses has increased flower production, plant vigor, and shoot elongation of both big and little rabbitbrush. Rabbitbrush resprouts following top removal, produce a large amount of seed, and its hairy achenes are readily carried by wind. Because of this large reproductive capacity, competition is very important in controlling rabbitbrush growth and reproduction [66]. On degraded plant communities in Nevada, only partial reduction in big sagebrush or green rabbitbrush populations by application of 2,4-D resulted in a rapid increase in seedling establishment of both species [129].

From studies at Benmore, Utah, it was concluded by Frischknecht [33] that big rabbitbrush was not competitive with seeded crested wheatgrass since the removal of rabbitbrush had little or no apparent effect on grass production. This lack of competition was attributed to big rabbitbrush having deep taproots rather than numerous lateral roots in the root zone of the grasses and having a later active growth period than crested wheatgrass. However, Cook et al. [24] concluded from studies in the same general area that big rabbitbrush was competitive with grasses for soil moisture since rabbitbrush competition reduced herbage yields, basal area, and number of seedheads of grasses as well as the root area occupied by grasses. Removal of little rabbitbrush in Nevada studies increased crested wheatgrass by eight to fourteen times [114].

Juniper-pinyon. This vegetation type is found in the Intermountain Region mostly at intermediate elevations in areas receiving mostly less than twenty inches of precipitation annually and is characterized by juniper (*Juniperus* spp.) and pinyon pine (*Pinus cembroides, edulis,* and *monophylla*). Juniper and pinyon trees are generally of minimal commercial value except for firewood, fence posts, and pinyon nuts. The juniper-pinyon type is estimated at about seventy million acres today.

Evidence is ample that juniper in most areas is climax only on rocky ridges and rimrocks where soil development is limited, but not originally in dense stands as generally found today [3, 13, 21, 120]. From their original habitat junipers have invaded and are continuing to invade areas of deeper soils on the valley slopes and bottoms previously occupied by grasslands and sagebrush-grass types (Figure 12). For example, juniper in the state of Texas was estimated in 1964

FIG. 12. *Juniper trees invading from ridge tops into sagebrush-grass type of slopes and valley bottoms in southern Idaho.*

to infest 21.5 million acres—an increase of 3.5 million acres in the preceding fifteen-year period (Figure 13) [102].

Probable causes of the invasion of juniper onto the deeper soils of the grass-sagebrush communities of the Intermountain Region include overgrazing, lack of recurring fires, greater seed dispersal, and climatic fluctuations [13]. Small trees of the nonsprouting junipers are readily killed by fire, and the reduction of fires in the sagebrush zone apparently allows juniper seedlings to establish in ever-increasing numbers. In southwestern Idaho the invasion of western juniper (*J. occidentalis*) was found to be directly related to the cessation of periodic fires resulting from active fire control, development of roads, and the reduced fuel left by grazing [14]. Range condition apparently had a negligible effect on juniper establishment, and the competitive effect of other vegetation during the establishment phase was insignificant.

Although drought favors the dominance of juniper, good years following drought also favor the establishment of new seedlings [49]. Sheep, deer, coyotes, many birds, and small mammals eat the juniper fruit and spread the seeds through the droppings [3, 2]. In fact, juniper seed passed intact by animals germinates faster than seeds not affected by digestion [49]. Juniper seed has been found to be long-lived and able to remain viable through several years of drought. Once seed sources become available, grazing versus protection seems to have little relation to increase in canopy cover of juniper [3].

The invasion of juniper into productive sites and the thickening of juniper canopy have reduced grazing capacity, increased the diffi-

FIG. 13. *Redberry juniper on grasslands of Briscoe County, Texas, before brush control was provided.* (Soil Conservation Service photo)

culty of handling livestock, reduced food for wildlife, and decreased water yields [13]. Pinyon-juniper control was found most beneficial in Arizona on sites with (1) dense overstory, (2) higher precipitation, (3) lower calcium carbonate percentage in the soil, and (4) medium texture soil [73]. On sites with inadequate ground cover, control of erosion often depends on juniper control and its replacement with herbaceous plants. Plummer et al. [84] have reported that even rocky slopes exceeding 50 percent were greatly improved in forage, ground cover, and retention of water from high intensity storms following juniper replacement. However, Williams et al. [125] concluded from their work with small plots that conversion practices currently used may not improve infiltration rates or reduce sediment on some juniper sites.

Alligator juniper (*Juniperus deppeana*), a sprouting species of juniper, occurs principally in Arizona. This sprouting characteristic is almost universal in young trees but less common in old trees. Sprouts may arise from the stem, roots, or especially from buds on the root crown [3, 48]. Mechanical control of alligator juniper requires that the root crown be removed from the soil.

Toxic agents in juniper and pinyon litter rather than root competition for soil moisture and nutrients were concluded in northern Arizona to account for the reduction of basal area and productivity of shallow-rooted grasses such as blue grama [46]. Extracts from juniper foliage have significantly decreased seed germination of blue

44

FIG. 14. *Gambel oak in central Utah on Uinta National Forest.*

grama (*Bouteloua gracilis*), crested wheatgrass, and sideoats grama (*Bouteloua curtipendula*) [57]. In related studies, extracts of Utah juniper (*Juniperus osteosperma*), ponderosa pine (*Pinus ponderosa*), and pinyon pine (*P. edulis*) decreased germination of squirreltail (*Sitanion hystrix*) and blue grama by 85 percent or more [47]. The extent and seriousness of these toxic juniper extracts remain uncertain. It has also been found that soil pH is higher under juniper trees than in the interspaces between juniper plants [120].

Gambel oak (*Quercus gambelii*). Gambel oak grows on several million acres in the states of Colorado, Utah, Arizona, and New Mexico. It is found principally at elevations of 5,000 to 9,000 feet and is associated with the semidesert shrub and grasslands at its lower extent and aspen and spruce at its higher extent. It frequently dominates sites in the mountain brush zone and often alternates with pinyon-juniper, sagebrush-grass, and ponderosa pine sites. Growth form varies from groves of small trees on favorable sites to dense shrub thickets on less favorable sites (Figure 14). Vegetative reproduction is prolific with sprouts growing from surface roots and from the base of the trunk and root crown.

Gambel oak generally spreads slowly, and its present range is probably not greatly different than in the late 1800s [18]. However, many Gambel oak stands have increased in density and canopy cover. Fire, cutting, or herbicides frequently kill the aboveground portions of the plant; but topkill only tends to thicken the stand and cause the merging of scattered stands where competition is minimal.

45

Since many plant stems die before reaching eighty years of age, lack of disturbance may be associated with thinning out and retreating of stands with age. Thick stands of Gambel oak are usually associated with a reduction in herbage production and increased water use [12]. However, Gambel oak is a source of emergency browse, provides mast, and reduces erosion on erosive sites. Where Gambel oak occurs on gentle slopes with productive soils, replacement with grass species or more palatable shrubs appears desirable.

Halogeton (*Halogeton glomeratus*). Halogeton is an annual forb accidentally introduced from the Old World. Following its initial discovery in northeastern Nevada in the 1930s, it has received widespread attention because of its poisonous properties and ability to rapidly invade denuded lands. It is now widely spread over the low rainfall areas of Nevada, Utah, Idaho, and Wyoming and probably occurs on more than twenty million acres. Overgrazed sites, abandoned farmlands, highway and railroad right-of-ways, and stock trails provide suitable habitat for halogeton. Vigorous perennial competition is probably its best control [25].

Halogeton is well adapted to invade dry, salty, open sites because of its low water loss from aerial parts, high tolerance of salinity and sodium in the soil through all phases of growth, and prolific seed production even under adverse conditions (Figure 15) [25]. Not only is halogeton capable of producing about seventy-five seeds per stem or up to 400 pounds per acre, but two types of seed are produced. Black seeds germinate readily and seldom remain viable over one

FIG. 15. *Halogeton, a poisonous plant, invading a dry site in Box Elder County, Utah. Note the absence of perennial-plant competition.*

year in the soil. Brown seeds break dormancy slowly and may persist in the soil for ten years or more, thereby providing a means for species survival during long periods of drought or control treatments.

Halogeton concentrates sodium in its tissues [51]. Subsequent leaching of the sodium into the surface soil makes the site much more favorable for halogeton than for grasses. The increased sodium greatly inhibits germination of seeds of other plant species and thus materially reduces competition. Tall wheatgrass (*Agropyron elongatum*) is least affected by the increased sodium, but halogeton areas are normally too xeric for this grass. Other plants showing considerable tolerance of sodium with possible implications for use in replacing halogeton are Russian wildrye (*Elymus junceus*), crested wheatgrass, and barley (*Hordeum vulgare*).

On established stands of crested wheatgrass in Utah, halogeton first invaded heavily grazed slick spots where soils contained higher total soluble salts and greater amounts of exchangeable sodium [34]. It was noted on these slick spots that the perennial grasses showed poor growth because of low soil moisture retention, frost-heaving of the grasses, and were shorter lived, less vigorous, and more heavily grazed by cattle and rabbits. It was concluded that the invasion of halogeton onto crested wheatgrass stands was favored by (1) heavy grazing, (2) grazing in the spring, (3) high soil salts, and (4) high summer rainfall one year followed by good moisture the following spring.

Cheatgrass brome (*Bromus tectorum*). Cheatgrass brome or downy brome, commonly referred to as cheatgrass, presents problems on western rangelands as serious as any of the woody perennials which it has often replaced [83]. This introduced annual appeared in western United States in about the 1890s and now occupies large areas from the moister part of the blackbrush and shadscale zones, through the sagebrush and juniper zones, and up into the mountain brush zone.

Cheatgrass has many of the same seed characteristics associated with other aggressive annuals [52, 103, 128]. It is a prolific seed producer, and seeds normally have high viability. Seeds of cheatgrass are readily spread by wind and water and have barbed spikelets that help in mechanical distribution by animals. Seeds germinate rapidly when rains come in late summer and early fall, but adequate numbers of viable seed are carried from one year to the next in the litter and soil to survive unfavorable years. Although naturally a winter annual, cheatgrass can germinate in the spring and produce seed even under unfavorable conditions. Replacement of cheatgrass requires control of seed production, germination, and establishment.

Natural or artificial seeding of perennial grasses combined with beneficial grazing practices are generally also required.

Cheatgrass rapidly invades sites where the natural perennial cover has been disturbed or lost through cultivation, grazing, repeated fire, or road construction [83, 52, 103]. It endures drought well and has the phenomenal ability to consume soil moisture on which perennial seedlings depend, thus enabling it to control vast areas for long periods. Wildfire greatly favors cheatgrass and allows it to extend its range. It is occasionally found to a limited extent in stable communities and may be considered a naturalized alien. Although its production is highly variable, cheatgrass does provide forage of medium quality for early spring use.

Medusahead (*Taeniatherum asperum*). Medusahead, an aggressive, introduced annual grass, was first established near Roseburg, Oregon, but has since spread through the Palouse region of Washington and Idaho, southern Idaho, and over the northern half and scattered southward in California [60]. In Idaho its distribution was estimated at 30,000 acres in 1952, 150,000 acres in 1955, and 700,000 acres in 1959 with the potential for invading Idaho's some 6 million acres of cheatgrass brome [110]. It is considered the worst range weed in southwestern Idaho because of its rapid migration, its vigorous competitive nature, and its extremely low forage value. Medusahead has reduced grazing capacity as much as 75 percent on some ranches in California [50]. Because it is largely ungrazed, it accumulates and creates a fire hazard.

In the western Great Basin, clay sites with well-developed soil profiles and sparse vegetation are most susceptible to medusahead invasion [27, 128]. Also, those sites receiving run-off water from adjacent areas favor medusahead invasion. Sagebrush or cheatgrass communities on well-drained, medium- to coarse-textured soils are quite resistant to medusahead invasion. Maintaining a good stand of perennial vegetation appears to be the best barrier to medusahead invasion into susceptible soils. It has been concluded that once medusahead invades a low seral community and fully occupies the site, the site is effectively closed to the establishment of native or seeded, introduced species unless intensive treatment is provided.

Medusahead produces a large amount of seed, which is of high viability even when the heads are still greenish in color [100]. Large numbers of seed carry over in the soil and remain viable for at least one year. Fire-damaged seeds do not germinate, but substantial numbers of seeds lying on the ground go through fire undamaged and readily germinate. Medusahead, normally a winter annual, is capable of delaying germination until spring, and this serves as a survival mechanism. The long awned spikelets are readily spread by animals, particularly by sheep in their fleece. This factor should be considered in efforts made to prevent its spread into uninfested areas. Dead litter from medusahead has inhibited early growth of seedlings of planted forage species. [70].

Macartney rose (*Rosa bracteata*). Macartney rose was introduced as a hedge plant into southeastern United States about 1870 and is one of several introduced roses that have become serious pests in the South. In 1964 Macartney rose was estimated to occupy 276,000 acres of Texas range, an increase of over 230,000 acres during the preceding fifteen-year period [102]. Estimates in 1965 placed the area of Macartney rose infestation in the U.S. at 500,000 acres [116].

Although providing some wildlife food and cover, range areas covered by established clumps of roses are nearly a total loss for grazing [116]. Besides seriously reducing range forage production, it also restricts livestock movement and seriously hinders effective management of the range. Where further spread is not prevented, it is capable of completely taking over fertile grasslands in a period of fifteen to twenty-five years after initial establishment. Macartney rose seed is readily spread by birds, small animals, and livestock, but spread from layering or mower cuttings also takes place. Ninety percent of seed recovered from the digestive tract of animals suffered no apparent damage, and it was concluded that the action of digestion greatly increased germination ability of intact seed [65].

Tarweed (*Madia glomerata*). Tarweed is a herbaceous annual that grows from Saskatchewan to northern Arizona and California, commonly in openings from 6,000 to 10,000 feet. It is a highly competitive plant and invades overgrazed mountain ranges that frequently have high potential for forage production. It is a prolific seed producer, but most of the seed germinate the first year. Control is based upon eliminating one year's seed production, but all plants must be killed to be effective [83]. Tarweed produces toxic agents that reduce the germination of grass seed and cause abnormal seedlings [16]. Leachate from amounts of tarweed commonly found in the field has reduced the germination of intermediate wheatgrass by 61 to 91 percent and prevented normal germination of this grass at higher levels of leachate.

Creosotebush (*Larrea tridentata*). Creosotebush is an evergreen shrub with an extensive lateral root system that dominates about forty-five million acres from west Texas to California and has apparently increased considerably from its original area [54]. It is nearly worthless for grazing, and sites dominated by creosotebush have little or no herbaceous cover or forage for grazing. Presumably it was originally found only on foothills and outlying well-drained knolls, while the level and gently rolling land was covered with grasses [118].

Creosotebush aggressively invades desert grasslands in Arizona, New Mexico, and west Texas [35]. It invades by frontal advances along the margins of the established creosotebush communities. As scattered plants establish in the margin of the grassland community, they provide a seed source which develops ever-expanding communities which gradually connect with the main creosotebush community. Seed production is high, but seed dispersion is generally limited to a few hundred feet on the leeward side of the seed plants, except where transported by rodents [118]. Seeds remain viable from four to seven years and readily establish seedlings in barren openings in good condition grass range where the soil has been disturbed by rodents. Where range is converted or returns to creosotebush type, the change becomes rather complete and the site deteriorates from wind and water erosion. Creosotebush produces germination- and growth-inhibiting substances that affect associated desert grasses [54].

Burroweed (*Haplopappus tenuisectus*). Burroweed is a poisonous, short-lived, perennial half-shrub found particularly on alluvial plains from the Big Bend region of Texas westward to southern Arizona and south into Mexico [112]. Establishment of burroweed is rapid in winters and springs of good moisture conditions. Burroweed stands usually mature in a few years after a favorable establishment season and then decline.

Burroweed invasion apparently results from continuous heavy grazing of grasses and the cessation of range fires [112]. Since removal of burroweed may expose protected grasses to overgrazing, little benefit can be expected from control unless grazing is regulated. Protection of grass plants from rabbits may also be necessary before grasses are able to improve in vigor sufficiently to prevent burroweed reinvasion.

Post oak and blackjack oak. Post oak and blackjack oak occur from eastern Texas, Oklahoma, and Kansas eastward to the Atlantic. About eleven million acres occur in Texas alone, with an additional sixteen million acres of live oak [102]. About half of these stands have over 20 percent canopy cover. Post oak and blackjack oak require considerable water and compete effectively with forage species for soil moisture.

The thickening and local spread of oak has converted the original open woodland or savannah into brushy or dense woodland stands. This has been associated with a material reduction in bluestems (*Andropogon* and *Schizachyrium* spp.) and associated forage grasses [28]. Heavy grazing, removal of the more desirable hardwoods, and reduction of ground fires appear to be mostly responsible for these changes. For satisfactory economic returns from forage production, it is recommended that oak canopy be reduced to 25 percent cover or less. Provision must be made for preventing the release or invasion of even worse species following oak control, i.e., winged elm in eastern Oklahoma.

FIG. 16. *Saltcedar establishing in a wet meadow near Springville, Utah, will eventually develop into dense stands which crowd out forage species unless controlled.*

Saltcedar. Saltcedar (*Tamarix* spp.) was introduced into the United States over one hundred years ago. Since about 1930 it has become a major problem in the southern half of the West on river flood plains, around lakes and reservoirs, and in irrigation systems [94]. Prior to 1935, it apparently did not occur in Salt Lake Valley, Utah Valley, or the Uintah Basin of Utah, but is widespread there today [19]. Saltcedar is estimated to have increased in the United States from about 10,000 acres in 1920 to over 900,000 acres in 1961, with further spread projected [94]. It is also greatly increasing in density as well as canopy on sites where it is found.

Saltcedar is a high water-consuming phreatophyte, is highly salt tolerant, crowds out forage plants, and impedes the passage of flood flows thereby increasing flood hazards and sediment deposits [94]. It has been cultivated for erosion control and windbreaks in some areas, but it readily escapes, naturalizes, and spreads from one valley to another (Figure 16). The problem is particularly severe in Arizona, New Mexico, Texas, and Oklahoma. In 1961 saltcedar was found in all seventeen western states except North Dakota and Washington, and its extent varied from about 1,000 acres in Idaho, Nebraska, and South Dakota each to over 450,000 in Texas.

Miscellaneous species.* Two introduced ornamental and shade trees that sometimes escape and become naturalized in moist pastures and meadows in some areas are Siberian elm (*Ulmus pumila*) and Russian olive (*Eleagnus angustifolia*) [20]. A third such species in more limited areas is Osage orange (*Maclura pomifera*).

Foxtail barley (*Hordeum jubatum*) is a native, short-lived perennial that frequently becomes a difficult pest in moist to wet meadows and irrigated pastures throughout the West. Leafy spurge (*Euphorbia esula*), whitetop (*Cardaria draba*), Canada thistle (*Cirsium arvense*), and musk thistle (*Carduus nutans*) are examples of weeds of cultivated lands that have almost unlimited ability to invade mesic rangelands and meadows (Figure 17). Shinnery oak (*Quercus harvardii*) and sand sagebrush (*Artemisia filifolia*) are two low shrubs that seriously reduce productivity of grasslands when allowed to develop into thick stands.

FIG. 17. *This dense stand of musk thistle has nearly excluded seeded grasses on a mesic foothill range site in central Utah.*

*Although emphasizing weeds of farmlands, many states and regions provide weed manuals including some undesirable plants of pasture and range [1, 32, 43, 71, 75, 105, 117]. A composite list of weeds which gives common and scientific names is recommended for reference [122].

Mulesear (*Wyethia amplexicaulus*) is a native, aggressive forb on many productive sites of mountain and foothill ranges. On sites where grass density has been reduced by livestock grazing, mulesear frequently increases to the virtual exclusion of other forage plants (Figure 18). Although undesirable and of low value on livestock range, mulesear provides fair to good deer and elk forage in the spring. However, scattered plants are more effectively utilized by big game than dense, extensive stands as commonly found.

Bitterweed (*Hymenoxys odorata*), orange sneezeweed (*Helenium hoopesii*), and many other poisonous forbs tend to spread under heavy grazing. Broom snakeweed (*Xanthocephalum* or *Gutierrezia sarothrae*), a short-lived, perennial half shrub, aggressively invades disturbed areas; but it is highly cyclic and is not considered a reliable indicator of overgrazing. Refer to the section on biological control for consideration of St. Johnswort or Klameth weed (*Hypericum perforatum*) and tansy ragwort (*Senecio jacobaea*).

FIG. 18. *Dense stand of mulesear in a dry mountain meadow in northern Utah nearly excluding other forage plants.*

Introduction to Plant Control

Eradication—complete kill or removal of noxious plants, including all plant structures capable of sexual or vegetative reproduction.

Control—manipulation and management for reduction of noxious plants. Control is a term of many degrees; it can

53

vary in use from only slightly limiting to nearly complete replacement of the undesirable species in question.

Prevention—avoiding contamination or infestation by a noxious plant still absent from the area.

Eradication is desirable and practical when the noxious plant is not yet prevalent in the locality, when the weed is particularly troublesome once established or allowed to spread, and where a satisfactory method is available. New, isolated infestations of halogeton or Canada thistle would normally justify rather high treatment costs per unit of land area to eradicate the plants in order to prevent spread to adjacent areas. On the other hand, the cost of eradication of such native and widely distributed species as big sagebrush and juniper from any sizeable area may not be desirable and would be prohibitive in cost even if adequate methods were available. Special emphasis, however, should be given to stopping the noxious plant in question before the initial establishment can be made.

In contrast to eradication and prevention, control may be the only practical approach where noxious plants are widespread and well established. Control implies keeping the undesirable species at a density that minimizes interference with the production and utilization of range forage. Control rather than eradication is the aim of most treatment programs on rangeland. After halogeton became widespread in western Utah, the only practical approach was to learn to live with it while managing desirable forage plants to provide maximum competition and indirect control.

Season of control treatment. Physiological stage of growth is a primary factor determining effective treatment periods. Since annuals complete their life cycle in one year, reproduction of annuals is dependent with few exceptions upon seed reduction. Thus, most control programs for annuals are aimed at the prevention of seed production. This is accomplished by killing the plants prior to or during early flowering. Preventing seed production is often an important consideration with perennial plants, also. Methods that completely remove plants from the soil, such as grubbing or removal by power equipment, effectively kill most plants at any stage of growth.

Perennial plant control methods that are based on the removal of top growth—cutting, mowing, grazing, or chemical defoliation—are normally most effective when applied when the plant's food reserves and ability to produce regrowth are at their lowest point [127]. Carbohydrate reserves accumulate in the roots and stem bases of deciduous perennials and also in the stems and branches of woody plants during the normal growing season. After reaching a peak after

54

maturation of leaves, there is a continued small loss of stored c[
bohydrates during the dormant season.

In the vegetative production of new tops in the spring, carboh[
drate reserves are rapidly depleted [127]. Depletion continues unt[
just before full-leaf stage in most species when the photosynthet[
area of the leaves becomes sufficient to provide carbohydrate needs[
and storage can be resumed. In nondeciduous perennials the carbohydrate depletion pattern may be reversed and the smallest reserves found just prior to winter dormancy. Single or repeated top
removal at these times of lowest storage generally gives the most effective plant kill.

Apical dominance is an important factor affecting the effectiveness of
top-growth removal as a control treatment. In many plants the apical meristem located at the tips of stems and upper branches produces hormones that dominate and control meristematic tissue or buds
located on the lower stem or underground on roots and rhizomes.
However, when the apical buds are removed, lateral buds are no
longer inhibited from breaking dormancy and producing new shoots.
Lateral buds may be *trace* buds (trace or connect to original vasculum) or *adventitious* buds (do not develop from original vasculum but
in an abnormal manner on roots, shoots, or callused cambium). Removal of apical dominance may thus result in a rapid increase in the
number of stems. Plants such as rabbitbrush and mesquite that readily sprout from buds on the root crown are ineffectively controlled by
top removal alone (Figure 19).

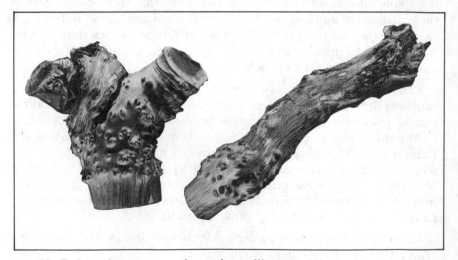

FIG. 19. *Buds on the root crown of mesquite readily sprout following top removal alone.*
These buds must be killed by mechanical or chemical treatment to prevent regrowth.
(Photo by Texas A&M University Research Station, Spur)

55

Herbicides such as 2,4-D enter the stomata after being applied as foliage sprays and are carried with the photosynthate stream throughout the plant. Most effective kill is obtained if phenoxy herbicides are applied when carbohydrate production and translocation rate is at the maximum, often near full-leaf stage. This effective period is generally shortly after the time of maximum depletion of carbohydrate reserves. The exact growth stage for maximum effectiveness varies with different plant species. Carbohydrate concentrations in honey mesquite roots were found to be closely coordinated with phenoxy herbicide translocation rates [126]. Both were low from bud break in early spring through leaf extension, good during rapid leaf growth once fully extended and through flower aging (May 15 to June 15), low again during seed development, but good again during the late seed stage (July 1 to July 15).

Other types of herbicides may have other periods of maximum effectiveness. Season of treatment is discussed in more detail with the individual plant control treatments.

Plant Control Considerations on Game Range

Most plant control techniques and principles used on livestock range are also applicable to game range or joint game-livestock range. Cattle and sheep have different forage plant preferences and adapt differently to different terrain and vegetation types. So also do big game species differ from livestock and even from each other in their habitat requirements. In general, ideal big game habitat has been equated with a greater mixture of forage species than needed for livestock, a mosaic of vegetation types, and greater availability of cover than needed for livestock.

Plant control, as well as other range treatment and development projects, directly affects the grazing habits of both big game and domestic livestock. Grazing animals generally move to places where treatment provides palatable forage, browse, and other attractive habitat features. However, in contrast to livestock management in which habitat use is greatly influenced by such practices as fencing, herding, and intensive grazing systems, the grazing of big game is much less regulated by direct control and more by habitat characteristics that attract or repel big game animals.

Lamb and Pieper [55] in New Mexico recommended that game range needed improvement whenever—

(1) It is infested with a solid stand of brush or trees of minimal value to wildlife.

(2) The stand is comprised of a single woody plant species that may furnish feed at one particular season but be a poor source of feed during the remainder of the year.

(3) Stands of shrubs or trees are too dense to allow ready access to game.

(4) Shrub stands are so tall that the edible browse is out of reach of the game.

(5) Stands of shrubs or trees are so dense as to crowd out all understory plants.

(6) Stands of woody plants are so dense that successful hunting is impossible.

A mosaic of treated and untreated areas is generally recommended for intensive woody plant control by mechanical, pyric, and chemical methods on big game range. Untreated areas are generally suggested for draws, ravines, rough ridges, and shallow, rocky sites; along permanent watercourses; and around watering places. Wildlife cover should be provided by leaving untreated areas, northeasterly exposures for summer use and southern exposures for winter use commonly being suggested. Untreated areas are also suggested for wildlife escape routes between water and food and between different stands of shelter.

The following categories of woody plant treatments for big game range emphasize somewhat different benefits, and have been proposed by Dasmann et al. [29]:

> **Browse rejuvenation***—projects where shrubs are crushed, burned, sprayed, mowed, chopped, or otherwise treated to encourage new growth of the shrubs in the form of sprouts and seedlings available for browsing. A secondary purpose is the natural establishment of grasses and forbs.
>
> **Browseway**—a lane built through a dense brush field by crushing with a dozer blade or other mechanical equipment to provide access by deer and man and to encourage browse production from sprouting shrubs. Differs from browse rejuvenation primarily in size and shape of treated areas.
>
> **Weed tree control**—reduction or elimination of undesirable woody vegetation, i.e. pinyon and juniper, while encouraging reestablishment and increased densities of desirable shrubs and herbaceous forage plants.

Since the opportunities and problems encountered in plant control on big game range differ not only between wildlife species but also between areas, the following discussions center around specific vegetation regions.

*Refer to Dimeo [30] for equipment for rejuvenating browse plants.

Texas shrublands. Brush control is a viable alternative to increase profits from ranching in south Texas [123], but in selecting a brush control program consideration must be given both to the livestock enterprises and to potential income from fee hunting of white-tailed deer and other wildlife and from exotic big game enterprises. Large block spraying of mature honey mesquite brushland reduced populations of white-tailed deer and wild turkey but not introduced nilgai antelope [5]. However, deer numbers were similar 27 months later after the forbs had returned to the understory.

Wildlife specialists in Texas commonly recommend that between 10 and 35 percent of brushlands be left untreated for wildlife cover and browse [85]. Possibilities raised have been to clear brush in strips, selective control to leave all plants of the more desirable woody plant species, and clearing brush only from the deeper soils (Figure 20). However, it was acknowledged that leaving scattered brush plants also leaves a seed source for rapid reinfestation. It was suggested that the nature of the individual plants and their reinvasion potential be carefully considered.

Aerial spraying leaving unsprayed strips has been researched in south Texas [5, 123] and on the northern Rio Grande Plain [104]. Aerially spraying 80 percent of an area in alternating strips did not reduce any of the wildlife populations, including white-tailed deer. It was concluded that the relative income from cattle or from white-tailed deer should be considered on private range in deciding wheth-

FIG. 20. *White-tailed deer are a highly profitable enterprise on this livestock ranch located on the Edwards Plateau of Texas. Much of the overstory of mesquite and live oak has been removed by chaining, but an adequate quantity of browse remains for deer.* (Soil Conservation Service photo)

er to completely treat or 80 percent treat in strips. It was noted in the northern Rio Grande Plain that deer did not rearrange their use to favor the untreated brush strips when the sprayed woody plants remained upright and provided the necessary cover.

Western shrublands. Short et al. [101] concluded that the benefits of widespread clearing of pinyon-juniper woodlands in Arizona are questionable, particularly on sites including low herbage production potential, and that large-block clearing should be restricted to areas of high potential for cattle production, little use by wildlife, and limited browse shrubs. They reported that large-scale clearing decreases both deer and elk use and recommended that woodland clearing be done on a localized, prescription basis. They suggested that clearings intended for wildlife habitat improvement and increased forage production be made in long, narrow strips 90 to 600 feet wide that conform to the terrain and are separated by strips left in suitable cover. For turkey habitat in Arizona pinyon-juniper, Scott and Boeker [98] recommended that cleared areas not be wider than 265 feet and that travel lanes be established between feeding areas and roosting areas.

Treatment prescribed by Lamb and Pieper [55] for improving pinyon-juniper habitat for big game in New Mexico was based on clearing strips not over ¼ mile wide, or only ⅛ mile wide on intensive deer management areas. This was equivalent to leaving 25 percent of the total area uncleared, or up to 50 percent uncleared in the intensive deer management areas. The suggested mosaic of cleared and uncleared strips and patches was considered as being beneficial to both livestock and game. This pattern was generally readily accomplished by following the meanderings of the most suitable soil type for intensive pinyon-juniper control and by leaving the rocky points, ridges, steep slopes, and areas of shallow soil in the uncleared portion.

Deer in Nevada were found to use chained pinyon-juniper areas 2.6 times more than unchained areas [113]. This increased use on the chained areas was associated with a 300 percent increase in bitterbrush leader length, in increased forb production, and in a tenfold increase in grass yield. In northern Arizona antelope and elk were observed to frequent the chained areas in pinyon-juniper, especially near their edges [3]. Although chaining released preferred game browse species such as winterfat, cliffrose, and shrubby eriogonums, deer movements were not noticeably affected.

Chaining pinyon-juniper areas in western Colorado improved deer habitat where continuous chained areas were not too large [67]. Deer droppings were higher in chained areas, particularly near the edge of the clearings. It was concluded that chained areas should be interspersed with undisturbed forest, so that continuous chained areas do not exceed one-quarter mile in width. However, other observations in Utah indicate that deer will graze uniformly over much larger chained areas where herbaceous plants and woody plant sprouts are plentiful [82]. Slash piles and stands of young junipers on pinyon-juniper chainings in the Intermountain Region have been observed to induce mule deer to graze at considerable distances from undisturbed cover [80]. It has also been noted that when the more favorable sites have been intensively treated, the less favorable areas left untreated generally benefit from reduced grazing and improve in range condition also [83].

In the Intermountain Region, leaving narrow strips of big sagebrush intermixed with cleared areas or striving for only partial control of big sagebrush has commonly resulted in a rapid reinvasion of sagebrush and a reduction in the effectiveness and life expectancy of grass seeding projects. In Montana it was found that the amount of sagebrush surviving control treatments was the most important factor related to its reinvasion [50]. Possibly, heavy winter deer use in local areas might delay this reinvasion, but a better practice would seem to be complete clearing of moderate acreages in blocks, leaving scattered, uncleared areas, as needed, for winter use by deer.

FIG. 21. *Chaining for browse rejuvenation and recovery of herbaceous understory in Utah.* (Utah Division of Wildlife Resources)

60

It has been noted that elk did not change their calving behavior or feeding habits on a sagebrush site where 97 percent of the big sagebrush cover had been killed by spraying with 2,4-D [121]. Chemical control of big sagebrush on the Gros Ventre elk range near Jackson, Wyoming, effectively influenced distribution of elk [124]. Sagebrush flats previously sprayed with 2,4-D attracted elk, particularly in late spring, and relieved some of the pressure on the already overused and abused ridge tops and steep slopes. Herbicide treatment of the flats increased grass production three times and decreased the brush to about 25 percent of its former stand. Herbicides were not applied to ridges and other areas grazed in midwinter by elk, since the sagebrush was an important forage at that time of year.

Browse production of mature and overmature curlleaf mountain-mahogany in Utah was approximately doubled by top pruning in spring or fall [108]. This method of rejuvenation was expensive in requiring about four man hours with a chain saw per acre. However, cost reportedly compared favorably with other game range improvement projects in the area on critical winter range. Top killing Gambel oak in old stands has also greatly proliferated sprouting. Although seldom preferred as browse, mule deer will browse the sprouts when they are the only feed available in abundance on winter or early spring range.

In managing chamise brushlands in California for game, special consideration is given to opening dense stands of woody plants to access by deer [8]. Alternatives suggested for opening chamise stands include bulldozing or disking, but also prescribed burning, spot burning, phenoxy sprays for sprouting species, or heavy browsing of sprouts following initial removal of brush.

Western aspen. Based on the finding that aspen understory in Arizona produced six times more herbage than conifer understory, Reynolds [91] recommended improving deer habitat by providing an interspersion of aspen groves in mixed conifer or ponderosa pine forest. Both deer and cattle made greater use of aspen groves than conifer areas. Thinning patches of aspen and removing associated coniferous reproduction increased the herbaceous understory by 250 percent and aspen sprouts by 400 percent.

Aspen provides browse for both big game and sheep, but frequently grows beyond the reach of browsing animals. Patton and Jones [76] concluded that 93 percent of the existing aspen stands in the Southwest are mature or overmature and produce only a minimum number of sprouts. They suggested management for browse production based on clearcutting or burning every 20 to 30 years to promote vigorous sprouting and reduce conifer invasion into the aspen stands.

Clear-cutting aspen in northern Utah resulted in prolific sucker production and marked increase in forage production [99]. A 30 percent canopy reduction gave a less marked response, while girdling produced few aspen suckers. Suckering of aspen in the Southwest has been induced by reducing or eliminating the aspen overstory and, less effectively, by protection from browsing [56].

Clearing dense stands of aspen by mechanical top removal and grass seeding in southern Utah increased the carrying capacity for sheep from less than one sheep month per acre to eight to ten sheep months per acre in two years [93]. Clearing also greatly stimulated native browse, forbs, and grass while stimulating aspen sprouting. This suggests that reduction of dense aspen canopy followed by partial retardation of aspen sprouting may be the most beneficial treatment on game-livestock range.

Western conifer. Grazing habits of cattle and big game have been studied in ponderosa pine forests in eastern Arizona in relation to the size and kind of open areas [90]. It was concluded that openings in the tree canopy around forty-six acres in size would best coordinate deer and elk habitat management with timber management. While elk and deer used openings up to forty-six acres in size more effectively than larger openings, cattle used all sized openings. Big game used cleared areas most effectively up to 800 feet from the forest borders. Deer used the smaller openings and adjacent timbered areas about equally but showed preference for forb areas. Elk and cattle used the openings most and preferred grassy areas. It was suggested that openings should be maintained against tree invasion and that larger openings be seeded to grass for cattle and elk and the smaller openings to forbs and browse for deer.

Slash clearing following logging in ponderosa pine in northern Arizona did not affect the production or composition of understory vegetation when compared with noncleared areas six years after logging [89]. Cattle use was greater in the slash cleared area, but deer use was greater in the areas where slash had not been removed. Herbage production was greater in thinned than unthinned Arizona ponderosa pine forests (257 pounds versus 157 pounds per acre) [23].

Clary [22] has prescribed the following guidelines for treating and maintaining big game habitat in Arizona ponderosa pine forests:

(1) Small openings to provide an abundant food supply.

(2) Forest densities between 40 and 80 square feet basal area per acre to provide adequate cover and also an adequate supply of herbaceous forage.

(3) Thickets of young trees to provide bedding cover.

(4) Preferred browse plants, at least 160 plants per acre.

(5) Moderate amounts of slash for additional cover.

Clear-cutting lodgepole pine and spruce-fir forest in strips in Colorado, while leaving alternating uncut strips of the same widths, doubled the use by mule deer ten years after logging [119]. The increase in use was mostly in the cut strips, where use as indicated by pellet-group counts was three times those on the uncut strips and on adjacent virgin forest. Clear-cut strips ranged from 66 to 396 feet in width, with the 198-foot strips receiving the maximum use. Lamb and Pieper [55] recommended spot clearing about one-fifth of dense spruce forest in New Mexico for game range improvement and that tree reproduction be controlled in the clearings. Deer forage benefits from clearcut logging of subalpine forests in Colorado reached a peak before 15 years but persisted beyond 20 years [86].

Developing and maintaining big game range must result from planning rather than be left to accident. In the Northern Rockies much of the new big game habitat is created as an unplanned by-product of logging and wildfire [59]. In this area a browse and herbaceous plant cover develops rapidly following tree removal by clear cutting or fire. However, under natural succession the forest vegetation returns and progresses toward mature tree communities. As a result, wildlife habitat gradually disappears under a closed canopy of trees, and many of the remaining shrubs grow beyond the reach of big game in a few years.

Logging and related silvicultural treatments can be used to improve key big game winter range [97]. Thinning overstocked timber stands is a means of improving both timber production and forage production by releasing understory plants. Thinning dense ponderosa pine stands in eastern Washington to 26-foot spacings greatly increased understory vegetation available to big game [64]. It was suggested that thinning as a range improvement practice be an adjunct to timber stand improvement and particularly on key range areas. Tree thinning for browse and herbaceous plant production alone is apt to be uneconomical and justified only when advantages accrue from timber production also. Thinning over-dense stands of young lodgepole pine (*Pinus contorta*) and larch (*Larix* spp.) has also increased the production of understory shrubs [97].

Clear-cutting and thinning of subsequent lodgepole pine regeneration in Montana provided a grazing resource for livestock and big game for an estimated 20 years or more [5]. Peak forage production of about 1,000 lbs. per acre occurred about 11 years after clear-cutting. It was concluded that because the palatability of the indigenous understory vegetation was low, the large acreages of lodgepole pine harvested annually warrant efforts to improve the quantity and quality of forage by reseeding.

Black et al. [9] concluded that maximum summer range capacity for mule deer and elk in the Blue Mountains of Oregon occurred when 40 percent of the area was left in timber cover and 60 percent of the area was clear-cut and maintained as forage areas. Cover recommendations included adequate provision for hiding cover, thermal cover, and calving or fawning cover and providing travel lanes across otherwise open slopes. Although slash in windrows or piles was found to provide some cover on critical cover-deficient areas, large amounts of slash or dead and down material materially reduced big game use. Roads also tended to lessen game use of immediately adjacent areas.

In the Coast Range of Oregon, it was concluded that only timber clear-cutting in blocks and slash burning was an adequate and practical means for perpetuating deer habitat [26]. The highest productivity of deer forage was in the 1- to 6-year age treatment classes. Nelson [72] concluded on the general principle that prescribed burning, and even wildfire in some cases, promoted healthier and more productive big game herds in Pacific Northwest forests.

Food production for white-tailed deer in the South is largely determined by timber stand conditions [38]. Forage production under the pines is inversely related to timber density. Pine stands periodically thinned to about seventy-five square feet of basal area per acre have produced good quantities of forage and fruit over an extended period. It was concluded that clear-cutting in units of 50 to 100 acres was a practical compromise between the larger units preferred for timber harvesting and smaller units of about 30 to 40 acres considered optimum for wildlife.

Blair and Brunett [10] concluded that the capability of all-aged loblolly-shortleaf pine-hardwood stands to sustain deer largely depends upon the intensity and frequency of timber cuttings. A diversified growth of palatable and nutritious forage depended upon frequent and substantial thinning. The deer carrying capacity was one deer per 12 to 18 acres for 3 to 4 years following logging but dropped to 40 acres per deer by the seventh year, with gradual reductions thereafter.

For improving white-tailed deer habitat in conjunction with timber production in loblolly-shortleaf pine forests in the South, Halls [39] recommended widely spacing and thinning trees, retaining hardwoods along drainageways, prescribed burning, regulating the size, shape, and distribution of cutting units, establishing and maintaining forest openings, and favoring desirable food plants.

Literature Cited

1. Alley, Harold P., and Gary A. Lee. 1969. Weeds of Wyoming. Wyoming Agric. Expt. Sta. Bul. 498.
2. Allred, B. W. 1949. Distribution and control of several woody plants in Texas and Oklahoma. J. Range Mgt. 2(1):17–29.
3. Arnold, Joseph F., Donald A. Jameson, and Elbert H. Reid. 1964. The pinyon-juniper type of Arizona: effects of grazing, fire, and tree control. USDA, Agric. Res. Serv. Production Res. Rpt. 84.
4. Basile, Joseph V., and Chester E. Jensen. 1971. Grazing potential on lodgepole pine clearcuts in Montana. USDA, For. Serv. Res. Paper INT-98.
5. Beasom, Samuel L., and Charles J. Scifres. 1977. Population reactions of selected game species to aerial herbicide applications in south Texas. J. Range Mgt. 30(2):138–142.
6. Bement, R. E. 1968. Plains pricklypear: relation to grazing intensity and blue grama yield on Central Great Plains. J. Range Mgt. 21(2):83–86.
7. Bentley, Jay R. 1967. Conversion of chaparral areas to grassland. USDA Agric. Handbook 328.
8. Biswell, H. H. 1954. The brush control problem in California. J. Range Mgt. 7(2):57–62.
9. Black, Hugh, Richard J. Scherzinger, and Jack Ward Thomas. 1976. Relationships of Rocky Mountain elk and Rocky Mountain mule deer habitat to timber management in the Blue Mountains of Oregon and Washington. Elk-Logging-Roads Symposium Proc., Moscow, Ida., pp. 11–31.
10. Blair, Robert M., and Louis E. Brunett. 1977. Deer habitat potential of pine-hardwood forests in Louisiana. USDA, For. Serv. Res. Paper SO-136.
11. Branscomb, Bruce L. 1958. Shrub invasion of a southern New Mexico desert grassland range. J. Range Mgt. 11(3):129–132.
12. Brown, Harry E. 1958. Gambel oak in west-central Colorado. Ecology 39(2):317–327.
13. Burkhardt, J. Wayne, and E. W. Tisdale. 1969. Nature and successional status of western juniper vegetation in Idaho. J. Range Mgt. 22(4):264–270.
14. Burkhardt, J. Wayne, and E. W. Tisdale. 1976. Causes of juniper invasion in southwestern Idaho. Ecology 57(3):472–484.
15. Cable, Dwight R., and S. Clark Martin. 1973. Invasion of semidesert grassland by velvet mesquite and associated vegetation changes. Ariz. Acad. Sci. 8(3):127–134.
16. Carnahan, Glenn, and A. C. Hull, Jr. 1962. The inhibition of seeded plants by tarweed. Weeds 10(2):87–90.
17. Carter, Meril G. 1964. Effects of drouth on mesquite. J. Range Mgt. 17(5):275–276.
18. Christensen, Earl M. 1949. The ecology and geographic distribution of oak brush (*Quercus gambelii*) in Utah. Master's Thesis, University of Utah, Salt Lake City.
19. Christensen, Earl M. 1962. The rate of naturalization of tamarix in Utah. American Midland Naturalist 68(1):51–57.
20. Christensen, Earl M. 1964. The recent naturalization of Siberian elm (*Ulmus pumila* L.) in Utah. Great Basin Naturalist 24(3–4):103–6.
21. Christensen, Earl M., and Hyrum B. Johnson. 1964. Presettlement vegetation and vegetational change in three valleys in central Utah. Brigham Young Univ. Sci. Bul., Biol. Series, Vol. IV, No. 4.
22. Clary, Warren P. 1972. A treatment prescription for improving big game habitat in ponderosa pine forests. Proc. 16th Annual Ariz. Watershed Symposium, pp. 25–28.
23. Clary, Warren P., and Peter F. Ffolliott. 1966. Differences in herbage-timber relationships between unthinned ponderosa pine stands. USDA, For. Serv. Res. Note RM-74.
24. Cook, C. Wayne, Paul D. Leonard, and Charles D. Bonham. 1965. Rabbitbrush competition and control on Utah rangelands. Utah Agric. Expt. Sta. Bul. 454.

25. Cronin, Eugene H., and M. Coburn Williams. 1966. Principles for managing ranges infested with halogeton. J. Range Mgt. 19(4):226–227.
26. Crouch, Glenn L. 1974. Interaction of deer and forest succession on clear-cuttings in the Coast Range of Oregon. *In* Hugh C. Black. Wildlife and Forest Management in the Pacific Northwest, School of Forestry, Oregon State Univ., Corvallis, pp. 133–138.
27. Dahl, B. E., and E. W. Tisdale. 1975. Environmental factors related to medusahead distribution. J. Range Mgt. 28(6):463–468.
28. Darrow, Robert A., and Wayne G. McCully. 1959. Brush control and range improvement in the post oak-blackjack oak area of Texas. Texas Agric. Expt. Sta. Bul. 942.
29. Dasmann, W., R. Hubbard, W. G. MacGregor, and A. E. Smith. 1967. Evaluation of the wildlife results from fuel breaks, browseways, and type conversions. Proc. Tall Timbers Fire Ecology Conf. 1967:179–193.
30. Dimeo, Art. 1977. An investigation of equipment for rejuvenating browse. USDA, For. Serv., Equip. Dev. Center, Missoula, Mon.
31. Fisher, C. E., C. H. Meadors, R. Behrens, E. D. Robinson, P. T. Marion, and H. L. Morton. 1959. Control of mesquite on grazing lands. Texas Agric. Expt. Sta. Bul. 935.
32. Frankton, Clarence, and Gerald A. Mulligan. 1970 (Rev.). Weeds of Canada. Canada Dept. Agric. Pub. 948.
33. Frischknecht, Neil C. 1963. Contrasting effects of big sagebrush and rubber rabbitbrush on production of crested wheatgrass. J. Range Mgt. 16(2):70–74.
34. Frischknecht, Neil C. 1968. Factors influencing halogeton invasion of crested wheatgrass range. J. Range Mgt. 21(1):8–12.
35. Gardner, J. L. 1951. The vegetation of the creosotebush area of the Rio Grande Valley in New Mexico. Ecol. Monogr. 21(4):379–403.
36. Glendening, G. E., and H. A. Paulsen. 1950. Recovery and viability of mesquite seeds fed to sheep receiving 2,4-D in drinking water. Bot. Gazette 111(4):486–491.
37. Glendening, George E., and Harold A. Paulsen, Jr. 1955. Reproduction and establishment of velvet mesquite as related to invasion of semidesert grasslands. USDA Tech. Bul. 1127.
38. Halls, Lowell K. 1970. Growing deer food amidst southern timber. J. Range Mgt. 23(3):213–215.
39. Halls, Lowell K. 1973. Managing deer habitat in loblolly-shortleaf pine forest. J. For. 71(12):752–757.
40. Heady, Harold F. 1954. Viable seed recovered from fecal pellets of sheep and deer. J. Range Mgt. 7(6):259–261.
41. Herbel, Carlton, Fred Ares, and Joe Bridges. 1958. Hand grubbing mesquite in the semidesert grassland. J. Range Mgt. 11(6):267–270.
42. Hoffman, G. O., and R. A. Darrow. 1964 (Rev.). Pricklypear—good or bad? Texas Agric. Ext. Serv. Bul. 806.
43. Holmgren, Arthur H., and Berniece A. Andersen. 1970. Weeds of Utah. Utah Agric. Expt. Sta. Spec. Rep. 21.
44. Houston, Walter R. 1961. Some interrelations of sagebrush, soils, and grazing intensity in the Northern Great Plains. Ecology 42(1):31–38.
45. Houston, Walter R. 1963. Plains pricklypear, weather, and grazing in the Northern Great Plains. Ecology 44(3):569–574.
46. Jameson, Donald A. 1966. Pinyon-juniper litter reduces growth of blue grama. J. Range Mgt. 19(4):214–217.
47. Jameson, Donald A. 1968. Species interactions of growth inhibitors in native plants of northern Arizona. USDA, For. Serv. Res. Note RM-113.
48. Jameson, Donald A., and Thomas N. Johnsen, Jr. 1964. Ecology and control of alligator juniper. Weeds 12(2):140–142.
49. Johnsen, Thomas N., Jr. 1962. One-seed juniper invasion of northern Arizona. Ecol. Monogr. 32(3):187–207.
50. Johnson, James R., and Gene F. Payne. 1968. Sagebrush reinvasion as affected by some environmental influences. J. Range Mgt. 21(4):209–213.

51. Kinsinger, Floyd E., and Richard E. Eckert, Jr. 1961. Emergence and growth of annual and perennial grasses and forbs in soils altered by halogeton leachate. J. Range Mgt. 14(4):194–197.
52. Klemmedson, James O., and Justin G. Smith. 1964. Cheatgrass (Bromus tectorum L.). Bot. Rev. 39(2):226–262.
53. Klingman, Dayton L. 1964. Research on control of weeds and brush on grazing land. ASRM, Abstract of Papers, 17th Annual Meeting, pp. 49–50.
54. Knipe, Duane, and Carlton H. Herbel. 1966. Germination and growth of some semidesert grassland species treated with aqueous extract from creosotebush. Ecology 47(5):775–781.
55. Lamb, Samuel H., and Rex Pieper. 1971. Game range improvement in New Mexico. New Mex. Inter-Agency Range Comm. Rep. 9.
56. Larson, Merlyn M. 1959. Regenerating aspen by suckering in the Southwest. Rocky Mtn. Forest & Range Expt. Sta. Res. Note 39.
57. Lavin, Fred, Donald A. Jameson, and F. B. Gomm. 1968. Juniper extract and deficient aeration effects on germination of six range species. J. Range Mgt. 21(4):262–263.
58. Lehrer, W. P., Jr., and E. W. Tisdale. 1956. Effect of sheep and rabbit digestion on the viability of some range plant seeds. J. Range Mgt. 9(3):118–122.
59. Lyon, L. Jack. 1966. Problems of habitat management for deer and elk in the northern forests. USDA, For. Serv. Res. Paper INT-24.
60. Major, J., C. M. McKell, and L. J. Berry. 1960. Improvement of medusahead-infested rangeland. California Agric. Ext. Serv. Leaflet 123.
61. Martin, S. Clark. 1966. The Santa Rita experimental range. USDA, For. Serv. Res. Paper RM-22.
62. Martin, S. Clark. 1970. Longevity of velvet mesquite seed in the soil. J. Range Mgt. 23(1):69–70.
63. Martin, S. Clark, John L. Thames, and Ernest B. Fish. 1974. Changes in cactus numbers and herbage production. Prog. Agric. Ariz. 26(6):3–6.
64. McConnell, Burt R., and Justin G. Smith. 1970. Response of understory vegetation to ponderosa pine thinning in eastern Washington. J. Range Mgt. 23(3):208–212.
65. McCully, Wayne G. 1951. Recovery and viability of Macartney rose seeds fed to cattle. J. Range Mgt. 4(2):101–106.
66. McKell, Cyrus M., and William W. Chilcote. 1957. Response of rabbitbrush following removal of competing vegetation. J. Range Mgt. 10(5):228–230.
67. Minnich, Don W. 1969. Vegetative response and pattern of deer use following chaining on pinyon and juniper forest. ASRM, Abstract of Papers, 22nd Annual Meeting, pp. 35–36.
68. Morris, Melvin S., and Charles P. Pase. 1963. Sagebrush ecology and management in western Montana. ASRM, Abstract of Papers, 16th Annual Meeting, pp. 28–29.
69. Mueggler, Walter F. 1956. Is sagebrush seed residual in the soil of burns or is it wind-borne? USDA, Intermountain Forest & Range Expt. Sta. Res. Note 35.
70. National Research Council. 1968. Principles of plant and animal pest control. II. Weed control. National Academy of Sciences Pub. 1597.
71. Nebraska Dept. of Agric. 1968 (Rev.). Nebraska weeds. Bul. 101-R.
72. Nelson, Jack R. 1976. Forest fire and big game in the Pacific Northwest. Proc. Annual Tall Timbers Fire Ecology Conf. 15:85–102.
73. O'Rourke, J. T., and P. R. Ogden. 1970. Pinyon-juniper control: where, why? Prog. Agric. in Arizona 22(6):12–15.
74. Parker, Kenneth W., and S. Clark Martin. 1952. The mesquite problem on southern Arizona ranges. USDA Cir. 908.
75. Parker, Kittie F. 1972. An Illustrated Guide to Arizona Weeds. Univ. Ariz. Press, Tucson.
76. Patton, David R., and John R. Jones. 1977. Managing aspen for wildlife in the Southwest. USDA, For. Serv. Gen. Tech. Rep. RM-37.

77. Paulsen, Harold A., Jr. 1950. Mortality of velvet mesquite seedlings. J. Range Mgt. 3(4):281–286.
78. Paulsen, Harold A., and Fred N. Ares. 1962. Grazing values and management of black grama and tobosa grasslands and associated shrub ranges of the Southwest. USDA Tech. Bul. 1270.
79. Pechanec, Joseph F., A. Perry Plummer, Joseph H. Robertson, and A. C. Hull, Jr. 1965. Sagebrush control on rangelands. USDA Agric. Handbook 277.
80. Phillips, T. A. 1977. An analysis of some Forest Service chaining projects in region 4, 1954–1975. USDA, For. Serv. Ogden, Utah.
81. Pieper, Rex D. 1971. Blue grama vegetation responds inconsistently to cholla cactus control. J. Range Mgt. 24(1):52–54.
82. Plummer, A. Perry. 1971. Personal communication.
83. Plummer, A. Perry, Donald R. Christensen, and Stephen B. Monsen. 1969. Restoring big game range in Utah. Utah Div. of Fish & Game Pub. 68-3.
84. Plummer, A. Perry, Donald R. Christensen, Richard Stevens, and Norman V. Hancock. 1970. Improvement of forage and habitat for game. Annual Conf. of Western Assoc. of State Game & Fish Commissioners, July 15, 1970.
85. Rechenthin, C. A., H. M. Bell, R. J. Pederson, and D. B. Polk. 1964. Grassland restoration. II. Brush control. USDA, Soil Cons. Serv., Temple, Texas.
86. Regelin, Wayne L., and Olof C. Wallmo. 1978. Duration of deer forage benefits after clearcut logging of subalpine forest in Colorado. USDA, For. Serv. Res. Note RM-356.
87. Reynolds, Hudson G. 1958. The ecology of the Merriam kangaroo rat (*Dipodomys merriami* Mearns) on the grazing lands of southern Arizona. Ecol. Monogr. 28(2):111–127.
88. Reynolds, Hudson G. 1959. Brush control in the Southwest. *In* Grasslands. Amer. Assoc. for Advancement of Sci., Washington, D.C.
89. Reynolds, Hudson G. 1966. Slash cleanup in a ponderosa pine forest affects use by deer and cattle. USDA, For. Serv. Res. Note RM-64.
90. Reynolds, Hudson G. 1966. Use of a ponderosa pine forest in Arizona by deer, elk, and cattle. Rocky Mtn. Forest & Range Expt. Sta. Res. Note 63.
91. Reynolds, Hudson G. 1969. Aspen grove use by deer, elk and cattle in southwestern coniferous forests. USDA, For. Serv. Res. Note RM-138.
92. Reynolds, Hudson G., and S. Clark Martin. 1968 (Rev.). Managing grass-shrub cattle ranges in the Southwest. USDA Agric. Handbook 162.
93. Robinson, M. E., and D. H. Matthews. 1955. Aspen cutting may increase feed resources in southern Utah. Utah Farm and Home Sci. 16(2):28–29, 44.
94. Robinson, T. W. 1965. Introduction, spread, and aerial extent of saltcedar (*Tamarix*) in the western states. U.S. Geol. Survey Prof. Paper 491-A.
95. Robison, E. D., C. E. Fisher, and P. T. Marion. 1968. Mesquite reinfestation of native grassland. Texas Agric. Expt. Sta. Prog. Rpt. 2586.
96. Sampson, Arthur W., and Arnold M. Schultz. 1957. Control of brush and undesirable trees. Unasylva 10(1):19–29, 10(3):117–128, 10(4):166–182, 11(1):19–25.
97. Schmautz, Jack E. 1970. Use of ecological knowledge to improve big-game range. *In* Range and wildlife habitat evaluation—a research symposium, pp. 217–218. USDA Misc. Pub. 1147.
98. Scott, Virgil E., and Erwin L. Boeker. 1977. Responses of Merriam's turkey to pinyon-juniper control. J. Range Mgt. 30(3):220–223.
99. Scotter, George W., and Calvin O. Baker. 1969. Response of aspen reproduction and forage production to silvicultural and soil treatments applied on aspen stands in northern Utah. ASRM, Abstract of Papers, 22nd Annual Meeting, p. 59.
100. Sharp, Lee A., M. Hironaka, and E. W. Tisdale. 1957. Viability of medusahead (*Elymus caput-medusae* L.) seed collected in Idaho. J. Range Mgt. 10(3):123–126.
101. Short, Henry L., Wain Evans, and Erwin L. Boeker. 1977. The use of natural and modified pinyon-juniper woodlands by deer and elk. J. Wildl. Mgt. 41(3):543–559.

102. Smith, H. N., and C. A. Rechenthin. 1964. Grassland restoration. I. The Texas brush problem. USDA, Soil Cons. Serv., Temple, Texas.
103. Stewart, George, and A. C. Hull. 1949. Cheatgrass (*Bromus tectorum* L.)—an ecologic intruder in southern Idaho. Ecology 30(1):58–74.
104. Tanner, G. W., J. M. Inglis, and L. H. Blankenship. 1978. Acute impact of herbicide strip treatment on mixed-brush white-tailed deer habitat on the northern Rio Grande plain. J. Range Mgt. 31(5):386–391.
105. Texas Agric. Ext. Serv. 1969. Weeds of the southern United States. Texas Agric. Ext. Serv. Misc. Pub. 897.
106. Thatcher, A. P., G. V. Davis, and H. P. Alley. 1964. Chemical control of plains pricklypear in southeastern Wyoming. J. Range Mgt. 17(4):190–193.
107. Thomas, Gerald W., and Robert A. Darrow. 1956. Response of pricklypear to grazing and control measures, Texas Range Station, Barnhart. Texas Agric. Expt. Sta. Prog. Rpt. 1873.
108. Thompson, R. M. 1970. Experimental top pruning of curlleaf mountain-mahogany trees. Range Improvement Notes 15(3):1–12 (U.S. For. Serv., Intermtn. Region)
109. Tisdale, E. W., M. Hironaka, and M. A. Fosberg. 1969. The sagebrush region in Idaho—a problem in range resource management. Idaho Agric. Expt. Sta. Bul. 512.
110. Torell, Paul J., Lambert C. Erickson, and Robert H. Haas. 1961. The medusahead problem in Idaho. Weeds 9(1):124–131.
111. Tschirley, Fred H., and S. Clark Martin. 1960. Germination and longevity of velvet mesquite seed in soil. J. Range Mgt. 13(2):94–97.
112. Tschirley, Fred H., and S. Clark Martin. 1961. Burroweed on southern Arizona rangelands. Arizona Agric. Expt. Sta. Tech. Bul. 146.
113. Tueller, Paul T. 1965. Deer use on the Spruce Mountain chaining project. Nevada Agric. Expt. Sta. Prog. Rpt. 13, pp. 29–30.
114. Tueller, Paul T., and Raymond A. Evans. 1969. Control of green rabbitbrush and big sagebrush with 2,4-D and picloram. Weeds 17(2):233–235.
115. Turner, George T., and David F. Costello. 1942. Ecological aspects of the pricklypear problem in eastern Colorado and Wyoming. Ecology 23(4):419–426.
116. USDA, Agric. Res. Serv. 1965. Weed roses—pests of pasture, orchard, and range. USDA, Agric. Res. Serv. 22–93.
117. USDA, Agric. Res. Serv. 1970. Selected weeds of the United States. USDA Agric. Handbook 366.
118. Valentine, K. A., and J. B. Gerard. 1968. Life-history characteristics of creosotebush, *Larrea tridentata*. New Mexico Agric. Expt. Sta. Bul. 526.
119. Wallmo, O. C. 1969. Response of deer to alternate-strip clear cutting of lodgepole pine and spruce-fir timber in Colorado. USDA, For. Serv. Res. Note RM-141.
120. Wambolt, Carl L. 1969. Size, density, growth rate, and age of Rocky Mountain juniper and interrelationships with environmental factors. ASRM, Abstracts of Papers, 22nd Annual Meeting, pp. 57–58.
121. Ward, A. Lorin. 1973. Sagebrush control with herbicide has little effect on elk calving behavior. USDA, For. Serv. Res. Note RM-240.
122. Weed Society of America. 1966. Report of the Terminology Committee. Weeds 14(4):347–386.
123. Whitson, Robert E., Samuel L. Beasom, and C. J. Scifres. 1977. Economic evaluation of cattle and white-tailed deer response to aerial spraying of mixed brush. J. Range Mgt. 30(3):214–217.
124. Wilbert, Don E. 1963. Some effects of chemical sagebrush control on elk distribution. J. Range Mgt. 16(2):74–78.
125. Williams, Gerald, Gerald F. Gifford, and George B. Coltharp. 1969. Infiltrometer studies on treated versus untreated pinyon-juniper sites in central Utah. J. Range Mgt. 22(2):110–114.
126. Wilson, Rodney T., Bill E. Dahl, and Daniel R. Krieg. 1975. Carbohydrate concentrations in honey mesquite roots in relation to phenological devel-

opment and reproductive condition. J. Range Mgt. 28(4):286–289.

127. Woods, Frank W. 1955. Control of woody weeds: some physiological aspects. USDA, Southern Forest Expt. Sta. Occasional Paper 143.

128. Young, James A., and Raymond A. Evans. 1970. Invasion of medusahead into the Great Basin. Weed Science 18(1):89–97.

129. Young, James A., and Raymond A. Evans. 1974. Population dynamics of green rabbitbrush in disturbed big sagebrush communities. J. Range Mgt. 27(2):127–132.

Chapter 3

Manual and biological plant control

Manual Methods of Control

Manual methods of noxious plant control have the disadvantages of being slow, normally adapted only to small areas, and costly unless noxious plants are in sparse stands. However, manual control is highly selective. Manual methods may be practical for clearing scattered plants just invading grasslands, for cleaning up following other control methods, for maintenance of cleared areas against reinvasions, or for removing limited stands of poisonous plants or new infestations of the more troublesome plants before they have an opportunity to spread further. Hand cutting or grubbing of chaparral is used in maintaining fuel breaks in California and in maintaining trails and right-of-ways where only minimum maintenance is required [2].

Simple hand tools such as the saw, axe, shovel, machete, mattock, and brush hook are easy to obtain, inexpensive, simple to operate, and easy to repair; but labor cost is high if the number of plants per acre is high. Machines available for use in manual control methods include the chain saw, tree girdler, and scythette.

Grubbing. Grubbing involves digging out plants with as much of their root system as necessary to prevent sprouting and regrowth. Grubbing hoes or mattocks (combination of axe and hoe) are frequently used to chop off herbaceous plants or small shrubs near the ground level or to grub out the roots. Axes can be used on any size brush to cut top growth or aid in grubbing roots.

In Arizona hand grubbing using mattocks was concluded to be the most inexpensive and most effective method of removing mesquite

71

when the trees are small (under one inch in diameter) and scattered [42]. Initial invasions while still widely scattered were controlled for a few cents an acre. The sprout buds on mesquite seedlings are rather shallow, and completely severing the root at three to four inches below ground removes the sprouting zone.

On the Jornada Experimental Range in New Mexico, hand grubbing of mesquite was recommended for controlling light stands of small mesquite to avoid further loss of valuable grassland [17]. Under these conditions, hand grubbing, flagging, and rechecking required about forty minutes per acre. Because of time considerations, only plants with crown diameters of thirty inches or less were considered grubbable. However, when only a few plants exceeded this size, it was suggested they be grubbed also.

Grubbing hoes were widely used in earlier days on pricklypear. Pricklypear is effectively killed by grubbing when the main root is cut two to four inches below the soil surface and the detached plants are piled for burning or chemical spraying to prevent rerooting. More recently a combination of manual and biological control has been used, i.e., singeing the spines to allow cattle to graze the plants during the winter, followed by grubbing the remaining fibrous trunks [19].

Cutting. Cutting plants above the root crown or trimming off green leaves and branches effectively kills many plants that do not root or basally sprout. Cutting of sprouting species must be followed by other treatments such as chemical or grazing if the plants are to be killed. The stems of small plants can be cut with machetes or brush hooks (swung like a scythe). The hand axe or chain saw is effective for clearing small areas of medium- to large-sized trees. Wheel-mounted circular power saws are infrequently used today because of the danger of hitting rocks and breaking the blade or tipping over on hillsides.

Cutting can be highly selective with little damage to desirable shrubs or herbaceous plants. Also, cutting can be done any time but often results in slightly less sprouting when done in spring or early summer [41]. However, hand cutting is used much less now than formerly because of the high cost of labor. In areas of the Southwest where unskilled labor is inexpensive during the off season, some use might still be practiced, particularly where some posts or Christmas trees can be salvaged, government cost-sharing payments may be received, or scattered stands can be eliminated (Figure 22).

The time required to fell alligator juniper in Arizona with a chain saw ranged from one minute to more than 2 hours per tree [33]. It was concluded that the average felling time of 2.5 minutes per tree

FIG. 22. *Excellent recovery of Indiangrass and bluestems near Johnson City, Texas, following hand cutting of nonsprouting juniper. Only juniper was removed; oaks were left for wildlife food and shelter.* (Soil Conservation Service photo)

could be expected for trees 15 inches in diameter and less. Felling the very large, basally branched trees was considered uneconomical because of high labor cost.

In clearing fuel breaks in California chaparral, brush is sometimes hand cut above the root crown and then piled or windrowed for burning. This procedure has ranged in price from $400 per acre for light brush up to $1,100 for very heavy brush, based on 1973 prices [43]. Grubbing the root crowns to prevent sprouting may increase costs by another 50 percent. Various individual tree treatment methods for removing small alligator juniper trees in Arizona were compared [4]. Comparative costs with labor at $1.25 per hour plus the use of a tractor and burning equipment for burning are given in Table 5.

Girdling. Girdling consists of removing a strip of bark and outer xylem completely around the tree. This practice is commonly used in forest and woodland areas. Girdling is generally more effective on larger trees since small trees are more apt to sprout; but not all woody species are readily killed by girdling. Girdles can be made with a hand axe or saw or with a mechanically powered girdler. Girdling may be more effective against sprouting species than clear cutting since girdling stops translocation of carbohydrates to the roots but permits greater exhaustion of root reserves than does cutting. Although girdling is frequently applied whenever labor use is most efficient, some studies suggest that spring and summer treatment is slightly more effective than winter or fall treatment of hardwood species.

Girdling in east Texas was determined to be efficient in thinning or removing light stands of hardwoods such as post oak and blackjack oak [6]. It was most effective when trunk diameter was at least eight to ten inches. However, girdling was found ineffective on many other hardwoods that sprout prolifically following girdling. Both girdling and cutting were generally ineffective on Gambel oak in Arizona [27]. Following winter top cutting, all stumps sprouted. Following winter girdling, 50 percent of the tops were killed but 93 percent of the plants sprouted.

Table 5. Cost per Acre of Individual Juniper Plant Control in Arizona, 1953*

No. of trees per acre	Costs per acre		
	Hand grubbing	Hand chopping	Individual tree burning
10	$ .44	$ .24	$ .38
20	.66	.36	.61
30	.92	.46	.81

*Source of data: [4].

Biological Control by Grazing

Both dual grazing and biological control of undesirable plants by grazing are based on selective grazing. Dual grazing increases actual grazing capacity by grazing a mix of animal species with different dietary preferences to uniformly harvest all potential forage plants. Biological control by grazing is effective when the right combination of grazing animals, season and system of grazing, and stocking rates results in heavy grazing of undesirable or less desirable plants to the competitive advantage of the favored plants. However, biological control by grazing often requires that the larger brush and trees first be brought down within reach of the grazing manipulators by mechanical or chemical means or by burning.

Selective grazing by livestock. It is well demonstrated that different species of grazing animals have different forage preferences. Cattle, while being primarily grass eaters, consume certain shrubs and forbs they find palatable. Sheep utilize numerous forbs and many shrubs and grasses. While deer select many forbs, both deer and goats tend to use large quantities of browse. On mountain summer range in the Intermountain Region, cattle grazing tends to reduce the grass component in the stand and increase the forbs and shrubs. Since sheep encourage many grass species by relatively heavier grazing of snowberry (*Symphoricarpos* spp.), geranium (*Geranium* spp.), dandelion (*Taraxacum officinale*), and butterweed (*Senecio serra*), moderate grazing of sheep on range tends to improve it as cattle range.

Small soapweed (*Yucca glauca*) has been controlled on Nebraska Sandhills grasslands by grazing with mature cattle during winter. Some ranchers report that winter feeding of protein supplements at more than the normal rate results in heavier grazing of soapweed. On the other hand, small soapweed in the Sandhills increases under continuous summer grazing by cattle. Sheep have been used with variable results in removing from dairy pastures weeds that give off flavors to milk.

Grazing cattle range with sheep and deer to reduce tall larkspur (*Delphinium* spp.), a plant poisonous to cattle but not to sheep and deer, has been suggested on subalpine grassland ranges. The lasting effectiveness of this approach has not been demonstrated. The presence of troublesome amounts of larkspur may indicate greater suitability of the area for grazing animals other than cattle rather than their use as a larkspur control measure.

Hubbard and Sanderson [20] in California found that grasses are competitive with bitterbrush (*Purshia tridentata*), a highly palatable and productive browse plant for deer. The reduced vigor of bitterbrush and the increased vigor of the grasses appeared to be associated with reduced cattle grazing. When the herbaceous competition was removed, bitterbrush previously in poor vigor increased average leader or twig length by 90 percent and average numbers of leaders by 223 percent. This and similar observations in other areas have suggested it is advantageous to graze some cattle on range managed primarily for deer when cattle grazing is managed to make maximum use of grasses and minimum use of bitterbrush and other shrubs palatable to deer.

Grazing cattle or sheep in the spring and early summer has shown promise in northern Utah for thinning the understory of forbs and grasses and increasing the vegetative output of desirable shrubs for winter browsing by elk [50]. Spring-grazing sheep on Utah foothill range later winter-grazed by mule deer resulted in a 2.3-fold increase in total range grazing capacity over deer alone [31]. The spring sheep grazing apparently increased deer browse production when the sheep stocking rates were carefully regulated and the sheep were removed from the range prior to July. It was considered that the complete elimination of livestock grazing from any Intermountain range is apt to encourage grass and forb production at the expense of shrubs [50]. Policies of managing state-owned big game range commonly includes provision for controlled cattle grazing to promote a better balance of forage species.

Spring deferment and heavy fall grazing by sheep (sixty sheep days per acre) on native sagebrush-grass range at Dubois, Idaho, improved range condition faster than total protection [29]. This system of grazing increased grasses and forbs but decreased sagebrush and raised range condition from poor to fair in seven years. Heavy fall grazing following spring rest for two or more years in succession was recommended as a range improvement practice on sagebrush-grass range. However, it was noted that successful use of this treatment required that sufficient perennial grass be present to respond and that the range must not be grazed in the spring if heavily grazed in the fall.

Sheep grazing in late fall was also found by Frischknecht and Harris [12] to effectively control big sagebrush on spring cattle range seeded to crested wheatgrass if applied before sagebrush became too dense. This treatment was effective when the density of big sagebrush was only about 1½ plants per 100 square feet and primarily resulted in a decrease in sagebrush plant size and in limiting reproduction. Cattle were found ineffective at any season in controlling big sagebrush, even when big sagebrush was present in minimal amounts.

FIG. 23. *Goats on range near Sonora, Texas, provide biological control of low-growing brush and woody plant sprouts.* (Soil Conservation Service photo)

Sheep grazing was found useful in California in reducing scattered stands of Klamath weed [36]. To be effective it was necessary to concentrate sheep grazing for short periods, avoiding long continued heavy grazing. The combination of sheep grazing during March and April to reduce vigor by grazing off new shoots and again in early summer to remove flowering tops was suggested. Selective grazing of leafy spurge in crested wheatgrass stands by sheep in Canada reduced the leafy spurge but was concluded to be an ineffective control method [28]. Leafy spurge was reduced in basal area from 3.14 percent to 0.22 percent by five years of selective early spring grazing, while crested wheatgrass increased from 17.7 to 22.0 percent. However, the remaining leafy spurge plants were considered potentially capable of reinfestation if the grazing treatment was discontinued.

Goating. The most effective control of undesirable woody range plants by livestock grazing has been from the use of goats. An FAO report [8] concluded that because of their general browsing tendencies, goats can be a potent factor in controlling woody plants and in preventing their return in areas of low and erratic precipitation throughout the world. It found that the adaptability of goats to desolate, semi-arid, exhausted, poorly watered sites was phenomenal, and they can often subsist after all other livestock have had to be removed.

Since goats will utilize large amounts of browse in their diets, they have been widely used in Texas and adjacent Mexico to control low-growing brush or sprouts resulting from previous control treatments (Figure 23). Repeated defoliation of woody species by goats has been found to either control the plant growth and spread or kill the plants if continued long enough. Goating has been effective on a rather wide range of shrubs in Texas including oaks, mesquite, sumac (*Rhus* spp.), and hackberry (*Celtis* spp.).

Unless carefully managed, goats will also graze out desirable forage species. The goat can cope with a variety of dietary alternatives. But when the shrubs available for browsing are highly unpalatable and unacceptable or unavailable, the goat often readily shifts to herbaceous species and may only worsen an already deteriorated situation [26]. Prior treatment by roller chopping or chaining often increases both accessibility and acceptability of woody plants for goat browsing.

Yearlong stocking rates commonly used to control sprouts in Texas are one goat for each two or three acres [6, 38, 41]. However, short-term grazing by larger numbers of goats, i.e., five to eight goats per acre for a thirty-day period, is more effective. The use of small pastures also increases shrub control. Grazing should be carried to the point of leaf defoliation of the shrubs while assuring that other preferred forage plants are not excessively grazed. A high degree of brush control usually takes about three years of intensive defoliation. Subsequent light stocking with goats grazed in common with other kinds of livestock will provide continuing maintenance control of the brush. In order to be an effective follow-up treatment, goating should immediately follow mechanical or chemical clearing in order to adequately control sprouts.

Goat enterprises have been added on many Texas ranches not only to control brush regrowth but also to provide additional ranch income. The economic returns from range goat enterprises have been similar to those from range cattle and sheep enterprises [32]. The Angora goat produces mohair, and the Spanish goat is readily marketed for meat. In a north central Texas study, goat enterprises were added without reducing the original number of cattle, thereby approximately doubling the number of animal units maintained [30]. The goat enterprises averaged 359 head and over a five-year period paid for their original purchase price, the added fencing and shelter costs, all year-to-year costs incurred in handling the goats, and the cost of mechanically clearing an average of 518 acres of range at an average price of $7.23 per acre.

Both the Spanish goat and the mohair-type Angora goat have been used successfully in biological brush control, but differences in physiology and nutritional requirements make the Spanish goat the more preferable for this use [32]. The latter has been found to be more rangy and can browse to heights of seven feet or more. It is readily available for purchase in the Southwest, is more prolific, is less vulnerable to extreme weather conditions, and includes more browse in its diet than the Angora goat.

Goat management problems have generally restricted the use of goating to Texas and the Southwest, but potential use may exist in other southerly areas. These management problems include the need for special goat-proof fences, sheds for protection from severe cold and rain, the shortage of winter forage where brush species are deciduous, predation by bobcats, coyotes, or dogs, particularly on the young kids, and the pulling of mohair by thorny plants [41]. Goats are highly competitive with deer and to a considerable extent with sheep, and stocking rates should be adjusted accordingly.

Goat grazing of Gambel oak sprouts following mechanical treatment in Colorado was found to provide up to 95 percent control [7]. The mechanical treatment was found necessary to allow the animals full access to all of the foliage, and roller chopping was suggested. Two defoliations of Gambel oak per year by concentrated goat grazing in short time periods under rotational grazing was found the most effective grazing treatment. No problems were observed from permitting the goats to mix freely with range cattle.

Following a wild fire about five years previous, the diet of Spanish goats in California chaparral under summer grazing consisted of about 80 percent of scrub oak and chamise, but little use was made of the manzanita and ceanothus [45]. In southern Utah, blackbrush normally provides little forage for range cattle because of its spiny, woody growth [40]. However, it was found that goats readily removed the woody branch stems by browsing, and the resulting new growth was more accessible, palatable, and nutritious for cattle grazing than the untreated plants. It was anticipated that goating would have to be repeated on a regular basis to provide these advantages for cattle grazing.

Selective grazing by deer. Selective grazing by deer on cattle range may result in reduced or limited competition with cattle. In fact, where many shrubs occur that are palatable to deer but not to cattle, deer grazing on cattle range beneficially controls such shrubs. On chained and seeded foothill areas in central Utah, heavy winter and early spring grazing by mule deer, in combination with competition from the herbaceous understory, prevented even juniper and pinyon from regaining dominance when present only as scattered small trees [46].

In California chaparral, small burns of four to five acres in size attract deer sufficiently to suppress sprouts and seedlings of chamise, oaks, and several other woody plants [2, 3, 44]. This localized heavy grazing by deer tends to maintain the herbaceous forage plants sought by cattle. On large burns, however, browse may be so abundant that brush control by resident deer is insufficient to be effective. On very large burns, sheep or goats may also be needed in suppressing sprouts. When grazed in midsummer after grasses and forbs have dried, sheep and goats may materially retard the reinvasion of chaparral shrubs.

Biological Control
by Insects

The biological control of noxious plants by insects has solved a few of the worst noxious plant problems, such as pricklypear in Australia. Since it deals with a complex balance of nature and with natural controls manipulated by man, its success is predicated upon detailed planning and preparation. Biological control of noxious plants by insects employs specific host plant-phytophagous insect relationships to reduce the population of the noxious plant to a status of little or no significance [11]. Successful biological control by insects in the past has generally been characterized as follows [1, 11, 39]:

1. A noxious plant has been accidentally introduced from a foreign land.
2. Its natural or potential predatory insects were left behind.
3. These natural enemies were identified, increased in population, and were released under conditions favorable to establishment in the new habitat.
4. Initial control was augmented by additional insect introductions or other control methods, when needed.

Before insect introductions can be made, a large amount of investigation must be carried out. Collections of the plant's natural and potential enemies are studied. After further investigations, the number of species being considered for introduction are further reduced. Prior to actual introduction, the insect must be determined to be:

1. Highly destructive of the noxious plant.
2. Highly specific as to its host plant and harmless to desirable plants.
3. Able to survive in the plant's habitat.
4. Free of its natural parasites.
5. Apparently subject to no new parasites in the host plant's habitat.

Before biological control by insects can be initiated, the plant in question must be widely accepted as an undesirable plant since the effects of this plant control method cannot be limited to a selected area as they can with other control methods. This is accomplished by an accurate appraisal of its net value made over the entire contiguous land mass where it is found. Although pricklypear programs in Australia have been highly successful, programs in the United States have been opposed largely out of deference to claims of emergency feed value for livestock, to considerations of its value in soil erosion control and habitat for non-game wildlife [22], and to certain other minor economic values.

Employment of biological control of noxious plants by insects has been reduced because of two principal factors [23]:

1. Conflict in acceptance of a plant as undesirable.
2. Fear that risks are too great compared with chances of success.

However, prior to introduction and release, the parasitic insects are carefully studied to assure they will not harm other plants or at least other useful plants. The purpose of biological control is not eradication but rather reduction of the noxious plant to a negligible status.

The emphasis thus far has mostly been targeted on alien plants, plants of natural areas, and those of a perennial nature [22]. Although these types of plants appear most promising there is hope that biological control by insects can be extended to other types of plants. However, the likelihood of finding natural enemies capable of controlling introduced weeds with a minimum of manipulation and attention is greater than is the case for indigenous or native weed species [1]. It is suggested that biological control by insects be considered as an alternative control method to consider rather than only as a last resort [23].

Insects have received primary attention thus far for the biological control of weeds [1, 11, 13]. Insects have the advantage of large size, high rates of reproduction, and high degree of host specificity. However, although their use has been negligible thus far, other natural enemies of potential control value include fungi, bacteria, viruses, parasitic higher plants, plant mites, nematodes, and an assortment of other small or microscopic animals [1].

The introduced enemies may destroy the individual plants through direct destruction of vital parts [22]. They may also destroy indirectly through creating a favorable environment for infection by other pathogenic agents or cancelling the competitive advantages possessed by the plant in its present environment. In most instances, the initial results have been indicative of a permanent control. In a few cases the initial effects have not been maintained and this partly because of an increase in the influence of natural enemies of the introduced control agents [23]. Additional control measures, providing competition by forage species through seeding competitive forages, and controlled grazing as to timing and intensity augments biological control by insects [37].

Pricklypear. The control of pricklypear in Australia has been the classic example of effective biological control by insects [5, 23]. This plant is not native to Australia but was introduced as an ornamental and hedge planting and as a potential fodder plant. By 1925 *Opuntia stricta* and six related species had escaped and spread over sixty million acres, of which about thirty million acres had become useless forage producing areas. Since the cost of chemical and mechanical control was generally greater than the value of the land, resort was made to biological control.

A total of twelve different insect species were introduced beginning in 1925 [5, 23]. After the introduction and striking success of an Argentine moth borer (*Cactoblastus cactorum*) in 1926, no further introductions were made. Although there were later fluctuations and adjustments of moth borer populations and pricklypear populations,

1935 to 1940 was a period of virtually complete control. The larvae of this insect species live inside the pricklypear plants and tunnel through and destroy the aboveground portions of the plants, penetrating even into the roots. This damage opens the plant tissues to the attack of secondary bacteria and fungi which furthers the destruction.

On Santa Cruz Island, located twenty-five miles south of Santa Barbara, California, biological control of pricklypear with insects has given considerable success [14]. Two species (*Opuntia littoralis* and *O. oriocola* and their hybrids) are native on the island. Rangelands on Santa Cruz Island, severely overgrazed by feral sheep, were invaded by these resident pricklypears until about 40 percent of the range was rendered useless for cattle grazing. In 1951 a cochineal insect (*Dactylopius* spp.) was introduced from Mexico by way of Hawaii. When combined with the removal of the sheep and restricted cattle management, the introduction and continuing distribution of the cochineal insect materially reduced pricklypear on the island. It was noted that two important predators of the insect on the Mexico mainland were not present on Santa Cruz Island.

St. Johnswort or Klamath weed (*Hypericum perforatum*). St. Johnswort is an aggressive perennial forb accidentally introduced into the United States from Europe. It was estimated in 1949 that the plant infested about five million acres in the United States with much of this area being heavily infested [23]. About half of this infestation occurred in northern California [25]. This poisonous plant later became established in other areas of the country, particularly in the Pacific Northwest states of Oregon, Washington, Montana, and Idaho. St. Johnswort was estimated in 1956 to infest 211,000 acres in Montana with an additional 20 million acres subject to infestation [34].

In the mid 1940s, two leaf-feeding beetles (*Chrysolina gemellata* and *C. hyperici*) were introduced into northern California [23]. By 1958 these beetles had reduced St. Johnswort to less than 1 percent of its former abundance in that area, and St. Johnswort was concluded no longer to constitute a problem there. Introductions of the *Chrysolina* beetles into the colder regions of Oregon, Washington, Idaho, and Montana have provided variable control. However, by 1971 the main St. Johnswort infestations in Oregon and many in Washington and Idaho had come under control [24]. The continuing predator-prey relationship established in Idaho has reduced and maintained St. Johnswort populations at about three percent of its 1948 levels [47]. Increases in plant populations in the intervening years have been followed within a year or two by corresponding *Chrysolina* beetle populations.

The larvae of *Chrysolina* beetles feed on the foliage of St. Johnswort and keep plants defoliated over a long period of time in the winter and early spring when food reserves in the plant are naturally low [23, 25]. Although adult feeding of the beetles alone extends over too short a period to give effective control, it has helped to control plants that greened up in the spring. Death of St. Johnswort is not caused directly by the loss of foliage but rather indirectly to inability to maintain sufficient root systems to survive low soil-moisture periods in the summer.

Chrysolina beetles provide the means of controlling St. Johnswort to the point that it is no longer a menace [36]. However, St. Johnswort seed is long lived and capable of germination and establishment after the original population has been decimated. Effective control requires that a few plants survive. Otherwise the beetle colonies become locally extinct and unavailable to control new St. Johnswort plants developing from seed.

Chrysolina beetles were initially distributed by placing a colony of about five thousand beetles in a ten-square-foot area where the St. Johnswort was thick [36]. More recently many colonies have been started with only a few beetles [24]. After taking three or more years to attain heavy attack densities, St. Johnswort populations have subsequently been reduced below toxic and competitive levels within a period of one or two years. One colony per 1,000 acres has been adequate where St. Johnswort is found in continuous stands [36].

On ranges in California the biological control of St. Johnswort has given marked increases in perennial grasses such as California oatgrass (*Danthonia californica*) and various winter annuals, particularly soft chess (*Bromus mollis*) and filaree (*Erodium* spp.) [21, 25]. However, grass seeding following St. Johnswort control has been required on many areas to achieve maximum forage production. The importance of good management practices following St. Johnswort has been emphasized if the invasion of other undesirable plants is to be prevented. Finding that medusahead, rather than perennial grasses, replaced St. Johnswort in some areas, Tisdale [47] concluded that more than biological control, such as reseeding and improved grazing practices, would be required to provide lasting range improvement.

Tansy ragwort (*Senecio jacobaea*). Tansy ragwort is a biennial or occasionally a winter annual or even perennial and is a native of Europe. It was accidentally introduced into Australia and New Zealand where it became a major problem. More recently it has been accidentally introduced on both the East and West Coasts of the United States. It has invaded and is now spreading from northern California and Oregon up to British Columbia [5]. Not only is it a competitive, unpalatable plant but it also is highly toxic to cattle and horses. It is not toxic to sheep and goats under range conditions, but symptoms have been produced experimentally.

Beginning in 1959 the cinnabar moth (*Tyria jacobaeae*) was introduced into California to biologically control tansy ragwort [15]. The larvae of cinnabar moth feed on all aboveground plant parts of tansy ragwort but prefer the foliage and young buds and flowers (Figure 24). These feeding habits greatly reduce plant vigor and seed production (Figure 25). The cinnabar moth has given partial control in parts of California and Oregon, but more effort has been needed in distributing the insects [16].

FIG. 24. *Larvae of cinnabar moth providing biological control by feeding on tansy ragwort in California.* (SEA—Agricultural Research)

To bolster the biological control of tansy ragwort, the ragwort seed fly (*Hylemya seneciella*) was brought to California and Oregon in 1966 [49]. The seed fly larvae consume many seeds and damage other seeds. However, survival has been limited under field conditions and the effectiveness of the ragwort seed fly is uncertain [10].

A third insect introduced for biological control of tansy ragwort is a European flea beetle (*Longitarsus jacobaeae*) [9]. The larvae of this flea beetle feed on the root crowns of tansy ragwort during the winter and spring while the adults feed on the plant foliage. Although this insect had been initially established in a field near Fort Bragg,

California, its longevity and effectiveness had not yet been determined in 1971 [10].

Miscellaneous. The mesquite twig girdler (*Oncideres rhodosticta*) has inflicted considerable damage on mesquite and shows promise in biological control programs [48]. In one locality in Texas, about 90 percent of the mesquite trees were attacked, and about 40 percent of all branches 0.2 to 0.8 inch in diameter were girdled. Biological control studies on mesquite are continuing [13].

A weevil (*Rhinocyllus conicus*), proved host specific to plants in *Carduus, Cirsium, Silybum,* and *Onopordum* thistle genera, has shown promise of controlling musk thistle (*Carduus nutans*) in Montana [18]. The larvae of this weevil feed on the developing seed and the weevil has

FIG. 25. *Tansy ragwort control by the cinnabar moth on grassland near Fort Bragg, California.* Top, *before control;* bottom, *after control.* (Courtesy of Kenneth E. Frick)

85

been found to greatly reduce the primary flowers of musk thistle. However, this 1976 report concluded that a secondary biological agent was needed to control seed production from the later seed heads. Augmenting biological control of musk thistle with 2,4-D or by mowing was suggested.

In addition to the successful applications of biological control discussed previously, current research in the United States and Canada is also exploring biological control by insects of such noxious plant species as Canada thistle (*Cirsium arvense*), musk thistle (*Carduus nutans*), bull thistle (*Cirsium vulgare*), Russian thistle (*Salsola kali*), Russian knapweed (*Centaurea repens*), halogeton (*Halogeton glomeratus*), hoary cress or whitetop (*Cardaria draba*), leafy spurge (*Euphorbia esula*), dalmatian toadflax (*Linaria dalmatica*), puncturevine (*Tribulus terrestris*), saltcedar (*Tamarix pentandra*), western ragweed (*Ambrosia psilostachya*), common ragweed (*A. artemisiifolia*), and silverleaf nightshade (*Solanum elaeagnifolium*) [13, 37].

Literature Cited

1. Andres, L. A., C. J. Davis, P. Harris, and A. J. Wapshere. 1976. Pages 481–499 *in* C. B. Huffaker and P. S. Messenger. Theory and Practice of Biological Control. Academic Press, New York.

2. Bentley, Jay R. 1967. Conversion of chaparral areas to grassland. USDA Agric. Handbook 328.

3. Biswell, H. H., R. D. Taber, D. W. Hedrick, and A. M. Schultz. 1952. Management of chamise brushlands for game in the north coast region of California. California Fish and Game 38(4):453–484.

4. Cotner, Melvin L. 1963. Controlling pinyon-juniper on southwestern rangelands. Arizona Agric. Expt. Sta. Rpt. 210.

5. Crafts, Alden S., and Wilfred W. Robbins. 1962 (Third Ed.). Weed Control. McGraw-Hill Book Co., Inc., New York City.

6. Darrow, Robert A., and Wayne G. McCully. 1959. Brush control and range improvement in the post oak-blackjack oak area of Texas. Texas Agric. Expt. Sta. Bul. 942.

7. Davis, Gary G., Lawrence E. Bartel, and C. Wayne Cook. 1975. Control of Gambel oak sprouts by goats. J. Range Mgt. 28(3):216–218.

8. French, M. H. 1970. Observations on the goat. Food and Agric. Organ., Rome, Italy.

9. Frick, Kenneth E. 1970. *Longitarsus jacobaeae,* a flea beetle for the biological control of tansy ragwort. I. Host plant specificity studies. Annals of the Entomological Society of America 63(1):284–296.

10. Frick, Kenneth E. 1971. Personal correspondence.

11. Frick, Kenneth E. 1974. Biological Control of Weeds: History, Theoretical, and Practical Applications. Pages 204–223 *in* Proceedings of the Summer Institute of Biological Control of Plant Insects and Diseases, Univ. Press of Miss., Jackson.

12. Frischknecht, Neil C., and Lorin E. Harris. 1973. Sheep can control sagebrush on seeded range. Utah Sci. 34(1): 27–30.

13. Goeden, R. D., L. A. Andres, T. E. Freeman, P. Harris, R. L. Pienkowski, and C. R. Walker. 1974. Present status of projects on the biological control of weeds with insects and plant pathogens in the United States and Canada. Weed Sci. 22(5):490–495.

14. Goeden, R. D., C. A. Fleschner, and D. W. Ricker. 1968. Insects control pricklypear cactus. California Agric. 22(10):8–9.

15. Hawkes, Robert B. 1968. The cinnabar moth, *Tyria jacobaeae,* for control of tansy ragwort. J. Econ. Entom. 61(2):499–501.

16. Hawkes, Robert B. 1971. Personal correspondence.

17. Herbel, Carlton, Fred Ares, and Joe Bridges. 1958. Hand-grubbing mesquite in the semidesert grassland. J. Range Mgt. 11(6):267–270.

18. Hodgson, J. M., and N. E. Rees. 1976. Dispersal of *Rhinocyllus conicus* for biocontrol of musk thistle. Weed Sci. 24(1):59–62.

19. Hoffman, G. O., and R. A. Darrow. 1964 (Rev.). Pricklypear—good or bad? Texas Agric. Ext. Bul. 806.

20. Hubbard, R. L., and H. R. Sanderson. 1961. Grass reduces bitterbrush production. California Fish & Game 47(4):391–398.

21. Huffaker, Carl B. 1951. The return of native perennial bunchgrass following the removal of Klamath weed (*Hypericum perforatum* Linn.) by imported beetles. Ecology 32(3):443–458.

22. Huffaker, C. B. 1957. Fundamentals of biological control of weeds. Hilgardia 27(3):101–157.

23. Huffaker, C. B. 1959. Biological control of weeds with insects. Annual Rev. of Entom. 4:251–276.

24. Huffaker, C. B. 1971. Personal correspondence.

25. Huffaker, C. B., and C. E. Kennett. 1959. Ten-year study of vegetational changes associated with biological control of Klamath weed. J. Range Mgt. 12(2):69–82.

26. Huss, Donald L. 1972. Goat response to use of shrubs as forage. USDA, For. Serv. Gen. Tech. Rep. INT-1, pp. 331–338.
27. Johnsen, Thomas N., Jr., Warren P. Clary, and Peter F. Ffolliott. 1969. Gambel oak control on the Beaver Creek Pilot Watershed in Arizona. USDA, Agric. Res. Serv. ARS 34–104.
28. Johnston, A., and R. W. Peake. 1960. Effect of selective grazing by sheep on the control of leafy spurge (*Euphorbia esula* L.). J. Range Mgt. 13(4):192–195.
29. Laycock, William A. 1961. Improve your range by heavy fall grazing. National Woolgrower 51(6):16, 30.
30. Magee, A. C. 1957. Goats pay for clearing Grand Prairie rangelands. Texas Agric. Expt. Sta. Misc. Pub. 206.
31. Malechek, John C. 1978. Animal production on rangelands. *In* Symposium: Agriculture, Everybody's Business, Brigham Young University, Provo, Utah, pp. 1–18.
32. Merrill, Leo B., and Charles A. Taylor. 1976. Take note of the versatile goat. Rangeman's J. 3(3):74–76.
33. Miller, Robert L. 1971. Clearing an alligator juniper watershed with saws and chemicals: a cost analysis. USDA, For. Serv. Res. Note RM-183.
34. Morris, Melvin S., and Joseph Meuchel. 1956. The problem of St. Johnswort, a noxious plant in western Montana. Montana Forest & Cons. Expt. Sta. Bul. 4.
35. Murphy, Alfred H. 1955. Vegetational changes during biological control of Klamath weed. J. Range Mgt. 8(2):76–79.
36. Murphy, Alfred H., R. Merton Love, and Lester J. Berry. 1954. Improving Klamath weed ranges. California Agric. Ext. Serv. Cir. 437.
37. National Research Council. 1968. Principles of plant and animal pest control. II. Weed control. National Academy of Sciences Pub. 1597.
38. Norris, Joe B. 1968. Biological control of oak. ASRM, Abstract of Papers, 21st Annual Meeting, p. 29.
39. Piemeisel, Robert L., and Eubanks Carsner. 1951. Replacement control and biological control. Science 113(2923):14–15.
40. Provenza, Fred D. 1978. Getting the most out of blackbrush. Utah Sci. 39(4):144–146.
41. Rechenthin, C. A., H. M. Bell, R. J. Pederson, and D. B. Polk. 1964. Grassland restoration. II. Brush control. USDA, Soil Conservation Serv., Temple, Texas.
42. Reynolds, Hudson G., and S. Clark Martin. 1968 (Rev.). Managing grass-shrub cattle ranges in the Southwest. USDA Agric. Handbook 162.
43. Roby, George A., and Lisle R. Green. 1976. Mechanical methods of chaparral modification. USDA Agric. Handbook 487.
44. Sampson, Arthur W., and Arnold M. Schultz. 1957. Control of brush and undesirable trees. Unasylva 10(1):19–29, 10(3):117–128, 10(4):166–182, 11(1):19–25.
45. Sidamed, A. E., S. R. Radosevich, J. G. Morris, and W. L. Graves. 1978. An assessment of goat grazing in chaparral. Calif. Agric. 32(10):12–13.
46. Stevens, Richard, Bruce C. Giunta, and A. Perry Plummer. 1975. Some aspects in the biological control of juniper and pinyon. *In* Proceedings of the Pinyon-juniper Symposium, Utah State Univ., Logan, pp. 77–82.
47. Tisdale, E. W. 1976. Vegetational responses following biological control of *Hypericum perforatum* in Idaho. Northwest Sci. 50(2):61–75.
48. Ueckert, Darrell N., Kenith L. Polk, and Charles R. Ward. 1971. Mesquite twig girdler: a possible means of mesquite control. J. Range Mgt. 24(2):116–118.
49. USDA, Agric. Res. Serv. 1966. Biological control of range weed. Agric. Res. 15(6):14.
50. Utah Agric. Expt. Sta. 1978. Grazing to make forage. Utah Sci. 39(4):150–151.

Chapter 4

Mechanical plant control

Choosing the best mechanical method to use in controlling undesirable range plants depends upon several factors:

1. *Characteristics of undesirable species present*—density, size of stem, brittleness, and sprouting ability.
2. *Need for seedbed preparation and revegetation.*
3. *Topography and terrain.*
4. *Kind of soil*—depth, amount and size of rock, erosiveness, and degree of compaction.
5. *Site potential*—cost of improvement must be consistent with expected productivity.

Bulldozing

The bulldozer consists of a crawler tractor equipped with a heavy-duty pusher blade which can be raised and lowered hydraulically or by cable. The standard straight blade consists of a solid, concave moldboard to which is attached a knifelike cutting edge at the bottom. The standard dozer is also frequently equipped with manual or power angling of the blade. Various special attachments and replacement blades are available. The bulldozer with straight blade is used to sever woody stems at or below the ground level and for uprooting and lifting larger brush and trees from the ground. It is also used for walking over and killing brittle brush species where minimum soil disturbance is desired and for crushing brush prior to burning. Other uses are for clearing and piling brush for burning or for windrowing on the contour to help in erosion control.

Bulldozing is best adapted to removing scattered stands of large brush or trees. It is less adapted to thick stands of brush, particularly sprouting species, and is more costly. Sprouting brush can be controlled by this method only if the root crown and sprouting zones of the roots are removed from the soil. Small trees in the open or under felled trees are commonly missed. Thus, control may be incomplete unless considerable follow-up is provided or complete blading of the ground surface is accomplished. Another problem with bulldozing is that much soil may be torn up and large pits left.

Pinyon and juniper grow back very slowly following bulldozing where clearing is thorough. However, small trees missed in the operation respond rapidly to release from removal of the larger trees [3]. Bulldozing is an effective but slower method than chaining for removing moderately dense stands of juniper and pinyon [53].

Blading with the dozer blade at ground level has also been used successfully in Utah following chaining in one direction to take out and windrow for burning the brush on sagebrush-juniper sites (Figure 26) [19]. Since some soil is scraped into the windrows along with the brush, placing windrows on the contour has been suggested.

Blading or scraping with a motor patrol at ground level has also been used to shear off small brush while leaving herbaceous vegetation mostly undisturbed. However, it has been effective on only a few undesirable plant species and is limited to nearly level land, fro-

FIG. 26. *Blading and windrowing sagebrush-juniper vegetation near Benmore, Utah, in preparation for drilling.* (Bureau of Sport Fisheries and Wildlife photo by Maurice F. Baker)

zen soil, and light blade pressure. Otherwise, grasses are damaged, and excessive amounts of soil end up in the windrows. Blading during ideal conditions in Wyoming gave nearly 100 percent control of pricklypear at a cost of $6 (1968) per acre [31]. But sprouting of pricklypear in the windrows had to be prevented by burning, chemical means, or additional mechanical control.

After evaluating BLM projects in the Intermountain Region, Aro [4] concluded that chaining-windrowing-drilling was the most costly but best method of converting pinyon-juniper sites to grassland. This method typically killed 95 percent or more trees, reduced shrub cover, and increased herbage production from negligible amounts to over 500 pounds annually. The following criteria were suggested for site selection for such intensive conversion: (1) deep soil (twenty-four inches minimum), (2) medium soil texture, (3) soil sufficiently rock free to allow seed drilling, and (4) slope not exceeding 15 percent.

The cost of bulldozing with individual plant removal increases as the density and size of brush and trees increase. The type of soil, amount of soil moisture, amount of rock, kind of terrain, size of tractor required and used, and experience of the operator also affect the cost of bulldozing. In Arizona studies reported in 1963, bulldozing pinyon and juniper cost from $.78 to $12.39 per acre with an average of $4.53 [11]. Based on contract charges of $12.00 per hour of bulldozer use, predicted costs of removing individual small trees ranged from $3.67 per acre with 50 trees per acre to $7.52 with 150 trees per acre. Corresponding costs predicted for removing large trees respectively were $4.38 and $9.66 per acre. Although cost per acre increases as tree density increases, the cost per tree decreases because of reduced travel time between trees. The average cost of bulldozing dense stands of juniper and pinyon trees into piles in southern Utah in 1962 was $14.00 per acre [24].

Bulldozing has been found to be well adapted for uprooting scattered stands of relatively large mesquite trees [43]. Dense stands of mesquite or oak in Texas cost $12 to $20 per acre to remove by bulldozing in 1964 [58]. Summer was concluded best for dozing mesquite, but control was considered adequate anytime. In a hardwood-pine area, bulldozing cost $11 per acre for a tree density of about 427 per acre if 90 percent of the trees were six inches or less in diameter at breast height [48].

The cost of bulldozing California chaparral in 1967 ranged from $20 per acre for light stands to $40 per acre for medium brush density [5]. It was concluded that bulldozing as a control practice on areas having coarse, erodible soils must be limited to gentle to moderate slopes and that topsoil disturbance must be minimized. Bull-

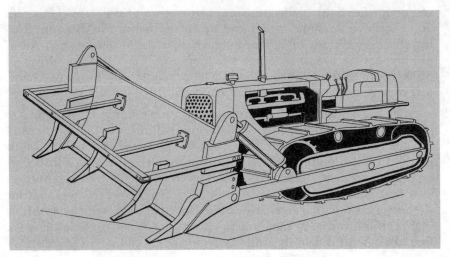

FIG. 27. *The hula dozer provides a means of individual tree or shrub removal but is used in many other phases of range development because of its versatility.*

dozing is recommended for controlling the early invasion of willow and aspen into Canadian fescue prairie [36]. Grass production under dense aspen in that area is only about one-third of that on adjacent fescue prairie (400 pounds versus 1,200 pounds per acre). Dense aspen production is mostly removed by bulldozing, piling, and later burning. Cleared areas in Canada are frequently cultivated for two years and then seeded to grass.

Hula dozer. The Hula dozer is a hydraulic power-controlled tilt dozer, generally equipped with four removable digging teeth spaced along the dozer blade (Figure 27) [57]. Both the side tilt and pitch of the blade are adjusted from the driver's seat and while the bulldozer is working under load. Most Hula dozers are equipped for manual angling or hydraulic angling of the blade by means of telescoping pusher arms. A hinged push-bar attachment is available for mounting above and in front of the blade.

The Hula dozer is a multipurpose dozer for use in brush and tree removal, rock clearing, digging contour trenches, installing gully plugs, excavating reservoirs and dugouts, and in trail and road construction. The teeth can be pitched downward for raking brush and roots. Pitching the teeth gives increased power advantage and results in less contact and disturbance at the ground surface than when the cutting edge of the blade is used for raking.

Tree dozer. The tree dozer is equipped with a *V*-shaped, solid dozer blade to sever the roots and remove the root crown and roots from the soil [61]. The *V*-blade pushes trees to the side while allowing the dozer to continue forward. A push bar mounted solid to the blade or

92

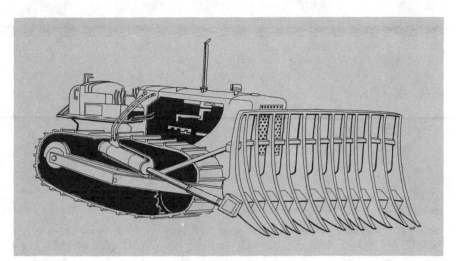

FIG. 28. *Brush rake (also called root rake or grubber) mounted on front end of crawler tractor.*

mounted and operated on separate arms extends out in front of the *V*-blade for knocking down trees while pushing the tops forward, exposing the roots, and protecting the equipment and driver. The tree dozer has been successfully used in eastern hardwoods, mesquite, oak, and juniper but is less versatile than the Hula dozer.

Brush rake. The brush rake (also called a stacker, root rake, or grubber) is a multitoothed adaptation of the standard dozer blade (Figure 28) [61]. The row of curved teeth projects slightly forward at the base, and a top guard or pusher bar joins the teeth at the top. This combination pushes the brush forward while allowing the dirt to clear between the teeth. The brush rake is used for uprooting small brush, for combing roots from the soil, and for piling after larger trees have been cut down and uprooted by other methods.

A rear-mounted brush rake is also adapted for brush removal (Figure 29). Removable brush rake attachments are also available for mounting directly in front of and slightly below straight dozer blades but are not as efficient as regular front-mounted brush rakes [60]. Welding a narrow cutting blade or root grubber across the bottom tips of the teeth is sometimes needed to uproot or cut off small, flexible shrubs that tend to slip through the rake teeth.

The brush rake has been used with considerable success on pricklypear and small brush in Texas [1]. In most cases pricklypear rapidly increases in density following root plowing, and brush raking as a follow-up treatment is often required. When following root plowing or whenever teeth are operated under the ground, the effects on perennial grasses are severe, and the area generally must be reseeded.

FIG. 29. *A rear-mounted brush rake equipped with broadcast seeder in use near San Angelo, Texas. Rake teeth are three feet long, spaced twenty-four inches apart, and operated ten to twenty inches deep.* (Soil Conservation Service)

FIG. 30. *Crawler tractor equipped with a stumper grubbing mesquite in Briscoe County, Texas.* (William K. Holt Machinery Co. photo)

The brush rake provides a better seedbed through soil breakup than does a solid blade.

The brush rake can be effectively used as a stacker. When compared with the straight dozer blade, the toothed brush rake moves little soil and still allows depth penetration. The soil is allowed to sift around the teeth rather than being pushed into windrows or piles. The brush rake works best in dry or sandy soil. On-site costs of clearing and piling in California chaparral in 1973 using a front-mounted brush rake varied from $23 to $68 per acre in lighter brush to $34 to $150 in the heaviest brush [60].

Miscellaneous attachments. A number of miscellaneous attachments such as the tree stumper, juniper bit, and stinger are available for replacing or attaching to the front and bottom of a straight dozer blade (Figure 30). These modifications have the advantage of concentrating the power of the crawler tractor on a narrow front in cutting, lifting, and uprooting individual trees and large shrubs with minimal soil disturbance. Farm tractors equipped with a stinger blade have been as effective and more economical for mesquite control than large crawler-type tractors [29].

Disk Plowing

Plows equipped with disks are widely used on plowable range sites for killing small, shallow-rooted plants and preparing a seedbed. Disk plows are useful on sagebrush, rabbitbrush, greasewood, low mountain brush, creosotebush, tarbush, and annual and perennial weed areas. Rabbitbrush and horsebrush are more difficult to kill and must be plowed about five inches deep for satisfactory kills [54]. Since the disk plowing destroys nearly all herbaceous vegetation, only those areas should be plowed where complete plant removal is desired and where subsequently seeded to desirable plants. Disk plowing has the advantage of leaving considerable mulch at or near the soil surface but it is not well adapted to very rocky or excessively gullied areas or where large brush is present [54].

The season for most effective plowing depends upon the species concerned, precipitation patterns, and seeding practices to be followed. Summer plowing in Nevada when the soil was dry and firm killed more mature big sagebrush plants than either spring or fall plowing [6]. Reestablishment of brush seedlings was negligible under summer plowing prior to seed maturity, moderate under spring plowing, and high on areas plowed after mid-October. Plowing near or after sagebrush seed maturation has commonly resulted in subsequent heavy infestation of seeded grass stands [35].

Plowing, disking, or root cutting in Utah is considered best in late spring or early summer when the ground is more easily worked [50]. With late spring treatment, moisture is conserved, loose soil tends to firm prior to late summer and fall seeding, and dead roots begin to decay. However, plowing just before or just after rain is less effective because many plants may reestablish and continue growing. Spring plowing and summer fallow followed by shallow fall tillage just before seeding to perennial grasses is considered the most effective way to rehabilitate cheatgrass range [49]. Mechanical control of chamise in California is suggested for late spring and early summer when plants are in a physiologically weak stage [40].

Standard disk plow. The standard disk plow has a single gang of a few to several disks on a frame supported by wheels. Each disk is moderately cupped, is slanted at an angle to the vertical, and has a separate bearing and frame attachment. The narrow cut, heavy draft and associated high power requirements, and breakage caused by embedded rocks make the standard disk plow generally too costly for range work [57]. However, a large two-way disk plow weighing over 1,000 pounds per foot of cut was effective in southern Arizona in killing 90 percent of sand-dune-type mesquite and creosotebush [37].

Wheatland plow. The wheatland plow (or one-way plow) is a wheeled implement consisting of a single gang of upright disks that are uniformly spaced on a common axle. The disks and axle rotate as a unit and at an angle of 35° to 50° from the line of travel. Wheatland plow models with disks twenty-four to twenty-eight inches in diameter and a heavy frame are best adapted to range work. In rocky soils and rough terrain breakage is common. Wheatland plows are best adapted to gentle terrain with fairly rock-free, moderately soft soil [57]. On firm soil the side draft may be excessive, and frequent adjustment is required.

The wheatland plow has given about 80 percent kill of big sagebrush with one treatment [9]. Two plowings are generally necessary to obtain a complete kill (over 95 percent), particularly where the sagebrush is tough and actively growing. When compared in their effectiveness in killing big sagebrush in Utah, plowing killed 87 percent of the brush while double pipe harrowing or double railing each killed only 64 percent [8]. Nine years later the sagebrush stand on the plowed land was 34 percent of the original while harrowing and railing had 57 and 60 percent, respectively, of the original stand.

The wheatland plow is frequently used on land previously cultivated and for light tillage as a second treatment. Costs on large acreages can often be reduced by pulling two or three plows in tandem behind a single tractor.

Brushland plow. The brushland plow is a heavily constructed three-wheeled implement and is similar to the standard disk plow in having no continuous, solid axle (Figure 31). Each pair of disks, the forward disk twenty-eight inches in diameter and the rear one twenty-six inches in diameter, has a single shaft and bearing assembly and is supported by a spring-loaded arm [57]. This allows individual disk sets to ride over boulders and stumps without interfering with the other disks.

The brushland plow can be used on rough, rocky sites without excessive breakage and is more effective on uneven terrain than the

FIG. 31. *Brushland plow has the ability to perform on hard soils, uneven terrain, and where rocks and stumps are prominent. The present model has a single rear wheel.* (Arcadia Equipment Co. photos)

wheatland plow. It is well adapted to sagebrush lands in the West and removes up to 90 or 95 percent of the sagebrush depending upon amount of rock present, travel speed, amount of slope, and size of brush. On very rough, rocky, uneven terrain with dense brush, plowing twice over may be necessary.

Although the initial cost is high, it has been purchased and widely used by federal land management agencies in the West and has also been made available to ranchers on a rental basis through soil and water conservation districts or federal land managing agencies. Many private contractors and a few ranchers have their own brushland plows. One disadvantage of the brushland plow is that it is not commercially available and must be custom built. It is also difficult to transport because of its bulk and weight.

Offset disk. The offset disk (also called tandem disk, disk harrow, Towner disk, and even brushland disk) consists of two gangs of disks, each gang having a separate axle and frame. One gang of disks moves the soil in one direction while the other gang moves it in the other direction (Figure 32). Offset disks are available commercially in a range of sizes. Only the heavy-duty models are generally adapted to range work, with some models ranging up to 1500 pounds per foot of cut.

Offset disks are well adapted to compact, dry, heavy soils and work effectively on moderately rocky or sticky soils (Figure 33). On loose, sandy soil it is difficult to regulate depth of plowing with wheelless models, and seedbeds are often left in a loose condition undesirable for seeding unless compacted with other equipment. Most older models do not have wheels, and the full weight rests upon the disks. However, providing wheeled models has helped greatly in controlling plowing depth and in transporting the equipment. The offset disk is adapted to small and medium brush, but is not effective on large brush and trees having large, woody stems and heavy root crowns.

When evaluated for California chaparral treatment, the offset disk was more effective than bulldozing where open stands of light brush or dense stands of brush sprouts were present [5]. Since the offset disk

FIG. 32. *The offset disk consists of two gangs which move soil in opposite directions.*

98

FIG. 33. *Offset disk (also known as Towner disk) preparing seedbed in light brush near Ukiah, California.* (Towner Manufacturing Co.)

cuts both ways, it normally kills a high percentage of brittle brush, grass, and forbs. But in the California studies double disking was needed for a high kill of most brush. It was further observed that care must be taken to avoid leaving dead furrows up and down slope and to avoid using the offset disk on slopes over 30 percent. The average cost of disking in California in 1973 varied from $25 to $75 per acre, depending upon the size of the shrubs and the need for single or double disking [60]. In other areas offset disking has varied recently from $12 to $25 per acre for once over. The rather high cost is related to the fact that approximately twice as much power is needed than for the brushland plow [57].

Root Plowing

Various planes, blades, grubbers, and cultivators have been used to sever the roots of brush and associated herbaceous perennials below the ground surface. The design most recommended for general range use is the root plow, which consists of a single large, shallowly V-shaped blade mounted on the rear of a large crawler tractor (Figure 34). The blade varies from eight to fourteen feet wide and is attached to two arms by means of two heavy upright shanks. The entire blade unit is raised and lowered hydraulically or by cable. Fins welded to the top of the blade help in severing horizontal rhizomes

99

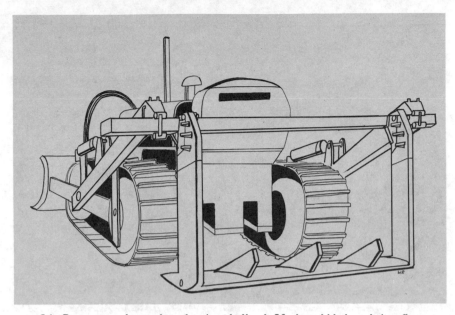

FIG. 34. *Rear-mounted root plow showing shallowly V-shaped blade and riser fins.*

and heaving roots and root crowns to the surface of the ground. The action of the fins or risers increases root kill through desiccation, reducing sprouting from surface roots, and breaking up large clods.

Root plowing is best adapted to large brush too dense for other types of mechanical treatment and to species not affected by herbicides. Root plowing is limited to deep soils that are fairly free of rocks and obstructions. High breakage may occur in rocky soils, and it is difficult to keep the blade at a satisfactory depth in rocky soil and on uneven ground. High kill is favored on dry, sandy soil. In wet, muddy, or particularly heavy soils, kill is reduced and stems and roots ball and hang up on the cutter blade. Swaths must be slightly overlapped so that live plants will not be left between the swaths. Using the dozer blade on the front of the crawler tractor facilitates passage of the root plow by pushing down or to one side the trees and large brush plants.

Root plowing kills most of the brush plants and also most of the non-rhizomatous perennial grasses and forbs. Although blading at deeper depths allows some rhizomatous grasses to survive, the use of riser fins to kill a maximum amount of brush also kills most of the herbaceous forage plants. Thus, root plowing should not be used in areas having a good grass stand under the brush [58]. Since root plowed areas must be reseeded, this operation should be carried out only in areas that have deep, fertile soils and ample rainfall to justify the cost. To be effective the blade must sever plants below the bud-

100

ding zone. Recommended depth is fourteen to sixteen inches for mesquite [58], eight to twelve inches for creosotebush [37], four to six inches for rabbitbrush and horsebrush [3, 49], twelve to eighteen inches for saltcedar [30], and twelve inches for Gambel oak and shrub live oak [33, 55]. Root plowing is expensive; cost comparisons with other methods of plant control in 1976 in Hidalgo Co., Texas, were as follows [16]:

Method	Ave. cost/acre
Tree dozing	$20
Rootplowing	$21
Rootplowing, reseeding	$33
Rootplowing, raking two ways, and reseeding	$55
Roller chopping	$13.50
Chaining	$ 9
Aerial or ground application, including herbicide	$12–$15

The root plow has been widely used in central and southern Texas on mesquite and associated mixed brush (Figure 35). It effectively kills established mesquite plants, but seedlings generally establish from residual seed in the soil and must later be controlled [32]. On mixed brush range in south Texas, the order of effectiveness in reducing brush cover was (1) root plowing plus raking, (2) front-end stacking plus disking, (3) surface blading and disking, (4) stacking, and (5) surface blading [14]. The root plowing plus raking reduced brush cover from the original 50 percent level to less than 5 percent and increased herbage production from 1,000 to 8,000 pounds per acre.

Root plowing in southern Arizona was considered capable of killing 95 percent of sand-dune mesquite and creosotebush when conscientiously applied, particularly in dry seasons [37]. Root plowing without grass seeding has seldom improved the productivity of rangeland because of the high damage to existing grass cover and rapid reinfestation of mesquite seedlings [17].

While decreasing mesquite materially and nearly eliminating coyotillo, root plowing in the Rio Grande Plains of Texas has often increased pricklypear by 100 to 300 percent [13]. Root plowing favors the spreading of pricklypear by removing brush competition and transplanting broken pricklypear plant parts [28]. Range sites to be root plowed should be relatively free of pricklypear or accompanied by effective pricklypear control measures such as herbicides, railing, mowing or piling, and burning [13, 56]. Railing followed by root plowing has decreased the density of pricklypear and other undesir-

101

FIG. 35. *Root plowing dense mesquite in Briscoe County, Texas.* (William K. Holt Machinery Co.)

able woody species and resulted in a relatively brush-free grassland [13]. A comparison of brush control methods in south Texas suggested that mechanical measures giving minimum soil disturbance during extended droughts were the most reliable in increasing forage production [56].

Root plowing in combination with herbicide application is the most effective method presently available to remove dense stands of saltcedar from water courses and flood plains [30]. This method consists of root plowing, piling and burning, seeding to grass, and then spraying with herbicides about six months later to kill saltcedar sprouts and seedlings. Root plowing gives highest kill when the soil is dry and the weather hot.

Root plowing was considered in Nevada as the only "one shot" method of killing rabbitbrush and horsebrush, but its use was considered limited because of rocky soil and high cost [59]. High soil moisture favored survival of the brush species as well as the grass understory. It was suggested that a chain be dragged behind the root plow in a *U*-shaped loop to knock over plants and prevent rerooting and that the operation be performed from spring to mid-September prior to rabbitbrush seed maturity.

Cutting off big sagebrush plants below ground with a Noble blade (medium-size blade mounted to a single shank) was more effective in the fall than in the spring (94 versus 47 percent kill) because under the spring treatment many sagebrush plants rerooted in the moist soil [8]. Root plowing Gambel oak in Colorado when soil was dry and rain did not follow was effective but only where the soil was deep and not rocky [33]. Root plowing killed 81 to 100 percent of shrub live oak in Arizona on fairly level areas of deep soil but was expensive [55].

Miscellaneous Plows and Tillers

Pipe harrow. The pipe harrow or Dixie drag consists of several spiked iron pipes, usually four inches in diameter, that trail behind a spreader bar (Figure 36). The pipes are swiveled so that they can rotate and clean themselves of trash. The pipe harrow is used for thinning low, brittle brush such as big sagebrush. Kill of 30 to 70 percent of big sagebrush is common; but two treatments, with the second treatment in the opposite direction, are often required for the higher level of control [50].

The pipe harrow gives low kill of young or willowy shrubs, brings many rocks to the surface, and is not effective on sprouting shrubs and annuals [50]. It normally uproots only 10 to 20 percent of the bunchgrasses and damages bitterbrush only slightly. The pipe harrow disturbs enough soil for seed coverage and is adapted to rocky ground and rough terrain, and its per-acre cost is low. Its use is mostly limited to portions of seeding projects that cannot be effectively plowed, on rocky scablands, burns, abandoned roads, and excavation scars [57].

Rockland tiller. This wheeled implement was designed specifically for scarifying rocky areas, killing weeds and annual grasses, and preparing a seedbed on such sites [57]. It consists of a series of teeth

FIG. 36. *The pipe harrow is used for thinning low, brittle brush and covering broadcast seed on rocky ground and rough terrain.* (Arcadia Equipment Co.)

103

mounted on three rotating main crossbeams. Teeth are kept in the ground by compression springs which allow the teeth to come out without breakage when an obstruction is hit. Although effective on rocky sites, this implement is no longer commercially available and is now seldom used [52].

Alpine cultivator. This small implement is designed for pulling by a small tractor or by horses. It consists of nine teeth staggered on two rows of shanks. Each shank is equipped with spring release and reset mechanisms to prevent breakage on rocky ground. Shanks may be equipped with either teeth or narrow duckfoot shovels. This implement is generally used only in small, inaccessible areas but is moderately effective on low weeds and small brush [57].

Moldboard plow. The moldboard plow is widely used on cultivated soils that are rock free, dense, and tight but has largely been replaced on range by the various disk plows. A series of furrows are opened by shears, and the soil is turned completely over by means of curved

FIG. 37. *Chaining is accomplished by dragging a heavy anchor chain between two crawler tractors traveling in parallel direction.*

moldboards. Characteristics of the moldboard plow limiting its use on range include slowness; high power per width of cut; high expense per acre; ineffectiveness on hard, sticky, or rocky soil; picking up roots; and being adapted only to very small brush [54, 66].

Chaining and Cabling

Chaining. Chaining is accomplished by dragging heavy anchor chain in a U-shape, half circle, or J-shape behind two crawler tractors traveling in a parallel direction (Figure 37). Chain lengths of 200 to 500 feet are most common. Anchor chains with links weighing 40 to 90 pounds are commercially available. However, chains with links weighing more than 70 pounds stay on the ground better and are more effective on young flexible trees and shrubs. Although the heavier chains are generally recommended, they may do more damage to desirable understory shrubs [53].

Chains must be dragged in a loose pattern rather than stretched taut, to maximize ground contact. The narrow U-shape provides more severe treatment and better pull-up of small trees and gives more windrowing effect. The half-circle shape increases swath width but decreases effect in the center section. A chain length to swath width ratio of 2:1 to 3:1 is recommended. When a 300-foot chain is used, a swath of 100 to 175 feet in width is common and generally gives good control. Higher traveling speed also gives more whip action and higher kill.

Chaining is adapted to varied terrain and is particularly useful on areas too rocky, rough, and steep for other equipment [53, 57]. Slopes up to 50 percent can be chained. However, rock outcrops that may catch the chain should be avoided. Steep slopes and sharp ridges can be treated by having one tractor travel along the ridge top and the other along the bottom of the slope. Chaining on the contour is preferable because of lower power requirements and the soil-disturbance furrows and brush windrows left on the contour.

Because of the weight of the chains, high power requirements, and special equipment required for loading and transporting, most chaining is done on commercial contract, particularly on privately owned range, but to a considerable extent on federal lands also.

Plant control from chaining. Chaining is an effective, low-cost, and widely applicable method for removing even-aged, mature, nonsprouting, single stemmed species such as most junipers, pinyon, and other conifers (Figure 38) [53]. Chaining is less effective on sprouting species because of regrowth, does not severely damage small trees or shrubs that bend easily without breaking, and gives little control of

105

FIG. 38. *Grass and browse recovery following chaining of juniper in central Utah.* (Utah Division of Wildlife Resources)

FIG. 39. *Juniper seedlings missed in a BLM chaining operation in San Juan County, Utah. Maintenance control such as broadcast burning should be considered.*

undesirable herbaceous plants in the understory (Figure 39). Chaining is useful in releasing valuable understory plants from brush and tree competition and in opening up shrub thickets. It can also be used to improve appearance and facilitate livestock movement on burned areas where many snags remain.

A major advantage of chaining is that large acreages can be treated at a lower cost per acre. Single chaining costs vary greatly from area to area but commonly ran from $7.50 to $25.00 per acre in light to moderate brush in 1973 [60]. The size and density of the brush and the size of the project affect costs. Two-way chaining costs about twice as much as a single chaining but is generally recommended under difficult conditions, where undesirable sprouting woody plants are present, for maximum uprooting of trees, and where broadcast seeding is planned. On BLM projects, single chaining killed from –28 to 95 percent of pinyon and juniper trees, depending upon the proportion of older, larger trees in the stand [4]. Single chaining killed an average of only 38 percent of the trees, while double chaining killed 60 percent.

Chaining does not materially increase runoff and erosion [51] but leaves debris and trash to protect the soil from erosion, and a light slash layer seems to favor the establishment of grasses and forbs. But excessive slash combined with the release of small trees missed in the chaining operation may pose problems [3, 4]. From a survey of Forest Service chaining projects in the Intermountain Region, it was concluded that pinyon and juniper trees surviving on the chained areas were mostly those that had survived the chainings rather than those since establishing by seed [51]. In order to minimize the number of small juniper and pinyon trees escaping chainings, Young and Evans [71] recommended selecting only areas for chaining alone that support closed-canopy, old growth pinyon and juniper trees.

Leaving debris in place following chaining in eastern Nevada without follow-up mechanical windrowing or debris burning has permitted a rapid return of juniper and pinyon [64]. Understory cover and production began a steady reduction beyond the fifth to eighth year following treatment, and trees appeared to redominate the chained sites in less than 15 years. The advantages of understory production resulting from the chaining was eliminated by the time the pinyon-juniper tree cover reached 12 to 15 percent. Twelve years after treatment only 1 percent of the trees were post-treatment seedlings. Chaining or cabling alone was concluded not sufficiently severe a disturbance to kill the many small established plants.

In Arizona about 55 percent of the control work in the pinyon-juniper type has been done by chaining and cabling [11]. In even-aged stands of trees fifteen to twenty-five feet high, chaining was found to pull out about 94 percent of the trees. However, in mixed-age stands when half of the pinyons and junipers were seedlings up to six feet high, only 43 percent of the trees were pulled from the ground. In northern Arizona the removal of large overstory trees resulted in an immediate release of small pinyons and junipers if they were missed in the control operation [3].

Based on his evaluation of completed BLM projects, Aro [4] concluded that chaining was an inefficient method of converting juniper sites to grasslands because of the problem of release of young trees. It was suggested that follow-up treatment under these conditions is generally necessary. Individual tree treatment such as bulldozing is more efficient and cheaper than cabling or chaining when trees are widely scattered or where large boulders and rock outcroppings make the terrain unsuited to chaining.

Chaining pinyon-juniper in northern Arizona increased forage production from 200 to 700 pounds per acre, reduced labor required for handling livestock, and released preferred game species such as winterfat, cliffrose (*Cowania mexicana*), and shrubby eriogonums (*Eriogonum* spp.) [3]. When present in the stand, shortgrasses and midgrasses responded very favorably to the chaining. However, undesirable shrubs and half-shrubs such as rabbitbrush and broom snakeweed were also released abruptly. Chaining juniper in Utah has also released bitterbrush (*Purshia tridentata*) and serviceberry (*Amelanchier alnifolia*) but has usually killed some of the mature cliffrose [52].

Anchor chaining gives a fair kill of old big sagebrush plants but kills few young, flexible plants. Late summer and early fall are most effective for chaining sagebrush as well as for other methods such as railing, harrowing, or rolling that uproot, crush, and break off the plants [50]. Dense stands of mature Gambel oak, chokecherry, or greasewood can be materially thinned and opened up by chaining [53]. This reduces competition so that herbaceous species can become established. However, maintenance control is necessary to prevent the rapid return of these shrubs.

Chaining has been found to be relatively economical but generally of temporary and limited effectiveness on Texas brush ranges [58]. It was found most effective on tree-type oaks and Ashe juniper. The benefits of chaining mesquite and creosotebush in the Southwest have generally been temporary in nature with other practices being necessary to control sprouts and seedlings [17, 37, 43]. Mesquites that are small at the time of chaining may be broken off or roughed up but not uprooted and may develop dense stands of sprouts. However, chaining was concluded to be the most economical method of retreating sprayed mesquite over five inches in diameter [68] and of removing the standing dead brush.

In south Texas mixed brush, chaining one way, double chaining, and double chaining-raking-stacking reduced live woody plant density by 39, 47, and 87% respectively [63]. The single chaining was considered ineffective, and the double chaining alone was less effective in promoting forage production than the combination method and did not improve livestock management efficiency.

Chaining has sometimes been used as a preliminary treatment to root plowing or in knocking down brush where followed by biological control with goat grazing. Double chaining followed by goating for five years at moderately heavy stocking rates at the Sonora Research Station [47] reduced liveoak by 91.5%, shinoak (*Quercus mohriana*) by 98.5%, mesquite by 77%, juniper by 97%, sacahuista (*Nolina texana*) by 72.5%, and pricklypear by 62.5%, with an average brush reduction of 83%.

Chaining semidesert grass-shrub range in Arizona opened up the brush, but the effects were considered temporary [62]. Chaining knocked down the old cholla plants but paved the way for many new cholla seedlings. Chaining in Arizona studies greatly increased all cacti species for one to two years, but this was not always permanent

FIG. 40. *Chain modification used on the Uinta National Forest, Utah.* Top, *showing axle segments welded to links;* bottom, *sagebrush range after two chainings and broadcast seeding between chainings.*

109

[44]. Chaining in west Texas has resulted in dense stands of prickly-pear because of the transplanting action of the chain [13].

Chain modifications. Various attempts have been made to modify anchor chains in order to obtain greater soil scarification and to hold down the chain for greater kill of small brush. Wrapping a second piece of chain around the center one-third of the main chain, using a heavier grade chain in the center section, or welding pieces of railroad rail to the center section links all add weight to the middle section of the chain and increase scarification and kill [27].

Providing a swivel near each end of an anchor chain has been helpful in allowing the chain to rotate and keep relatively free of trash. Connecting the swivel to the tractor by a 15- to 20-foot smooth-link lead chain provides room for the tractor to maneuver. A third swivel installed in the center of the chain, allowing each half to turn independently, has not increased effectiveness appreciably but does keep the chain from occasionally becoming twisted [60].

The use of full double loops or a double loop on the midsection only has been tried but was generally found less effective than making two passes in opposite directions with a single loop chain [67]. The double loop accumulates trash on the back loop which is difficult to clean, requires tractors to operate in lower gear, excessively destroys the sod and roughens the ground surface, and reduces the effectiveness of the front loop. The double loop gives little or no whip action on the young trees and has generally been abandoned because of these problems. The Johnson chain drag has been proposed and used to some extent on dense Hawaiian brush [42]. It consists of two or three loops of chain attached to a beam pulled behind a single crawler tractor in a *U*-shape.

Links in the anchor chain have been variously modified to increase chopping, digging, and pulling out brush and for scarifying soil for reseeding. The Ely chain [7] modification consists of welding a short section of railroad rail (about 18 inches long) across each link and extending about four inches on either side of the link.

It was suggested that the Ely chain not exceed 210 feet when used in dense stands of pinyon-juniper or 300 feet in open stands of pinyon-juniper and/or big sagebrush [7]. Using three-inch malleable angle iron in place of the railroad rail was found unsatisfactory because of excessive wear and twisting. The use of eighteen to twenty-inch lengths of auto axles has been effective, but does not add as much weight to the chain as do railroad rail segments (Figure 40) [53]. Using crossbars of harder steel or hardsurfacing the tips of softer iron crossbars is suggested to reduce wear.

A chain developed on the Dixie National Forest in Utah has utilized short segments of railroad rail as diggers when welded to all links except the terminal fifteen links at each end of the chain [34]. An eight-inch digger weighing about twenty pounds was attached to each side of the link. Instead of welding the rail sideways, the top of each segment was welded to one side of the link on the outer margin. This reduced the loss of rail sections. The Dixie chain worked particularly well in sagebrush and scattered juniper. It provided more soil scarification and more kill of juniper seedlings than the regular chain, but it was most effective in the spring when the soil was plowable. It did not work satisfactorily in thick pinyon-juniper since the diggers hooked the trees and carried them along.

In Canada anchor chain for removing brush has been modified by attaching two 80- to 130-foot chain segments to a four-ton ball [18]. The loose ends of the chain are tied to two crawler tractors the same as with the regular chain. The heavy ball causes the chain to advance in an inverted V-shape which tends to give uniform brush treatment and pile trees for later burning. Chaining in Canada where sloughs, muskegs, and soft ground are found is suggested for the winter season, but most of the roots are left in the frozen ground at this season.

A ball-and-chain technique of chaining has been used successfully in California on slopes too steep for using regular chaining techniques [60]. One end of a 120- to 150-foot chain is attached to a crawler tractor, which operates near the top of the slope; the other end is attached to a 5-foot marine buoy filled with water or other dense material to add weight. The round buoy holds the chain in working position by rolling down slope and keeping the chain tight while moving across slope parallel to the tractor. Site treatment costs have varied (1976) from $22 to $50 per acre in light brush and from $40 to $70 per acre in heavier brush where three to four passes were required.

In Australia a 100-foot chain was developed into a 100-foot-long one-way plow without wheels [2]. Steel disks twenty-two inches in diameter and three-quarter-inch thick were welded to alternate links of chain. The chain was further modified by placing a walking disk at each end of the chain and one in the center. This adaptation was considered useful were soil is deep and plowing is necessary to establish new stands of grass.

Cabling. Steel cables 1.5 to 2 inches in diameter and 200 to 600 feet long (or occasionally 1,000 feet) may be used in the same manner as the anchor chain. Although the cable works similarly to a chain on large trees, it tends to slip over small trees and often merely tips over intermediate-size trees. The efficiency of cabling on young flexible juniper and pinyons has been increased somewhat by operating at the highest possible speed to get maximum whip action and by tying railroad rails to the cable. Cable is less flexible than anchor chain and does not hug the ground as well. Cabling is generally ineffective for opening brush thickets [3, 53]. But it is a preferred treatment where only thinning of the brush is desired [52].

A study in Arizona demonstrated that size of trees, number of trees, and size of project affect per acre costs of cabling juniper [11]. On the 100 projects studied and reported in 1963, cabling costs ranged from $.62 to $10.67 per acre with an average of $2.06. Calculated costs of cabling small trees ranged from $1.01 for 50 trees per acre to $1.56 for 150 trees per acre. Cost calculations of cabling large trees were $1.30 and $2.33 per acre, respectively, for 50 and 150 trees per acre. The per-acre time required for cabling pinyon-juniper of average height and density was calculated at 0.2 hours per acre for 500-acre projects, 0.15 hours for 1,000-acre projects, and 0.075 hours for 3,000-acre projects. Similar relationships are presumably also true with chaining.

Railing

Railing (or dragging) consists of dragging railroad rail, *H*-beam, or channel iron along the ground to break off or uproot brittle shrubs to allow herbaceous forage plants to increase. The three most recommended designs are the *A-rail* (a rigid frame pulled with the apex forward), the *Supp rail* (a flexible, three-section straight rail mounted on post skids), and the *drag rail* (rails arranged in three tandems) [57]. Two-way railing is generally required for satisfactory kill. To be effective the brush must be brittle and the terrain relatively flat.

Railing has the advantage of being cheap. It has most commonly been used on mature big sagebrush, where a kill of 30 to 80 percent is obtained [50]. However, kill on sagebrush seedlings and other willowy species is quite low. Root sprouters such as rabbitbrush and horsebrush are not effectively killed by railing. Railing generally does little damage to grasses. It is generally more effective when the soil and vegetation are dry. Although balling up with trash is a problem, this is circumvented somewhat by the *A*-rail and Supp rail. In a Wyoming study, railing and rotary beating of big sagebrush returned 4.5 percent on total costs compared to 10.5 percent with chemical control (based on no cost-sharing assistance and an a.u.m. of forage valued at $2.50) [38].

Railing has been used with considerable success in Texas in killing pricklypear. Railing twice over in opposite directions crushes most of the cactus stems and pads. However, railing at six to twelve-month intervals in a series is generally required for high kill. Railing in a series followed by a final treatment of 2,4,5-T or silvex has given nearly 100 percent kill [28, 58]. Although found moderately effective in New Mexico in dry seasons, mechanical methods have generally proven inferior to hand grubbing and piling pricklypear because of the scattering of joints and pads [23].

Rolling
Brush Cutter

The rolling brush cutter (also called roller-cutter, rotary brush chopper, or Marden brush cutter) consists of a cylindrical roller or drum equipped with several full-length cutting blades (Figure 41). The drum is usually hollow and filled with water to bring total weight up to desired weights, often ten to twenty-five thousand pounds. The rolling brush cutter may be pulled straight away or at a diagonal to increase the chopping action. Two or more rolling brush cutters are often pulled in tandem and set at slight contrasting angles. One large self-propelled cutter utilizes three cutter-equipped drums and cuts a sixteen-foot swath [58].

The heavy weight of the cutter crushes the brush, and the cutter blades chop up the brush and scarify the ground surface. This combination crushing-chopping action effectively knocks down and cuts up small brush with stem diameters up to five inches and kills

FIG. 41. *Rolling brush cutter used for crushing and chopping up brush and for ground disturbance for range seeding on a ranch near Paducah, Texas.* (Soil Conservation Service photo)

113

brittle, nonsprouting species. However, since a low percentage of the sprouting plants are actually killed, the treatment is generally only temporary unless combined with other treatments. The cutter is also used for crushing brush prior to burning or grazing by goats or in compacting loose seedbeds. The cutter blades are quickly dulled by large rocks in the soil.

When the rolling brush cutter was used to control mesquite in Texas, retreatment thereafter about every five years was necessary, or annually with lightweight cutters [17]. The rolling brush cutter was ineffective on both mesquite and creosotebush in southern Arizona [37]. It has been used in Canada to break down and cut up poplar and willow up to four inches in diameter when the soil is frozen or hard [18]. A single treatment with a tandem brush cutter cuts the horizontal stems and terminal buds of saw-palmetto in the Southeast resulting in an 80 percent control when done during the winter [70]. Forage production from creeping bluestem increased from 1,600 to 6,000 pounds per acre as a result of the saw-palmetto control. In south Florida both rootplowing (referred to as webbing) and double treatment with a rolling brush cutter (referred to as cross-chopping) were effective in reducing ground cover of all shrub species, especially saw-palmetto [41]. The rolling brush cutter was generally more effective, particularly on moist sites.

Mowing, Beating, and Shredding

Rotary mower. Mowers with one or more blades that rotate in a horizontal plane are the most widely used type of mowers in range and pasture improvement. The blade is attached in the center to a power source and provided with cutting edges at each end on the leading edge. Heavy-duty rotary mowers are capable of cutting weeds and brush stems up to four inches in diameter. One heavy-duty, trailing rotary mower equipped with two revolving blades cuts a swath of twelve feet, is capable of mowing four acres per hour, and has been used effectively on uniform stands of big sagebrush and other low, nonsprouting brush (Figure 42).

Sickle mower. Sickle mowers may be used on brush stems up to 1.5 inches in diameter if provided with a heavy, relatively short cutter bar, stub-nosed guards, a heavy, smooth section, and a double set of clips to hold the sections snugly against the blunt guards. Even with this heavy-duty construction, breakage from large stems and rocks is relatively high, tractor speed must be slow, and frequent stops and starts are generally required. When use is limited to herbaceous plants and browse seedlings, breakage may be kept down. A principal advantage is that most ranches have sickle mowers that can be adapted to range and pasture use. Although capable of cutting most sand sagebrush plants, the stems of big sagebrush are too thick for cutting with a sickle mower [50].

Rotobeaters and shredders. The rotobeater knocks down and shreds small brush and weeds by means of a series of flails attached to a horizontal shaft which revolves at high speed. Several large, massive implements have been developed capable of chopping, shredding, and masticating brush and trees up to several inches in diameter and converting slash into mulch. These machines have variously been called *bushwhackers, tree eaters* or *shredders, brush cutter-chippers,* and *tree crushers.* The tree eaters and similar equipment are generally unadapted to rocky or muddy soils or steep slopes, and costs have been excessive. In Texas, shredding in May was the most effective time for mesquite control [69]. Shredding to near ground level in May killed 75 percent of the plants with little damage to the grass understory.

Results of mowing and beating. Mowing was formerly used much more frequently than now for range and pasture improvement. Mowing nearly always improves the appearance of an area but often fails to kill many perennial plants. Mowing is effective on upright annuals. Repeated mowing will give temporary control of tall growing perennial weeds and small-stemmed brush by depleting the underground food supply. However, mowing is ineffective on prostrate

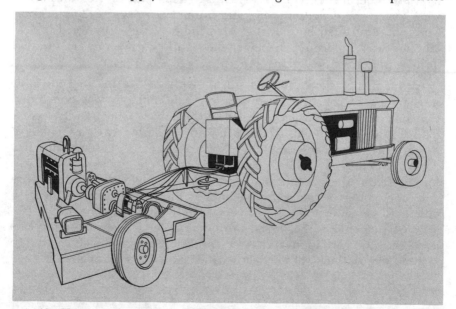

FIG. 42. *Heavy-duty, trailing mower with two revolving blades. Effective for mowing sagebrush or other low, nonsprouting brush.*

115

and low-growing perennials and sprouting species. Mowing is also used as a follow-up treatment on sprouts but often must be repeated. Mowing and beating are not adapted to rough, rocky land and do not prepare a seedbed. Rotary mowers and brush beaters, but not sickle mowers, scatter the plant residues and avoid smothering plants underneath except where the foliage is very dense.

In Nebraska studies, three annual sprayings of 2,4-D at one pound acid equivalent per acre reduced weed stands by 70 percent, while mowing in the same three successive years reduced the weed stand only 30 percent [39]. Mowing was ineffective on perennial ragweed, hoary vervain (*Verbena stricta*), goldenrod (*Solidago* spp.), and common yarrow (*Achillea millefolium lanulosa*) [46]. Annual mowing of western ironweed for ten successive years had little effect on plant numbers [45]. The Nebraska work showed that mowing decreased the vigor of most of the weeds, did not satisfactorily control any species of perennial pasture weeds, and reduced total forage production [20].

Broad-leaved annual weeds such as sunflower (*Helianthus* spp.), marshelder (*Iva xanthifolia*), common ragweed (*Ambrosia artemisiifolia*), and mustard (*Sisymbrium* spp.) have been controlled in South Dakota when mowed after weeds had made considerable growth but prior to seed production [12]. Biennials such as gumweed (*Grindelia squarrosa*) and mullein (*Verbascum thapsus*) were controlled by mowing before seed production the second year. Grassy weeds such as foxtail barley, wild barley (*Hordeum leporinum*), and annual bromegrass (*Bromus* spp.) were not controlled by mowing. Mowing of western ragweed in Oklahoma was ineffective because the shallow rhizomes resprouted quickly and grew new plants [15].

Although two successive years of mowing or beating in June gave satisfactory kill of sand sagebrush in Oklahoma, it was concluded that mechanical control was more costly, slower, and usually less effective than 2,4-D [46a]. Rotobeating in Wyoming gave a fair kill of big sagebrush at a cost of $4.76 (1967) per acre while increasing forage production by 133 percent [38]. In Oregon rotobeating was found effective on mature big sagebrush but ineffective on seedlings and young plants under twelve inches high [26]. The rotobeating killed 84 percent of the big sagebrush on fair condition range but only 61 percent on the poor condition range. However, eight years later the fair condition and poor condition range had 88 and 110 percent, respectively, of the pretreatment amounts of sagebrush.

Rotobeating at a cost of $5 per acre in 1965 killed 90 percent of the pricklypear in eastern Wyoming, increased forage production by two times, and increased the uniformity of utilization [31]. Mowing silverberry (*Eleagnus commutata*) as many as three times in Canada was less effective than a single application of one pound a.e. of 2,4-D per acre made in early summer [10]. Rotary mowing in Canada was found good for cutting and shredding willow, silverberry, pasture sage, and snowberry especially after being killed with herbicides [18]. But mowing for two or three years was generally required to halt regrowth. Mowing costs were reported at $3 to $6 per acre.

Mowing may be used to clip weeds in new seedings. However, care must be taken that tops of grasses are not clipped off sufficiently to damage the young grasses. Mowing is advantageous only if the forage grasses recover more rapidly than the weeds. Delayed mowing of broad-leaved weeds appears to help suppress weedy grasses.

Debris Disposal

Light, dispersed debris from mechanical brush control is often merely left in place, where it rapidly deteriorates. In situations where it does not greatly interfere with the use of the area, the debris can be important in providing erosion control and later in adding organic matter to the soil and possibly providing wildlife protection. However, debris left by chaining, bulldozing, rootplowing, and tandem disking may be excessive and interfere with the subsequent development and use of the area. This problem is common on sites with dense, robust stands of juniper, scrub pines, oaks and maples, mesquite, saltcedar, or aspen.

Debris from brush and small trees may have to be crushed, dispersed, or disposed of before subsequent reseeding and effective livestock grazing can take place. On public lands the interim problems of interference with other multiple uses, access by men and equipment, and general effects on esthetics must be considered.

The following checklist indicates some of the techniques for excessive slash disposal available for use on rangelands. Some have been developed for slash disposal following timber harvesting [25] or road construction [21], and some will be too costly for range use except in critical areas.

(1) Crushing and dispersal with rolling brush cutters or tree crushers.
(2) Broadcast burning, often preceded by brush crushing.
(3) Individual tree burning, with or without prior bulldozing.
(4) Windrowing and later burning windrows.
(5) Windrowing and piling without further disposal.
(6) Placing in gullies or ravines or burying.
(7) Treatment and dispersal with tree eaters, chippers, or hammermills.
(8) Removal from site for wood products processing.

117

Literature Cited

1. Allison, D. V., and C. A. Rechenthin. 1956. Root plowing proved best method of brush control in south Texas. J. Range Mgt. 9(3):130–133.
2. Anonymous. 1963. The widest plow in the world. Western Lvst. J. 41(15):28.
3. Arnold, Joseph F., Donald A. Jameson, and Elbert H. Reid. 1964. The pinyon-juniper type of Arizona: effects of grazing, fire, and tree control. USDA, Agric. Res. Serv. Prod. Res. Rpt. 84.
4. Aro, Richard S. 1971. Evaluation of pinyon-juniper conversion to grassland. J. Range Mgt. 24(3):188–197.
5. Bentley, Jay R. 1967. Conversion of chaparral areas to grassland. USDA Agric. Handbook 328.
6. Bleak, A. T., and Warren G. Miller. 1955. Sagebrush seedling production as related to time of mechanical eradication. J. Range Mgt. 8(2):66–69.
7. Cain, Don. 1971. The Ely chain. USDI, Bur. Land Mgt., Ely, Nevada.
8. Cook, C. W. 1958. Sagebrush eradication and broadcast seeding. Utah Agric. Expt. Sta. Bul. 404.
9. Cook, C. Wayne. 1966. Development and use of foothill ranges in Utah. Utah Agric. Expt. Sta. Bul. 461.
10. Corns, William G., and R. J. Schraa. 1965. Mechanical and chemical control of silverberry (*Eleagnus commutata* Bernh.) on native grassland. J. Range Mgt. 18(1):15–19.
11. Cotner, Melvin L. 1963. Controlling pinyon-juniper on southwestern rangelands. Arizona Agric. Expt. Sta. Rpt. 210.
12. Derscheid, Lyle A., and Raymond A. Moore. 1965. Improving "worn out" pastures. South Dakota Agric. Ext. Serv. Fact Sheet 256.
13. Dodd, J. D. 1968. Mechanical control of pricklypear and other woody species on the Rio Grande Plains. J. Range Mgt. 21(5):366–370.
14. Drawe, D. Lynn. 1977. A study of five methods of mechanical brush control in south Texas. Rangeman's J. 4(2):37–39.
15. Elder, W. C. 1951. Controlling perennial ragweed to make better pastures. Oklahoma Agric. Expt. Sta. Bul. 369.
16. Everitt, J. H., and J. A. Cuellar. 1976. Use and management of the rangelands of Hidalgo County, Texas. Rangeman's J. 3(5):155.
17. Fisher, C. E., C. H. Meadors, R. Behrens, et al. 1959. Control of mesquite on grazing lands. Texas Agric. Expt. Sta. Bul. 935.
18. Friesen, H. A., M. Aaston, W. G. Corns, et al. 1965. Brush control in Canada. Canada Dept. Agric. Pub. 1240.
19. Frischknecht, Neil C. 1971. Personal communication.
20. Furrer, John, and Don Burzlaff. 1960. Pasture weed control. Nebraska Agric. Expt. Sta. Campaign Cir. 171.
21. Gallup, Robert M. 1976. Clearing, grubbing, and disposing of road construction slash. USDA, For. Serv. Equip. Dev. Center, San Dimas, Calif.
22. Gay, Charles W., Jr. 1965. Control cholla cactus by mechanical and chemical methods. New Mexico Agric. Ext. Serv. LB 804.
23. Gay, Charles. 1965. Mechanical and chemical methods of controlling pricklypear cactus. New Mexico Agric. Ext. Serv. LB 801.
24. Gray, James R., Thomas M. Stubblefield, and N. Keith Roberts. 1965. Economic aspects of range improvements in the Southwest. New Mexico Agric. Expt. Sta. Bul. 498.
25. Harrison, Robin T. 1975. Slash: equipment and methods for treatment and utilization. USDA, For. Serv. Equip. Dev. Center, San Dimas, Calif.
26. Hedrick, D. W., D. N. Hyder, F. A. Sneva, and C. E. Poulton. 1966. Ecological response of sagebrush-grass range in central Oregon to mechanical and chemical removal of *Artemisia*. Ecology 47(3):432–439.
27. Hoffman, Darrel C. 1968. Double anchor chain. Range Improvement Notes 13(3):7.
28. Hoffman, G. O. 1967. Controlling pricklypear in Texas. Down to Earth 23(1):9–12.
29. Hoffman, G. O. 1971. Mesquite control. Texas Agric. Ext. Serv. Misc. Pub. 386.

30. Horton, J. S. 1960. Use of a root plow in clearing tamarisk stands. USDA, Rocky Mtn. Forest & Range Expt. Sta. Res. Note 50.
31. Hyde, Robert M., Arlowe D. Hulett, and Harold P. Alley. 1965. Chemical and mechanical control of plains pricklypear in northeastern Wyoming. Wyoming Agric. Ext. Cir. 185.
32. Jaynes, C. C., E. D. Robison, and W. G. McCully. 1968. Root plowing and revegetation on the rolling and southern High Plains. Texas Agric. Expt. Sta. Prog. Rpt. 2585.
33. Jefferies, Ned W., and J. J. Norris. 1965. Management and improvement of Gambel oak ranges, San Juan Basin Branch Station, Hesperus, 1965. Colorado Agric. Expt. Sta. Prog. Rpt. 147.
34. Jensen, Frank. 1969. Development of the Dixie-sager. Range Improvement Notes 14(1):1-10.
35. Johnson, James R., and Gene F. Payne. 1968. Sagebrush reinvasion as affected by some environmental influences. J. Range Mgt. 21(4):209-213.
36. Johnston, A., and S. Smoliak. 1968. Reclaiming brushland in southwestern Alberta. J. Range Mgt. 21(6):404-406.
37. Jordan, Gilbert L., and Michael L. Maynard. 1970. The San Simon Watershed: shrub control. Prog. Agric. in Arizona 22(5):6-9.
38. Kearl, W. Gordon, and Maurice Brannan. 1967. Economics of mechanical control of sagebrush in Wyoming. Wyoming Agric. Expt. Sta. Sci. Monogr. 5.
39. Klingman, Dayton L., and M. K. McCarty. 1958. Interrelations of methods of weed control and pasture management at Lincoln, Nebraska, 1949-55. USDA Tech. Bul. 1180.
40. Laude, Horton M., Milton B. Jones, and William E. Moon. 1961. Annual variability in indicators of sprouting potential in chamise. J. Range Mgt. 14(6):323-326.
41. Lewis, Clifford E. 1972. Chopping and webbing control saw-palmetto in south Florida. USDA, For. Serv. Res. Note SE-177.
42. Lyman, Robert A., and Walter E. Sykes. 1954. The Johnston chain drag for clearing brush from rangeland. J. Range Mgt 7(1):31-32.
43. Martin, S. Clark. 1966. The Santa Rita Experimental Range. USDA, For. Serv. Res. Paper RM-22.
44. Martin, S. Clark, and Fred H. Tschirley. 1969. Changes in cactus numbers after cabling. Prog. Agric. in Arizona 21(1):16-17.
45. McCarty, M. K., and D. L. Linscott. 1962. Response of ironweed to mowing and 2,4-D. Weeds 10(3):240-243.
46. McCarty, M. K., and C. J. Scifres. 1968. Pasture improvement through effective weed control. Nebraska Quarterly 14(4):23-26.
46a. McIlvain, E. H., A. L. Baker, W. R. Kneebone, and Dillard H. Gates. 1955. Nineteen-year summary of range improvement studies at the U.S. Southern Great Plains Field Station, Woodward, Oklahoma, 1937-1955. USDA, Agric. Res. Serv. Mimeo.
47. Merrill, Leo B., and Charles A. Taylor. 1976. Take note of the versatile goat. Rangeman's J. 3(3):74-76.
48. National Research Council. 1968. Principles of plant and animal pest control. II. Weed control. National Academy of Sciences Pub. 1597.
49. Parker, Karl G. 1968. Range plant control modernized. Utah Agric. Ext. Cir. 346.
50. Pechanec, Joseph F., A. Perry Plummer, Joseph H. Robertson, and A. C. Hull, Jr. 1965. Sagebrush control on rangelands. USDA Agric. Handbook 277.
51. Phillips, T. A. 1977. An analysis of some Forest Service chaining projects in region 4, 1954-1975. USDA, For. Serv., Ogden, Utah.
52. Plummer, A. Perry. 1971. Personal communication.
53. Plummer, A. Perry, Donald R. Christensen, and Stephen B. Monsen. 1968. Restoring big game range in Utah. Utah Div. of Fish and Game Pub. 68-3.
54. Plummer, A. Perry, A. C. Hull, Jr., George Stewart, and Joseph H. Robertson. 1955. Seeding rangelands in Utah, Nevada, southern Idaho, and western Wyoming. USDA Agric. Handbook 71.

55. Pond, Floyd W., D. T. Lillie, and H. R. Holbo. 1965. Shrub live oak control by root plowing. USDA, For. Serv. Res. Note RM-38.
56. Powell, Jeff, and Thadis W. Box. 1967. Mechanical control and fertilization as brush management practices affect forage production in south Texas. J. Range Mgt. 20(4):227–236.
57. Range Seeding Equipment Committee. 1965 (Rev.). Handbook of range seeding equipment. USDA and USDI.
58. Rechenthin, C. A., H. M. Bell, R. J. Pederson, and D. B. Polk. 1964. Grassland restoration. II. Brush control. USDA, Soil Cons. Serv., Temple, Texas.
59. Robertson, Joseph H., and H. P. Cords. 1957. Survival of rabbitbrush, *Chrysothamnus* spp., following chemical, burning, and mechanical treatments. J. Range Mgt. 10(2):83–88.
60. Roby, George A., and Lisle R. Green. 1976. Mechanical methods of chaparral modification. USDA Agric. Handbook 487.
61. Sampson, Arthur W., and Arnold M. Schultz. 1957. Control of brush and undesirable trees. Unasylva 10(1):19–29, 10(3):117–128, 10(4):166–182, 11(1):19–25.
62. Schmutz, Ervin M., Dwight R. Cable, and John J. Warwick. 1959. Effects of shrub removal on the vegetation of a semidesert grass-shrub range. J. Range Mgt. 12(1):34–37.
63. Scifres, C. J., J. L. Mutz, and G. P. Durham. 1976. Range improvement following chaining of south Texas mixed brush. J. Range Mgt. 29(5):418–421.
64. Tausch, Robin J., and Paul T. Tueller. 1977. Plant succession following chaining of pinyon-juniper woodlands in eastern Nevada. J. Range Mgt. 30(1):44–49.
65. Tueller, Paul T. 1965. Deer use on the Spruce Mountain chaining project. Nevada Agric. Expt. Sta. Prog. Rpt. 13, pp. 29–30.
66. USDA, Agric. Res. Serv. 1961. Equipment for clearing brush from land. USDA Farmers' Bul. 2180.
67. USDA, Forest Service. 1958. Brush removal by chain drag. Range Improvement Notes 3(2):1–5.
68. Weddle, Jon P., and Henry A. Wright. 1970. An evaluation of five methods to retreat sprayed mesquite. J. Range Mgt. 23(6):411–414.
69. Wright, Henry A., and Kenneth J. Stinson. 1970. Response of mesquite to season of top removal. J. Range Mgt. 23(2):127–128.
70. Yarlett, Lewis L. 1965. Control of saw-palmetto and recovery of native grasses. J. Range Mgt. 18(6):344.
71. Young, James A., and Raymond A. Evans. 1976. Control of pinyon saplings with picloram or karbutilate. J. Range Mgt. 29(2):144–147.

Chapter 5

Herbicidal plant control

Advantages of Herbicidal Control

Chemical control of undesirable range plants by the use of herbicides has developed rapidly in recent years and is now the most widely used means of removing noxious shrubs and weeds from grazing lands. A major impetus to the use of herbicides on rangelands was the discovery of 2,4-D in 1942 and its availability at reasonable cost following World War II. Since that time many new herbicides have been introduced and are now available for range improvement. Still others are in the experimental stage.

About eight million acres of pasture and range in the U.S. were treated with phenoxy herbicides in 1966, and phenoxy herbicides accounted for 98 percent of the herbicides used in 1966 [140]. In 1971 7.9 million pounds of phenoxy herbicides were applied to U.S. privately owned pastures and rangelands [25]. The largest portion was 2,4-D, which was applied to six million acres. About one million pounds of 2,4,5-T was applied to 1.1 million acres of land, mostly rangelands. In 1971 five percent of the 100 million acres of cultivated pasture in the U.S. received phenoxy herbicides, while 2.1 million, or 0.4 percent, of the 530 million acres of rangelands were treated with phenoxy herbicides.

The use of herbicides to control undesirable range plants has the following advantages over other control methods [96, 80]:

1. Cheaper than most mechanical control methods, but may cost more than fire.
2. Can be used where mechanical methods are impossible, such as on steep or rocky slopes.

121

3. Reduces labor requirements for treatment.
4. Provides a selective means of killing certain weed and brush species such as root sprouters that cannot be efficiently controlled by other methods.
5. Grasses usually escape damage when directions for use are followed.
6. Maintains a grass and litter cover and does not expose soil to erosion.
7. Provides a rapid control method from the standpoint of both plant response and acreage covered.
8. Safer than fire when proper safeguards are used.

These advantages explain the current widespread use of herbicides on range and pasture. However, disadvantages include: (1) no chemical control has yet proven effective for some species, (2) herbicides provide a desirable seedbed for grass seeding only under certain conditions, (3) costs of control may outweigh expected benefits on low-potential range, (4) the careless use of chemicals is hazardous to cultivated crops and may contaminate water supplies, and (5) herbicides may kill associated forbs and shrubs important for grazing.

Benefits of
Herbicidal Control

The benefits of mesquite control in experimental studies on the Rolling and Red Plains at Spur, Texas, in 1968, included the following: (1) increased forage production by five times, (2) increased calf weaning weights by forty pounds per animal, (3) increased stocking rate by 30 percent, (4) increased net returns per acre by $.81 annually, and (5) increased weaning weights of calf per acre from 49.8 to 65.8 pounds [61]. Based on interviews made with ranchers in north central Texas in 1965, the spraying of herbicides to control mesquite gave 26.75 percent return on upland sites and 15.75 percent on bottomland sites [150]. These returns were based on $3.00 cost per acre for the spraying operation, a grazing rental of $3.35 per animal unit month (a.u.m.), and government cost-sharing through the ACP Program of 50 percent of application and herbicide costs. In the semidesert area of southern Arizona, spraying 2,4,5-T in two successive years approximately doubled forage production of the native grasses by the end of the second growing season [21]. Forage production was tripled when Lehmann lovegrass (*Eragrostis lehmanniana*) was broadcast seeded the first year.

Wide use has been made of 2,4-D for control of big sagebrush on native ranges and in renovating seeded grass ranges rendered unproductive by big sagebrush reinvasion. Kearl [77] reported that from

1952 to 1964 a total of 350,000 acres of private lands and 155,000 acres of Forest Service and BLM lands were sprayed in Wyoming to kill sagebrush. Since big sagebrush is highly competitive with grasses for soil moisture and probably nitrogen, control makes additional soil moisture available for grass production.

Big sagebrush control with 2,4-D in Wyoming reduced the rate of soil moisture withdrawal by 14 percent, with 75 percent of this reduction in the three-to-six-foot zone [127]. However, the herbicide application increased soil moisture withdrawal in the one-to-two-foot zone, indicating that the increase in grass herbage production is most strongly reflected in that zone. Sagebrush control in the Red Desert and in the Big Horn Mountains of Wyoming increased soil moisture by July the following year [124]. Fisser [44] concluded that grasses are able to withdraw more water from the soil, over a longer period of time, than sagebrush. On mesic sites in Wyoming, sagebrush control reduced soil moisture by late summer, while on arid sites it had little effect.

FIG. 43. *Response of sideoats grama, blue grama, and buffalograss near Erick, Oklahoma, following mesquite control with 2,4,5-T and two seasons of deferred grazing. Note unsprayed range in background.* (Soil Conservation Service photo)

Snow-holding capacity was not affected by chemical control of sagebrush in Wyoming, and uncontrolled areas retained snow for a shorter time in the spring [124]. The latter was attributed to a "black body" radiator effect by the live sagebrush plants in uncontrolled areas. Various widths of sagebrush strips left uncontrolled, in regions where drifting usually occurred, did not measurably affect the snow-holding capacity.

Herbicidal brush control in higher rainfall areas of the Southeast has also given beneficial results. On oak-grass range in Oklahoma sprayed with 2,4,5-T and deferred from grazing for two years, grass production per acre was 6,000 pounds compared to 400 to 600 pounds on untreated areas [39]. Spraying hardwoods in the Missouri Ozarks increased the annual herbage production, principally little bluestem (*Schizachyrium scoparius*), from 220 to 1,210 pounds, while burning increased annual herbage production by only 50 pounds because of slight brush kill [38]. Repeated application of 2,4,5-T over two to three years in Arkansas killed 95 percent of the woody plants and increased steer gains per acre from about 13 to 45–50 pounds per acre under summer grazing [33]. When undesirable hardwoods were killed out of pine-grass stands in central Louisiana, bluestem forage averaged 2,415 pounds, compared to 675 pounds per acre on the check [54].

Spraying Macartney rose in Texas in 1968 returned $3 in a cow-calf operation for each $1 spent on 2,4-D [62]. Before spraying with 2,4-D, the range was in poor condition and produced 2,800 pounds of herbage per acre annually. After two annual applications of 2,4-D, Macartney rose was reduced by 95 percent, the range increased to good condition and produced 5,400 pounds of herbage per acre, and carrying capacity increased from one animal unit (a.u.) yearlong per fifteen acres to one a.u. per five acres. In Utah, 70 percent reduction in mulesear increased grass forage production from 280 to 1,353 pounds per acre [131].

Herbicides have been used to manipulate habitat for wildlife. Browse species commonly grow so tall that the forage they produce becomes unavailable to deer and elk. Herbicides have been utilized, as in the case of quaking aspen, to obtain top kill and accelerate sprout production at ground level. Herbicides such as 2,4,5-T have been frequently used in California brushland ranges to lower the browse level, stimulate palatable regrowth, and increase the production of forbs and grasses for winter and spring use [7]. Silvex application resulting in topkill of Gambel oak and a 44 percent increase in grass production two years after treatment resulted in 73 percent increase in elk use and 16 percent increase in mule deer use in northwestern Colorado [81], but respraying every three years was recommended to maintain the improved situation.

On big game range with a variety of browse plants of different palatabilities and susceptibilities to herbicides, evaluation of long-term benefits of herbicidal treatment is difficult. On a shrub range in the northern Rockies, 2,4,-D, 2,4,5-T, and a mixture of both were applied at the rate of three pounds per acre and evaluated five years later [85]. The translocated herbicides apparently did not improve the range. Redstem ceanothus (*Ceanothus sanguineus*), the most desirable browse plant tested, was killed by all treatments. The herbicides used did not effectively lower the crowns of Rocky Mountain maple (*Acer glabrum*) and were ineffective in stimulating sprouts of Scouler willow (*Salix scouleriana*). However, it was speculated that if contact herbicides instead of the translocated herbicides had been used, a high proportion of the aerial crowns might have been killed without inhibiting the roots and stems from sprouting.

Big sagebrush control in Idaho with 2,4-D resulted in a high top-kill of snowbrush ceanothus (*Ceanothus velutinus*), downy rabbitbrush (*Chrysothamnus puberulus*), quaking aspen (*Populus tremuloides*), chokecherry (*Prunus virginiana*), willows (*Salix* spp.), and snowberry (*Symphoricarpos oreophilus*), but a high proportion of the plants sprouted profusely [8]. Bitterbrush, a particularly valuable forage species, was unharmed or only slightly damaged.

In Oregon and California considerable 2,4-D damage has been reported on bitterbrush. Based on Oregon experiments, 2,4-D should be applied as early as good big sagebrush kill can be obtained in order to reduce damage to bitterbrush [70, 122]. The best time for ap-

FIG. 44. *Nearly complete kill of big sagebrush from aerial application of 2,4-D by the U.S. Forest Service near Richfield, Utah. Bitterbrush was initially damaged, but survival of individual plants was almost 100 percent.*

125

plication was when new leaves were present on sagebrush and bitter-brush and Sandberg bluegrass was in late boot or early head. Spraying at the time of leaf origin, and before the appearance of distinct twig elongation or flowering in bitterbrush, left only a small amount of dead tissue on large plants; but bitterbrush less than twelve inches high was frequently killed. After flower eruption in bitterbrush, significant increase in crown damage and mortality was realized.

Maintenance Control

Eradication of large acreages of undesirable range plants by herbicides is seldom possible. Areas treated with herbicides tend to become reinfested by sprouting of brush species, by seedlings arising from seeds present in the soil or brought in by wildlife or livestock, or by seedlings coming from plants missed in the initial treatment (Figure 45). At Spur, Texas, all 213 mesquite plants were removed by hand grubbing from a small pasture in 1940 [61]. However, over the next twenty-five years periodic maintenance removed an additional 476 trees without eradicating the mesquite from the pasture.

FIG. 45. *Big sagebrush reinvasion into seeded created wheatgrass range near Eureka, Utah, was controlled on the right with a maintenance application of 2,4-D.* (Utah State University photo)

Maintenance control recommended for areas previously receiving chemical or mechanical mesquite control treatments included the following steps: (1) remove plants escaping original treatment, (2) remove individual invading plants, (3) defer grazing during the first growing season and periodically thereafter, (4) avoid overuse of forage grasses, (5) avoid moving in cattle from areas of mesquite infestations, (6) prevent mesquite from maturing seed, and (7) respray to control mesquite and secondary species after a period of a few years [61]. Other possibilities are burning or goating.

When pasture and range units are only partially sprayed, livestock may concentrate on the sprayed area and overuse the forage there. Spraying big sagebrush range in strips leaving alternate check strips has contributed to selective grazing and sagebrush reinvasion [76]. These situations should be avoided by spraying the entire unit or fencing off and managing the sprayed area separately for a period of a few years.

Many land management agencies and agricultural technicians encourage or require deferment of grazing for one or more grazing seasons following herbicide application. Grazing deferment for at least one growing season following spraying is required in some states before ranchers are eligible to receive cost-sharing payments. Although deferment for one or more growing seasons favors maximum grass improvement, this practice may place a temporary hardship on the ranch operator or reduce the benefits of biological control that browsing can provide.

Hoffman [59] recommended that all weed controlled range areas in Texas be deferred from grazing until the key management plants have matured seed. The National Research Council [96] recognized that alternating use and nonuse periods and growing season deferment often are particularly helpful on range sprayed for brush and weeds but suggested that benefits in the short run were often greater than in the long-term.

Robertson [105] simulated deferment for two years after spraying in Nevada and found this benefited the crested wheatgrass but may not have been essential since good grass recovery was obtained in either case under proper utilization. Smith [119] reported that continuous moderate utilization (30 to 40 percent) did not retard the revegetation process on sagebrush range after spraying when compared to one to three years of deferment and did not influence subsequent reinvasion of sagebrush. Following herbicide treatment of subalpine tall forb-grass communities to control tall larkspur in Utah, delayed but not early grazing permitted the return of desirable decreaser grasses [28].

Although good management may prolong the benefits of herbicidal plant control on range, the effects are seldom permanent. Retreatment will usually be needed after a few years depending upon the regenerative potential of the noxious plants [96]. In Wyoming, where at least 75 percent of the big sagebrush is killed by spraying, the range can be expected to remain relatively free of sagebrush seedlings for at least four years after treatment [145].

In other studies in Wyoming, the density of young and mature sagebrush plants began to increase within five years after spraying whether the range was grazed or left ungrazed [76]. Seventeen years after treatment, sagebrush on grazed range was somewhat denser than before spraying and was about the same as before spraying on ungrazed range. In this study, life expectancy of brush control was about fourteen years. However, retreatment at an earlier stage of reinvasion would appear desirable.

In Oregon, range condition affected the degree of big sagebrush kill from 2,4-D and its subsequent reinvasion [56]. On fair condition range, two pounds of 2,4-D ester killed 98 percent of the big sagebrush, increased the production of perennial grasses by 2.5 times, and was associated with no increase in brush by eight years later. On comparable sites in poor range condition 85 percent of the big sagebrush was killed, herbage production increased about 4 times but was composed mostly of annual grasses, and sagebrush had increased slightly by eight years after herbicide application.

On Ozark hardwood range, spraying with 2,4,5-T to improve the range for cattle was recommended at less than eight-year intervals, but for deer habitat less frequently and in an alternate block pattern [53]. Young hardwood plants, seedlings, and sprouts began to interfere with grazing within five years after treatment [33].

Adequate initial control of many woody plants requires two or more herbicidal treatments in consecutive years. Similar initial control may result from paired treatments in alternate years as in successive years unless regrowth becomes excessive. Sprouts in chaparral species allowed to reach thirty-six inches high become less susceptible to hormone herbicides [7]. Third-year sprouts of chaparral species are generally too old for effective kill. The best kill is obtained when chamise sprouts are twelve inches high and scrub oak sprouts are eighteen to twenty-five inches high. Respraying Gambel oak sprouts during the second growing season in Utah has been more effective than during the first growing season or in later years.

Safeguards in Herbicide Use

The use of pesticides including herbicides in relation to the environment has recently received wide public attention. Pesticide critics have frequently based their appeals for discontinued use of pesticides on emotionalism rather than scientific fact. Pesticides are commonly criticized as a group rather than being recognized as including many chemicals of highly varied classification, origins, uses, toxicities, and other properties. The beneficial uses of pesticides have often been ignored.

A special commission under the Department of Health, Education, and Welfare included the following warning in their 1969 report [141]: "Reporting and exaggerating the danger of pesticide usage without equal treatment of the beneficial aspects of pesticides threatens to retard the advancing technology required to meet food and public health demands around the world." This report further concluded, "Herbicides, as presently used, do not present serious and widespread hazards to nontarget organisms. With few exceptions, most herbicides have a low order of toxicity to aquatic and terrestrial animals."

After extensive reviews of the published literature, Scifres [117] concluded that herbicides, when properly applied to range, provide excellent levels of weed and brush control without undue hazard to sensitive crops; do not endanger man, his livestock or wildlife; and, in most cases, are dissipated from the ecosystem during the growing season in which they are applied. There is no evidence of biological magnification with herbicides in food chains.

Although the problem has since been corrected, a carcinogenic impurity called dioxin was discovered during the mid 1970s to be contaminating 2,4,5-T at times during the manufacturing process. In support of continued use of 2,4,5-T and related products, the Society for Range Management [123] in 1978 issued the following position statement: "The Society for Range Management supports the continued production and currently registered uses of 2,4,5-T and related products. Although many questions about this herbicide's safety have been surfaced through news reports, the news media have failed to report the preponderance of reliable scientific research which clearly and conclusively documents for safety of 2,4,5-T. This vast body of scientific evidence demonstrates that 2,4,5-T as it is presently manufactured and used poses no harm to humans, wildlife, or any other form of animal life."

USDA concluded in 1970 [140] that a moratorium on the use of phenoxy herbicides in the U.S. would (1) reduce net farm income by $290 million annually, (2) require about twenty million more hours of family labor, (3) reduce pasture and range income $69 million annually, and (4) leave the range producer with no economically feasible alternative treatment on many sites. Based on USDA estimates Burnside et al. [20] calculated in 1970 that pesticides in total save U.S. farmers and ranchers $2.5 billion annually in production costs. They further concluded that the 16.5 percent of annual income then being spent for food in the U.S. would rise by 50 to 75 percent if pesticides were eliminated.

Registration of pesticides. All pesticides must be registered by the Pesticides Registration Division, Office of Pesticides Programs, Environmental Protection Agency, before entering into interstate or intrastate commerce. Authority has been given EPA by the Federal Insecticide, Fungicide, and Rodenticide Act (FIFRA) of 1972 and subsequent amendments to approve all uses of pesticides including herbicides, regulate instructions on pesticide labels, set tolerances in animal feeds and human foods, seize any raw agricultural commodities not complying with these tolerances, and punish violators using nonregistered pesticides or making unapproved use of registered herbicides.

EPA registration of herbicides insures that they are released for public sale and use only after detailed research and thorough testing. Tolerance levels set for human intake of pesticides include rather large safety factors and are generally set at one percent or less of the highest level causing no adverse effect in the most sensitive animal species [20, 141]; but zero tolerance is mandatory in some cases. The total cost of herbicide development by chemical companies is now (1979) about $15 million per chemical eventually registered. This cost includes basic research and chemical screening; development of promising chemicals, including manufacturing and analytical methodology; and the technical services of evaluating the effects and impacts of using the herbicide.

EPA Compendium of Registered Pesticides. Volume I [142] is the official source of new uses and changes in old uses of federally registered herbicides, defoliants, dessicants, plant regulators, and algaecides. Basic information is provided about each chemical, including the information on the herbicide label. Four of the eight group headings under which the approved uses for each chemical are listed are (1) agricultural crops (including fallow land, pastures, and rangeland), (2) agricultural premises, (3) forest, chaparral, non-agricultural land, and wasteland, and (4) aquatic areas.

In addition to the EPA, one lead agency within each state is designated by its governor to participate in pesticide regulation. Individual states may have special registration and use requirements for pesticides. Also, the designated state agency is charged with certifying pesticide applicators. Only certified pesticide applicators are permitted to purchase and use restricted use pesticides, including paraquat and picloram or those on emergency exemption.

In addition to the regular federal registration of pesticide uses, three special registrations are provided for additional pesticide use approval:

(1) *Experimental label.* This special federal label permits new products or old products being considered for removal of registration to be further researched and evaluated before final approval is given.

(2) *Emergency exemption.* The federal administrator of EPA may exempt any federal or state agency so requesting unapproved pesticide usage provided that the emergency requires such exemption.

(3) *Special state label.* A state may provide registration for additional uses of federally registered pesticides within the state, if such uses have not previously been denied, disapproved, or cancelled by EPA. Final approval of the state label is given by EPA unless such use is known to have definite detrimental effects on humans or the environment.

Animal toxicity. Herbicides such as 2,4-D, 2,4,5-T, MCPA, and silvex are not poisonous to livestock, wildlife, or man at recommended application methods and rates [98, 108, 138]. In fact, livestock need not be removed from areas to be sprayed except when poisonous plants are present. However, milk cows should be removed for one week from sprayed areas as a safeguard against any possible contaminants in the milk. Toxic amounts of these herbicides are obtained only if animals have access to spray tanks, herbicide containers, or herbicide in the concentrate form. These characteristics generally apply to other herbicides used on rangelands. Any exceptions are noted in *Table 8, Properties of Herbicides Used on Rangeland or Proposed for Range Use.*

The relative degree of toxicity of the various herbicides to warm-blooded animals has been determined experimentally. The relative degree of toxicity is expressed as the acute oral LD_{50} (the single dosage by mouth that kills 50 percent of the test animals expressed as mg./kg. of body weight). The LD_{50} for each herbicide is given in Table 8. The toxicity class, the LD_{50}, and lethal dosage for man have been related as shown in Table 6. On the basis of these ratings, 2,4-D, 2,4,5-T, and silvex are classified as mildly to moderately toxic; paraquat, diquat, and PCP as moderately toxic; fenac, TBA, amitrole, atrazine, ammate, monuron, and dicamba as mildly toxic; bromacil, picloram, and dalapon as nontoxic; and only DNBP as highly toxic. For comparative purposes, both aspirin (LD_{50} at 1,240 mg./kg.) and common salt (LD_{50} at 3,320 mg./kg.) are classified as mildly toxic.

Some herbicides such as 2,4-D and related compounds temporarily increase the palatability of affected plants. This is serious when poisonous plants are present. A general recommendation is that areas with poisonous plants should not be grazed until after affected plants

Table 6. Toxicity Class, LD$_{50}$, and Lethal Dosage for Man*

Class	LD$_{50}$	Lethal dosage for 150 lb. man
Highly toxic	50 or less	few drops to 1 teaspoon
Moderately toxic	50–500	1 teaspoon to 1 oz.
Mildly toxic	500–5,000	1 oz. to 1 pt.
Nontoxic	above 5,000	1 pt. to over 1 qt.

*Source: [138].

begin to dry and lose their palatability (generally three weeks or more after application). Following application of two pounds a.e. per acre of 2,4,5-T or silvex, levels of miserotoxin, a toxic glucoside in timber milkvetch (*Astragalus miser*), decreased rapidly and was only one-third of the controls after four weeks [147]. A thoroughly bleached condition of timber milkvetch was recommended before grazing in the area was resumed.

Herbicides may also increase levels of toxic principles in poisonous plants [148]. Silvex and 2,4,5-T increased the alkaloid content of leaves and stems of tall larkspur (*Delphinium barbeyi* and *occidentale*). Alkaloid levels were doubled in some years. The alkaloid content of

FIG. 46. *Tall larkspur controlled in foreground with 2,4,5-T. Hormone herbicides per se such as 2,4,5-T are safe to livestock but have increased the palatability and alkaloid content of some poisonous plants such as tall larkspur.*

FIG. 47. *Specimen label for herbicide as approved by EPA, giving active ingredients. Use directions and precautions and special instructions, also printed on the container, are not shown here.*

falsehellebore (*Veratrum californicum*) was increased by 2,4,5-T but not silvex. No herbicide tested prior to 1966 affected the oxalate content of halogeton or the alkaloid content in low larkspur (*Delphinium nelsoni*).

Proper handling. Proper handling and application of herbicides are necessary to prevent possible injury to the operator and handler, to livestock, desirable plants, fish and wildlife, and equipment. In order to insure that herbicides are safely handled and used, the following safeguards should be carefully followed:

1. Read the label on each container before using the contents. Follow instructions; heed all cautions and warnings. (See Figure 47.)
2. Apply only as directed, to the crops specified, and in amounts specified. Follow grazing or crop feeding restrictions.
3. Store in original labeled containers and keep out of reach of children, pets, livestock, and irresponsible people.
4. Avoid prolonged contact with the skin, breathing vapors or dust, or splashing herbicide solutions into eyes and mouth.
5. Clean spraying equipment immediately after application of herbicides to prevent contamination of future spray mixtures and to prevent damage to the equipment.
6. Do not clean spraying equipment or dump excess spray material near lakes, streams, ponds, or recreational areas.
7. Dispose of excess chemicals and empty herbicide containers at sanitary land-fill dumps or by burying at least eighteen inches underground in an isolated place that cannot contaminate water supplies. Do not reuse pesticide containers.
8. Confine chemicals to the property and crop being treated. Avoid drift of spray onto susceptible crops or ornamentals.

Chemicals such as dalapon, TBA, picloram, and fenac are not strongly absorbed on the soil colloids and are relatively free to move with percolating water into ground water or stream flow. However, limited use of these herbicides in areas where leaching occurs and their low toxicity reduce the hazard [96].

The problem of herbicide drift and volatization must be guarded against in order to prevent damage to susceptible crops near the target area. *Drift* is the horizontal displacement of airborne spray particles following application or after volatization. *Volatization* is the change of an herbicide after application from the liquid to the vapor form. When in the vapor form, herbicides are highly subject to drift.

Volatization of 2,4-D esters and related herbicides is affected by the size and structure of the alcohol moiety [79]. The high volatile esters include methyl, ethyl, isopropyl, butyl, and amyl; the low volatile esters are butoxyethanol, butoxyethoxypropanol, ethoxyethoxy-propanol, isooctyl, and propylene glycol butyl ether (PGBE) [78]. The more volatile esters quickly volatize and are carried downwind as soil or plant surfaces reach high temperatures. The amine, Na, K, and NH_4 salts of 2,4-D have little or no volatility.

The direction, distance, and amount of spray drift that result before the herbicide reaches the ground are influenced principally by the size of droplets, specific gravity, evaporation rate, height of release, direction and velocity of wind, vertical air movements, and type of application equipment used. Herbicide drift is a greater problem in aerial spray application because release is made farther from the ground and because of air turbulence generated. However, drift can be a problem with ground application as well.

Spray droplets should be large enough to minimize drift hazards and yet be sufficiently small and properly distributed to give good coverage. Finely atomized spray drops may drift from the target area or evaporate before reaching foliage. Larger droplets reduce spray drift but provide less uniform coverage. Smaller droplets (below 0.1 mm in diameter) of all herbicidal formulations used in one study were more effective than larger droplets (0.3 mm in diameter) [41]. Even if average droplet size is kept at a desirable level, lack of uniformity in droplet size will be associated with high numbers of fine droplets.

In applying phenoxy herbicides to mesquite, it has been established that about seventy drops per square inch are needed for minimum coverage [96]. Efficient spray distribution and plant coverage is especially critical when low rates of herbicide and carrier are used. Factors influencing spray distribution include swath width, height of spray release, nozzle placement as well as size and orientation, wind, and physical characteristics of the spray.

Methods of maintaining sufficient droplet size and uniformity to reduce drift include [1, 78, 96]:

1. *Use low nozzle discharge pressures*—not more than 35 psi for boom sprayers or 100 psi for spray guns.
2. *Use nozzles with larger orifices (not less than one-eighth inch) and that have no internal spray directional mechanism.*
3. *Direct nozzles in aerial application with the slipstream or a maximum of 10° downward.*
4. *Place no nozzles within three feet of the wing tips* because of high air turbulence in that area.
5. *Use invert emulsions* (water in oil).
6. *Use polymeric thickeners or particulating agents.*

135

Other means of reducing drift of herbicides include [1, 40, 78, 139, 138]:

1. *Increase specific gravity of droplets.* Oil-spray droplets are lighter and drift farther.

2. *Retard evaporation by adding anti-evaporants.* Water evaporates more rapidly than diesel oil.

3. *Reduce height of release*—particularly in aerial application.

4. *Use positive liquid shutoff systems in aerial application.*

5. *Use granular herbicides in soil application.*

6. *Avoid spraying on windy days*—not over 10 mph for aerial application nor over 15 mph with ground rigs, and preferably not over 5 mph for both aerial and ground application.

7. *Select days with a slight, continuous wind movement from one direction and away from susceptible crops.*

8. *Align flight paths so that no airplane turns are made near susceptible crops.*

9. *Avoid vertical air movement*—temperature differential is least in early morning, late evening, and night.

10. *Spray on days when atmospheric humidity is high*—this reduces evaporation.

11. *Keep rates and acreage minimum when susceptible crops are downwind.*

Classification and Application of Herbicides

Herbicide—a chemical that kills plants (syn. *phytocide*).

Contact—an herbicide that kills only plant parts directly exposed to the chemical and is directly toxic to living cells.

Translocated—an herbicide applied to one part of a plant but which is spread throughout the plant where effects are produced (syn. *synthetic hormone herbicide, systemic herbicide,* or *growth regulator*).

Selective—an herbicide that kills or damages a particular plant species or group of species with little or no injury to other plant species. (Are often nonselective at heavy rates.)

Nonselective—an herbicide that kills or damages all plant species to which applied (syn. *general weed killer*).

Soil sterilant—an herbicide which kills or damages plants when present in the soil. The effect may be temporary or permanent and selective or nonselective.

Refer to Table 8 for properties of herbicides used or proposed for use on rangelands. Chemical control recommendations for individual noxious range plants are summarized in Table 9.

The following classification of herbicides used on range considers method of application, selectivity, and type of action:

I. Foliage spray application—broadcast or individual plant.
   A. Selective
      1. Translocated—2,4-D, 2,4-DB, 2,4-DP or dichlorprop, 2,4,5-T, silvex or 2,4,5-TP, MCPA, dalapon, dicamba, picloram.
      2. Contact—paraquat, dinoseb salt (unimportant in range improvement).
   B. Nonselective
      1. Translocated—AMS, amitrole, glyphosate.
      2. Contact—diquat and paraquat (effect generally low on broadleaf plants), AMS, kerosene, and diesel oil.
II. Foliage dust application—unimportant in range improvement.
III. Individual trunk treatment (special woody plant applications).
   A. Trunk base spray—2,4,5-T, picloram, silvex.
   B. Frills—AMS, 2,4,5-T, 2,4-D amine.
   C. Trunk injection—2,4,5-T, picloram, 2,4-D amine.
   D. Cut-stump treatment—AMS, 2,4,5-T, 2,4-D amine, picloram.
IV. Soil application (temporary sterilization)—including spray, pellet, or granular application or by soil injection—atrazine, bromacil (nonselective), dicamba, diuron, fenuron*, karbutilate (nonselective), monuron, fenac, 2,3,6-TBA (nonselective), picloram, tebuthiuron (nonselective).

Herbicides used on rangeland may also be classified on the basis of chemical groupings:

1. Alaphatics—dalapon, glyphosate.
2. Benzoics—dicamba, 2,3,6-TBA.
3. Carbamate—karbutilate.
4. Dypiridylium—paraquat, diquat.
5. Inorganics—AMS.
6. Organics—diesel oil, kerosene.
7. Phenoxys—2,4-D, 2,4-DB, 2,4-DP or dichlorprop, 2,4,5-T, silvex or 2,4,5-TP, MCPA.
8. Substituted phenols—dinoseb or DNBP.
9. Substituted uracil—bromacil.
10. Substituted urea—monuron, fenuron*, diuron, karbutilate, tebuthiuron.
11. Triazine—atrazine.
12. Triazole—amitrole.
13. Unclassified—fenac, picloram.

Foliage spray application. Broadcast spraying is the application method most commonly used on rangelands. Since the herbicide is

*Fenuron was discontinued in 1971 for an unknown period of time.

applied on all plants, desirable as well as undesirable, selective herbicides are generally required. Broadcast spray applications can be made either by ground rigs or by aerial application. When herbicides are applied by ground rigs, a spray volume of ten to twenty gallons or occasionally up to fifty gallons per acre is used. With aerial application, total spray volume generally ranges from two to five or sometimes as low as one gallon per acre. The advantages of using ground versus aerial broadcast spray application are as follows:

Advantages of ground application	Advantages of aerial application
1. Adapted to small acreages.	1. Faster coverage.
2. No landing strip required (also true of helicopter application).	2. Adapted to wet, rough, or rocky ground.
3. Less drifting and less subject to fog or wind.	3. Lower cost per acre often on larger acreages (200 acres or more).
4. Commercial equipment not required.	4. No mechanical disturbance of soil or vegetation.
5. Safer for applicators.	5. Better coverage of tall, dense brush stands.

Fixed-wing aircraft are commonly used to apply herbicidal sprays to large areas (Figure 48). Fixed-wing aircraft with payload capacities from 100 to 750 gallons are available for herbicide application. Helicopters generally are equipped to carry from 30 to 120 gallons and are more commonly used now than formerly (Figure 49). Comparative advantages include:

Advantages of fixed-wing aircraft	Advantages of helicopters
1. Greater lifting power in thin, warm air.	1. No landing strip required.
2. Greater payload (may be offset by travel to landing strip).	2. Trees, snags, and steep terrain less troublesome.
3. Cheaper on large projects.	3. Greater accuracy resulting from slower airspeed and greater maneuverability.
4. More available.	

Boom-type ground sprayers utilize nozzles evenly spaced along a horizontal boom. The boom can be mounted on a truck, pickup, wheel or track tractor, trailer, or aircraft. Nozzles giving a flat, fan-shaped spray are more commonly used with ground rigs since they give a more uniform coverage and more forceful application than nozzles giving a hollow, cone-shaped spray. Boomless or wide-angle nozzle-type ground rigs should generally be used only on fence rows, roadsides, ditches, or crops where uneven spray pattern is less impor-

FIG. 48. *Fixed-wing aircraft for herbicide application.* A, *bi-wing agricultural appli-cator;* B, *converted torpedo bomber with large payload capacity;* C, *one pound of 2,4-D being applied near Woodward, Oklahoma, which killed up to 80 percent of the sand sagebrush, increased forage production 60 percent for the first three years, and doubled net returns per acre.* (Photo C courtesy of USDA, SEA–AR)

tant and in areas where drift is not a dangerous factor. A comparison of boom and boomless application with ground rigs indicates:

Boom advantages	**Boomless advantages**
1. More even distribution of spray.	1. Equipment often less expensive.

139

FIG. 49. *Helicopter being refilled with herbicide mix while resting on landing provided on top of mixing truck near Post, Texas.* (Soil Conservation Service photo)

2. Spray less subject to drift.
3. Spray less affected by wind.

2. Easier to maneuver equipment.
3. Convenient for brush, fence rows, or rough terrain.
4. Less nozzle stoppage.

One self-propelled ground sprayer developed by the Agricultural Research Service in Oklahoma has sprayed sixty acres per hour at a 1964 application cost of $.28 per acre (Figure 50a and 50b) [6]. The sprayer utilizes a self-propelled combine chassis. The 100-foot boom has an application height of 6 to 18 feet at the outer ends and folds into five sections. The machine readily maneuvers on level or rolling sandhills terrain and sprays shinnery oak over ten feet high. Total spray volume can be varied from 1.5 to 40 gallons per acre.

A small track tractor equipped with a 44-foot boom and two 170-gallon tanks has been developed by the Range Seeding Equipment Committee in conjunction with the Arcadia Equipment Co. (Figure 50c) [103]. It has been used principally on California chaparral and has given very satisfactory performance on front grades of 0 to 45 percent (some slippage on grades 45 to 55 percent) and side slopes of 0 to 32 percent. It is also equipped with a spot sprayer which will reach up to 150 feet. The boom can be raised or lowered for use on brush stands of different heights.

Annual application of low, defoliation levels of phenoxy herbicides have been investigated by USDA at Woodward, Okla. [93]. Summerlong suppression of shinnery oak, sand sagebrush, skunkbush, and perennial ragweed was achieved by applying 1/16 pound of 2,4,5-T or 1/8 pound of 2,4-D a.e. per acre; two to three sprayings per summer defoliated saltcedar. Spraying for 2 to 3 years and

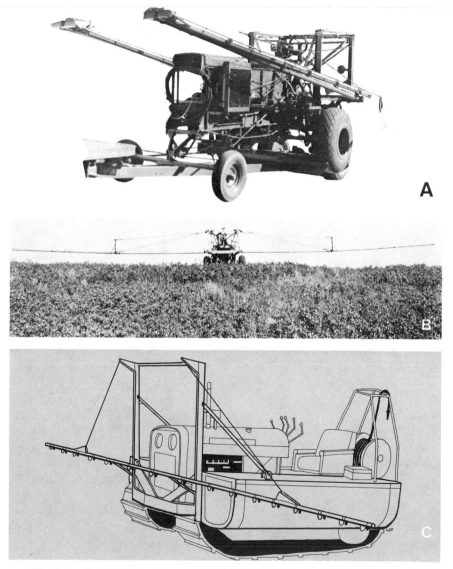

FIG. 50. *Ground rigs for broadcast application of herbicide.* A, *"Goliath" spray rig developed on self-propelled combine chassis;* B, *"Goliath" spraying sand sagebrush at sixty acres per hour;* C, *ground rig developed for spraying chaparral on steep slopes.*

thereafter as needed for desired suppression was suggested. These light rates were applied by mistblower sprayer mounted on a pickup following predetermined spraying trails at 10 m.p.h. A crosswind of 5 to 12 m.p.h. permitted spraying of 100-foot strips at a rate of 120 acres per hour and resulted in an application cost of 10¢ and chemical cost of 10–20¢ per acre.

Proper swath widths are important in preventing skips or overlapping swaths and in obtaining complete coverage of the foliage in broadcast application. However, effective swath widths with aerial application may be less than half of the overall swath width and will require considerable marginal overlap [139]. Swath width varies primarily with altitude above the vegetation from which the spray is released, and swath width checks should be made periodically to insure proper coverage. Flagging is essential in aerial application (Figure 51). A flagman is required at each end of the flight run with additional flagmen on very long runs, on varied terrain, or in tall brush or trees. Colored paper strips laid on the ground or colored smoke are sometimes used to supplement flagging; and radio communication between the flagmen and the pilot is desirable. Mechanical or other means of swath marking is also required in broadcast ground application of herbicides. Many aircraft are now equipped with automatic flaggers which dispense strips of wet paper.

Individual plant foliage spraying may have advantages over broadcast foliage spraying on spot infestations of undesirable plants,

FIG. 51. *Flagging, as being used in this Texas spray project on mesquite, is essential in the aerial application of herbicides.* (Soil Conservation Service photo)

142

widely scattered plants, along roads and trails, where the terrain is too rough for wheeled machinery, and in forests where only a small proportion of the plants are to be removed. Individual plant treatment allows nonselective herbicides to be used selectively through positive control of spray stream. Also, complete wetting is sometimes required and can be provided easier in individual plant application. Disadvantages include a high cost per plant, excessive costs where plant density is high, high labor demand, difficulty of wetting plants over six feet high, and slow job completion.

The most commonly used herbicide for individual plant wetting sprays for brush is 2,4,5-T. Picloram at three pounds acid equivalent

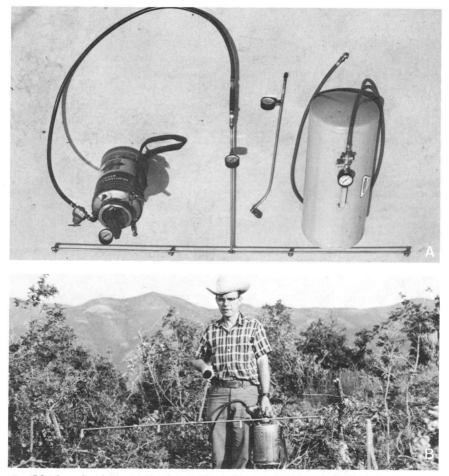

FIG. 52. A, *a hand sprayer developed for plot or individual plant spraying with single-nozzle and four-nozzle boom applicators and portable air tank [43]; B, same being used on Gambel oak control research plots in Utah.*

per 100 gallons (a.e.h.g.) as a wetting spray is effective on all oaks, huisache, mesquite, pricklypear, sagebrush, willow, and sumac [128].

A combination of broadcast spraying followed by individual plant treatment is often the best method, such as on hard-to-kill chaparral plants [102]. Individual plant spraying can be accomplished by hand sprayers or auxiliary hand booms or mist blowers on power sprayers.

Individual trunk treatment. The treatment of trunks or stems of individual woody plants is generally practical only where there is a sparse stand of single stem trees or where small areas of rangeland are to be treated [96]. The three special methods applicable to woody plants are trunk base spray; frills, notches, or trunk injection; and cut-stump treatment.

Trunk base spray is applied to the lower 12 to 18 inches of the tree trunk until it runs down the bark channels to the groundline. This method is particularly effective against hardwood trees up to five inches in diameter where the bark is not too thick. A basal spray of eight pounds acid equivalent of 2,4,5-T ester per 100 gallons of diesel oil or kerosene is effective on most hardwoods [65]. Sixteen pounds a.e.h.g. is recommended for hard-to-kill species. One gallon should treat from fifteen to twenty four-inch trees. The 2,4,5-T spray can be applied either in summer or winter.

Application of 2,4,5-T or silvex in a water carrier is less effective, and 2,4-D has been found less effective than 2,4,5-T or silex in basal sprays. Kerosene or diesel oil has been effective on mesquite even when not fortified with additional herbicide. Basal spraying of plants under five inches in diameter, in one study, was the most economical of five methods used to retreat sprayed mesquite [143]. Picloram has shown particular promise as an effective basal trunk spray.

Frill or tree injection is recommended for killing most woody plants six inches in diameter or larger. Frills are made by overlapping downward axe cuts through the bark around the tree trunk near the groundline; the herbicide is applied in the frills. Tools for tree injections have a hole in a chisel point through which herbicide is injected after the chisel is driven into the stem. Injections are commonly made in cuts about two inches apart at base of stem.

For frill cuts, 2,4,5-T ester at eight and sixteen pounds a.e.h.g. of diesel oil is recommended for average and resistant hardwoods. Apply herbicide until it bubbles out of the freshly cut frills [65]. Undiluted 2,4-D amine has also been used with 95 percent effectiveness on oak in California [55]. For tree injection, undiluted 2,4,5-T ester or diluted at 32 pounds a.e.h.g. in diesel oil has been effective on hardwood species more than one inch in diameter [128]. Frill or injection treatments have generally been effective during all seasons of the year.

Treating recently cut stumps with herbicide effectively controls most sprouting species. Herbicides can be applied by a large oil squirt can, a squeeze bottle, or a paint brush to entire cut surface of small stumps or to a strip two to four inches wide around the outer

FIG. 53. *Herbicide application by individual trunk treatment.* A, *trunk base spraying;* B, *frill spraying; and* C, *pour-on method being used in Texas to control mesquite with diesel oil.* (Photo C courtesy of Soil Conservation Service)

145

circumference of cut surfaces of large stumps. Treatment can be made at any time of the year but is most effective soon after cutting and when trees are cut close to the ground. Herbicides and rates used in basal stem and frill application are generally adapted to cut stump applications.

Shredding followed by stump application of herbicides provided a safe and economical means of controlling mesquite on small areas in Texas [19]. Picloram-2,4,5-T mixtures in a water carrier at 3 pounds a.e.h.g., applying about 1/10 gallon per stump, resulted in root mortality averaging 57 percent and was effective even during the winter months providing soil moisture content was adequate.

Ammonium sulfamate (AMS) can also be used in individual trunk treatment with frill or stump application. For cut-stump treatment sprinkle crystals directly on the cut surface. A spray comprised of 3.5 to 5 pounds of crystals per gallon of water can be applied either to cut surfaces or to frills. AMS is also effective as an individual plant foliage wetting spray when applied after the brush has reached the full leaf stage and until color changes in the fall. Concentrations of fifty pounds AMS active per 100 gallons in oil or sixty pounds active per 100 gallons in water are used in foliage sprays [144, 63].

Soil application. Applying herbicides directly to the soil is another means of killing undesirable plants. Soil-active herbicides are more commonly applied as dry granules or pellets than as spray since vegetation intercepts some or most of the spray. The advantages and disadvantages of granular or pellet forms of soil sterilants are as follows [96, 83].

Advantages
1. Reduced drift.
2. Not intercepted by foliage.
3. Controlled release through selection of carrier.
4. Ease of handling and application.
5. Premixing reduces mixing errors.
6. Simpler application equipment generally.
7. Generally remains active longer.

Disadvantages
1. More expensive generally to make, package, and ship.
2. Some herbicides are less effective in granular form.
3. Generally more expensive per unit of killing power.
4. Difficult to spread evenly if particle size is variable.
5. Application equipment lacks versatility.
6. Moved by surface winds or water action after application.
7. Needs rainfall to activate.
8. Difficulty of handling and storing bags.

Soil surface application is less dependent on stage of growth than foliage sprays but does require rainfall to dissolve and penetrate into the soil. In some areas excessive losses may result from leaching or adsorption on soil colloids. Soil sterilants with high solubility in water normally are faster acting but persist in the soil for a shorter period of time than related sterilants with low solubility. Herbicide granules or pellets may be broadcast, placed in grid fashion, or placed around the base of individual plants or clumps of brush. Applying granular or pelleted herbicides in continuous narrow bands 6 to 10 feet apart has been equal or superior to broadcast application for brush control and may offer a means of reducing damage to herbaceous, understory vegetation by some herbicides [45].

Soil sterilants in liquid solution can also be placed at the base of brush plants by soil injectors but require placement in contact with underground parts of the plant for maximum effectiveness. Eight pounds a.e.h.g. of 2,4,5-T in diesel oil or six to eight pounds active ingredients per 100 gallons of water have been effective when applied at the rate of two ounces per inch of trunk base diameter [90]. An experimental herbicide applicator designed to apply liquid herbicides in continuous narrow bands underground eliminated spray drift and significantly reduced forage plant injury [12]. Although this application method offers the possibility of year-round use in brush control, it had difficulties in rocky and excessively wet soils.

Formulating Herbicide Sprays

The formulation of a herbicide is a major factor determining its effectiveness when applied to noxious plants. Since the formulation of a herbicide considers a multiplicity of factors and ingredients, the ultimate effect of a herbicide such as 2,4-D can be markedly altered by changing any of these factors. The combination of formulation, rate of application, and method of application generally determine whether a recommended herbicide will be highly selective and effective or not. Formulation requires an understanding of the following terms:

> **Formulation**—the manner in which the toxicant, carrier, and other ingredients are mixed. Considerations include the kind, chemical and physical form, and concentration of the toxicant; kind and amount of carrier; whether solution, emulsion, suspension, granules, or pellets in final form; and kind and amount of surfactants added.
>
> **Toxicant**—the herbicide or chemical agent that causes a toxic effect on plants.

Carrier—the diluent in which the toxicant is mixed to provide greater bulk for more effective application.

Commercial product—the herbicide as purchased from the store, usually in a concentrated form.

Spray mix (or spray liquid)—herbicide formulation prepared in liquid form for spray application.

Solution—a homogeneous spray mixture in which the toxicant (solute) is distributed in molecular or ionic form within the carrier (solvent).

Emulsion—a mixture in which the toxicant in liquid form (discontinuous phase) is dispersed as globules in a liquid carrier (continuous phase), i.e., oil in water.

Invert emulsion—a water-in-oil emulsion, i.e., the toxicant in aqueous form is dispensed in an oil carrier. Generally has a distinct mayonnaise consistency requiring special application equipment.

Suspension—a mixture in which the toxicant consisting of solid, finely divided particles is dispersed in a liquid carrier.

Surfactant (surface active agent)—materials used in herbicide formulations to facilitate or accentuate emulsifiability, spreading, wetting, sticking, dispersibility, solubilization, or other surface-modifying properties.

Active ingredient—that part of a commercial product or spray mix which directly causes the herbicidal effects.

Acid equivalent (a.e.)—the amount of active ingredient expressed in terms of the parent acid or the amount that theoretically can be converted to the parent acid.

The active ingredient of phenoxy herbicides is expressed in terms of acid equivalent. This is a relative term relating esters and salts to the pure acid, a form which is seldom available but may occur in minor amounts mixed with the other forms. The acid equivalent is a more precise measurement than the actual amount of the particular chemical form. However, acid equivalent measures toxicity only indirectly since other factors in the formulation also affect toxicity to plants.

The tag on one chemical product of silvex available for purchase provides the following information about the herbicide:

1. 58.9 percent trichlorophenoxy propionic acid, butoxy ethanol ester.
2. 41.1 percent inert ingredients.
3. 42.9 percent trichlorophenoxy propionic acid equivalent.
4. 4 pounds a.e. per gallon by weight.

The 58.9 percent represents the total chemical form of the silvex. Since 42.9 percent is the acid equivalence, the difference (16 percent) represents approximately the butoxy ethanol ester fraction. The 41.1 percent is principally inert ingredients but includes additives such as surfactants.

Calculations*. Herbicide recommendations are generally given in one or more of the following ways:

1. Pounds of active ingredients (or acid equivalent) per acre or per square rod for broadcast application.
2. Pounds of active ingredients (or acid equivalent) per 100 gallons of mix or percent concentration for wetting sprays, frill or cut-stump application, or plant or soil injection.
3. Weight (grams or ounces) or volume (tablespoons or cups) of actual herbicide product per plant or clump of plants.

The following are examples of calculations frequently used in mixing and applying herbicides:

1. *Rate per acre for liquid formulation*—if two pounds a.e. per acre is recommended and a commercial product containing four pounds a.e. per gallon is purchased,

   $$\frac{2\text{ lb. a.e./acre}}{4\text{ lb. a.e./gal. product}} = \text{½ gal. (or 4 pts.) or product is required per acre.}$$

   Calculate as above or obtain from Table 10.

2. *Rate per acre for granular form*—if three pounds active per acre is recommended and a commercial product containing 10 percent active ingredients in granular form is purchased,

   $$\frac{3\text{ lb. a.e./acre} \times 100\%}{10\%} = \text{30 lb. of granules is required per acre.}$$

3. *Weight of commercial product per gallon*—if a commercial product of silvex includes four pounds a.e. per gallon by weight and is 42.9 percent trichlorophenoxy propionic a.e.,

   $$\frac{4\text{ lb. a.e./gal.} \times 100\%}{42.9\%\text{ a.e.}} = \text{9.3 lb. per gal. is the weight of the commercial product.}$$

   The weight per gallon of contents may or may not be given on the herbicide container.

4. A wetting spray containing 8 pounds a.e.h.g. is recommended and a commercial product containing 4 pounds a.e. per gallon is purchased.
   a. Ounces of commercial product required per gallon of spray liquid is 3.2. (Obtain directly from Table 11 providing chemical product weighs 10 pounds per gallon. Correct for

---

*A table of weights, measures, and equivalents is found in *Appendix I.*

other weights of commercial products as shown in footnote.)
   b. Tablespoons of commercial product required per gallon of
      spray liquid is 5.1. (Obtain directly from Table 11.)

5. A 100-square-foot trial plot is to be sprayed at the rate of one
   pound active per acre. If a liquid product containing two
   pounds active per gallon is used, mix 5.2 gm. or 0.88 teaspoons
   of commercial product in the desired carrier. (Obtain directly
   from Table 12; correct if necessary.)

6. A 1 percent a.e. (by weight) wetting spray is to be applied to
   individual plants. If a commercial product containing 20 percent a.e. is purchased and diesel oil is used as a carrier,

$$\frac{1\% \text{ (desired conc.)} \times 7 \text{ (wt./gal. diesel)} \times 453.6 \text{ (gm./lb.)}}{20\% \text{ (a.e. in commercial product)}}$$

$$= 158.8 \text{ gm. of commercial pro-}$$
duct to be included in each
gallon of diesel oil carrier.

7. Two commercial products of 2,4,5-T are available for spraying
   sixty acres of mesquite. Source A has four pounds a.e. per gallon and sells for $9.20 per gal. while Source B has two pounds
   a.e. per gallon and sells for $5.50 per gal. The 2,4,5-T is to be
   applied at the rate of two-thirds pound a.e. per acre. Spray mix
   is to include commercial product plus one-half gallon of diesel
   oil and sufficient water for a total application rate by airplane
   of three gallons per acre.
   a. Compare cost of herbicide from the two sources:

	Source A	Source B
Lbs. a.e. required	40	40
Gals. of C.P. required	10	20
Cost of herbicide	$92.00	$110.00

   b. Calculate ingredients necessary to formulate the spray mix
      required.

	Source A (gal.)	Source B (gal.)
Total spray	180	180
Diesel oil	30	30
Commercial product	10	20
Water	140	130

Surfactants. Surfactants may increase, decrease, or have no effect on
the herbicidal effect. The use of surfactants to increase herbicidal activity is a desirable practice but must not be used indiscriminately.
Surfactants are selective for herbicides, carriers, and even plant species. Although normally harmless per se, surfactants are capable of

150

increasing the toxicity of sprays or reducing their selectivity sufficiently to cause injury to desirable plants normally resistant to the herbicide.

Surfactants include such common products as soaps, detergents, and shampoos. They combine two opposing characteristics, lipophilic (nonpolar) and hydrophilic (polar) tendencies, and reduce the interfacial tension between oil and water or between the herbicide spray and plant surfaces. Surfactants, functioning as emulsifying agents, are necessary to stabilize emulsions by preventing or retarding the dispersed droplets from coalescing to form larger droplets and settling out.

The effects of surfactants in herbicides have been proposed as follows: (1) improve foliage coverage, (2) remove air films between spray and foliage surface, (3) reduce interfacial tension, (4) induce stomatal entry, (5) increase the permeability of the plasma membrane, (6) facilitate movement through cell walls, (7) act as cosolvents, (8) interreact directly with herbicide in some manner, and (9) prevent drying of spray and thereby maintain herbicide in an absorbable condition for a longer time [31]. When water is used as the carrier, a surfactant or both a surfactant and oil frequently improve herbicidal activity on woody plants. Emulsifiers are generally added as needed at the factory.

Carriers. Water, oil, or water-oil mixtures are all used as herbicide carriers on rangeland. Diesel oil is the most common oil used as a carrier, but kerosene and fuel oil are sometimes used. Water and oil as carriers are contrasted as follows:

Water
1. Water molecule is polar and very good to ionize salts.
2. Many nonionizing herbicides go into solution due to solubility in water.
3. High surface tension often results in differential wetting.
4. Good driving force through foliage to reach lower branches or understory.
5. Penetrates mostly through stomata.
6. Water is nontoxic.
7. Higher evaporation during application.

Oil
1. Oil molecule is nonpolar, thus nonpolar materials readily dissolve in oil.
2. Low surface tension and spreads evenly.
3. Low driving force through foliage.
4. Lower nozzle pressure required to reduce mist and drift.
5. Penetrates cuticle (nonpolar, waxy) as well as through stomata.
6. Oil is also toxic and may provide additive control.
7. Adds cost to the spraying program.

8. Greater difficulty of
handling an oil carrier and
cleaning equipment.

For contact herbicides, complete spray coverage of plant foliage is required. Adequate quantity of carrier must be used to obtain a good coverage of the foliage sprayed. Different carriers have often been reported to be equally effective. When emulsifiable concentrates of 2,4-D esters were sprayed on big sagebrush, water, diesel, and diesel-water emulsions as carriers gave similar results [69]. Similar results were observed with western ragweed [15], Russian olive [11], and sand sagebrush [9]. On saw-palmetto (*Serenoa repens*) in Florida, increasing amount of oil in the carrier from 0 to 100 percent decreased kill from 2,4,5-T and silvex but increased kill by 2,4-D and dalapon [87].

Recommendations for the aerial application of herbicides to kill brush species commonly include a recommendation of one gallon diesel oil, sufficient commercial product to provide the acid equivalence, and enough water to total five gallons of spray mix per acre. Diesel oil alone used as a carrier for 2,4,5-T and silvex on oaks only slightly increased leaf defoliation over diesel-water emulsion but increased drift [39]. Adding 10 percent diesel oil in the 2,4,5-T spray mixture applied at 50 g.p.a. greatly increased effectiveness on broom snakeweed in Arizona [111]. Schuster [113] concluded that pricklypear was not satisfactorily controlled by herbicides applied in water carriers; at low concentrations, effectiveness increased with the proportion of diesel oil in the carrier. Roots generally absorb polar agents, such as water, better than nonpolar agents.

Chemical form and rate. General characteristics of the different chemical forms of 2,4-D are presented in Table 7. The ester forms are the most toxic per unit of acid equivalence for most plants, are the most volatile, and must be emulsified before they will mix with water. Herbicidal action by esters is favored by warm weather, apparently through more rapid cuticle penetration. Although esters make some entry into leaves in a vaporous form, the amount of volatility associated with an ester is questionable relative to effectiveness. On big sagebrush in Oregon, the isopropyl ester (less volatile) of 2,4-D was slightly less effective than butyl and ethyl esters (more volatile) at low herbicide application rates but was similar at practical rates [68]. In studies with rabbitbrush (*C. viscidiflorus*), the low volatile and high volatile esters of 2,4-D were equally effective [82]. Emulsifiable concentrates of phenoxy herbicides are commonly used on rangeland in broadcast spray application. These commercial products are prepared by dissolving the ester forms and an emulsifying agent in an organic solvent and may be used with either water or oil as the diluent.

The amine and sodium salts of 2,4-D have no volatility hazard, are directly soluble in water, but are most effective when foliage is moist from dew. The amine salt is generally less toxic than the ester forms but may be similar to esters under ideal conditions. It is used less frequently than the esters on rangeland but is preferred when susceptible crops are nearby.

The amount of herbicide required to provide adequate control varies greatly with kind of herbicide, plant species, and method of application. Herbicide rate recommendations primarily consider optimum toxic effects. Higher rates are rarely more effective and may prove detrimental. Many translocated herbicides may be reduced in effectiveness when high rates are used because leaves may die so quickly that the toxicant is not absorbed. When multiple herbicide

FIG. 54. *Anatomical symptoms of 2,4-D effects on susceptible plants include bending and twisting of stems, petioles, and blades (epinasty) and the accentuation of leaf sinuses. Picture shows results of spraying Gambel oak sprouts with 2,4-D.*

153

Table 7. General Characteristics of Different Forms of 2,4-D*

Form of 2,4-D	Soluble in water	Soluble in oil	Appearance when mixed with water	Precipitate formed in hard water	Volatility	Toxicity ranking	Comments
2,4-D acid (wettable powder)	no	no	milky	yes	no hazard	3	seldom used
Amine salt	high	no	clear	yes	no hazard	2	widely used
Sodium salt	medium	no	clear	yes	no hazard	4	less frequently used
High volatile ester	no**	yes	milky	no	volatile	1	hazardous around susceptible crops
Low volatile ester	no†	yes	milky	no	medium to low volatility	1	widely used

*Source: Adapted from Klingman [79].
**But can be emulsified. Includes methyl, ethyl, propyl, butyl, pentyl, amyl, and isopropyl esters.
†But can be emulsified. Includes isooctyl, butoxyethanol, butoxyethoxypropanol, ethoxyethoxypropanol, tetrahydrofurfuryl, and propylene glycol butyl ether (PGBE) esters.

applications are required for satisfactory kill, a higher rate of herbicide should not be expected to replace the need of the repeat treatment. However, reducing rates below recommended levels to save chemicals may sharply reduce kills, particularly when less than ideal conditions are encountered.

Mixing herbicides has remained relatively uncommon in range improvement until recently except for "brushkiller" which includes equal parts of 2,4-D and 2,4,5-T on an acid equivalent basis. Herbicide mixtures have often resulted in no added benefits, and contact herbicides in mixture with phenoxy herbicides are capable of interfering with their translocation within the plant and reducing the permanent kill. However, mixtures of phenoxy herbicides with picloram are now sold commercially in some states. Mixtures of 2,4,5-T and picloram have given the best control of brush species such as Gambel oak. The following three reasons for mixing herbicides have stimulated active interest and research into its possibilities [48]:

1. Control a greater number of weed species than individual herbicides, i.e., brushkiller.
2. Synergism—when the benefits of mixing herbicides are greater than the summation of advantages of both herbicides used singly, i.e., picloram and 2,4,5-T.
3. Partial replacement of a more costly herbicide by one cheaper in cost with little loss in effectiveness, i.e., adding 2,4-D to picloram.

Herbicide Selectivity and Plant Factors

Herbicide selectivity is the result of a toxicant reaching and disrupting a vital function in one plant but not in another plant. Selectivity is relative and depends on the presence and effectiveness of barriers that prevent herbicidal action. Barriers can prevent or reduce herbicidal action when effective at any of these steps [96]:

1. Achievement of surface contact by herbicide with plant or plant parts.
2. Penetration of the plant.
3. Translocation to a site of toxic action.
4. Disruption of some vital function.

Penetration. Foliar absorption of herbicides occurs most rapidly at open stomata, hydathodes (epidermal structures allowing exudation of water), lenticels, and natural fissures; insect punctures and other imperfections of the cuticle; glandular and nonglandular epidermal hairs; and directly over veins [96]. Manipulating the factors that favor or prevent penetration of a herbicide and thereby influence selectivity depends upon an understanding of the factors that influence penetration of herbicides through foliage, roots, and epidermis of the stem [46, 31, 96]:

1. *Leaf size and shape.* Leaves that are small, narrow, waxy, corrugated, or pubescent (such as many of the grasses) are less easily intimately contacted and thereby penetrated by sprays.
2. *Stomata.* The location of large, numerous stomata on the upper as well as the lower surface of the leaves favors penetration.
3. *Leaf orientation.* Plants with leaves oriented horizontally rather than vertically are more easily reached by herbicides applied from above.
4. *Cuticle.* Thick, waxy cuticles without cracks and natural perforations are less easily penetrated, particularly by herbicides in a water carrier. The cuticles of leaves tend to become less easily penetrated as they mature. Rotary hoeing of pricklypear has effectively increased herbicide penetration and control through puncturing the pads and reorienting the pads into a horizontal position [146].
5. *Canopy.* Plants with dense canopies tend to shield lower branches and leaves as well as the understory plants.
6. *Growing point location.* Plants with growing points located near the ground or otherwise protected from direct contact are less subject to herbicidal damage.
7. *Water balance.* Plants having a water balance favorable to rapid leaf and root absorption are more readily penetrated by herbicides.

Translocation. Translocated foliar herbicides are carried in the phloem with the assimilate stream from the foliage to regions of active growth, reproduction, and storage [96]. By contrast, herbicides applied to roots move through the xylem in the transpiration stream with water and soil nutrients to sites of photosynthetic and metabolic action. The rate and extent of translocation and associated herbicidal effects depend on several factors:

1. *Physiological differences.* Different species have different enzyme systems, different chemical constituents in tissues, and different cell characteristics. Plants may have chemical and biophysical barriers to translocation (i.e., inactivation by adsorption, enzymatic detoxification, high death loss of phloem cells by herbicide toxicity, differential tolerances to toxicants, or chemical reaction)` [46, 96].
2. *Stage of growth.* The most effective stage of growth for herbicidal application must be determined experimentally for each spe-

156

cies. Growth stage is more reliable than calendar date. Rabbitbrush in Colorado studies was most effectively controlled during the growing season prior to flowering when the new growth was three to six inches long. Two pounds of 2,4-D a.e. per acre killed 82 percent of the plants when the new shoots were three to six inches long but only 65 percent up to to three inches long [100a]. In Oregon the ideal time to spray big sagebrush was based on growth stage of Sandberg bluegrass (*Poa secunda*) beginning with early heading and ending when the green head color was half gone [69].

In Nevada little rabbitbrush sprayed with 2,4-D at the half-leaf stage in early June increased because unaffected or resprouted plants increased in vigor and production as a result of forb reduction [82]. Whereas application at the full-leaf stage in late June was most effective, 2,4-D application in August was ineffective in killing rabbitbrush. Variability in growth stage susceptibility is less of a problem in soil-applied herbicides [96].

3. *Age of plant.* Seedlings of broadleaf species, as well as grasses, are generally more susceptible to herbicides than older plants. This has variously been attributed to changes in leaf characteristics, translocation differences, or other physiological or enzyme factors. Old sagebrush plants have commonly been reported more susceptible to phenoxy herbicides than young vigorous plants aged two to six years, but exceptions have been noted.

4. *Leaf damage.* Plants previously injured by frost, hail, insects, or grazing may not fully respond to herbicides. Defoliation prior to foliage application of translocated herbicides interferes with herbicidal action apparently by removing photosynthetic tissue and thereby curtailing translocation.

Environmental Factors

Environmental factors, principally climatic and edaphic, materially influence the effectivenss of herbicides. These factors principally affect herbicide penetration or translocation or both:

1. *Soil moisture.* Plants under moisture stress do not readily translocate herbicides. Cook [22] in Utah studies reported unsatisfactory big sagebrush kills when soil moisture in the upper two feet of a silty clay loam soil was below 12 percent. This minimum soil moisture level was interpreted to be equivalent to 15.5 percent on clay soils, 13.5 percent on loam soils, 12.5 percent on clay loams, 10.5 percent on silt loams, and 5.5 percent on sandy loams. When soil moisture is deficient following seasonal or prolonged drought, do not apply phenoxy herbicides. Soil sterilants should be applied when there is a rea-

157

sonable chance of rainfall and being carried into the soil. Otherwise losses of herbicide though photodecomposition and volatization may be excessive [96].

2. *Other soil factors.* Rapid growth associated with ideal soil texture, structure, and fertility appears to increase herbicidal effects. On soils high in nitrogen, 2,4-D has caused more root damage of broadleaf plants than on deficient soils [46]. Light soil types, low adsorption of the herbicide, and high rainfall favor deep penetration of soil sterilants [96]. Reduced control on bottomland soils with soil sterilants may be associated with increased herbicide adsorption. Less silvex has also been required to control tree-type mesquite in Texas aerially applied on sandy soil sites than on heavy clay sites [64]. The higher mesquite kill from 2,4,5-T on upland and sandy sites compared to bottomland and clay sites has been attributed to higher soil temperature [32]. Plants on waterlogged sites have also been found more resistant to herbicides than plants of the same species on well-drained sites [149].

3. *Rainfall.* Rain four hours or more after foliar application of a herbicide seldom reduces herbicidal effects. However, rain shortly after application may wash herbicide off the foliage and reduce control. Excessive rainfall may also leach soil sterilants from the soil profile before maximum root absorption takes place. In low rainfall zones, herbicide application often becomes impractical. In Wyoming big sagebrush control in the seven-to-nine-inch rainfall areas is generally associated with very small grass response [76].

4. *Temperature.* Warm but not excessive air temperature promotes the entry and translocation of herbicides, respiration, protoplasmic steaming, and plant growth and thereby herbicide effectiveness [96]. Temperatures of 70° to 85° F favor maximum herbicidal effects in most plants. Low temperatures during the preceding week tend to slow plant growth and retard herbicide activity. Cook [22] in Utah concluded that unsatisfactory kill of big sagebrush could be expected when minimum daytime temperatures are below 70° F or minimum nighttime temperatures were below 40°. Temperatures above 85° to 90° are generally less satisfactory for spraying because of reduced plant growth rate and increased losses of herbicides from volatization [138]. However, best control of mesquite with 2,4,5-T was found when soil temperature at the eighteen-inch zone was over 80° F compared to low kills at lower temperatures [32]. Hot, dry winds also result in thicker, less penetrable cuticle [96].

5. *Humidity.* High relative humidity increases the effectiveness of most herbicide sprays by reducing plant water stress, by delaying drying of droplets, favoring stomatal opening, and probably increasing the hydration and thus permeability of leaves to polar substances

[46]. Low humidity effects can partially be overcome by using oil or oil-water emulsions instead of water as a diluent [138].

6. *Wind.* Wind above minimum levels frequently causes improper distribution of sprays and increases drift of herbicides onto susceptible crops. (Refer to section on safeguards in herbicide use.) In many areas of the Great Plains, aerial application of herbicides is best done in early morning and secondarily late evening because of typically reduced wind velocities, higher humidity, and lower temperatures.

7. *Light.* Light generally promotes herbicide penetration by stimulating stomatal opening and accentuates photosynthesis and translocation [31]. However, night spraying of plains pricklypear (*Opuntia polyacantha x phaecantha*) with silvex increased kill over daytime application by 20 to 59 percent [113]. This difference was attributed to the finding that the stomata of pricklypear are open only at night.

Table 8. Properties of Herbicides Used on Rangeland or Proposed for Range Use*‡

Common name (Chemical name)	Group and type of herbicide	Forms** and trade names	Animal toxicity, LD_{50} in mg./kg.†
Amitrole (3-amino-S-triazole)	Triazole; foliage, nonselective, translocated.	WSP, WML; Amizol, Weedazol.	LD_{50} 24,600; no human hazard.
AMS (ammonium sulfamate)	Inorganic; contact and translocated, nonselective, foliage.	SP; Ammate.	LD_{50} 3900; nonflammable; not injurious as normally used.
Atrazine (2-chloro-4-(ethylamino)-6-(isopropylamino)-S-triazine)	Triazine; selective, soil sterilant.	WP, WML, G; AAtrax.	LD_{50} 3080.
Bromacil (5-bromo-3-sec-butyl-6-methyluracil)	Substituted uracil; nonselective, soil sterilant.	WP; Hyvar.	LD_{50} 5200. May cause skin irritation.
Brushkiller††			
Dalapon (2,2-dichloropropionic acid)	Alaphatic; translocated, selective; foliage.	Sodium salt, WP, SP; Basfapon B, Dowpon.	LD_{50} 7570 to 9330. Nearly harmless to fish. Skin and eye irritations to some people.
Dicamba (3,6-dicloro-o-anisic acid)	Benzoic; selective, translocated; foliage or soil.	G, WSL (amine salt); Banvel.	LD_{50} 2900; low toxicity to fish, wildlife, or livestock.

*Registration of herbicides for range and pasture uses and the accompanying restrictions are subject to continual change. Current clearance and restrictions at both state and federal levels should be checked and complied with. (Literature citations refer to Chapter V.)

**Abbreviations for forms registered and available for use:

C = crystals	Pr = pressurized container	WS = water soluble
EC = emulsifiable concentrate	S = solution	WSP = water soluble powder
G = granules	SP = soluble powder	WML = water miscible liquid
Pl = pellets	Tech = technical material	WSL = water soluble liquid
	WP = wettable powder	

Federally approved uses and restrictions*	Plant physiological effects	Range and pasture uses; comments*
Noncropland use principally.	Inhibits chlorophyll formation; plants turn white, red, or brown; desiccation; kills in 2-3 weeks; absorbed through foliage and roots	Used on Canada thistle, horsetail, leafy spurge, whitetop, cattails, poison ivy. Water usual carrier. Persists 2-4 weeks in soil. Some use in England for chemical fallow when applied at 2-4 lb. active/A.
For forest, range, and pasture. For individual plant treatment.	Absorbed through foliage and green stems.	Woody plant control on noncrop, range, and pasture. Foliage (30-100 lbs. actual per acre), frill, stump or notch. Use greatest in humid regions. Persists 1-3 weeks in soil. Water or oil-water carrier. Nonvolatile.
Noncropland use; experimental on range.	Inhibits CO_2 fixation; other effects; absorbed by roots.	Kills annual grasses and shows promise for chemical fallow on range. Persists for over 1 year in soil. Established wheatgrasses damaged with 1 lb. a.e./acre in Nevada [37]. Water usual carrier. Has increased protein content in perennial grasses [66].
Use not approved for forage crops.	Inhibits photosynthesis; absorbed principally by roots.	Experimental only on range. Shows promise as brush killer at higher rates but damages perennial grasses at 2-4 lb. active/A. Very persistent in soil. Spread dry or in water carrier.
Pasture or noncropland use.	Interferes with pantothenic acid formation. Absorbed by roots and leaves.	Foliage spray on emerged aquatics such as cattails and rushes, also medusahead and foxtail barley. Nonvolatile. Water carrier only. Persists in soil up to 2-6 weeks. Some use in chemical fallow at 10 lb./acre. Little effect on broadleaf weeds.
Cleared for pasture and range at rates up to 8 lb. active/A.	Interferes with plant enzyme system; absorbed through roots and foliage.	Controls difficult plants such as Russian knapweed, Canada thistle, leafy spurge. Also useful in brush control. Persists in soil for up to a few months. Nonvolatile. Water carrier.

†For rats unless otherwise indicated.
††A 1:1 ratio of 2,4-D and 2,4,5-T. Gives control of a wider range of plants than either chemical used alone but their effects are not fully additive [96].
‡The manufacture of fenuron has been discontinued by DuPont and its future is uncertain.

Table 8. (Continued)

Common name (Chemical name)	Group and type of herbicide	Forms** and trade names	Animal toxicity, LD_{50} in mg./kg.†
2,4-D ((2,4-dichlorophenoxy) acetic acid)	Phenoxy; selective, translocated.	EC, S, WS salt; several trade names.	LD_{50} 300-1000; not harmful to wildlife or livestock.
2,4-DB (4-(2,4-dichlorophenoxy) butyric acid)	Phenoxy; selective, translocated.	EC, WML, WS salt; Butoxone, Butyrac.	LD_{50} 300-1000. May irritate skin and eyes of humans.
2,4-DP or dichlorprop (2-(2,4-dichlorophenoxy) propionic acid)	Phenoxy; selective, translocated.	EC, WS salt; Weedone 2,4-DP.	LD_{50} 400-800.
Dinoseb or **DNBP** (2-sec-butyl-4,6-dinitrophenol)	Substituted phenol; contact, nonselective (except DNBP salt).	WML, E.C., G; several.	LD_{50} 5-60; very toxic to fish; yellows skin. Do not inhale or contact spray. Skin absorbed.
Diquat (6,7-dihydro-dipyrido (1,2-a: 2', 1'-c) pyrazinediium ion)	Dipyridylium; foliage, nonselective, mostly contact.	EC, S, WS salt; Ortho Diquat, Reglone.	LD_5 400-440. May irritate skin. Nontoxic to fish at recommended rates.
Diuron (3-(3,4-dichlorophenyl)-1,1-dimethylurea)	Substituted urea; soil sterilant.	WP, WML, G; Karmex.	LD_{50} 3400-7500.
Fenac ((2,3,6-trichlorophenyl) acetic acid)	Unclassified; translocated by roots, selective, temporary soil sterilant.	WS salt, WSP, EC, G; Fenac.	LD_{50} 1780.

Federally approved uses and restrictions*	Plant physiological effects	Range and pasture uses; comments*
Pasture and range. Do not graze poisonous plants for 3 weeks after treatment.	Affects respiration, food reserves, and cell division. Causes bending and twisting of stems and petioles. Foliar absorbed, also root.	Highly effective as foliage spray on many broadleaved herbaceous plants and many shrubs. 2,4-D amine used in frill cuts. Persists in soil for 1-4 weeks. Volatility depends on chemical form. Water, diesel oil, or combination carrier.
Forage legumes at maximum of 2 lb. a.e./A.	Most weeds convert readily to 2,4-D; legumes convert slowly and effects slight (except sweetclover).	Used for weed control in legumes (½-2 lb. a.e./A) except sweetclover. Water carrier. Persists in soil 3-6 weeks.
Range use in some states.	Foliar absorbed similar to 2,4-D.	Used on sand sagebrush, locoweed, and shinnery oak. Water or water-oil carrier.
Do not use on forage crops.	Coagulates protein; uncoupler of oxidative phosphorylation; inhibits respiration.	Formerly used extensively as an oil fortifier. Gives high foliage kill through acute cell necrosis. Water or oil-water carrier. Persists 3-5 weeks in soil. Top kills only.
Nonforage crop weed killer.	Rapid, acute action. Damaged chemical radicals released within cells. Foliage absorbed.	Gives top kill of perennials only but controls most annuals at rates of 1.5-2 lb. active/A. Ineffective on shrubs. Nonvolatile. Inactivated rapidly by contact with soil. Water carrier. Experimental on range in chemical fallow.
Use for spot treatment. Do not plant treated area for one year.	Inhibits photosynthesis; results in accumulation of toxic by-products; root absorbed.	Used as local soil sterilant. Persists in soil 6-18 months. Spread dry or sprayed in water carrier. Less important on range than monuron.
Spot treatment on range.	Causes epinasty, bud necrosis, and bud inhibition. Root absorbed, also foliage.	Used on Canada thistle, leafy spurge, Russian knapweed, and woody plants. Persists one year or longer in soil. Water carrier.

Table 8. (Continued)

Common name (Chemical name)	Group and type of herbicide	Forms** and trade names	Animal toxicity, LD_{50} in mg./kg.†
Glyphosate (N-(phosphono-methyl)glycine)	Alaphatic; mostly non-selective, translocated, foliage.	WML; Roundup.	LD_{50} 4320.
Karbutilate (m-(3,3-dimethyl-ureido)phenyl tert-butyl-carbamate	Carbamate and substituted urea; non-selective, soil applied.	WP, G, EC; Tandex.	LD_{50} 3000.
MCPA (((4-chloro-o-tolyl)oxy) acetic acid)	Phenoxy; selective, trans-located, foliage.	EC, WML, WS salt; several.	LD_{50} 800. May irritate eyes and skin.
Monuron (3-(p-chlorophenyl)-1,1-dimethyurea)	Substituted urea; partially selective, soil sterilant.	WP, EC, WML, G; Telvar.	LD_{50} 3600; does not harm fish up to 18 p.p.m.
Paraquat (1,1'-dimethyl-4,4'-bipyridinium ion)	Dipyridylium; selective or nonselective, contact, foliage.	EC, WML; Ortho Paraquat, Gramoxone.	LD_{50} 150; low fish and wildlife tox-icity.

Federally approved uses and restrictions*	Plant physiological effects	Range and pasture uses; comments*
Experimental on range; broad spectrum herbicide.	Causes wilting, yellowing to browning. Interferes with amino acid synthesis.	Shows promise in brush control; but also kills desirable grasses and forbs. Shows promise in killing undesirable grasses such as foxtail barley or saltgrass. Persists 1-3 weeks in soil. Water carrier.
Experimental on range; noncrop herbicide.	Prevents photosynthesis. Mostly absorbed by roots.	Effective on many plant species; injurious to forage plants; persists several months in soil. Water carrier.
Pastures use.	Similar to 2,4-D. Foliar and root absorbed.	Use is similar to 2,4-D but more costly and generally less effective on most broadleaf weeds. Persists in soil 3-10 weeks. Water usual carrier.
Spot treatment only.	Inhibits photosynthesis. Root absorbed, less so through foliage.	Used in soil sterilization to control all vegetation. At high rates persists 6 to 20 mo. Ureabor contains 94% disodium tetraborate and 4% monuron. Water carrier or dry.
Use as spot treatment on noncropland only; pasture renovation.	Breaks down pyradine ring and liberates toxic peroxide radical within cells. Foliar absorbed but not by trunks.	Experimental on range and pasture. Major interest in grass seedbed preparation by application at ¼-1 lb./A just prior to seeding. Rapid acting, nonvolatile. Soil contact inactivates. Water carrier. Has minor effect on broadleaf perennials. Low rates somewhat selective on range clovers. Low rate (0.2 lb./A) chemically cures but does not kill perennial grasses [120, 121].

Table 8. (Continued)

Common name (Chemical name)	Group and type of herbicide	Forms** and trade names	Animal toxicity, LD_{50} in mg./kg.†
Picloram (4-amino-3,5,6-trichloropicolinic acid)	Unclassified; selective, translocated, foliage or soil.	WML, Pl, G; Tordon, Amdon.	LD_{50} 8200; no apparent hazard to wildlife, livestock, or fish. Fish safe to prolonged exposure to 100 to 320 p.p.m.
Silvex or 2,4,5-TP (2-(2,4,5-trichlorophenoxy) propionic acid)	Phenoxy; selective, translocated, foliage.	EC esters, WS amine salt, G; Kuron, Weedone.	LD_{50} 650. May irritate eyes.
2,4,5,-T ((2,4,5-trichlorophenoxy) acetic acid)	Phenoxy; selective, translocated, foliage.	EC, WML, WS salt; several.	LD_{50} 300; hazard to livestock and wildlife is negligible.
2,3,6-TBA (2,3,6-trichlorobenzoic acid)	Benzoic; nonselective, soil sterilant.	EC, G, WML, WS salt; Benzac, Trysben.	LD_{50} 750–1000; may cause skin and eye irritation. Toxic to fish and small marine life.
Tebuthiuron (N-[5-(1, 1-dimethylethyl)-1,3,4-thiadiezol-2-yl]-N, N^1-dimetylurea)	Substituted urea; nonselective, translocated, soil sterilant.	WP; Pl; Spike.	Very low.

Federally approved uses and restrictions*	Plant physiological effects	Range and pasture uses; comments*
Noncropland, spot treatment. Limited clearance in some states for range use.	Interferes with hormone systems. Symptoms are bent plant tips, rolled leaves, chlorosis, swollen tips. Absorbed by foliage and roots.	Very effective on leafy spurge, Russian knapweed, low and tall larkspur, whorled milkweed, and also many shrubs such as rabbitbrush and oaks. Nonvolatile. Most perennial grasses tolerate 1-2 lb. a.e./A, some tolerate 4 lb. per acre. Rates over 1 lb. may persist for 2 or 3 years. Synergic with phenoxy herbicides often. Water carrier.
Pasture and range clearance. Do not use on newly seeded pasture or range. Do not graze for 3 weeks when poisonous plants present or until plants have dried.	Similar to 2,4-D.	Plant control including oaks, maples, yucca, cholla, pricklypear, tall larkspur, saltcedar, and Dalmatian toadflax. Persists 2-5 weeks in soil. Water or water-oil carrier for foliage; oil carrier for basal stem or stump treatment.
Rangeland clearance. Do not graze poisonous plants for 3 weeks after treatment or until plants have dried.	Interferes with plant hormone systems. Absorbed through roots and foliage.	Foliage spray on woody plants including oak, maple, mesquite, elm, ceanothus, cholla, roses, huisache, pricklypear, saw-palmetto, yucca, and yaupon. Also used in oil carrier in basal trunk spray, frills, and stump treatment; persists 4-8 weeks in soil. Water, oil, or combination carrier on foliage.
Noncropland or spot treatment on range. Not for food or feed crops.	Interferes with plant hormone systems. Absorbed best as foliage-soil application.	Used on leafy spurge, Canada thistle, Russian knapweed. At high rates persists 18-24 months. Water carrier. Benzabor granules contain 90% disodium tetraborate and 8% 2,3,6-TBA.
Experimental on range-lands.	Inhibits photosynthesis. Absorbed through roots.	Holds promise for controlling mixed brush in Texas at 2 lb. a.e./A [118] while increasing grass production. Persists up to several months.

Table 9. Chemical Control of Noxious Range Plants*

Common name Scientific name	Herbicide	Rate, volume, carrier
Arrowgrass (*Triglochin maritima*)	2,4-D	2 lb. a.e./A** [78]†
Aspen, quaking (*Populus tremuloides*)	2,4-D ester	2 lb. a.e./A [18, 47]; aerial—4 gal. including 1 gal. diesel; ground—20 gal. water.
	2,4,5-T	12 lb. a.e.h.g.†† [5]; diesel oil.
Aster, heath (*Aster ericoides*)	2,4-D	1 lb. a.e./A [78].
Barley, foxtail (*Hordeum jubatum*)	Dalapon	48 lb. a.e./A [104]; 50 gal. water.
Bitterweed (*Hymenoxys odorata*)	2,4-D	1 lb. a.e./A [126].
	2,4-D	2-4 lb. a.e.h.g. in water [24].
Brome, cheatgrass (*Bromus tectorum*)	Atrazine	1 lb. a.e./A [49, 78]; 10 gal. water (ground).
	Paraquat	½-1 lb. a.e./A [24]; 20-30 gal. water (ground).
Buckbrush (*Symphoricarpos orbiculatus*)	2,4-D ester	2 lb. a.e./A [97, 138]; use sufficient water to insure good coverage.
	2,4,5-T or brushkiller	3-4 lb. a.e.h.g. [58]; water.
Burroweed (*Haplopappus tenuisectus*)	2,4-D	1-2 lb. a.e./A [133]; water and oil emulsion.
Cattail (*Typha latifolia*)	Dalapon	15 lb. a.e./A [2, 107]; 50 gal. water/A.
	2,4-D ester	6 lb. a.e. and 10 gal. diesel oil in 200 gal. water/A [2, 75].
Ceanothus, wedgeleaf (*Ceanothus cuneatus*)	2,4,5-T	2-3 lb. a.e./A [84]; water (include ½ gal. diesel oil).

*Recommended chemicals, rates, and methods of application vary from locality to locality and are continually being modified as research progresses. For these reasons, the recommendations given in this table, although documented, should be considered only as tentative. Further information and use clearance should be obtained locally.

Method and season of application	Special considerations
Foliage spray.	Fair control; repeat treatment may be necessary.
Foliage spray; leaves fully expanded.	Top kill at ½ lb./A used for browse rejuvenation.
Trunk base spray; leaves fully expanded.	Spray trunk to height of 2 feet.
Foliage spray.	Good control.
Foliage spray; spring, during rapid growth.	Kills desirable forage grasses; reseed desirable grasses. Practical only in limited situations.
Foliage spray; prebloom to early flowering.	Excellent control.
Foliage wetting spray; prebloom	
Foliage-soil spray; fall or early spring.	Used in chemical fallow; wait one year before grass seeding.
Foliage spray; spring after emergence.	Add ½ lb. 2,4-D ester if broadleaf weeds present. Spray prior to deep-furrow grass drilling [24].
Broadcast foliage spray; spring, in early full foliage stage.	Some resprouting from roots and stem base; repeat spraying in subsequent year often necessary.
Plant wetting spray; full leaf to early bloom.	
Foliage spray; spring, when leaves are succulent.	
Foliage spray; sprouts 2-3 feet, flowering to fruiting.	Do not graze or cut for feed for one year. May need to repeat treatment.
Foliage spray; early head stage and repeat on regrowth.	Saturate plants; three applications over two years often needed.
Foliage spray; rapid growth.	Retreatment may be needed. 2 lb. 2,4-D or 2,4,5-T adequate for 1- or 2-year-old seedlings.

**Acid equivalent per acre.
†Numbers in brackets refer to literature citations at the end of Chapter V.
††Acid equivalent per 100 gal.
‡1:1 ratio of 2,4-D and 2,4,5-T.

Table 9. (Continued)

Common name Scientific name	Herbicide	Rate, volume, carrier
Chamise (*Adenostema fasciculatum*)	Brushkiller‡	5 lb. a.e./A [101]; water-oil emulsion.
	Brushkiller	8 lb. a.e.h.g. [7]; water-oil emulsion.
	Picloram	0.5-2 lb. a.e./A [50]; water plus wetting agent.
Chokecherry (*Prunus virginiana*)	2,4,5-T or brushkiller	2 lb. a.e./A [58]; water-oil emulsion.
	2,4,5-T or brushkiller	3-4 lb. a.e.h.g. [58]; water.
Cholla (*Opuntia* spp.)	2,4,5-T or silvex	8 lb. a.e.h.g. [128]; diesel.
	Picloram	2 lb. a.e.h.g. [104]; water.
Clubmoss (*Selaginella densa*)	Monuron	2 lb. a.e./A [109]; water.
Cocklebur (*Xanthium* spp.)	2,4-D	1 lb. a.e./A [78].
Coneflower (*Rudbeckia* spp.)	2,4-D	1 lb. a.e./A [78].
Coyotillo (*Karwinskia humboldtiana*)	2,4,5-T	24 lb. a.e.h.g. [128]; diesel oil.
	2,4,5-T or silvex	1-2 lb. a.e./A [78].
Creosotebush (*Larrea tridentata*)	2,4,5-T	2 lb. a.e./A [110].
	Picloram	0.5-1 lb. a.e./A [110]; water.
Croton, Texas (*Croton texensis*)	2,4-D	1 lb. a.e./A [78].
Currant, wax (*Ribes cereum*)	2,4-D	1-2 lb. a.e./A [17]; water.

Method and season of application	Special considerations
Foliage application; early spring.	2,4-D ester at 4 lb./A effective on sprouts first or second spring after fire or mechanical top removal.
Foliage-stem, individual plant spray; spring is best.	Saturate plants. Allows high selectivity.
Foliage application; spring.	Trial basis only. High kill.
Foliage application; early summer, after leaves fully formed.	Repeat 1 or 2 years. Not practical on large scale; spray when necessary to prevent livestock poisoning.
Plant wetting spray; full leaf to early bloom.	
Foliage-stem, individual plant spray; during rapid growth.	Thoroughly wet plants.
Foliage-stem, individual plant spray; rapid growth.	High kill.
Foliage-soil spray; spring or fall.	Mechanical treatments and nitrogen fertilizer have decreased clubmoss and increased forage production in northern mixed prairie.
Foliage spray; rapid growth.	Insure complete coverage.
Foliage spray; rapid growth.	Good kill.
Stem base spray; spring during rapid growth.	
Foliage spray, rapid growth.	Good to fair kill.
Foliage spray; full flowering to mid-fruiting.	
Same.	Good kill.
Broadcast foliage spray; rapid growth in spring.	High kill possible.
Broadcast spray application; spring.	Use sufficient carrier to assure good coverage.

Table 9. (Continued)

Common name Scientific name	Herbicide	Rate, volume, carrier
Dandelion, common (*Taraxacum officinale*)	2,4-D	1 lb. a.e./A [78].
Deathcamas (*Zygadenus* spp.)	2,4-D ester	2 lb. a.e./A [138, 2, 104]; water or oil.
Elm, Siberian (*Ulmus pumila*)	2,4,5-T	16 lb. a.e.h.g. [128]; diesel oil or kerosene.
Elm, winged (*Ulmus alata*)	2,4,5-T	16 lb. a.e.h.g. [128]; diesel oil or kerosene.
	Picloram	2 lb. a.e./A [52].
Falsehellebore (*Veratrum californicum*)	2,4-D or silvex	2 lb. a.e./A [149]; 20-50 gal. of water.
Gallberry, common (*Ilex glabra*)	2,4,5-T	2 lb. a.e./A [104]; water.
	2,4,5-T	16 lb. a.e.h.g. [104]; diesel oil.
Goldenrod, rayless (*Haplopappus heterophyllus*)	2,4-D ester	4 lb. a.e.h.g. [49, 125]; water.
	Picloram	4 lb. a.e.h.g. [125]; water.
Greasewood (*Sarcobatus vermiculatus*)	2,4-D	1-2 lb. a.e./A [24]; water.
Grounsel, threadleaf (*Senecio longilobus*)	2,4-D	1-2 lb. a.e./A [128].
Gumweed (*Grindelia squarrosa*)	2,4-D ester	1 lb. a.e./A [78].
Halogeton (*Halogeton glomeratus*)	2,4-D ester	2 lb. a.e./A [2, 138]; water.
	Monuron or 2,3,6-TBA	8 lb. active/A or 4 lb. active/A respectively [99].

Method and season of application	Special considerations
Broadcast spray application; rapid growth.	Highly susceptible when rapidly growing.
Broadcast spray application; 3-5 leaf stage and before flower buds appear.	Spray only when necessary to prevent livestock poisoning; do not graze for two weeks after treatment.
Frill, stump, or basal spray; winter or summer.	Foliage applications of phenoxy herbicides generally ineffective because of root sprouting.
Frill, stump, or basal spray; winter or summer.	Refer to Siberian elm.
Soil application as spray or granular; spring or early summer.	Noncropland use generally.
Broadcast foliage spray; after all leaves have expanded but before blooming.	Use sufficient carrier to assure good coverage; retreat in following year for 95% kill. Complete control on wet sites may require subsequent individual plant treatment.
Broadcast foliage spray; August.	Use sufficient carrier to assure good coverage.
Basal trunk spray; August.	
Foliage spray; saturate individual plants during rapid growth.	Recommended for controlling poisonous plants in limited areas.
Foliage spray; individual plant.	High kill.
Broadcast foliage spray; spring or early summer during rapid growth.	High kill but benefits usually low. Sometimes resprouts.
Foliage spray; prebloom, spring.	
Foliage spray; early June for second year growth; late June if seedlings present.	
Broadcast foliage spray; early branching, prebloom stage.	One-year control; reinfests from seed. Summer fallowing, grass seeding in early spring, and spraying 0.5 lb. 2,4-D after grass emergence gives high control of halogeton.
Broadcast spray or pellet application.	Use for eradication of spot infestations by temporary soil sterilization.

Table 9. (Continued)

Common name Scientific name	Herbicide	Rate, volume, carrier
Horsebrush (*Tetradymia* spp.)	2,4-D ester	2.5 lb. a.e./A [99].
Houndstongue (*Cynoglossum officinale*)	2,4-D	2 lb. ae./A [99]; water.
Huisache (*Acacia farnesiana*)	2,4,5-T ester	8 lb. a.e.h.g. [128]; diesel oil.
	2,4,5-T and picloram	.5 + .5 lb. a.e./A [128]; oil-water emulsion.
Iris, Rocky Mountain (*Iris missouriensis*)	2,4-D	2 lb. a.e./A [35]; water.
Ironweed, western (*Vernonia baldwini*)	Picloram	0.5-1 lb. a.e./A [89]; water.
	2,4-D	2 lb. a.e./A [97]; diesel oil.
Juniper (*Juniperus* spp.)	2,3,6-TBA	8-12 lb. a.e.h.g. [73]; diesel oil.
	Picloram	2-4 lb. a.e./A [114] equivalent.
	Picloram	.5-1 lb. a.e.h.g. [104, 114]; water.
Knapweed, Russian (*Centaurea repens*)	Picloram and 2,4-D	1 + 2 lb. a.e./A [2, 75, 107].
	2,3,6-TBA	20 lb. active/A [2, 107].
Larkspur, low (*Delphinium* spp.)	2,4-D ester	2 lb. a.e./A [138, 71]; oil or water.
	Picloram	.25-0.5 lb. a.e./A [2]; water.

174

Method and season of application	Special considerations
Broadcast foliage spray; spring.	Treating large areas impractical. For spot application, use wetting spray of 20 lbs. a.e.h.g. Root-sprouting controlled by mechanically severing 4-6 inches deep.
Broadcast foliage spray; spring, prior to blooming.	Biennial; repeat spraying the following spring may be necessary. Provide perennial competition.
Stem-foliage spray, individual plant treatment; any time soil is dry.	Kerosene or diesel oil are effective in stem base application.
Spring; 90 days after bud break; foliage spray.	
Spring during late vegetative to late bloom stage.	Has given 91-100% control in mountain meadows and increased grass and grasslike plant production by 58 to 360% [35].
Foliage spray; spring during rapid growth.	
Foliage spray; spring during rapid growth.	
Trunk base spray.	Chaining or burning is effective on dense, extensive, mature stands except sprouting species (*J. deppeana* or *pinchotii*).
Applied as pellets under canopy cover only.	Broadcast spraying not recommended but picloram shows promise at 1-2 lb. a.e./A. Pelleted TBA also shows promise with
Individual plant wetting spray; has given 95% control when applied April to June in Texas [114].	individual trees. Individual tree burning or cutting on nonsprouting species. Basal spray effective with dicamba, karbutilate, and picloram.
Foliage spray, spot application; fall or spring on vigorous growth.	Also ½-¾ lb./sq. rod of 2% picloram granules [75].
Spot treatment as foliage-soil spray or soil granules.	Also 4-6 lb. active/A of dicamba [75].
Foliage application; fully emerged but before flowerstalks appear.	Spray only where necessary to prevent livestock losses. Do not graze for at least 3 weeks following application.
Foliage application; growing season.	Spot treat for hard to kill species such as *D. geyeri*.

175

Table 9. (Continued)

Common name Scientific name	Herbicide	Rate, volume, carrier
Larkspur, tall (*Delphinium* spp.)	2,4,5-T or silvex	4 lb. a.e./A [27]; insure good coverage.
	Picloram	2 lb. a.e./A [135].
Loco (*Astragalus* spp., *Oxytropis lambertii*)	2,4-D ester	2 lb. a.e./A [2,138].
	2,4-D ester	4 lb. a.e.h.g. [138].
Lupine (*Lupinus* spp.)	2,4-D ester	2 lb. a.e./A [2, 104].
Manzanita (*Arctostaphylos* spp.)	2,4-D ester	3-4 lb. a.e./A [104, 84].
	2,4-D ester	2 lb. a.e.h.g. [84]; water.
Maple (*Acer* spp.)	Silvex or 2,4,5-T	2 lb. a.e./A [78].
Medusahead (*Taeniatherum asperum*)	Dalapon	3 lb. a.e./A [132]; 10 gal. or more; water.
Mesquite (*Prosopis juliflora*)	2,4,5-T ester	2/3 lb. (running type) or 0.5 lb. (tree type) a.e./A [128]; 5 gal., oil-water emulsion.
	2,4,5-T ester	8 lb. a.e.h.g. [128]; diesel.
	2,4,5-T ester	2 lb. (ester) a.e.h.g. [64]; water.
	Diesel oil or Kerosene	About 1 pint per tree [86], or up to ½ gal. on large trees [64].
	Picloram + 2,4,5-T	⅓ + ⅓ a.e./A [128]; oil-water emulsion.

176

Method and season of application	Special considerations
Foliage spray; late vegetative stage (8-10″) but before flower production; repeated application the following year suggested for high kill [27].	Treat dense patches to prevent livestock poisoning. Remove livestock until plants have dried. Grubbing out isolated plants at an 8-inch depth is effective.
Soil application of granular form.	
Foliage application; bud to early bloom stage.	Spray local areas necessary to control livestock poisoning. Do not graze for 3 weeks until foliage has dried.
Foliage application; individual plant wetting spray.	
Foliage application; early bud stage.	Spraying often impractical except on dense stands or unless also for simultaneous kill of big sagebrush. Repeat application when necessary.
Foliage spray; during rapid spring growth.	Repeat applications for up to two additional years when necessary to kill sprouting species.
Individual plant wetting spray; rapid spring growth.	1 lb. a.e. of picloram/acre highly effective [104].
Foliage application; early full leaf stage.	Try 16 lb. a.e.h.g. as basal spray. Repeated applications generally required.
Foliage spray; from beginning of active spring growth and prior to boot stage.	Refer to chemical seedbed preparation in Chapter VIII.
Broadcast foliage spray; 40 to 90 days after leaf growth begins.	Repeat when necessary to control sprouts and new seedlings. Increase 2,4,5-T rate to 1 lb./A for dense trees on heavy clay soils [64].
Frill, stump, or basal trunk spray; when soil is dry.	Frill before basal spraying when stems over 5″ in dia.
Wetting foliage spray, individual plant treatment; full leaf development.	
Basal trunk spray; saturate bark and soil at base of tree.	Not recommended where sprout buds are buried by silt or for running forms without definite trunks; most effective on porous, dry soils.
Broadcast foliage spray; rapid growth following full leaf stage.	Consistent high kill.

Table 9. (Continued)

Common name Scientific name	Herbicide	Rate, volume, carrier
Milkvetch, timber (*Astragalus miser*)	Silvex or 2,4,5-T	2 lb. a.e./A [29, 138]; use ample carrier for complete coverage.
Milkweed, common (*Asclepias syriaca*)	Amitrole	4 lb. active/A [107]; 20 gal.; water.
	Picloram + 2,4-D	1 + 2 lb. a.e./A [49, 107]; water.
Milkweed, whorled (*Asclepias subverticillata*)	Picloram	1 to 2 lb. a.e./A [88]; water.
	Picloram + 2,4-D	1 + 2 lb. a.e./A [49]; water.
Mulesear (*Wyethia amplexicaulus*)	2,4-D ester	1-2 lb. a.e./A [114]; water.
Oak, blue (*Quercus douglasii*)	Silvex	2 lb. a.e./A [84]; water (include 1 gal. diesel).
	2,4-D amine	Undiluted liquid concentrate [55].
Oak, California scrub (*Quercus dumosa*)	Brushkiller	4 lb. a.e./A [84]; diesel oil or water-diesel emulsion.
Oak, Gambel (*Quercus gambelii*)	Silvex or 2,4,5-T	2 lb. a.e./A [74]; water-oil emulsion.
	Silvex + picloram	2 + 0.5 lb. a.e./A
	Silvex or 2,4,5-T	8 lb. a.e.h.g.; diesel oil.
Oak, interior live (*Quercus wislizenii*)	Silex or brushkiller	4 lb. a.e.h.g. [84]; diesel oil or diesel-water emulsion.
Oak, live (*Quercus virginiana*)	2,4,5-T ester	8 lb. a.e.h.g. [128]; diesel oil.
	Picloram	2 lb. a.e./A [13].

Method and season of application	Special considerations
Foliage spray; full flower.	Do not graze treated plants with cattle or horses until fully dried.
Broadcast foliage spray; bud to bloom stage.	Chemical control often impractical.
Foliage application; after emergence in the spring.	Spot treatment only.
Foliage application; early growth to bud stage.	1 lb. has given good control; 2 lbs. has eliminated stand. Spot treatment only.
Foliage application; early growth to bud stage.	Spot treatment only.
Broadcast foliage spray; rapid growth before bloom.	Spray where dense stands depress forage production.
Broadcast foliage spray; when growth is most active.	Repeat in two years.
Applied in overlapping frill cuts made in trunk at ground level.	Winter and spring.
Foliage application; during active growth.	Follow up with one or more individual plant treatments using 4 lb. a.e.h.g. Broadcast spray second year after burn.
Foliage spray; full leaf stage.	Retreat at 1-lb. rate in each of one or two successive years; single application results in incomplete root kill.
Foliage spray; full leaf stage.	High top kill; retreat one or two years later for sprout and root kill.
Basal stem application; growing season.	Spot treatment only; effective on stems up to 6″ dia. Basal stem spray with picloram at 4 lb. a.e.h.g. effective at any season.
Foliage spray, individual plant treatment; when leaves are plentiful.	Retreatment usually necessary.
Individual plant treatment by frill, stump, or basal trunk spray; winter or summer.	
Broadcast foliage spray; rapid growth in spring.	

Table 9. (Continued)

Common name Scientific name	Herbicide	Rate, volume, carrier
Oak, post and blackjack (*Quercus stellata* and *marilandica*)	2,4,5-T ester	2 and 1.5-2 lb. a.e./A in two successive years [128]; oil-water emulsion.
	2,4,5-T ester	16 lb. a.e.h.g. [128]; diesel oil.
	Picloram	2 lb. a.e./A [52]; water.
Oak, shinnery (*Quercus harvardii*)	Silvex or 2,4,5-T ester	0.5 to 1.0 lb. a.e./A [128]; diesel-oil emulsion.
Oak, shrub live (*Quercus turbinella*)	Silvex or 2,4,5-T	2 lb. a.e./A [112]; use sufficient carrier to insure good coverage.
Olive, Russian (*Eleagnus angustifolia*)	Brushkiller	2 lb. a.e./A [107]; sufficient carrier to insure good coverage.
Osageorange (*Maclura pomifera*)	2,4,5-T	2-4 lb. a.e.h.g. [97]; water
Pingue (*Hymenoxys richardsonii*)	2,4-D ester	3-4 lb. a.e./A [49]; 15 to 25 gal.; water.
Pinyon (*Pinus* spp.)	Karbutilate or picloram	1.4 or 0.7 gram active per tree [151].
Poisonhemlock (*Conium maculatum*)	2,4-D	2 lb. a.e./A [104]; use sufficient carrier to insure good coverage.
Pricklypear (*Opuntia* spp.)	2,4,5-T or silvex ester	2 lb. a.e./A [128]; oil-water emulsion; ground application at 25 gal./A in dense, upright stands.
	2,4,5-T or silvex ester	8 lb. a.e.h.g. [128]; diesel oil.
	Picloram + 2,4,5-T	.5 + .5 lb. a.e./A [128]; oil-water emulsion.

Method and season of application	Special considerations
Broadcast foliage spray; spring, after leaves fully developed and before summer dormancy.	
Individual plant treatment in frill, stump, or basal stump spray; winter or summer.	
Foliage spray.	Gave similar control to repeated application of 2,4,5-T.
Broadcast foliage spray; full leaf and growing rapidly.	Repeat if needed. Addition of 0.5 a.e./A of picloram synergically increases plant kill [115]. Pelleted picloram or tebuthiuron also shows promise.
Broadcast foliage spray; spring, during rapid growth.	Repeat the following year to kill sprouts. 8 lb. per acre of TBA effective on sprouts.
Broadcast foliage application.	Repeat treatment the following year for complete control. Diesel oil or picloram basal spray also effective.
Foliage drench application; active growth, full-leaf stage.	Also 2,4,5-T at 10 to 20 lb. a.e.h.g. in diesel oil for basal stem or stump application [97].
Foliage application; from full leaf to early flower stage.	Use 2,4-D ester at 2-4 lb. a.e.h.g. as wetting spray on spot infestations [104].
Granular form applied to soil under canopy.	Effective on saplings up to seven feet high.
Foliage spray; early growth to early flowering stage.	1 lb. a.e./A of silvex gives excellent control also [78].
Foliage spray; active growth in the spring, early bloom.	Rail or rotary hoe to puncture pads just prior to herbicide application; repeat herbicide following year, if necessary.
Wetting spray applied to individual plants; when plants are actively growing.	Same as above. High degree of control with rates as low as 1 lb. a.e.h.g. have been obtained with night application [113].
Foliage spray; spring, 40-90 days after bud break.	

Table 9. (Continued)

Common name Scientific name	Herbicide	Rate, volume, carrier
Rabbitbrush, big (*Chrysothamnus nauseosus*)	2,4-D ester	3 lb. a.e./A [24, 138]; oil, oil-water emulsion, or water.
	Picloram	1 lb. a.e./A [23]; water.
Rabbitbrush, little (*Chrysothamnus* *viscidiflorus*)	2,4-D ester	3 lb. a.e./A [36, 72, 138].
	Picloram	1 lb. a.e./A [23].
	Picloram + 2,4-D	½ + 2 lb. a.e./A [42].
Ragweed, western (*Ambrosia psilostachya*)	2,4-D ester	2 lb. a.e./A [138]; sufficient carrier for complete wetting.
Ragwort, tansy (*Senecio jacobaea*)	2,4-D ester	2 lb. a.e./A [78, 138].
Rose, Macartney (*Rosa bracteata*)	2,4-D ester	4 lb. a.e./A [128]; 5-15 gal. (aerial) or 25-30 gal. (ground).
	2,4-D	4 lb. a.e.h.g. [128]; water plus surfactant.
	Picloram + 2,4,5-T	.5 + .5 lb. a.e./A [116]; 5 gal./A in water or oil-water emulsion.
Rose, multiflora (*Rosa multiflora*)	Brushkiller	2 lb. a.e./A [136].
Rose, Wood's wild (*Rosa woodsii*)	2,4,5-T	2 lb. a.e./A [136].
	2,4,5-T	3-4 lb. a.e.h.g. [58]; water.
Sage (*Salvia* spp.)	2,4-D ester	2 lb. a.e./A [78].
Sagebrush, big (*Artemisia tridentata*)	2,4-D ester	2 lb. a.e./A [2, 138]; water or oil.

182

Method and season of application	Special considerations
Broadcast foliage spray; new twig growth 3″ and until *Poa secunda* loses green color in herbage.	For high kill, apply 2 lb./A for 1 or 2 years after initial spraying, burning, or mechanical top removal. On dry, sandy soil, root plowing or plowing at 5″ depth gives good kill.
Foliage spray.	High kill. May not give adequate control of big sagebrush.
Foliage spray; 3-6″ new growth.	Better simultaneous control when applied towards end of sagebrush susceptibility period.
Foliage spray; full leaf stage.	May not give adequate kill of big sagebrush [134].
Same.	Gives high kill of both green rabbit-brush and big sage even in a droughty year.
Foliage spray; spring during rapid growth.	
Foliage spray; early bolting stage in spring or rosette stage in fall.	See section on biological control by insects in Chapter III.
Foliage spray; full leaf to bud stage in spring.	Repeat at 1 lb./A when necessary. Burning or mowing followed by spraying the following year effective at 2 lb. a.e./A.
Individual plant wetting spray; rapid growth in fall or spring.	
Broadcast foliage spray; full leaf to bud stage.	More effective on plants not disturbed by previous shredding or spraying.
Foliage spray; full leaf to bud stage.	Repeat at 1 lb. a.e./A when necessary. Treat whenever escapes cultivation.
Foliage spray; full leaf to bud stage.	Repeat at 1 lb. a.e./A when necessary to control sprouting.
Individual plant wetting spray; full leaf to early bloom.	
Foliage spray; rapid growth in spring.	
Broadcast foliage spray; spring after first new leaves reach maximum length, or from early heading to marked loss of color in *Poa secunda* heads.	Burning or plowing are also effective. Increase rate to 3 lb. or use 2 lb. silvex to control reinvasion into seeded grass stands.

Table 9. (Continued)

Common name Scientific name	Herbicide	Rate, volume, carrier
Sagebrush, California (*Artemisia californica*)	2,4-D	2 lb. a.e./A [78].
Sagebrush, low (*Artemisia arbuscula*)	2,4-D ester	2 lb. a.e./A [138].
Sagebrush, sand (*Artemisia filifolia*)	2,4-D ester	1-2 lb. a.e./A [91, 97]; use sufficient volume to insure good coverage; oil-water emulsion suggested.
Sagebrush, silver (*Artemisia cana*)	2,4-D ester	2 lb. a.e./A [2, 84].
Sagewort, fringed (*Artemisia frigida*)	2,4-D ester	2 lb. a.e./A [94].
Sagewort, gray (*Artemisia ludoviciana*)	2,4-D ester	2 lb. a.e./A [94].
Sagewort, green (*Artemisia biennis or campestris*)	2,4-D ester	3 lb. a.e./A [95]; water.
St. Johnswort or Klamath weed (*Hypericum perforatum*)	2,4-D ester	2-3 lb. a.e./A [104]; use sufficient carrier to give thorough coverage.
Saltcedar (*Tamarix* spp.)	Silvex	2-4 lb. a.e./A [138]; use sufficient carrier to insure good coverage.
Saw-palmetto (*Serenoa repens*)	2,4,5-T	3-4 lb. a.e./A [104]; 50 gal.; water.
Silktassel (*Garrya* spp.)	Brushkiller	4 lb. a.e./A [84]; water-oil emulsion.
Silverberry or silver buffaloberry (*Eleagnus commutata*)	2,4-D	1 lb. a.e./A [26].
Snakeweed, broom (*Xanthocephalum sarothrae*)	2,4-D ester	1-2 lb. a.e./A [97, 128].

Method and season of application	Special considerations
Broadcast foliage application; during rapid growth.	
Broadcast foliage spray; spring, during rapid growth, or late boot stage to preanthesis in *Poa secunda*.	Top removal by burning or mechanical means is also effective.
Broadcast foliage spray; spring, when growing rapidly and having 6 to 8″ of new twig growth.	Repeat if necessary. One lb. silvex has given nearly complete control [9]. Mowing or burning also effective; root sprouts occasionally. One lb. of 2,4-DP effect similar to 2,4-D.
Broadcast foliage spray; 3-4 inches of new twig growth and growing rapidly.	
Broadcast foliage spray; spring, when new growth is 2″ long or longer.	Often impractical for control of fringed sagewort alone.
Broadcast foliage spray; spring.	Repeat in late summer, if necessary.
Broadcast foliage spray; when plants are 4-8″ high and rapidly growing.	Repeat in two years.
Foliage spray; early bud stage, when plants are about 6″ tall.	
Broadcast foliage spray.	Resprouts from roots and stem base. More effective on young plants than old plants. Mechanical removal followed by chemical treatment of regrowth more effective than herbicide alone. Soil application of dicamba or picloram looks promising.
Foliage spray; summer.	Roller chopping also effective.
Broadcast foliage spray; spring.	Treat regrowth in spring following burn.
Broadcast foliage spray; early summer (Canada).	Repeat in second year.
Broadcast foliage spray; first leaves fully expanded, new twig growth 3 5″, but before flower buds appear.	Control not always practical since growth is cyclic. Thin stands do not justify herbicide use.

Table 9. (Continued)

Common name Scientific name	Herbicide	Rate, volume, carrier
Sneezeweed, orange (*Helenium hoopesii*)	2,4-D ester	3 lb. a.e./A [138]; use sufficient water carrier to give complete wetting.
Soapweed, small (or yucca) (*Yucca glauca*)	Silvex ester	0.75 to 2 lb. a.e./A [10, 97, 128]; water or water-diesel oil emulsion.
	2,4,5-T ester	8 lb. a.e.h.g. [128]; diesel oil.
Spurge, leafy (*Euphorbia esula*)	Dicamba	6-8 lb. active/A [2, 75]; water; high volume.
	2,3,6-TBA	20 lb. active/A [2, 75]; water.
	Picloram	2 lb. a.e./A [2, 75]; water.
Sumac, skunkbush (*Rhus trilobata*)	2,4,5-T, silvex	2 lb. a.e./A [83]; diesel oil-water emulsion.
	Picloram + 2,4,5-T	.5 + .5 lb./A [128]; oil-water emulsion.
Tarbush (*Flourensia cernua*)	2,4,5-T ester	2 lb. a.e./A [110].
Tarweed (*Madia glomerata*)	2,4-D amine or ester	1 lb. a.e./A [104]; water.
	2,4-D ester	0.5 lb. a.e./A [67]; water.
Thistle, Canada (*Cirsium arvense*) continued on page 188	Picloram	1 lb. a.e./A [2, 75].
	Picloram + 2,4-D	1 + 2 lb. a.e./A [2, 75]; water.
	Picloram	0.5 lb. actual of 10% granules/sq. rd. [2, 75].

186

Method and season of application	Special considerations
Foliage spray; prebloom during formation of flower buds.	Respray following year when necessary.
Broadcast foliage spray; rapid growth in spring but prior to full bloom.	Winter grazing or mowing and spring-summer cattle grazing is effective [97].
Foliage spray for individual plants; spring or summer.	Saturate root crown at ground level for complete kill.
Foliage-soil spray; bud to bloom stage.	Nebraska recommends 10 lb. per acre [107]. Granular form also effective. Do not feed or graze within 30 days of slaughter.
Foliage-soil spray, or granular TBA to soil; fall or early spring.	Apply granular TBA at 1.5 lb. actual/sq. rod.
Spot treatment as foliage spray; rapid growth in fall or spring.	Picloram + 2,4-D at 1 + 2 lb./A also recommended for spot treatment [107], or use picloram granules.
Broadcast foliage spray; active growth.	Repeat as required.
Foliage spray; spring, 40-90 days after bud break.	
Foliage spray; rapid growth.	Often impractical for large scale programs.
Broadcast foliage spray; spring, prior to 4-leaf stage of tarweed.	Respray second growing season for kill of new seedlings. Summer fallowing also effective if cultivated in spring after leaf development but before seed set.
Broadcast foliage spray; early spring; no damage to grass seedlings in 1- to 4-leaf stage.	
Foliage spray; fall or spring on vigorous growth.	Spot application only; gives nearly complete kill. Do not plant to grass crops for 2 years.
Foliage spray; fall or spring on vigorous growth.	Spot application only; gives nearly complete kill. Do not plant to grass crops for 2 years.
Soil application to small plots; just prior to expected growing season precipitation.	Spot application. Do not plant to grass crops for 2 years.

Table 9. (Continued)

Common name Scientific name	Herbicide	Rate, volume, carrier
Thistle, Canada (cont.)	2,3,6-TBA	20 lb. active/A [2, 107]; water.
Thistle musk (*Carduus nutans*)	2,4-D ester	1.5-2.0 lb. a.e./A [2, 107].
Toadflax, Dalmation (*Linaria dalmatica*)	Silvex	2 lb. a.e./A [2].
	Picloram	1 lb. a.e./A [2].
Waterhemlock (*Cicuta douglasii*)	2,4-D ester	2 lb. a.e./A [138]; 10 gal.; water.
	2,4-D	4 lb. a.e.h.g.; water [75].
Weed seedlings (broadleaf) in new grass seedings.	2,4-D amine	(1) 0.25-0.5 lb. a.e./A [80, 107]. (2) 0.5-0.75 lb. a.e./A [80, 107]. (3) 1 lb. a.e./A [80, 107].
Whitethorn, chaparral (*Ceanothus leucodermis*)	2,4,5-T ester	3 lb. a.e./A [84]; diesel oil.
Whitetop or hoary cress (*Cardaria draba*)	2,3,6-TBA	20 lb. active/A [107].
	Amitrole	4 lb. active/A [2, 75]; 80 gal./acre; water.
Willow (*Salix* spp.)	2,4-D ester	2 lb. a.e.h.g. [104]; water.
	2,4-D ester	20 lb. a.e.h.g. [75]; diesel.

Method and season of application	Special considerations
Foliage-soil spray application; spring or fall.	Application of 1.5 lb. actual of TBA granules (10%)/sq. rod also effective. Also 6-8 lb. dicamba/A [75].
Foliage spray; late fall on rosettes or spring before bolting.	Annual treatment for a few years may be necessary to kill new seedlings. Use same treatment for Scotch thistle.
Broadcast foliage spray; rapid growth in spring, bud to bloom stage.	Repeat treatment required.
Spring or fall.	
Foliage spray; bud to early bloom stage in spring.	Treat spot infestations. Do not graze for at least 3 weeks to prevent livestock poisoning.
Wetting foliage spray; early growth to early bud stage.	Treat spot infestations. Do not graze for at least 3 weeks to prevent livestock poisoning.
(1) Foliage spray, warm-season grass seedlings in 2- to 4-leaf stage.	Light injury generally outgrown by grasses. But heavy rates may kill a large proportion of small grass seedlings.
(2) Foliage spray, cool-season grass seedlings in 2- to 4-leaf stage.	
(3) Foliage spray when grass seedlings beyond 4-leaf stage.	
Foliage spray; spring.	Most effective on sprouts.
Foliage spray; fall or spring.	Repeated annual treatments with 2,4-D for several consecutive years also effective. Picloram also effective in spot treatments.
Foliage spray; bud to early bloom stage.	Some retreatment may be necessary. Do not plant to forage plants for 8 months after treatment.
Individual plant wetting spray; spring, after leaves fully leaf out.	2 lb. 2,4-D a.e./acre as broadcast foliage spray effective if complete coverage obtained.
Individual basal stem spray; anytime.	Frill chop large trees near base before spraying.

Table 9. (Continued)

Common name Scientific name	Herbicide	Rate, volume, carrier
Yaupon (*Ilex vomitoria*)	2,4,5-T ester	2 lb. a.e./A [51].
	2,4,5-T ester	8 lb. a.e.h.g. [128]; diesel.
	Picloram	2 lb. a.e./A [51]; water.

Table 10. Pints of Commercial Product Needed per Acre

Pounds of active ingredients per gal. of commercial product	Pounds active ingredients per acre					
	1/4	1/2	1	2	3	10
1.0	2	4	8	16	24	80
2.0	1	2	4	8	12	40
3.0	2/3	1 1/3	2 2/3	5 1/3	8	26 2/3
3.34	3/5	1 1/5	2 2/5	4 4/5	7 1/5	24
4.0	1/2	1	2	4	6	20
6.0	1/3	2/3	1 1/3	2 2/3	4	13 1/3

Method and season of application	Special considerations
Broadcast foliage spray; spring, during rapid growth.	Annual treatments for 3 successive years generally required. Sprouts from roots and stem base.
Frill, stump, or trunk base spray; summer or winter.	
Broadcast foliage spray; spring, during rapid growth.	Picloram applied as granules also highly effective.

Table 11. Commercial Product Required per Gallon of Spray Liquid for Wetting Sprays

Herbicide concentration in commercial product	Pounds active ingredients per hundred gallons (a.e.h.g.) recommended															
	1	2	3	4	6	8	16	32	1	2	3	4	6	8	16	32
Liquid (lb. active/gal.)	ounces of product*								tablespoons of product**							
1.0	1.6	3.2	4.8	6.4	9.6	12.8	25.6	51.2	2.6	5.1	7.7	10.2	15.4	20.5	41.0	81.9
2.0	.80	1.6	2.4	3.2	4.8	6.4	12.8	25.6	1.3	2.6	3.9	5.1	7.7	10.2	20.5	41.0
3.0	.53	1.1	1.6	2.1	3.2	4.3	8.5	17.1	.85	1.7	2.6	3.4	5.1	6.8	13.7	27.3
3.33	.50	1.0	1.4	1.9	2.9	3.8	7.7	15.4	.77	1.5	2.3	3.1	4.6	6.2	12.3	24.6
4.0	.40	.80	1.2	1.6	2.4	3.2	6.4	12.8	.64	1.3	1.9	2.6	3.8	5.1	10.2	20.5
6.0	.27	.53	.80	1.1	1.6	2.1	4.3	8.5	.43	.85	1.3	1.7	2.6	3.4	6.8	13.7
Wettable powder, granules, or pellets (% active)	ounces of product**															
25	.64	1.3	1.9	2.6	3.8	5.1	10.2	20.5								
50	.32	.64	.96	1.3	1.9	2.6	5.1	10.2								
75	.21	.43	.64	.85	1.3	1.7	3.4	6.8								
80	.20	.40	.60	.80	1.2	1.6	3.2	6.4								
85	.19	.38	.56	.75	1.1	1.5	3.0	6.0								
95	.17	.34	.51	.67	1.0	1.3	2.7	5.4								

*Weight of commercial product assumed to be 10 lb./gal. If weight of commercial product is materially different, correct as follows:

$$\frac{\text{factor from table} \times \text{lbs./gal. weight of commercial product}}{10} = \frac{\text{ounces of commercial product/gal.}}{\text{of spray liquid}}$$

**No corrections needed for other weights per gallon of commercial product.

Table 12. Commercial Product Required per 100 Square Feet *

Herbicide concentration in commercial product	Pounds active ingredients per acre recommended													
	1/4	1/2	3/4	1	2	3	10	1/4	1/2	3/4	1	2	3	10
Liquid (lb. active/gal.)	grams of product**							teaspoons of commercial product†						
1.0	2.6	5.2	7.8	10.4	20.8	31.2	104.0	.44	.88	1.32	1.76	3.52	5.28	17.6
2.0	1.3	2.6	3.9	5.2	10.4	15.6	52.0	.22	.44	.66	.88	1.76	2.64	8.8
3.0	.87	1.7	2.6	3.5	6.9	10.4	35.0	.15	.29	.44	.59	1.17	1.76	5.9
3.33	.78	1.6	2.3	3.1	6.2	9.4	31.0	.13	.26	.40	.53	1.05	1.58	5.3
4.0	.65	1.3	2.0	2.6	5.2	7.8	26.0	.11	.22	.33	.44	.88	1.32	4.4
6.0	.43	.85	1.3	1.7	3.5	5.2	17.5	.07	.17	.22	.29	.59	.88	2.9
Wettable powder, granules, or pellets (% active)	grams of product†													
2	12.5	26.3	38.8	52.0	103.8	156.3	520.0							
8	3.1	6.6	9.7	13.0	25.9	39.1	130.0							
10	2.5	5.3	7.8	10.4	20.8	31.3	104.0							
20	1.3	2.6	3.9	5.2	10.4	15.6	52.0							
25	1.0	2.1	3.1	4.2	8.3	12.5	41.6							
50	.50	1.1	1.6	2.1	4.2	6.3	20.8							
75	.33	.70	1.0	1.4	2.8	4.2	13.9							
80	.31	.66	.97	1.3	2.6	3.9	13.0							
85	.29	.62	.91	1.2	2.4	3.7	12.2							
95	.26	.55	.82	1.1	2.2	3.3	10.9							

*Multiply factor in table times appropriate factor for spot treatments larger or smaller than 100 sq. ft., i.e., on square rod plots (272.2 sq. ft.) use factor of 2.7.

**Weight of commercial product assumed to be 10 lb./gal. If weight of commercial product is materially different, correct as follows:

$$\frac{\text{factor from table} \times \text{lbs./gal. weight of commercial product}}{10} = \frac{\text{grams of commercial product/gal.}}{\text{of spray liquid}}$$

†No corrections needed for other weights per gallon of commercial product.

Literature Cited

1. Akesson, Norman B. 1955. A statement concerning drift problems in the application of 2,4-D by aircraft. Down to Earth 10(4):16–18.
2. Alley, H. P., and A. F. Gale. 1976 (Rev.). Wyoming weed control guide, 1976. Wyo. Agric. Ext. Bul. 442R.
3. Alley, Harold P. 1965. A promising future for the control of perennial weeds. Down to Earth 21(1–2):8–10.
4. Alley, Harold P. 1967. Some observations on Tordon-2,4-D herbicide combinations. Down to Earth 23(1):2, 35–36.
5. Arend, John L. 1953. Controlling scrub aspen with basal sprays. Down to Earth 9(1):10–11.
6. Armstrong, C. G., M. R. Gebhardt, and E. H. McIlvain. 1964. Goliath—a long-boom range sprayer. J. Range Mgt. 17(6):340–342.
7. Bentley, Jay R. 1967. Conversion of chaparral areas to grassland. USDA Agric. Handbook 328.
8. Blaisdell, James P., and Walter F. Mueggler. 1956. Effect of 2,4-D on forbs and shrubs associated with big sagebrush. J. Range Mgt. 9(1):38–40.
9. Bovey, R. W. 1964. Aerial application of herbicides for control of sand sagebrush. J. Range Mgt. 17(5):253–256.
10. Bovey, R. W. 1964. Control of yucca by aerial application of herbicides. J. Range Mgt. 17(4):194–196.
11. Bovey, R. W. 1965. Control of Russian olive by aerial application of herbicides. J. Range Mgt. 18(4):194–195.
12. Bovey, R. W., T. O. Flynt, R. E. Meyer, J. R. Baur, and T. E. Riley. 1976. Subsurface herbicide applicator for brush control. J. Range Mgt. 29(4):338–341.
13. Bovey, R. W., S. K. Lehman, H. L. Morton, and J. R. Baur. 1969. Control of live oak in south Texas. J. Range Mgt. 22(5):315–318.
14. Bovey, R. W., S. K. Lehman, H. L. Morton, and J. R. Baur. 1968. Control of running live oak, huisache and mesquite in Texas. Texas Agric. Expt. Sta. Prog. Rpt. 2592.
15. Bovey, R. W., M. K. McCarty, and F. S. Davis. 1966. Control of perennial ragweed on western Nebraska rangeland. J. Range Mgt. 19(4):220–222.
16. Bovey, R. W., H. L. Morton, J. R. Baur, J. D. Diaz-Colon, C. C. Dowler, and S. K. Lehman. 1969. Granular herbicides for woody plant control. Weed Sci. 17(4):538–541.
17. Bovey, Rodney W. 1977. Response of selected woody plants in the United States to herbicides. USDA Agric. Handbook 493.
18. Bowes, G. G. 1975. Control of aspen and prickly rose in recently developed pastures in Saskatchewan. J. Range Mgt. 28(3):227–229.
19. Boyd, W. E., Ronald E. Sosebee, and E. B. Herndon. 1978. Shredding and spraying honey mesquite. J. Range Mgt. 31(3):230–233.
20. Burnside, Orvin C., John D. Furrer, and Robert E. Roselle. 1970. Pesticides—heroes or villains? Nebraska Quarterly 17(2):4–7.
21. Cable, Dwight R., and Fred H. Tschirley. 1961. Responses of native and introduced grasses following aerial spraying of velvet mesquite in southern Arizona. J. Range Mgt. 14(3):155–159.
22. Cook, C. Wayne. 1963. Herbicide control of sagebrush on seeded foothill ranges in Utah. J. Range Mgt. 16(4):190–195.
23. Cook, C. Wayne, Paul D. Leonard, and Charles D. Bonham. 1965. Rabbitbrush competition and control on Utah rangelands. Utah Agric. Expt. Sta. Bul. 454.
24. Cords, H. P., and J. L. Artz. 1976 (Rev.). Rangeland, irrigated pasture, and meadows weed control recommendations. Nev. Agric. Ext. Cir. 148.
25. Council for Agric. Sci. and Tech. (CAST). 1975. The phenoxy herbicides. Weed Sci. 23(3):253–263.

26. Corns, William G., and R. J. Schraa. 1965. Mechanical and chemical control of silverberry (*Eleagnus commutata* Bernh.) on native grassland. J. Range Mgt. 18(1):15–19.
27. Cronin, E. H. 1974. Evaluation of some herbicide treatments for controlling tall larkspur. J. Range Mgt. 27(3):219–222.
28. Cronin, E. H. 1976. Impact on associated vegetation of controlling tall larkspur. J. Range Mgt. 29(3):202–206.
29. Cronin, Eugene H., and M. Coburn Williams. 1964. Chemical control of timber milkvetch and effects on associated vegetation. Weeds 12(3):177–179.
30. Currie, Pat O. 1976. Recovery of ponderosa pine-bunchgrass ranges through grazing and herbicide or fertilizer treatments. J. Range Mgt. 29(6):444–448.
31. Currier, H. B., and C. D. Dybing. 1959. Foliar penetration of herbicides—review and present status. Weeds 7(2):195–213.
32. Dahl, B. E., R. B. Wadley, M. R. George, and J. L. Talbot. 1971. Influence of site on mesquite mortality from 2,4,5-T. J. Range Mgt. 24(3):210–215.
33. Davis, A. M. 1967. Rangeland development through brush control in the Arkansas Ozarks. Arkansas Agric. Expt. Sta. Bul. 726.
34. Davis, E. A., P. A. Ingebo, and C. P. Pase. 1968. Effect of a watershed treatment with picloram on water quality. USDA, For. Serv. Res. Note RM-100.
35. Eckert, Richard E., Jr., Allen D. Bruner, Gerald J. Klomp, and Frederick F. Peterson. 1973. Control of Rocky Mountain iris and vegetation response on mountain meadows. J. Range Mgt. 26(5):352–355.
36. Eckert, Richard E., Jr., and Raymond A. Evans. 1968. Chemical control of low sagebrush and associated green rabbitbrush. J. Range Mgt. 21(5):325–328.
37. Eckert, Richard E., Jr., Gerald J. Klomp, Raymond A. Evans, and James A. Young. 1972. Establishment of perennial wheatgrasses in relation to atrazine residue in the seedbed. J. Range Mgt. 25(3):219–224.
38. Ehrenreich, John H., and John S. Crosby. 1960. Forage production on sprayed and burned areas in the Missouri Ozarks. J. Range Mgt. 13(2):68–70.
39. Elwell, Harry M. 1964. Oak brush control improves grazing lands. Agron. J. 56(4):411–415.
40. Ennis, W. B., Jr. 1965. The new research in pesticides. J. Range Mgt. 18(6):297–300.
41. Ennis, W. B. Jr., and Ralph E. Williamson. 1963. Influence of droplet size on effectiveness of low-volume herbicidal sprays. Weeds 11(1):67–72.
42. Evans, Raymond A., and James A. Young. 1975. Aerial application of 2,4-D plus picloram for green rabbitbrush control. J. Range Mgt. 28(4):315–318.
43. Fenster, C. R., and L. R. Robison. 1965. A hand sprayer for herbicides. Nebraska Ext. Cir. 65–166.
44. Fisser, Herbert G. 1968. Soil moisture and temperature changes following sagebrush control. J. Range Mgt. 21(5):283–287.
45. Flynt, T. O., R. W. Bovey, R. E. Meyer, T. E. Riley, and J. R. Baur. 1976. Granular herbicide applicator for brush control. J. Range Mgt. 29(5):435–437.
46. Freed, V. H., and R. O. Morris (Editors). 1967. Environmental and other factors in the response of plants to herbicides. Oregon Agric. Expt. Sta. Tech. Bul. 100.
47. Friesen, H. A., M. Aaston, W. G. Corns, J. L. Dobb, and A. Johnston. 1965. Brush control in Canada. Canada Dept. Agric. Pub. 1240.
48. Fryer, J. D., and S. A. Evans (Editors). 1968 (Fifth Ed.). Weed control handbook. Volume II. Recommendations. Blackwell Scientific Publications, Oxford and Edinburg.
49. Gerard, Jesse B. 1976. Chemical control of noxious range weeds. New Mex. Agric. Ext. Guide 400 B-811.
50. Goodin, J. R., L. R. Green, and V. W. Brown. 1966. Picloram—a promising new herbicide for control of woody plants. California Agric. 20(2):10–12.
51. Haas, R. H. 1968. Controlling yaupon—a component of the east Texas woodlands. Texas Agric. Expt. Sta. Prog. Rpt. 2598.

52. Haas, R. H., and R. L. Watson. 1968. Post oak and mixed hardwood control using 2,4,5-T and picloram. Texas Agric. Expt. Sta. Prog. Rpt. 2599.
53. Halls, L. K., and H. S. Crawford. 1965. Vegetation response to an Ozark woodland spraying. J. Range Mgt. 18(6):338–340.
54. Halls, L. K., R. H. Hughes, R. S. Rummell, and B. L. Southwell. 1964. Forage and cattle management in longleaf-slash pine forests. USDA Farmers' Bul. 2199.
55. Harvey, W. A., W. H. Johnson, and F. L. Bell. 1959. Control of oak trees on California foothill range. Down to Earth 15(1):3–6.
56. Hedrick, D. W., D. N. Hyder, F. A. Sneva, and C. E. Poulton. 1966. Ecological response of sagebrush-grass range in central Oregon to mechanical and chemical removal of *Artemisia.* Ecology 47(3):432–439.
57. Heikes, E. G. 1964. Tordon and other herbicides field testing for the control of deep rooted perennial weeds in Colorado. Down to Earth 2(3):9–12.
58. Heikes, P. Eugene. 1978. Colorado weed control handbook. Colorado State University, Ft. Collins. Mimeo. (Updated looseleaf.)
59. Hoffman, Garlyn O. 1961. Economics of chemical weed control. Down to Earth 17(3):3–6.
60. Hoffman, G. O. 1967. Controlling pricklypear in Texas. Down to Earth 23(1):9–12.
61. Hoffman, Garlyn O. 1968. Maintenance control for mesquite. Texas Agric. Ext. Serv. Fact Sheet 766.
62. Hoffman, Garlyn O. 1968. Range improvement with the use of herbicides. ASRM, Abstract of Papers, 21st Annual Meeting, pp. 28–29.
63. Hoffman, G. O. 1969 (Rev.). Brush control with AMS—hand (ground) application. Texas Agric. Ext. Serv. Leaflet 413.
64. Hoffman, G. O. 1971. Mesquite control. Texas Agric. Ext. Serv. Misc. Pub. 386.
65. Hoffman, G. O. and B. J. Ragsdale. 1967 (Rev.). Brush control with 2,4,5-T—hand (ground) application. Texas Agric. Ext. Service Leaflet 414.
66. Houston, W. R., and D. H. van der Sluijs. 1973. Increasing crude protein content of forage with atrazine on shortgrass range. USDA Prod. Res. Rep. 153.
67. Hull, A. C., Jr. 1971. Spraying tarweed infestations on ranges newly seeded to grass. J. Range Mgt. 24(2):145–147.
68. Hyder, D. N., W. R. Furtick, and F. A. Sneva. 1958. Differences among butyl, ethyl, and isopropyl ester formulations of 2,4-D, 2,4,5-T, and MCPA in the control of big sagebrush. Weeds 6(2):194–196.
69. Hyder, Donald N., and Forrest A. Sneva. 1955. Effect of form and rate of active ingredient, spraying season, solution volume, and type of solvent on mortality of big sagebrush. Oregon Agric. Expt. Sta. Tech. Bul. 35.
70. Hyder, D. N., and Forrest A. Sneva. 1962. Selective control of big sagebrush associated with bitterbrush. J. Range Mgt. 15(4):211–215.
71. Hyder, Donald N., Forrest A. Sneva, and Lyle D. Calvin. 1956. Chemical control of sagebrush larkspur. J. Range Mgt 9(4):184–186.
72. Hyder, D. N., Forrest A. Sneva, and Virgil H. Freed. 1962. Susceptibility of big sagebrush and green rabbitbrush to 2,4-D as related to certain environmental, phenological, and physiological conditions. Weeds 10(4):288–295.
73. Jameson, Donald A., and Thomas N. Johnsen, Jr. 1964. Ecology and control of alligator juniper. Weeds 12(2):140–142.
74. Jefferies, Ned W., and J. J. Norris. 1965. Management and improvement of Gambel oak ranges, San Juan Basin Branch Station, Hesperus, 1965. Colorado Agric. Expt. Sta. Prog. Rpt. 147.
75. Jensen, Louis A., John O. Evans, J. LaMar Anderson, and Alvin R. Hamson. 1978. Chemical control guide for Utah, 1977 (plus 1978 addendum). Utah Agric. Ext. Cir. 301.
76. Johnson, W. M. 1969. Life expectancy of a sagebrush control in central Wyoming. J. Range Mgt. 22(3):177–182.
77. Kearl, W. Gordon. 1965. A survey of sagebrush control in Wyoming, 1952–1964. Wyoming Agric. Expt. Sta. Misc. Cir. 217.

78. Klingman, D. L., and W. C. Shaw. 1975 (Rev.). Using phenoxy herbicides effectively. USDA Farmers' Bul. 2183.
79. Klingman, Glenn C., Floyd M. Ashton, and Lyman J. Noordhoff. 1975. Weed science: principles and practices. John Wiley and Sons, New York.
80. Klomp, G. J., and A. C. Hull, Jr. 1968. Effects of 2,4-D on emergence and seedling growth of range grasses. J. Range Mgt. 21(2):67–70.
81. Kufeld, Roland C. 1977. Improving gambel oak ranges for elk and mule deer by spraying with 2,4,5-TP. J. Range Mgt. 30(1):53–57.
82. Laycock, William A., and Thomas A. Phillips. 1968. Long-term effects of 2,4-D on lanceleaf rabbitbrush and associated species. J. Range Mgt. 21(2):90–93.
83. LeGrand, F. E. 1965. 1965 guide to safe and effective chemical weed control. Oklahoma Agric. Ext. Serv. Cir. 763.
84. Leonard, O. A., and W. A. Harvey. 1965. Chemical control of woody plants. California Agric. Expt. Sta. Bul. 812.
85. Lyon, L. Jack, and Walter F. Mueggler. 1968. Herbicide treatment of north Idaho browse evaluated six years later. J. Wildl. Mgt. 32(3):538–541.
86. Martin, S. Clark. 1966. The Santa Rita Experimental Range. USDA, For. Serv. Res. Paper RM-22.
87. McCaleb, J. E., E. M. Hodges, and C. L. Dantzman. 1961. Effect of herbicidal control of saw-palmetto on associated native forage plants in peninsular Florida. J. Range Mgt. 14(3):126–130.
88. McCarty, M. K., and C. J. Scifres. 1968. Western whorled milkweed and its control. Weed Science 16(1):4–7.
89. McCarty, M. K., and Charles J. Scifres. 1969. Herbicidal control of western ironweed. Weed Sci. 17(1):77–79.
90. McCully, Wayne. 1956. Soil injection of herbicides for controlling individual brush plants. Texas Agric. Expt. Sta. Prog. Rpt. 1869.
91. McIlvain, E. H., and C. G. Armstrong. 1963. Progress in shinnery oak and sand sage control at Woodward. USDA, Agric. Res. Serv. Mimeo.
92. McIlvain, E. H., and C. G. Armstrong. 1965. Siberian elm—a tough new invader of grasslands. Weeds 13(3):278–279.
93. McIlvain, E. H., and G. C. Armstrong. 1974. Brush and weed suppression with mistblowers. USDA, Agric. Res. Serv., Woodward, Okla.
94. Mitich, Larry W. 1965. Pasture renovation with 2,4-D in North Dakota. Down to Earth 20(4):26–28.
95. Morrow, L. A., and M. K. McCarty. 1975. Control of green sagewort in the Nebraska sandhills. Weed Sci. 23(6):465–469.
96. National Research Council. 1968. Principles of plant and animal pest control. II. Weed control. National Academy of Sciences Pub. 1597.
97. Owensby, Clenton E., and J. L. Launchbaugh. 1976. Controlling weeds and brush on range land. Kan. Agric. Ext. Leaflet 465.
98. Palmer, J. S., and R. D. Radeleff. 1969. The toxicity of some organic herbicides to cattle, sheep, and chickens. USDA, Agric. Res. Serv. Prod. Res. Rpt. 106.
99. Parker, Karl G. 1968. Range plant control—modernized. Utah Agric. Ext. Serv. Cir. 346.
100. Parker, Robert, and Lyle A. Derscheid. 1970. Chemical weed control in pasture and hayland. South Dakota Agric. Ext. Serv. Fact Sheet 426.
100a. Paulsen, Harold A., Jr., and John C. Miller. 1968. Control of Parry rabbitbrush on mountain grasslands of western Colorado. J. Range Mgt. 21(3):175–177.
101. Perry, Chester A., Cyrus M. McKell, Joe R. Goodin, and Thomas M. Little. 1967. Chemical control of an old stand of chaparral to increase range productivity. J. Range Mgt. 20(3):166–169.
102. Plumb, T. R. 1967. Brushkiller to control scrub oak sprouts. USDA, For. Serv. Res. Note PSW-146.
103. Range Seeding Equipment Comm. 1965 (Rev.). Handbook of range seeding equipment. USDA and USDI, U.S. Govt. Printing Office, Washington, D.C.

104. Range Seeding Equipment Comm. 1966. Chemical control of range weeds. USDA and USDI, U.S. Govt. Printing Office, Washington, D.C.
105. Robertson, J. H. 1969. Yield of crested wheatgrass following release from sagebrush competition by 2,4-D. J. Range Mgt. 22(4):287–288.
106. Robison, E. D. 1968. Chemical control of mesquite with 2,4,5-T and combinations of chemicals. Texas Agric. Expt. Sta. Prog. Rpt. 2589.
107. Robison, Laren, Dave Nuland, and John Furrer. 1970. Chemicals that control weeds. Nebraska Agric. Ext. Serv. Cir. 70–130.
108. Rowe, V. K., and T. A. Hymas. 1954. Summary of toxicological information on 2,4-D and 2,4,5-T type herbicides and an evaluation of the hazards to livestock associated with their use. Amer. J. Vet. Res. 15(57):622–629.
109. Ryerson, D. E., J. E. Taylor, L. O. Baker, Harold A. R. Houlton, and D. W. Stroud. 1970. Clubmoss on Montana rangelands: distribution, control, range relationships. Mon. Agric. Expt. Sta. Bul. 645.
110. Schmutz, Ervin M. 1967. Chemical control of three Chihuahuan desert shrubs. Weeds 15(1):62–67.
111. Schmutz, Ervin M., and David E. Little. 1970. Effects of 2,4,5-T and picloram on broom snakeweed in Arizona. J. Range Mgt. 23(5):354–357.
112. Schmutz, Ervin M., and David W. Whitham. 1962. Shrub control studies in the oak-chaparral of Arizona. J. Range Mgt. 15(2):61–67.
113. Schuster, Joseph L. 1971. Night applications of phenoxy herbicides of plains pricklypear. Weed Sci. 19(5):585–587.
114. Schuster, Joseph L. 1976. Redberry juniper control with picloram. J. Range Mgt. 29(6):490–491.
115. Scifres, C. J. 1972. Herbicide interactions in control of sand shinnery oak. J. Range Mgt. 25(5):386–389.
116. Scifres, C. J. 1975. Fall application of herbicides improves Macartney rose-infested coastal prairie rangelands. J. Range Mgt. 28(6):483–486.
117. Scifres, C. J. 1977. Herbicides and the range ecosystem: residues, research, and the role of rangemen. J. Range Mgt. 30(2):86–90.
118. Scifres, C. J., and J. L. Mutz. 1978. Herbaceous vegetation changes following applications of tebuthiuron for brush control. J. Range Mgt. 31(5):375–378.
119. Smith, Dixie R. 1969. Is deferment always needed after chemical control of sagebrush? J. Range Mgt. 22(4):261–263.
120. Sneva, Forrest A. 1967. Chemical curing of range grasses with paraquat. J. Range Mgt. 20(6):389–394.
121. Sneva, Forrest A. 1973. Nitrogen and paraquat saves range forage for fall grazing. J. Range Mgt. 26(4):294–295.
122. Sneva, Forrest A., and D. N. Hyder. 1966. Control of big sagebrush associated with bitterbrush in ponderosa pine. J. For. 64(10):677–680.
123. Society for Range Management. 1978. SRM response for the Environmental Protection Agency's rebuttable presumption against registration of 2,4,5-T (WH-569). Rangeman's J. 5(4):132–133.
124. Sonder, Leslie W., and Harold P. Alley. 1961. Soil-moisture retention and snow-holding capacity as affected by the chemical control of big sagebrush (Artemisia tridentata Nutt.). Weeds 9(1):27–35.
125. Sperry, Omer E. 1967. Experimental studies on the control of rayless goldenrod and perennial broomweed. Texas Agric. Expt. Sta. Prog. Rpt. 2456.
126. Sperry, O. E., J. W. Dollahite, G. O. Hoffman, and B. J. Camp. 1964. Texas plants poisonous to livestock. Texas Agric. Expt. Sta. Bul. 1028.
127. Tabler, Ronald D. 1968. Soil moisture response to spraying big sagebrush with 2,4-D. J. Range Mgt. 21(1):12–15.
128. Texas Agric. Expt. Sta. and Ext. Serv. 1974 (Rev.). Suggestions for weed control with chemicals: range and timber lands, and pasture, and forage crops. Texas Agric. Ext. Misc. Pub. 1060.
129. Thilenius, John F., Gary R. Brown, and C. Colin Kaltenbach. 1975. Treating forb-dominated subalpine range with 2,4-D: effects on herbage and cattle diets. J. Range Mgt. 28(4):311–315.
130. Thomson, W. T. 1967. Agricultural chemicals. Book II. Herbicides. Simmons Pub. Co., Davis, California.

131. Tingey, D. C., and C. Wayne Cook. 1955. Eradication of mulesear with herbicides. Utah Agric. Expt. Sta. Bul. 375.

132. Torell, Paul J. 1967. Dowpan—an aid to reseeding medusahead-infested rangeland. Down to Earth 23(1):6–8.

133. Tschirley, Fred H., and S. Clark Martin. 1961. Burroweed on southern Arizona range lands. Arizona Agric. Expt. Sta. Tech. Bul. 146.

134. Tueller, Paul T., and Raymond A. Evans. 1969. Control of green rabbitbrush and big sagebrush with 2,4-D and picloram. Weed Sci. 17(2):233–235.

135. Tueller, Paul T., and J. H. Robertson. 1965. Control of tall larkspur (*Delphinium stachydeum* A. Gray) by pelleted 4-amino-3,5,6-trichloro-picolinic acid. *In* Research in range management. Nevada Agric. Expt. Sta. Rpt. 13.

136. USDA, Agric. Res. Service. 1965. Weed roses—pests of pasture, orchard and range. USDA, Agric. Res. Serv. 22–93.

137. USDA, Agric. Res. Serv. 1969. Chemical control of brush and trees. USDA Farmers' Bul. 2158.

138. USDA, Agric. Res. Serv. 1973. Guidelines for weed control. USDA Agric. Handbook 447.

139. USDA, Agric. Res. Serv. 1976 (Rev.). Aerial application of agricultural chemicals. USDA Agric. Handbook 287.

140. USDA, Econ. Res. Serv. and Agric. Res. Serv. 1970. Restricting the use of phenoxy herbicides—costs to farmers. USDA Agric. Econ. Rpt. 194.

141. U.S. Dept. of Health, Education, and Welfare. 1969. Report of the Secretary's Commission on Pesticides and their relationship to environmental health. Part I and II. U.S. Govt. Printing Office, Washington, D.C.

142. U.S. Environmental Protection Agency. 1975. EPA compendium of registered pesticides. Volume I. Herbicides and plant regulators. Technical Services Division, Office of Pesticides Programs, U.S. Env. Prot. Agency, Washington, D.C. (Updated by supplements periodically.)

143. Weddle, Jon P., and Henry A. Wright. 1970. An evaluation of five methods to retreat sprayed mesquite. J. Range Mgt. 23(6):411–414.

144. Weed Sci. Soc. Amer. 1974 (3rd Ed.). Herbicide handbook of the Weed Science Society of America. WSSA, Champaign, Ill.

145. Weldon, L. W., D. W. Bohmont, and H. P. Alley. 1958. Re-establishment of sagebrush following chemical control. Weeds 6(3):298–303.

146. Wicks, G. A., C. R. Fenster, and O. C. Burnside. 1969. Selective control of pricklypear in rangeland with herbicides. Weed Sci. 17(4):408–411.

147. Williams, M. Coburn. 1970. Detoxification of timber milkvetch by 2,4,5-T and silvex. J. Range Mgt. 23(6):400–402.

148. Williams, M. Coburn, and Eugene H. Cronin. 1966. Five poisonous range weeds—when and why they are dangerous. J. Range Mgt 19(5):274–279.

149. Williams, M. Coburn, and L. B. Kreps. 1970. Chemical control of western false hellbore. Weed Science 18(4):481–483.

150. Workman, D. R., K. R. Tefertiller, and C. L. Leinweber. 1965. Profitability of aerial spraying to control mesquite. Texas Agric. Expt. Sta. Misc. Pub. 784.

151. Young, James A., and Raymond A. Evans. 1976. Control of pinyon saplings with picloram or karbutilate. J. Range Mgt. 29(2):144–147.

Chapter 6

Range improvement by burning

The Role of Fire on Rangeland

Fire is a natural factor on wildlands, and probably no range site with its associated plant community has developed without being influenced by fire. Fires have occurred historically when fuel accumulation and weather conditions made ignition and burning possible. Biswell [13] has considered fire as nature's way of keeping the primitive forests open and parklike, in a stable equilibrium, and immune to extensive crown fires. Biswell [15] further concluded that fire exclusion is something new and unnatural in forest and brushland environments.

Sauer [144] proposed that the primary cause of grassland development and maintenance was fire and not climate. He suggested that firesetting activities of man, along with ignition by lightning, historically brought about a lasting modification of the climax. Although this view may be extreme, fire is obviously an important part of the climax grassland environment. Many ecologists credit fire with a major role in maintaining desert grasslands, grass-sagebrush communities, and midgrass prairies against major encroachments of shrubs and trees [86, 152].

Burning is the oldest known practice used by man to manipulate vegetation on grazing lands. For many thousands of years both accidental and deliberate burning by aboriginal populations greatly increased the frequency of fires over those resulting from natural causes in inhabited regions [7]. Deliberate burning was used with hunting to increase visibility for finding game animals, to drive game animals, and to attract game animals after the burning by vegetative resprouting. It was also used to reduce woody plants and improve herbaceous pasture for livestock grazing, for clearing and planting of food plants, and even for military attack and defense. Although often used improperly in the past, and obviously not a cure-all for all range problems, burning can be an effective and practical tool in range improvement. Whether burning can be recommended for a particular situation depends upon the site, the vegetation composition, multiple use considerations, and the objectives of management. Prescribed burning should be planned so as not to add materially to air pollution around large urban centers [127].

Because haphazard or accidental burning can be harmful or even disastrous, burning as a wildland improvement practice has only recently gained favor with many range users, technicians, and researchers. This increased standing of burning has resulted from additional knowledge of the benefits of using this tool along with improved practices in how to control it. Wildfire remains a major destroyer of natural resources and a threat to life and property. However, burning frequently benefits all land users and resources when properly planned and executed. Public relations is important in giving the general public an understanding of the role of fire in the development and management of natural communities [127].

When adapted for use, prescribed burning has several advantages over other noxious plant control methods. Excluding erosion hazard on highly erodible sites, it is widely applicable regardless of rockiness of soil, steepness of slope, or irregularity of terrain providing there is sufficient ground fuel to carry the fire [129]. Since it requires machines only in protection and patrolling, fire can be used on many sites where mechanical treatment is not feasible and at a moderately low cost. It also has the advantage of giving fast coverage. Prescribed burning is a low-cost and low petroleum-consuming technique for rangeland vegetation manipulation for which costly alternatives are often impractical [153].

The following terms are suggested for use with range burning, but only prescribed burning considers all factors conducive to maximizing benefits [142, 143]:

Convenience burning—considers only the time and place of firing with little or no attempt at control; hazardous, and benefits—if any—often of short duration.

Controlled burning—planned application and confinement of fire to a preselected land area. Considers only time, place, and fire control required.

Prescribed burning—systematically planned firing of selected land areas when weather and vegetation favor a particular method and intensity of burning that can be expected to accomplish planned benefits. Considers all known factors affecting burning effectiveness and is more definitive than controlled burning.

Purposes of Range Burning

Reasons for burning rangeland generally include one or more of the benefits expected from other noxious plant control methods: (1) kill or suppress undesirable brush plants, (2) prevent invasion of inferior species in the understory, and (3) increase forage production and thus grazing capacity. However, depending upon the geographical area and vegetation type under treatment, burning generally includes one or more of the following additional purposes:

1. Increase palatability of forages and remove old, dead material and thereby increase utilization by grazing animals.
2. Improve grazing use and distribution of undergrazed areas.
3. Improve access and availability of forage to grazing animals in dense brush areas.
4. Provide access for hunting and facilitate the movement of men and machines.
5. Initiate herbaceous plant growth one to three weeks earlier on fresh burns.
6. Release plant nutrients in the soil and litter for plant use.
7. Increase temporarily the nutrient content of forage, particularly protein and phosphorus early in the growing season.
8. Rejuvenate woody plants for browse production.
9. Increase legumes for wildlife in the Southeast.
10. Reduce heavy needle and debris mats that inhibit grass and other desirable plants.
11. Reduce consumption of water by brush and tree species, at least temporarily.
12. Reduce temporarily the amount of litter and vegetation that intercepts precipitation from light rains.
13. Remove accumulated debris, brush, and logging slash to reduce or prevent wildfire (Figure 55); i.e. presuppression burning.
14. Prepare seedbeds for artificial seeding of forage species.
15. Prepare ash seedbeds for natural or artificial reproduction of desirable trees as a silvicultural practice.
16. Reduce growth stagnation in stands of dense tree reproduction by thinning tree seedlings as a silvicultural practice.
17. Control undesirable hardwoods in southern pine plantations.
18. Reduce the incidence of insect and fungal disease attack, such as brown spot needle blight of longleaf pine in the Southeast.

FIG. 55. *Damage caused by wildfire in 1965 to the Nebraska State 4-H Club Camp located in the Nebraska National forest. Excessive accumulation of debris fuel materially contributed to the destruction.*

FIG. 56. *Although burning is an improvement practice commonly used on bluestem range, this wildfire on the University of Nebraska North Platte Station consumed forage planned for winter grazing.*

Burning increases forage production on many grasslands, with some exceptions on shortgrass and other semiarid and arid grasslands. Gains in herbage yield from prescribed burning are economical only when they are not offset by forage losses resulting from the following [43]:

1. Grazing must be reduced for one or more seasons in advance of burning on some sites to allow sufficient litter to accumulate to carry the planned burn.
2. Grazing must be delayed following burning on some sites to prevent excessive grazing use and allow recovery of forage plants.
3. Standing, matured grass consumed by fire in areas where grass cures well represents an actual forage loss (Figure 56).
4. Some sensitive but desirable plant species may not recover.

Increased livestock gains from burned grasslands have often been attributed to the superior nutritive quality of the forage produced on recently burned areas. However, these increased gains may also result from nearly complete availability of the regrowth on burned areas, less grazing effort and reduced energy expenditure, and consumption of forage undiluted by weathered material [43]. In wet, warm areas burning annually may be necessary to remove unused grass that has lost its food value and is a hindrance to grazing and to maintain new herbage in satisfactory nutritive condition. However, repeated burning can seriously deplete the root reserves of desirable perennials and even cause their death, particularly in semiarid areas.

Plant control. Several undesirable plants are selectively controlled by burning at the proper season if soil moisture conditions are satisfactory and adequate precautions are observed. Big sagebrush and many other woody, nonsprouting plants are readily and effectively killed by fire. In a Canadian study, understory plants found susceptible to fire were those with fibrous roots or having stolons or rhizomes which grew above the mineral soil [119]. Plants resistant to fire were those having rhizomes which grew more deeply and in the mineral soil, those with taproots able to regenerate below their root crowns, and those which quickly reestablished from seed following fire.

Plants with perennating buds at or above the ground level are more likely to be injured or even killed by fire than those with buds underground at the time of burning [86]. While the growing point of dormant grasses is close to the ground, shrubs have their growing points exposed on the ends of branches or in the cambium layer under the bark. Killing the cambium at the base of the stem kills the entire shrub above the killed zone, but allows sprouting shrubs to sprout from lower depths. However, anatomical differences alone often fail to explain the differential effects of fire on plants.

Wright [169] found that natural tolerance of herbage removal and the density of dead plant material accumulating at the base of the plant were important factors in perennial grass response to burning. Squirreltail was less susceptible to fire than needle-and-thread, apparently because it was more tolerant of herbage removal, was less leafy and more stemmy, and thus accumulated less flammable material at the plant base. It was concluded that fuel within the plant can generate enough heat upon ignition to kill it regardless of the total heat generated by a sweeping fire. Heavy grazing tends to reduce fire damage to perennial grasses by reducing fuel accumulating at the plant bases [167].

Season of burning and phenological stage of growth directly affect plant response to burning, with each plant species responding somewhat differently [43]. Perennials that have already broken spring dormancy and made considerable new growth are more damaged by intense fire than plants still dormant. For this reason early spring burning favors warm-season grasses over cool-season grasses. Plants subjected to fire during or immediately prior to extended drought are more severely affected than under normal weather conditions. But burning following a rain reduces damage to plants by increasing the moisture content of the plant base and litter around the base of the plant [167].

Thermal brush control is a means of killing nonsprouting woody plants with blasts of hot air while preventing ignition [10]. This control technique has given up to 100 percent kill of big sagebrush while leaving plant residues in place and eliminating fire control problems. It has also readily induced sprouting in sprouting shrub species.

A propane burner with a directional air blast has been used to sear the plants while keeping flame from reaching the plants [10]. In sagebrush the results of treatment have been rapid and readily apparent from the loss of natural coloration of the foliage. A change from a natural light green to a silvery tint occurs rapidly. The only serious obstacle in widespread use of this control technique has been the conservation and high cost of fuel. Burning equipment developed by 1973 enabled 7.5 acres per hour to be treated with fuel consumption for heat generation at about 16 gallons per acre.

Even when some undesirable species are not directly killed by fire, temporary suppression by burning may justify its use. Periodic fires kill only a portion of mesquite plants, but repeated topkill prevents maturity and seed production of surviving plants and thus prevents spread [86]. Other sprouting species such as Gambel oak may actually be stimulated by reduced competition to develop even denser thickets than before the burn [24]. But where there is a good understory in the burned areas or it is provided by artificial seeding, Gambel oak may be suppressed somewhat as the result of fire [132].

Cost of burning. Cost of burning big sagebrush based on 1962 figures varied from $1 to $4 per acre on tracts of 1,000 acres or more [129]. Planned burning was considered most useful on fairly level tracts of 1,000 acres or more. The average cost of controlled burns in 1956 on California brushlands was $3 per acre compared to $97 per acre for wildfire suppression [107]. It was concluded the cost of controlled burning was justified on the basis of wildfire hazard reduction alone.

The largest cost of burning is the preparation of fire lines and fire-control trails. Other costs include the cost of alternative grazing if required, and actual costs of burning and patrolling. The actual cost of burning per acre varies depending upon the size of the burn and the amount of preparation required. Estimates of burning costs ranged in 1971 from as low as $.50 to $1.00 per acre on tobosa and $1.00 to $2.00 per acre of dozed juniper in Texas [168] to $10.00 to $36.00 in California chaparral where site preparation costs are much higher. The costs of 190 burns in California in 1947 and 1948 were related to the number of acres in the project [141]. Costs were $3.00 per acre for 40-acre burns, about $.75 for 320-acre burns, and $.45 for 480-acre burns.

Wildlife hazard. Most prescribed burns are not directly destructive to wild mammals [82, 105]. Most vertebrates manage to escape the heat by flying or running away, going below ground a few inches, hiding in rock outcrops, or seeking islands missed by fire. Few mature big game animals are injured by fire, but some very young may be trapped by fast moving wildfires. Prescribed burning in contrast to wildfire provides many avenues of escape by means of fuelbreaks, firebreaks, and reduced size of block burning and can actually reduce wildlife hazard from wildfire [70].

Opening up and rejuvenating dense brush stands and the stimulation of herbaceous understory plant growth by fire is generally very beneficial to game animals [124]. Following wildfires in timbered areas, deer and elk use has been observed to decline temporarily but to increase several fold over preburn use after regrowth [108]. If the size of the burn is restricted, desirable habitat features can often be obtained more easily by fire than other methods [146]. But fire can be devastating to ground nesting birds by destroying the nests, by removing the productive cover, and reducing insect food resources [43]. Burning prior to or after the nesting season is suggested. Although mice and other small rodents suffer little direct death loss from fire, the loss of essential food and cover may materially reduce these populations temporarily.

Burning Practices and Safeguards

Preparing for burning. Prescribed burning includes four basic steps: (1) locating the limits of the area to be burned, (2) preparing the fire-control lines, (3) preparing the fuel if necessary, and (4) conducting the burn [143]. State regulatory agencies should be consulted to assure legal conformity to burning regulations. They can also help assure more effective burns and safe use of fire through on-the-ground consultation and often even direct assistance in making and controlling the burn.

The following terminology has been recommended for use in wildland fire control and management [65]:

> **Fireline:** a narrow line, 2 to 10 feet wide, from which all vegetation is removed down to mineral soil by sterilization of the soil, by yearly maintenance, or by clearing just ahead of firing out from the line.
>
> **Firebreak:** a fireline wider than 10 feet, frequently 20 to 30 feet wide, prepared each year ahead of the time it may be needed for use in controlling a fire.
>
> **Fuelbreak:** a strategically located block or strip (minimum width of 200 feet suggested) on which a cover of dense, heavy, or flammable vegetation has been changed to new vegetation of lower fuel volume or flammability and subsequently maintained as an aid to fire control.
>
> **Fuel modification:** fire control practices including cleanup of fuel hazards, permanent fuel reduction on limited areas, or periodic fuel reduction on large acreages.

The location of boundary and other firelines and firebreaks should consider both natural and artificial features. Such should be made to serve dually as barriers against uncontrolled fire and as ignition points for prescribed burning. Natural physical barriers that can be used include recent burns, streams, roads, rock ledges, ridge tops, and ravines. Fuelbreaks that can provide equally effective barriers include sparse vegetation, green grass, and mesic brush types on north slopes.

Artificial firelines and firebreaks include plowed or dozed strips. Backfiring into the wind to preburn a strip while vegetation is still green in the spring is used in some areas such as the South. Another method of developing firebreaks or widening existing ones is to con-

FIG. 57. *Grazed fuelbreaks provide forage as well as a means of fire control.* Top, *chain-wide fuelbreak in south Georgia seeded with Coastal bermudagrass and common lespedeza for summer grazing and common oats and white clover for winter forage and grazed at a rate of more than one cow per acre [71]; bottom, chaparral ridgetop with productive soil on Los Padres National Forest cleared for grazed fuelbreak.* (U.S. Forest Service photos)

209

struct double parallel lines up to 200 feet apart and to burn the area between. However, using fire as a preburning treatment requires that intensive precautions be taken to prevent a breakout. Dry fuels such as piles of dead juniper give off dangerous firebrands when burning [167]. Their hazard is considerably dependent upon the wind velocity. One approach is to burn out the brush piles when surrounded by green grass and later backfire the strips under more moderate burning conditions.

In areas with light fuels, wetlines using water applied with standard firefighting equipment have provided effective, rapid, low-cost, temporary firing lines without permanent disturbance to the landscape [112]. A single wetline can be developed into a fireline by short headfiring into the wetline, or by using parallel, double wetlines and burning out the fuel between them.

Grazed fuelbreaks are now in common use in many parts of the United States and have a twofold value (Figure 57) [71]. First, they help to slow down the spread of forest fire, provide a strip from which to backfire when making prescribed burns or attacking wildfires, and serve as access roads for deployment of men and equipment. Second, grazed fuelbreaks benefit livestock and big game by producing good forage during a part of the year and by providing a means of getting increased use of native plants in adjacent forest areas.

After the brush is cleared from appropriate areas, grazed fuelbreaks in southern forests are developed by selecting and planting forage grasses and legumes which are palatable, are adapted to close grazing, are highly productive under fertility, retard fire through long growing periods, and remain green under drought [71]. Low growing, low volume, slow burning shrubs in discontinuous mixtures with grasses and forbs have also been recommended for fuelbreak plantings on cleared chaparral sites in California [125]. Grazed pasture fuelbreaks are less likely to erode than plowed lines and are less dangerous to install than burned firebreaks. If closely grazed, grazed fuelbreaks have less flammable fuel than burned breaks and also provide valuable grazing. Fuelbreaks can also be made to provide safety zones for firefighting personnel and equipment, an interior strip of fireline, and water for fire fighting and for grazing animals [66].

Provision must be made to insure that fires do not get out of control. In addition to prearranged firebreaks and firelines, this requires trained manpower on hand for both ignition and control, the necessary equipment and tools and judgment as to when burning should be undertaken, and when and where it should be stopped. The use of desiccating chemicals to permit more latitude in selecting the time

210

FIG. 58. *Preparation for prescribed burning of sagebrush at the Benmore Experimental Area by the U.S. Forest Service. A, fire line constructed by blading; B, widening fire lines by backfiring with a drip torch; C, adequate fire crew with equipment for ignition and fire control; D, using mobile weather unit to provide continuous weather information.* (Bureau of Sports Fisheries & Wildlife photos by Maurice F. Baker)

and place of burning shows promise. Chemical desiccants now under investigation include quick-acting agents such as pentachlorophenol in weed oils or paraquat or slow-acting agents such as the phenoxy herbicides [11]. Other preparations for burning include getting weather information and forecasts, inspecting the fuel, notifying adjacent landowners, and clearing burning plans with the proper fire control authorities (Figure 58) [72].

FIG. 59. *Preparing brush fuel for prescribed burning in California chaparral by crushing with a heavy roller. This roller can also be used to cover broadcasted seed where site is unadapted for drilling and/or breaking down brush skeletons left by fire or herbicide treatment.* (U.S. Forest Service photo)

Spotty burns often result on heterogeneous vegetation and soil areas. Where a complete burn is not obtained on the first attempt, reburning soon afterward is generally undesirable and often impossible because of the lack of fuel necessary to carry the fire. Adequate amounts, dispersion, and compaction of the ground fuel supply is very important in broadcast burning in carrying fire from plant to plant. In an Ashe juniper community in Texas it was concluded that 900 pounds of fine fuel per acre was needed to carry a fire to kill juniper seedlings and burn piles of dozed juniper [165]. Grazing is sometimes deferred to allow a fuel supply to collect. Another practice commonly used to give a more complete burn in California chaparral, but with promising results in other areas as well, is brush crushing or mashing (Figure 59).

Crushing or mashing brush prior to burning on California brushlands has had the following advantages: (1) fuel is compacted, (2) fire is easier to handle since burning time is shortened, (3) burns are possible under higher humidity and lower temperature when danger is reduced, (4) mashing in the spring prior to summer burning kills many plants, and (5) burning leaves a cleaner seedbed of good white ash from the burn [111]. Mashing in chaparral also enables burns hot enough to kill many old plants and undesirable seed in the soil and makes firing easier [27]. Brush mashing followed by burning after two or three years kills the first crop of sprouts as well as removing the smashed debris.

Brush crushing and compacting in chamise is commonly accomplished by a bulldozer with the blade raised about twelve inches above the ground [139]. The blade knocks down the brush or breaks it off at ground level, and the tractor tracks provide crushing action. Pulling a rolling brush cutter or other roller or crusher behind the bulldozer provides additional crushing. The bulldozer can also be used to fell individual trees into the crushed brush. Chaining is sometimes used in crushing and windrowing but is less effective on young, open stands [11].

Since brush smashing adds to burning costs, the goal is the cleanest possible burn with the smallest amount of mechanical preparation [56]. Brush condition and brush density are given primary consideration. When the brush is young, and vigorously growing, with few dead twigs and litter, or when the litter on the ground will not carry a fire, 100 percent of the area should be smashed. Complete smashing is also desirable where seeding requires the area be relatively free of stubs and snags or where high risk requires spring burning. When brush is dense or overmature, or has a high amount of litter on the ground, smashing 25 to 35 percent of the area may be adequate.

Burning techniques. Although the ignition of the foliage of individual trees is useful for eliminating scattered trees, broadcast burning requiring sufficient flammable fuel to carry the fire is most commonly used. Rapid, controlled ignition along predetermined fire lines is important. Drip torches using kerosene under pressure are light and portable, provide fast ignition, and give a residual flame. Propane torches and diesel flamethrowers are adequate for igniting most range fuels but are heavy and awkward unless truck-mounted. Flares and grenades are available which can be thrown by hand, shot by gun, or strategically placed on the ground for initiating small burns but require high flammability to be effective. Flying drip torches suspended from helicopters have also been used in slash burning.

A headfire is ignited on the windward side of the area to be burned and carried leeward by the wind. A backfire is started on the lee side of the area and allowed to burn slowly into the wind. These single ignition methods may be partly or completely replaced by area ignition, in which numerous individual fires are lit rapidly and close enough together to cause each to burn more intensely and rapidly. This permits the several fires to join into one intense fire with a

213

strong indraft which keeps the fire from running. However, such area ignition should be attempted only by an experienced person because of the hazards involved. Area ignition has resulted in more complete burns under conditions where single ignition spreads slowly and burns incompletely [56].

Headfires are generally required when ground fuel is minimum, where weather conditions are less than ideal, or when a maximum kill of woody plants is desired. Backfiring is frequently used where selective burning is required, i.e., removing hardwoods fom conifer stands. Many desirable tree species receive little or no damage from backfires under proper conditions but are readily killed by headfires that develop into crown fires [43]. On pine needle fuel beds, backfires were found to spread more slowly, had longer burning time, had less flame depth, burned much deeper into the fuel beds, and were less affected by wind velocity than headfires [9]. Since favorable conditions for backfiring seldom persist long, difficulty may be experienced in backing fires into the wind for long distances before the prescribed conditions cease. Dividing larger areas into numerous strips using fire lines and firing these simultaneously reduces burning time but increases burning costs somewhat [47].

Wind direction and speed should be given special attention. Wind changes on level ground may develop an intense, fast-running headfire which will crown. On slopes where fire is made to burn downhill, shifting of wind is not likely to cause crowning. Fire travels much faster uphill than downhill. Burning should not be done when the wind is calm since flames tend to go straight up. Wind velocity of three to ten mph has been recommended for burning in the Southeast [12]. Ten to 15 mph winds have generally been required for effective sagebrush-juniper burns in Utah.

·Fire burns hotter and kills more plants when fuel humidity is low, the amount of fuel is high, and the wind velocity is high [143]; but hazards are also greatest then. Conditions must be dry enough to carry a good fire without being hazardous. Adequate soil moisture is important in reducing damage to desirable species; but if conditions are too wet, spotty brush kill and litter removal may result. Fires may be started late in the day so that little time will remain before temperatures drop, humidity goes up, and the wind dies. However, in determining the best time for burning the seasonal fire susceptibility of both the undesirable and desirable plant species must be considered as well as weather factors. [56].

In burning studies with mesquite in Texas, fuel moisture content was the most important factor affecting ignition, with wind speed and amount of total fine fuel the next important factors [23]. Low relative humidity and high air temperature materially reduced the moisture content of the fire fuels which greatly increased ignition. Low relative humidity, and thus low fuel moisture levels, was the most important factor affecting mesquite burndown, followed by increasing windspeed and higher levels of total fuel. Low fuel moisture and wind, continuous as to speed and direction, were important in preventing spotty burns.

Fuel accumulation and the direction of fire spread in relation to wind direction affects the intensity of burning. Where woody fuels .are present in high amounts such as in down timber, slash, or windrowed brush, considerable damage may result to the herbaceous plant understory. Since prescribed burning in such areas implies good soil moisture and ideal weather conditions, the heat and severity of the burns can be modified. By contrast, wildfires often occur under conditions least beneficial to the desirable herbaceous species. Adequate soil moisture and short duration exposure to the high temperatures generated by fires reduce damage to the root crowns of most herbaceous species.

Management after burning. Prescribed burning for range improvement must be followed by appropriate grazing practices, but the best grazing practices will depend upon the locality, the nature of the resident vegetation, the kinds of grazing animals, and the specific objectives sought from burning. Livestock and big game generally concentrate on areas recently burned, apparently because the forage there is more palatable or more completely available [49, 100, 113, 116]. For this reason, spot burning of undergrazed areas is sometimes practiced to increase grazing use.

On many arid and semiarid ranges, heavy concentration of livestock grazing on burned areas during the early part of the subsequent growing season results in soil loss and damage of the forage plant cover. Localized heavy grazing is common when only smaller portions of range accessible to livestock have been burned. In order to prevent this uneven distribution of grazing, every effort should be made to burn the entire pasture or range unit uniformly. However, when wildfire or planned burning cover only a portion of the grazing unit, deferment of grazing until after the end of the subsequent growing season reduces the differential grazing.

Management recommendations after prescribed or accidental burning of sagebrush-grass areas include (1) protection of burned areas from trailing by livestock during at least the first fall, (2) protection of burned areas from grazing for one full year, and (3) light grazing for the second year and at proper rates thereafter [130]. Where these practices were applied following sagebrush burns and checked nine years later, the perennial grass production and grazing capacity had increased by 106 and 83 percent, respectively, and there were only five sagebrush plants per 100 square feet. On similar range where these practices were not applied following burning, perennial grass production was reduced by 23 percent, carrying capacity was increased by only 4 percent, and there were fifty-five sagebrush plants per 100 square feet.

On range sites in the subhumid and humid regions, deferment of grazing prevents realization of expected benefits from grazing. Burning in the Southeast is used to establish a rotational grazing scheme, and grazing utilizes the lush growth that follows proper burning practices [49]. Burning tallgrass range in eastern Kansas stimulates earlier growth of the fire-tolerant grasses; delaying the grazing does not benefit the range and reduces the value of the new forage to livestock [4]. Since burning of tobosa grass greatly improves the palatability of this rather unpalatable species, adequate utilization results only when grazed as the new shoots are growing [166].

Prescribed burning in California chaparral is practiced partly to rejuvenate woody plants and to bring them within reach of deer (Figure 60) [61]. However, the sprouting woody species often grow beyond the reach of deer when allowed a complete year of protection. In this situation continuous browsing is recommended to maintain the browse for a longer time in an available form. Burns on wildlife range in any one year should be limited in size to provide a degree of biological control, since deer are unable to fully utilize sprouts and forb growth on large continuous burns. However, it was found in California that different chaparral species required different browsing pressures to maintain them in vigorous, available forms following burning [61].

Where follow-up treatments to burning are required, they should be applied promptly in order to be effective. Where spraying of brush sprouts or forbs is required as a follow-up to burning, it should normally be performed the first or second growing season following burning [123]. When artificial seeding is to follow burning, seeding should be made as rapidly as seasonal climate permits in order to take advantage of the reduced competition and to provide ground cover and forage as soon as possible.

Emergency treatments often required following wildfires may be quite different from management provided following a prescribed burn [127]. Management objectives after wildfires are often limited to minimizing expected damage from flooding and erosion, repair or replacement of range improvements, providing for alternative livestock or big game grazing, and so on.

FIG. 60. *Vigorous sprouting of chaparral in California during the second year after burning. Although this woody plant rejuvenation provides deer browse, follow-up treatments are generally required for adequate woody plant control.* (U.S. Forest Service photo)

Burning Effects on Soil

The widespread soil damage earlier attributed to burning rangelands is now being questioned. Although wildfire or breakouts are more apt to damage soil properties, fire suppression and presuppression, including prescribed burning, has aided in preventing damage to sites subject to soil damage. Under conditions of prescribed burning where sites for burning are carefully selected, it has been concluded that prescribed burning has had little permanent damage to the soils of pine-grasslands of the South [160], of grassland prairies in Iowa [53], of mesquite-tobosa grass communities [156], of ponderosa pine types of the West [163], or of sagebrush-grass types [129]. However, burning should not ordinarily be used on steep slopes or on soils that blow or wash readily.

Humus and mulch. Fire removes all or part of the mulch on the soil depending upon the severity of the fire. On Douglas fir sites in the Pacific Northwest, complete destruction of the duff layer was accompanied by a loss of twenty-five tons (or 89 percent) of the organic matter per acre [89]. Light fires in forests consume only the up-

217

permost part of the mulch and litter, but severe fires or repeated light fires are capable of exposing nearly all of the mineral soil [109]. Attempts are frequently made to burn so that a portion of the mulch remains [59]. Burning when soils and lower portions of the mulch are moist removes less mulch, minimizes erosion potential, and assures better plant response [127].

Most herbaceous perennials on forested or mesic grassland sites are stimulated by the removal of litter and decline in vigor as the litter again accumulates to a high level [43]. In tallgrass areas the normal litter cover is naturally restored in two to five years [53]. In the more arid grasslands, the destruction of the protective soil mulch may be associated with reduced forage production [80].

Although somewhat variable, the humus within grassland soil is not greatly affected by burning [43]. Annual controlled burning in longleaf pine in the Southeast has increased the organic matter in the soil [76]. Burning in the sagebrush-grass type has reduced the organic matter in the top half inch of the soil but only temporarily [18]. On chaparral sites in California, the reduction of soil humus was minimal except under intensive burns, and then only in the surface inch of soil [44]. It was concluded that forest fires generally destroy soil humus only in the immediate soil surface [109, 58].

Soil chemistry. Burning of litter, debris, and live plant material in forests and dense brushlands releases the minerals tied up therein and increases the supply of plant nutrients available in the surface soil. Soluble salts from the ashes left on these sites are leached into the surface soil causing an increase in pH, total P and N, exchangeable bases such as Ca and K, and total soluble salts, and a decrease in the carbon to nitrogen ratio [44, 58, 59, 109, 158]. Combustion volatizes part of the nitrogen and sulphur in the duff, but these losses are offset by increased availability of the remaining N and S [43]. The remaining N is readily available as ammonia nitrogen or through microbial mineralization as nitrate nitrogen [44].

Where the ashes left from burning are high, the fertility effects are generally sufficient to materially improve reseeding and establishment of forage plants [115]. This has led to the suggestion that under these situations it might be better to apply fertilizer, where recommended, the second year rather than the first season after burning [158]. In California chaparral, the higher nitrate content of the upper soil layer on fresh burns was largely lost in the second and third years but probably accounted for the characteristic robust growth of invading plants established the first year [140]. The benefits of increased fertility have been mostly temporary in forest soils [58].

Slash burning in Douglas fir in the Pacific Northwest changed the duff reaction from pH 4.95 to 7.6 while releasing about 435 pounds of nitrogen per acre to the atmosphere [89]. Soil acidity in Alaskan coniferous forests was raised by burning up to levels of pH 6.5 to 7.5 [109]. Burning increased pH in Arizona brushland soils by 0.5 from slightly acidic to slightly basic [115] but increased pH in California chaparral soils only under heavy ash [140]. On forest soils, burning destroys the organic acids and liberates bases which further neutralize acids in the soil. This improves soil conditions for biological processes, including bacterial activity, and accelerates nitrification [58, 59, 109].

Annual burning in longleaf pine stands in the Southeast was found to increase pH by 0.15 to 0.48 units, double the replaceable calcium, and slightly increase total nitrogen in the surface soil [76]. The release of soil nutrients in grasslands is much less than in forest burning [43]. The pH is raised slightly because of Ca, Mg, and K release, but this is considered of little significance. Nitrogen content in grassland soils is measurably increased when burning increases the proportion of legumes on the site. Available P was increased by burning in Iowa prairie [53]. In sagebrush-grass burns, nitrogen was reduced in the surface soil, but this was considered only temporary [18].

Soil moisture. The effect of burning on soil moisture is variable but commonly results temporarily in a decrease as a result of higher soil temperatures, increased evaporation, reduced snow catch, and possibly reduced soil wettability and water infiltration following rainfall [43, 44]. Where burning materially thins the vegetation and mulch, reduced transpiration and interception may partially offset these losses in soil moisture. Net losses of soil moisture from burning may be unimportant when followed by favorable rainfall, but may accentuate drought when followed by low rainfall.

Burning in coniferous forest on steep slopes, particularly southern exposures, was found to increase surface runoff and reduce soil moisture, but the effect was small on level sites [109]. Annual burning of tallgrass prairie in the Kansas Flint Hills reduced soil moisture at all depths whether burning was done in winter, early spring, or late spring [3]; but the reduction was least under late spring burning (about May 1), generally reducing the inches of available water in the top five feet of soil by only 15 to 25 percent [74]. However, in later Kansas studies, burned pasture had more inches of soil moisture throughout the growing season on ordinary upland than in unburned pasture, with late burning maintaining the highest levels [4]. The lower moisture levels in the unburned pasture was attributed to higher water use by plants.

Wink and Wright [165] have emphasized that soil moisture should be given primary consideration before prescribed burning. Following spring burns at the time of good soil moisture, forage plants recover rapidly and begin growth and soil erosion is minimal. Without good soil moisture at the time of the burn, which is usually the case when wildfires are common, burning increases the drought stress of plants, and herbaceous yields are reduced. Also, the soil may remain bare for considerable time and be subjected to severe wind and water erosion.

Soil temperature. During burning the temperature of the surface soil increases sharply for a brief period [1]. However, below two inches in the soil even on forest sites, temperatures do not rise greatly during fire. Maximum soil temperatures in California chaparral burns were up to 700°C at the surface but only up to 190°C at a one-inch depth; and heat transfer was found least into moist-wet, heavy soil with good organic matter [44]. Following burning, the temperature of blackened and unshaded soil, both forest and grassland, are appreciably warmer than unburned areas during the daytime [43, 1]. This presumably is the major cause of the appearance of grass shoots on fresh burns one to three weeks earlier in the growing season than on unburned areas [43]. Although the earlier production of forage is an advantage, postburn microclimate may favor frost damage to seedlings by hastening spring germination.

Soil erosion. Erosion following burning apparently is not accelerated in the Southeast pine-grass or in tallgrass areas, excepting possibly on very steep terrain, but has been reported from drier grassland habitats following wildfire [43]. On shortgrass range wildfire has been found to reduce percent basal cover and increase soil losses through erosion [80]. Burning in the sagebrush-grass areas of the Intermountain Region is recommended only where the soils are fairly firm and the slopes less than 30 percent [130]. Under these conditions soil losses from planned burning have been slight, and soil movement has been arrested almost completely by the end of the first spring. However, on sandy loam soils in Idaho, burning was noted to increase wind erosion [18]. This has also been a problem on sandhills range in Nebraska.

On forest and dense brushland sites, intense burns may increase erosion by removing organic matter down to the mineral soil, decreasing soil aggregates and porosity and increasing bulk density, and reducing wettability, penetration, and water retention in the soil [1, 59, 127, 39]. However, these effects of soil exposure and surface modification generally do not last beyond one or two years. The severity is accentuated on steep slopes. Prescribed burning following

220

dozing on central Texas brushlands did not affect erosion losses, runoff, and water quality on level areas, but had some adverse effects lasting 9 to 15 months on moderate slopes and 15 to 30 months on steep slopes [173]. Burning as late in the dormant season as possible provides less potential for erosion, since it minimizes the period the soil is without vegetation cover and generally destroys less organic matter.

Hot fires, including wildfire, frequently increase water yields in Arizona chaparral, but the charred stems and thin layer of ashes remaining were found incapable of controlling erosion on highly erodible granitic soil [29, 62]. In southern California chaparral, wildfire increased dry creep erosion markedly on steep terrain and almost immediately after the fire [96]. The dry season debris movement was observed generally to considerably exceed the wet season movement. However, it has been concluded that light, flashy burns in grassland or mixed grass-shrub areas seldom modify surface soil characteristics or seriously inhibit infiltration as do intense burns in forest, woodland, or chaparral [127].

Burning Effects on Vegetation

Sagebrush-grass range. Prescribed burning gives a complete kill of young and old plants of big, low, and black sagebrush (*Artemisia nova*) (Figure 61) [129]. However, since black sagebrush is a valuable forage plant on big game and sheep winter range, it should be protected from fire. This is also true of big sagebrush on critical deer winter range. Kill of three-tip sagebrush (*Artemisia tripartita*) is high, but a small proportion of the plants basally sprout. Silver sagebrush kill is low because it readily sprouts.

Sagebrush burning is recommended only when the following can be met [130]:

1. Where big and three-tip sagebrush is dense and forms more than half the plant cover.
2. Where fire resistant perennial grasses and forbs form more than 20 percent of the plant cover or will be seeded following burning.
3. Where the principal use of the area is livestock grazing; but only under certain conditions on watersheds, big game winter range, or sage grouse habitat.
4. In the late summer or early fall.
5. Not earlier than ten days after perennial grass seed is ripe and scattered, and after leaves are nearly dry.

221

FIG. 61. *Prescribed burning of big sagebrush in central Utah. Nearly complete kill in foreground where fuel was adequate to carry fire but with reduced kill on ridgetops.* (U.S. Forest Service photo by Neil C. Frischknecht)

In one burning experiment on sagebrush range in southeastern Idaho, grazing capacity was increased 69 percent, associated with a doubling of perennial grass production; 25 percent increase in perennial forbs; and 64 to 93 percent increase in forage availability [130]. Much of this increase had developed by the second growing season. In related studies, fire and rotobeating each increased production of desirable and moderately desirable grasses by two-thirds when evaluated three years later, while spraying and railing each increased production by only one-third [121]. Burning increased availability of grasses and forbs by 75 to 98 percent in dense sagebrush stands. In still other studies grass production was reduced the first year, was maximized at three years, but declined somewhat subsequently [18].

From the several studies in the sagebrush zone in the northern Intermountain Region, range plant species are classified as to damage from late summer burning as follows [18, 121, 129, 130, 169, 175]:

Severely damaged	Slightly damaged to favored	Greatly favored
Big sagebrush	Bluebunch wheatgrass	Arrowleaf balsamroot
Bitterbrush	Indian paintbrush	Cheatgrass
Broom snakeweed		Crested wheatgrass

Cliffrose
Curlleaf mountain-
  mahogany
Eriogonum
Idaho fescue
Pussytoes
Threadleaf sedge
Three-tip sagebrush

Indian ricegrass
  Needle-and-thread
Nevada bluegrass
Penstemon
Prairie Junegrass
Squirreltail
Tailcup lupine
Tapertip hawksbeard
Thurber needlegrass
Timber milkvetch

Douglas sedge
Foothill deathcamas
Green ephedra
Horsebrush
Lambstongue grounsel
Medusahead
Mulesear
Purple pinegrass
Rabbitbrush
Sandberg bluegrass
Serviceberry
Snowberry
Snowbrush ceanothus
Sticky currant
Subalpine needlegrass
Thickspike wheatgrass
True mountain-
  mahogany
Velvet lupine
Western wheatgrass
Yarrow

Damage by fire to desirable forage plants in the sagebrush zone is fairly low, excepting Idaho fescue, curlleaf mountainmahogany, cliffrose, and bitterbrush. Idaho fescue, in contrast to bluebunch wheatgrass, is very leafy and compacted at the base where dead material accumulates as fuel [158]. In one study in eastern Oregon in which the results of an accidental burn in sagebrush-grass was evaluated eleven months later, Idaho fescue had suffered a 27 percent death loss and a 50 percent total reduction compared to 1 and 29 percent for bluebunch wheatgrass [36]. On California perennial range, Idaho fescue had been reduced over 75 percent when evaluated five years after severe burning [41]. Burning nearly eliminates mature, live plants of cliffrose and curlleaf mountainmahogany [132].

Bitterbrush has been severely damaged at some locations, but it often stem sprouts with light burning and on moist soils [129]. Bitterbrush sprouts originate from dormant buds wholly or partially encircling the stem at ground level or from a callus or meristematic tissue encircling the stem formed beneath the bark after treatment [19]. In Idaho studies heavy burns prevented 50 to 75 percent of the plants from sprouting while light burns prevented 0 to 20 percent [17]. It was concluded in Idaho that controlled burning followed by good management should not permanently decrease bitterbrush. However, severe burns in California and the western Great Basin have been reported to be more injurious to bitterbrush [41, 19]. Layering forms of bitterbrush survive fire much better than nonlayering forms [132].

Burning for seedbed preparation in central Utah was found to be the most effective means of controlling big sagebrush prior to seeding [37, 38]. Fall burning reduced big sagebrush by 95 percent and held the sagebrush cover to 9.6 percent of the original stand by seven years after burning and drilling to wheatgrasses. However, in order to be this effective the brush had to be burnt to the ground and a uniform burn obtained over the area. Fall burning gave higher initial kill and lower reinvasion than spring burning. Burning has given very erratic kill of rabbitbrush, but burning on August 1 and spraying with 2,4-D the following spring has given a 94 percent kill of big rabbitbrush [138]. Broom snakeweed, although killed by fire, readily reestablishes by seed on disturbed range and may be increased by burning [34].

Burning is sometimes used to control fire-susceptible shrubs invading into crested wheatgrass. In North Dakota, April burning reduced yield of crested wheatgrass over the next two years [106]. Fall burning reduced yield the first season but increased yield the second season without changing crested wheatgrass density. This indicates that fire must be used with caution in removing brush from seeded grass stands.

Fires that occur after the seeds have matured and dropped kill some seeds but are ineffective in killing annual grasses because the remaining seed is released [86, 175]. Cheatgrass brome matures in early June and quickly dries out and becomes very flammable. On the other hand, perennial grasses mature later and dry out slowly. Burning may be effective in killing cheatgrass and preventing natural reseeding the following year only when made after the seeds are ripening but before they fall [133, 94].

Repeated burning, or burning in late summer, strongly favors cheatgrass and may be a major cause of its spread and establishment. If cheatgrass rapidly stabilizes at a new high level following burning, the site will be mostly closed to seedlings of perennial grasses [175]. Where cheatgrass brome is more than half of the understory of the area or cannot be protected against subsequent accidental burns, big sagebrush sites should not be burned [130]. Summer burning was found to favor cheatgrass even over sand dropseed (*Sporobolus cryptandrus*) and perennial threeawn (*Aristida* spp.) [34].

Medusahead has a seedhead moisture content above 30 percent for approximately a month after the leaves and stems begin to dry and provide a sufficient volume of fuel to carry a fire. While the seed moisture content is still high (soft dough stage), medusahead seed is readily killed by high temperatures. Burns at this stage of growth are recommended in the afternoon when burning slowly into a mild wind [118]. However, burning after medusahead seeds shatter and fall to the ground is very ineffective.

In California annual grasslands, filaree seed shatter and fall early and soft chess matures at least twenty days earlier than medusahead [122]. Burning when soft chess seed is shattering and before medusahead begins to shatter favors soft chess and filaree over medusahead. Medusahead seed is not viable until awns begin to curl. However, this delayed burning may injure early maturing cool-season grasses present, if any. In other studies on California annual grassland, July burning did not affect total herbage production but did reduce annual grass production by 25 percent while increasing bur clover (*Medicago hispida*) and filaree [75].

Juniper range. Burning has been an effective, economic control of nonsprouting juniper where a uniform burn has been obtained (Figure 62). Broadcast burning of live juniper trees requires a dense stand of trees (400 or more per acre) or a flammable understory to be effective [6]. Burns in pinyon-juniper have often burned clean on flat or gently rolling terrain but left islands of unburned trees on hills and ridges where the junipers grow in a thin stand with few understory plants. Large juniper trees often provide their own firebreak zone through intensive competition, and grazing further removes the flammable understory. Leaving juniper on the steep, shallow soil ridges may be desirable for watershed protection as well as big game cover.

Grasslands being invaded by junipers should be broadcast burned to control the junipers before the herbaceous understory is reduced to the degree it will not carry a uniform burn. Nonsprouting juniper seedlings and saplings are readily killed under these conditions, and even sprouting junipers may be materially suppressed. Following intense wildfires in the pinyon-juniper type in central Utah, big sagebrush often became dominant by 11 years later, with juniper beginning to appear shortly thereafter [8]. The juniper often became prominent after about 46 years and dominant at about 70 years, the latter two making broadcast burning difficult except under hazardous high-wind, high-temperature, dry conditions.

Burning a black grama-galleta range in Arizona invaded by juniper reduced the grasses the first year, but they recovered by the end of the second growing season [90]. Galleta being rhizomatous was found resistant to fire; black grama being stoloniferous was considerably damaged by spring burns. Fire in this study killed 70 to 100

225

FIG. 62. *Burning for range improvement in pinyon-juniper type in Utah. A, broadcast burning very effective on dense stands where fire will carry from tree to tree; B, individual tree ignition as follow-up treatment (also useful on open stands as initial treatment); C, manyfold increase in grass production following burning; D, burning out windrows left in chaining-windrowing-drilling treatment.* (Photo B courtesy of U.S. Forest Service)

percent of the juniper trees under four feet high but only 30 to 40 percent of trees five and six feet high. However, when the larger trees had tumbleweeds and other debris under them, the kill increased to 60 to 90 percent.

In Arizona oak-juniper woodland, fire killed 76 percent of oneseed juniper (*Juniperus monosperma*) while 10 percent of the living trees sprouted from the base [92]. However, alligator juniper had a low mortality from the fire, and 42 percent of the living trees sprouted. The fire also considerably suppressed beargrass (*Nolina texana*), an undesirable range plant, by killing 10 percent of the plants and damaging all except 13 percent of the plants.

Burning is also used following chaining, bulldozing, or juniper piling to get rid of unwanted slash and to kill additional juniper plants. Slash burning in December in Arizona removed the least slash and killed the fewest trees, while August burning removed the most slash and killed the most trees with temporary damage on the blue grama [6]. April burning was intermediate in killing junipers left from mechanical clearing.

Burning in pinyon-juniper in northern Arizona greatly benefited cattle forage and was considered acceptable for deer [116]. Fire greatly increased the grasses and forbs but extensively killed the cliffrose. However, deer grazing following burning was greater on the burned area up to one-quarter mile from the edge of the burn than in the unburned areas, and deer diets went heavily to forbs and grasses in all seasons of the year.

In Texas, spring burning damaged the herbaceous vegetation under Ashe juniper less than summer burning [161]. In the Texas studies, juniper cutting decreased total forage density by 9 percent by four years after treatment; juniper cutting and burning increased it by 345 percent; and cutting, piling, and burning increased the total forage density by 645 percent. Juniper reinvasion by the fourth year after treatment was only 21 percent as great on cut and burned areas as on those cut only. In later studies in the Ashe juniper community, prescribed burning on sites previously bulldozed readily killed small junipers, increased grass production except rhizomatous sideoats grama, and removed the dead brush without accelerating erosion [167].

March burning of shallow glade range in Missouri under good moisture conditions killed many small trees of the invading eastern juniper (*Juniperus virginiana*) [114]. However, when followed by drought and unrestricted grazing, it was destructive on the thin forage and soil cover. When followed by severe drought, burning alone was more severe than overgrazing alone on the forage cover.

Juniper in open stands can be individually ignited and burned by the use of propane or oil burners with fuel tanks mounted on pickups, trailers, or wheel tractors. Trees are ignited from the base up as a means of getting the canopy enveloped in flame. It was concluded in northern Arizona that 100 percent kill of nonsprouting junipers could be obtained by scorching 60 percent of the crown in individual tree burning [91]. This method was not appropriate for trees over ten feet tall, because of high cost, or on sprouting species such as alligator juniper and redberry juniper. Hot, dry, still days were found best for individual tree burning. A disadvantage of broadcast burning in the juniper zone is that valuable associated shrubs on deer and sheep winter range may be killed [132].

In Arizona in 1963, costs per acre for individual tree burning for small trees were $1.97, $3.61, and $5.81 for 50, 100, and 150 trees per acre. For large trees, respective costs per acre were $3.17, $6.00, and $8.76 [40].

Coniferous forest. The benefits of ground fires in ponderosa pine have been listed as (1) increased water production, (2) increased forage production for livestock and game, (3) providing room for camping and space for hunting, (4) preventing total destruction by wildfire, and (5) improving the forest aesthetically [5, 163, 171]. It has been concluded that the moderately open savannah stand of ponderosa pine, essentially free of understory trees and shrubs and supporting a good stand of grasses and forbs, has been maintained by natural fire [86]. These recurrent natural fires largely prevented dense thickets of pine reproduction and understory brush and kept fires relatively cool by preventing large accumulations of fuel.

Protection, regeneration, and continued development of ponderosa pine is dependent to a considerable extent on the use of fire under proper control [162]. Biswell [13] has reported that fire exclusion, timber cutting, and other land management practices in ponderosa pine have allowed dangerous fuel situations to build up and wildfires to result. He recommended prevention, suppression, and fuel reduction as the three essentials in adequate fire protection. Biswell et al. [16] concluded that controlled burning every 6 to 7 years maintained reduced fuels while meeting the ecological requirements of ponderosa pine. Burning in northern Arizona has been recommended in the fall when the air temperature is low, humidity is higher than usual, and preferably following a rain when the soil is damp [59]. September and October burns here have killed most saplings under four feet high, with damage to an occasional tree above sapling size. Grass density was reduced about 50 percent the first year with some reduction even into the second year [60].

Burning ponderosa pine types in the Pacific Northwest also reduced the heavy needle and debris mats that inhibit grass and other forage plants, thereby increasing grasses such as Kentucky bluegrass (*Poa pratensis*), Idaho fescue, needlegrasses (*Stipa* spp.), bluebunch wheatgrass, and pinegrass (*Calamagrostis rubescens*) [163]. Bitterbrush was reduced but returned by seedlings when grazing pressures were reduced. The following shrubs were topkilled back to the ground but then produced palatable sprouts for big game the following year: chokecherry, willow, serviceberry, and snowberry.

Burning is used in the Douglas fir and spruce-fir regions to clean up slash and debris after logging and pulpwood harvests while reducing wildfire hazards and preparing timber sites for regeneration. In south central Idaho, fire materially rejuvenated big game habitat through the greater availability, release, and stimulation of maple, willow, snowbrush ceanothus, serviceberry, snowberry, and sticky currant (*Ribes viscosissimum*) [110]. However, within ten years after burning Douglas fir begins to overshade the shrubs; by twenty years most of the wildlife potential is gone; and by forty years the vegetation is similar to prefire composition. From this it is concluded that wildlife habitat values can be expected to reach a maximum during the first fifteen years following fire and then begin a slow decline unless a second burn takes place on Douglas fir and spruce-fir sites.

Burning in northern Idaho coniferous forests increased elk winter range browse production mainly by (1) reduction in crown height of existing shrubs, (2) addition of new browse plants by seed germination, and (3) increased palatability of young growth [101]. Increased browse production and availability primarily came from Rocky Mountain maple (*Acer glabrum*), willow (*Salix scouleriana*), and serviceberry. But normal grazing pressures by elk did not prevent many plants from growing out of reach of the elk within two growing seasons. Spring burning was considered advantageous over fall burning because of less expense in manpower and fire lines required in the spring, more sprouting from spring treatment, and current winter browse being removed by fall burning.

Both ponderosa pine and aspen are considered fire climax trees on many Douglas fir and spruce-fir areas [96, 163]. Although quaking aspen is readily topkilled by hot fires, abundant root sprouts favor aspen over spruce and fir. Fire in a spruce-fir forest in New Mexico increased aspen sprouts from 100 to 14,000 per acre and was considered a potential tool in producing browse for deer and elk [128]. However, after about six years a majority of the fire-stimulated sprouts had grown out of reach of the big game. In Canada a moderate burn that killed the tree canopy and undergrowth and eliminated the litter and part of the duff was more effective in stimulating aspen sprouting than less intensive burns [81].

Arizona chaparral (brushlands). The use of fire to improve range conditions in Arizona chaparral shows some promise, plus considerable difficulties (Figure 63) [157]. Small burns about five acres in size have received about twice as much deer use as adjacent unburned areas. Other studies indicate that fires in mixed oak-mountainmahogany types considerably increased the water yield but also the sediment yield [62]. The rough terrain and erosive soils on much of the Arizona oak chaparral region make soil stabilization a major concern. Seeding to grasses such as Lehmann lovegrass or weeping lovegrass (*Eragrostis curvula*) immediately following fire has been successful in providing cover and increasing herbaceous forage production [154]. Burning-grass seeding-herbicide treatment has further delayed the return of brush.

However, the partial conversion of oak chaparral to herbaceous plants is only temporary, and the crown cover generally approaches that of adjacent unburned areas in four to seven years [29, 135]. Grazing was found to have no significant effect on shrub recovery [131]. Burning to remove shrub live oak seedlings from areas pre-

FIG. 63. *Prescribed burning of Arizona chaparral has often given only partial and temporary conversion to herbaceous plants but shows promise when used in conjunction with other treatments.* (University of Arizona photo by R. R. Humphrey)

viously burned and seeded to grass was found ineffective. Since October burning gave less sprouting than earlier treatment, fall burning followed by chemical spraying in the spring was suggested [104]. However, the delayed sprouting of many shrubs, particularly in dry years, complicated the use of herbicides as a follow-up to burning or cutting [131].

Burning reduced shrub live oak stems in number only after annual burning for five successive years—a practice considered impractical if not impossible [134]. Skunkbush sumac (*Rhus trilobata*) was similar to shrub live oak in requiring repeated burnings to kill and in displaying delayed sprouting. Wright silktassel and hollyleaf buckthorn (*Rhamnus crocea*), two desirable browse plants, were completely killed by two annual burns or two burns spaced two years apart. Both desert ceanothus (*Ceanothus greggii*), a desirable browse plant for deer, and pointleaf manzanita, an undesirable plant, were killed by a single burn. However, desert ceanothus has responded favorably to prescribed burns in Utah [132].

California chaparral. Controlled burning is reported to be the most widely used tool for range improvement in the California chaparral, being used to increase forage supplies for livestock and big game, improve watersheds, and reduce the hazards of wildfire [107]. Chamise brushlands are generally considered valuable primarily for game habitat and watersheds or, when partially or completely converted to herbaceous vegetation, as livestock range. Frequently, little management has been applied to chaparral except to attempt total fire exclusion, which has been only partly successful and often results in large wildfires [14]. The benefits of burning alone have generally been rather temporary in nature including (1) a slightly prolonged period of more nutritious vegetation, (2) temporarily greater variety and abundance of accessible forage, and (3) greater accessibility of the area by animals and man [140].

Complete conversion of chaparral to herbaceous range has generally required burning, grass seeding, and herbicidal control of brush sprouts and seedlings [107]. Prescribed burning has been recommended over bulldozing in preparation for seeding where it is considered safe and effective because it (1) is cheaper, (2) does not disturb the soil, (3) leaves a clean seedbed with few viable seed of competitive annual grasses, and (4) raises the availability of soil nutrients for better establishment and growth of new grass cover [11]. Both fall and spring burning have been used successfully. When burning only is used to improve chaparral range, late fall burning results in less sprouting than spring burning [25]. However, other research has suggested that spring burning gives more latitude, is less

231

dangerous, and fits better into a complete conversion program in which seeding is to follow [28]. Spring burning of chaparral mashed the preceding spring or summer permits spring burning when herbaceous vegetation is green and relatively fire resistant. In high risk and high liability areas, mashing followed by winter or spring burning is recommended.

Only chaparral areas in California receiving seventeen inches or more average annual precipitation are believed capable of being converted to and managed as perennial grass range [25]. Repeated burning in initial clearing of brush is not recommended except in maintenance of annual grass cover, and conditions necessary for a successful reburn are not easily obtained [11]. However, burning after four or five years can be used successfully to control brush sprouts and seedlings but may damage seeded grasses. Such a delayed repeat burn did not damage harding-grass (*Phalaris tuberosa* var. *stenoptera*) but killed nearly all of the smilograss (*Oryzopsis mileacea*) growing in heavy fuel. Rose clover (*Trifolium hirtum*) declined slightly from burning, as some seed was killed by the fire, while a mixture of wheatgrasses including tall, crested, pubescent, and intermediate wheatgrass was not damaged by a hot fire at the end of the third growing season [93]. In fact, by the end of the fourth growing season, the mixture had improved as a result of the rhizomatous nature of intermediate and pubescent wheatgrass.

Spot burning areas of about five acres in size and scattered evenly through the chaparral has been recommended for initiating black-tailed deer habitat improvement in chamise chaparral [14]. Larger size openings were suggested where the deer numbers were sufficient to suppress the sprouts in the clearing and prevent them from growing beyond reach of the deer. However, leaving at least 30 percent of the area in dense brush as cover for game and seeding some of the openings to herbaceous plants for spring variety was suggested. Black-tailed deer have been studied under three conditions of chamise brushland management. The results included [14]:

1. *Heavy brush*—deer numbers stable at 10 to 30 per section; fawn crop, 60 to 85 percent; low herbaceous plant use in the spring.

2. *Wildfire burn*—deer numbers from 5 to 160 per section, large number moved in when sprouts were young and tender but moved out during cold weather; fawn crop, 100 to 110 percent; medium herbaceous plant use in the spring.

3. *Opened brush consisting of interspersion of grass with patches of dense brush*—deer numbers relatively stable at 40 to 100 per section; fawn crop, 115 to 140 percent; high herbaceous plant use in the spring.

Southwest desert and semidesert grasslands. Fire has never been an important factor in the Chihuahuan and Sonoran Desert, except probably locally in the sacaton flats and tobosa swales [87]. However, in the desert grasslands it is considered important in maintaining a brush-free, open grassland. Although fuel to kill large brush is often inadequate today in many former desert grasslands, repeated burning at five- to ten-year intervals still offers promise as an effective and low-cost control method in some areas. The effectiveness of fire in arid and desert regions is dependent upon sufficient grass understory and proper weather conditions to carry a fire [86]. Schmutz [145] has proposed a special four-pasture, four-year grazing system for year-long Southwestern ranges that provides one year of rest to permit accumulating enough fuel to carry a fire followed by two years of deferred grazing.

Burning is effective in killing burroweed if there is 500 pounds of fuel or more per acre and burning is done in the dry season (June) just prior to summer rains [31, 85, 113, 137]. However, where there is no competitive grass understory or the grass is damaged by the fire, burroweed may rapidly reinvade since it is an opportunist and a prolific seed producer. Because the benefits of burning burroweed are only temporary unless a desirable, competitive cover can be provided, excluding livestock from the area for a year preceding the burn to build up ground fuel may not be economical. Also, small burned areas attract both rodents and big game as well as livestock, and heavy overgrazing often results under unrestricted grazing.

Wildfire in an upper desert grassland type in Arizona was found to be very damaging to larchleaf goldenrod (*Haplopappus larcifolius*), moderately damaging to mesquite, creosotebush, ocotillo (*Fouquieria splendens*), and Wheeler sotol (*Dasylirion wheeleri*), and to have little effect on false-mesquite (*Calliandra eriophylla*) and velvetpod mimosa (*Mimosa dysocarpa*) [32, 164]. Burning has reduced cholla by up to 50 percent and pricklypear up to 25 percent and killed broom snakeweed by up to 95 percent [136, 88]. In New Mexico a June burn killed 96 percent of broom snakeweed, but only 35, 25, and 45 percent of the broom snakeweed plants were killed respectively by October, January, and April burns [50]. However, repeated burning or low range condition may encourage broom snakeweed to rapidly reseed back into the stand. Many seeds in the soil are not killed by burning, and new seedlings readily establish from seed after fire where perennial grass competition is low.

Although fire may completely kill nonsprouting shrubs in the semidesert grasslands, individual plants of sprouting species such as mesquite readily resprout from the stem base and roots. Although periodical fires apparently greatly limited the spread of mesquite originally by killing or suppressing the plants while still small, the practical use of burning to control mesquite today is probably limited to areas having adequate fuel, where forage loss is not detrimental, and where the mesquite stand is mostly of small trees and seedlings invading grasslands [63].

Young mesquite trees are more readily killed by fire than older trees. In an Arizona study fire killed 60 percent of the mesquite saplings less than one-half inch in diameter but only 20 percent and 11 percent of the trees one to two inches and over five inches in diameter respectively [63]. In Texas studies [172] honey mesquite trees up to 1.5 years of age were easily killed by fire, severely harmed when 2.5 years of age, but very tolerant of fire after 3.5 years of age. Many of the larger mesquite trees are readily topkilled by fire but later sprout.

Kill of mature trees has been found accelerated by previous topkill by herbicides, drought, rodent and insect damage, and competition from grass [172]. When sprayed with 2,4-D four years prior to burning, larger mesquite trees in Texas were found easier to kill because of greater dead fuel and an extended period of burning [23]. From a Texas study it was concluded that by using fire on upland sites at 5- to 10-year intervals, 50 percent of the older trees can be killed [172]. Burning of mesquite range in Texas has been more effective in March when hot fires can be obtained and when grasses are more tolerant of fire [174]. In Arizona the most effective time for burning mesquite and other shrubs has been in June, also the hottest and driest time of the year [32, 113].

Although tobosa (*Hilaria mutica*) and sacaton (*Sporobolus wrightii*) are not generally damaged by fire, other grasses such as black grama (*Bouteloua eriopoda*), curlymesquite (*Hilaria belangeri*), and Lehmann lovegrass may be extensively damaged by a hot fire [166]. Spring burning (late March) of tobosa flats in a wet year in west Texas increased yield of tobosa by about two-thirds. Although the resulting ash had a slight fertilizing effect, litter removal stimulated the production of tobosa more than any other factor. Mechanical removal of litter was nearly as effective but more costly than fire. Sharrow and Wright [146] concluded that tobosa should not be burned more frequently than every five to eight years in order to prevent depletion of soil nitrogen reserves.

Although winter burning of weeping lovegrass in Texas increased production by only 14 percent, it increased utilization by cattle by 53 percent and crude protein levels from 3.5 to 7.6 percent during the summer [95]. Wildfire in one study in Arizona killed 90 percent of the black grama and 98 percent of the Lehmann lovegrass, although the latter actually greatly increased in plant density because of the many new seedlings following the burn [30].

Wildfire during dry periods, or even when drought follows prescribed burning, often deleteriously affects the desirable bunchgrasses in the semidesert grasslands. Although these effects often last only one or two seasons, long-range benefits in perennial grass production from woody plant suppression is not always realized because of failure to regulate grazing [31]. Another factor is that reduced shade and the resulting higher surface soil temperatures and evaporation rates may cancel out much of the expected improvement in soil moisture from brush removal [32]. Annual grasses may be greatly stimulated the year following fire if the fire did not destroy the previous year's seed crop. Burning brushlands in the Rio Grande Plains and in south central Texas has given some brush suppression but without apparent lasting brush kill and with variable effect on herbaceous forage species [20, 21, 79].

Midgrass and shortgrass range. A low intensity experimental burn in April in New Mexico shortgrass with a scattered overstory of juniper reduced herbage production by 30 percent the first year, but not the second growing season, and increased percent composition of blue grama [51]. In related studies, an April burn increased blue grama production by 56 percent, but burning at other seasons increased blue grama production an average of only about 10 percent. April burning increased blue grama crude protein content by 1.3 percent (16.1 versus 14.8 percent) when sampled in early June. Suppressing the scattered juniper overstory probably released some soil moisture for grass use.

In the Texas High Plains, burning in the fall, spring, or summer significantly reduced total forage production by about 25 percent the year after burning [155]. Fire appeared to increase the vigor of blue grama at the expense of sand dropseed and red threeawn. Burning shortgrass in North Dakota reduced production during at least the first four years after burning [45]. In the shortgrass area of southern Alberta, spring burning reduced forage production by 50 percent the first year, 15 percent the second year, with recovery completed by the third year [35]. Fall burning was less serious and reduced production by 30 percent the first year with recovery complete by the end of the second year.

Pasture burning by wildfire in western Kansas has been found detrimental on both shortgrass and midgrass range [80]. Burning on shortgrass habitat reduced the basal cover of buffalograss (*Buchloe dactyloides*) and blue grama by 48 and 67 percent, respectively. On

235

the midgrass habitat, fire reduced the basal cover of sideoats grama, big bluestem (*Andropogon gerardi*), and little bluestem by 16, 10, and 34 percent, respectively. First-year forage production on fall burned and spring burned (March) shortgrass areas were 1,918 and 635 pounds per acre, respectively, compared to 2,704 pounds per acre on the unburned. Shortgrass areas with heavy litter were severely damaged by burning, based on basal cover and forage production, compared to light damage on areas with light litter. Broom snakeweed was largely killed by burning while goldenrod received light kill.

In a later study near Hays, Kansas, March burning decreased first-year yields in a buffalograss-blue grama mixture by 65 percent and in a western wheatgrass-shortgrass type on two locations respectively by 82 and 48 percent [98]. By the third growing season, production differences were no longer significant. The reduced production following the fire was attributed to (1) partial killing of the forage grasses, (2) reduction of plant vigor of remaining forage plants and (3) reduced moisture penetration associated with the reduced ground cover.

The results of these studies suggest that burning semiarid shortgrass sites is undesirable in reducing both production and ground cover. However, it should be noted that much of the evidence is based largely on wildfires or burns made without relation to ideal weather conditions. Wright [168] has concluded that burning during dry years is apt to give negative results on shortgrass range and semidesert grasslands. However, his work in Texas suggests that if burning is done during good moisture periods, infrequent burning should not harm the grasses but may provide little benefit unless an excessive litter buildup has occurred.

Prescribed burning of mixed prairie in west Texas reduced rhizomatous sideoats grama but greatly increased the bunch form of the species [167, 170]. Fire did not greatly affect the vigor and production of buffalograss, blue grama, and sand dropseed, but increased vine-mesquite, Arizona cottontop (*Tricachne californica*), little bluestem, plains bristlegrass (*Setaria macrostachya* or *leucopila*), tobosa, and Texas cupgrass (*Eriochloa sericea*) for one to two years.

Tallgrass range. Fire is considered a natural factor in establishing and maintaining bluestem prairie in the Flint Hills of eastern Kansas [22], and these tallgrass ranges have a long history of applied burning. Since being settled over 100 years ago and recognized for its adaptation to spring and early summer grazing, this area has commonly been burned each spring during normal and above-normal rainfall periods as a means of providing early, palatable forage [149]. After initial years of research, Aldous [2] in 1934 recommended that

Flint Hills ranges be burned only in years having a large carry-over of grass. He reported that burning in early May decreased forage yield by only 14 percent compared to a 65 percent reduction in early December, that late spring burns killed the most brush and weeds, and that the least damage was caused when soils were wet.

Beginning in 1950, experimental pastures near Manhattan, Kansas, were burned annually at set calendar dates [4, 120]. Forage production during the 1958–1965 period are given in Table 13. May 1 was less detrimental to forage production than the other burning dates but was advantageous over the unburned in (1) increasing big bluestem and leadplant (*Amorpha canescens*), a desirable legume shrub, (2) controlling Kentucky bluegrass, Japanese brome (*Bromus japonicus*), and snowberry, and (3) more rapid beef gains.

Annual burning about May 1 increased the botanical composition of decreaser grasses from 68.5 to 87.1 percent and total perennial grasses from 88.7 to 96.4 percent [120]. Results of this study also suggested that attempts should be made to burn only when the soil surface was wet in order to minimize damage to the perennial grasses. It was concluded that the prairie was indestructible by annual fire except when in association with heavy grazing. All spring burning dates increased summer steer gains over the unburned controls, but late spring burning (May 1) was the most effective [4]. Steer gains were higher on all burning treatments than the control pastures during May, but during September steer gains on the unburned pasture exceeded gains on all burning treatments.

In related studies in the Kansas Flint Hills, burning in late spring (about May 1) has resulted in slightly reduced or nearly equal forage production to that on unburned range, caused earlier spring growth, increased crude protein content in forage by 0.5 to 2 percent and

Table 13. Effects of Different Burning Treatments on Flint Hills Range, Kansas*

	Lbs. forage/acre		Weed prod./acre
	1960	1958–1965	1958–1965
Check (no burning)	3,960	3,919	300
Late spring (May 1)	3,449	3,529	161
Mid spring (April 10)	3,536	3,238	289
Early spring (March 20)	2,770	2,612	335
Winter (December 1)	2,667	—	—

*Source: [4, 120]. Annual burning treatments initiated in 1950.

slightly increased phosphorus content and protein, crude fiber, and NFE digestibility [2, 3, 74, 126, 148, 149]. It is generally recommended that burning be used no more than necessary for good management and livestock husbandry. When deemed necessary, the burning should be delayed until late spring and preferably when the soil is moist. However, when burning is discontinued, the rate of woody plant invasion is often rapid on lowland, lower-slope, and steep rocky soil sites and moderate on heavy, upland soils [22].

Burning Iowa prairie in late winter initiated green growth two to three weeks earlier but caused earlier maturity also, gave no apparent reduction in total herbage production through the third year after burning, reduced the growth of Kentucky bluegrass in favor of the warm-season native grasses, and stimulated seedstalk production the first year [52, 53].

Annual burning in late March in central Missouri prairie increased grass yields to June 15 by 33 percent, to July 15 by 58 percent, and to September 15 by 111 percent [97]. This increase in yields was associated with earlier growth by seven to ten days due to warmer soil temperature in early spring, less shading, and greater availability of nutrients. Big bluestem and Indiangrass under annual burning constituted about 75 percent of the total plant mass. Burning increased flower stalk production of big bluestem, little bluestem, and Indiangrass by 270, 1,200, and 400 percent, respectively, which accounted for at least part of the additional herbage production.

Annual spring burning about April 1 for three successive years was used near Stillwater, Oklahoma, to improve poor condition, loamy prairie range [64]. Burning increased desirable forage species, decreased undesirable annuals, particularly prairie threeawn (*Aristida oligantha*), and stimulated earlier growth; but the final composition was not too different from the unburned sites. Spring burning had been included in the Oklahoma studies to control the expected invasion of cool-season species, particularly Japanese brome, because of nitrogen fertilizer application.

Although shinnery oak is very fire hardy and slightly increases under spring burning, the combined use of April burning and June spraying with herbicide was recommended on sandy soil in western Oklahoma [117]. This combination practice controlled shinnery oak while maintaining high quality and quantity of forage production on a sustained basis. Little bluestem slightly decreased while sand bluestem (*Andropogon hallii*) and switchgrass (*Panicum virgatum*) usually increased. Burning greatly increased the availability and palatability of range forage. It was suggested that on sandy soil burning be done when the soil and litter moisture is high in order to leave 25 to 50 percent of the fuel unburned.

Burning has also been used on tallgrass-hardwood savannahs to control the brush while maintaining or improving the forage [159]. Burning a brush-prairie savannah in northwestern Wisconsin protected from burning for twenty-five years increased green herbage production of grasses and forbs threefold over the control during the first and second growing seasons postburn. It was suggested that burning once every four to six years would be the most effective in maintaining a ground layer of herbaceous plants, but that burning once every ten years would maintain a brush-prairie savannah and keep it from reverting to forest.

Spring burning of partially wooded areas in the Missouri Ozarks was ineffective in controlling either the large overstory trees, principally oaks, or the small tree reproduction [54, 55]. A combination of burning and girdling was more effective than burning alone. However, 2,4-D increased herbage production 4.5 times over burned areas and 5.5 times over the control. Little bluestem increased 40 percent from burning but 1,600 percent from 2,4-D spraying. A combination of prescribed burning and proper follow-up management of livestock was concluded to show promise for hardwood control in the post oak-blackjack type in Texas [42]. Where there was an adequate supply of fuel, ground fires killed small-diameter oaks and some underbrush, while the bluestems and associated grasses were quite tolerant of fall or early spring burning.

Southeastern pine-bluestem range. Burning in the coastal plain, pine-hardwood forests favors a subclimax association where pine is the dominant species [147]. Where fuel is adequate, most hardwoods under four inches in diameter are killed by prescribed burns in east Texas. Pine reproduction there withstands prescribed burning once it reaches ten years of age or two inches in diameter. Winter burns every two or three years using backfires or strip headfires are recommended to reduce natural hardwood plant succession.

In the South Carolina coastal plain, prescribed burning retained the understory vegetation under loblolly pine (*Pinus taeda*) in a subclimax condition favoring both domestic and wild animals [103]. Burning every two or three years during the winter was recommended for providing good conditions for both cattle and whitetailed deer and wild turkey. However, annual winter burning produced the most legumes and other forbs for quail and the highest forage yields for cattle.

Hardwoods under 1.5 inches in diameter in immature loblolly pine stands in Texas were controlled more readily by burning than those of larger sizes [57]. Fire was found not sufficiently selective where the pines and the hardwoods were about the same size, although fires during the growing season were more effective than during the nongrowing season. Prescribed burning every five years permitted adequate slash pine (*Pinus elliottii*) reproduction in Louisiana and was required to prevent most scrub hardwoods and shrubs from reaching a size uncontrollable by fire and to prevent pine litter from eliminating herbaceous plants from the understory [69].

Burning in southern pine forests also helps keep browse plants low and available to deer [70]. In southeast Texas, burning improved browse quality for fall and winter use by deer [99]. Spring and summer burning increased protein content of browse as much as 43 percent and phosphorus content as much as 78 percent during rapid growth, but less so by winter. Since the increased nutrient levels were lost within a year or two, it was suggested that burning be practiced as often as good forestry permitted. But summer burns or too frequent winter burns may kill understory species [70]. In Texas studies, repeated burning reduced mass production for wildfire use by 70 percent. Thus, it may be desirable in limited areas to exclude burning sufficiently to allow large shrubs and midstory hardwoods to reach fruit-bearing size [100].

Burning has long been used for range improvement in the longleaf pine-bluestem region extending from Texas to western Florida. This area consists of an overstory of longleaf pine (*Pinus palustris*), a desirable lumber tree, and various undesirable hardwood trees and a grass understory including little bluestem and slender bluestem (*Andropogon tener*). As early as 1939 researchers reported that annual burning maintained more favorable composition, quality, and quantity of forage; increased cattle gains by 37 percent during the seven summer months; and retarded growth of longleaf pine trees only in the sapling stage [160].

Under moderate to heavy grazing on longleaf pine-bluestem range or on cutover lands maintained predominantly as open grassland, periodical burning is not required to maintain high production [46, 48]. However, under dense pine canopy, grasses become so mixed with pine needles that cattle avoid grazing there. Burning in such cases removes old grass and litter and increases utilization of grasses from less than 20 to over 80 percent. When burned in patches, cattle concentrate on the burned patches, particularly when grazed during the spring following burning; this requires that grazing capacity be calculated mainly on the burned patches [150].

A rotation burning program has been recommended for both cutover and timbered longleaf pine-bluestem ranges in central Louisiana [49]. It consists of burning each year in winter or early spring one-third of a range unit without interior fencing. The purpose of this program on yearlong range is to improve distribution of grazing and to increase summer and fall cattle gains sufficiently, by means of increased forage value, to withstand winter losses. Grass utilization under this program averaged 78, 31, and 18 percent in the first, second, and third seasons after burning. Close grazing of the new growth the first year kept the vegetation palatable and nutritious, while the lighter grazing the last two years of the cycle restored vigor in the grasses. The burning also topkilled hardwood brush and aided herbage growth by removing the pine litter. It was suggested that regenerated pinelands be burned in early March by low intensity backfires after waiting until the trees were beyond the damage stage. Burning cutover lands in early May by a free-running headfire best maintained the area in open grassland.

Winter burning in central Louisiana increased protein content in young leaves of bluestem grasses from 9 to 13 percent crude protein but only slightly by time of full leaf and mature leaf stages [33]. In related studies on longleaf pine-bluestem range in Louisiana, spring burning materially improved the protein content of forage in June without reducing yield [83]. Mid-July burning yielded herbage available for winter grazing of higher protein content than spring or winter burning. Burning different parts of the range at intervals from late winter until midsummer provided herbage of relatively high quality during most of the growing season. However, since midsummer burning reduces the quantity of herbage available for winter use, only a portion of the winter range should conceivably be burned in any one year [68].

Although burning on pine-bluestem range improves the quality of herbage and may increase yield, removal of the litter is the major cause of such benefits [67]. Burning destroys the excess litter, but close grazing largely prevents it from accumulating except under dense pine stands. Burned plots and closely mowed and raked plots did not differ significantly in quantity and nutrient content of herbage. Periodical burning generally has had little effect on long-range litter accumulation, vegetational cover, and botanical composition of grazed range [48]. Burning removes old growth and stimulates new growth high in crude protein and phosphorus similar to the effect of close grazing.

Southeastern pine-wiregrass range. This range type, also referred to as the pine flatwoods, occurs in Georgia and Florida and has an overstory of longleaf pine and slash pine (*Pinus elliotii*). Utilization of the understory wiregrasses, pineland threeawn (*Aristida stricta*) and Curtiss dropseed (*Sporobolus curtissii*), is almost entirely limited to fresh burns; otherwise it is tough and unpalatable [151]. Burning is recognized as the oldest and cheapest method of improving wiregrass range and is widely used (1) to improve quality and productivity of forage, (2) to aid distribution of cattle, (3) to stimulate native grasses to produce seed, and (4) to check growth of undesirable shrubs (Figure 64) [77].

Burning wiregrass range is recommended every two to three years, but fire should be excluded until the pines are eight to ten feet high [73, 77, 78]. Periodical burning thereafter helps to control brown spot disease in pines, increases the availability and usefulness of early growing wiregrasses in winter and spring before other species have made substantial growth, improves wildlife habitat, and reduces wildfire hazard. Under burning pineland threeawn maintains dominance but is grazed well the first two to three months following the burn.

Burns are made in wiregrass from October to May to extend the nutritious forage season [77]. May burns in Florida have produced 400 percent as much palatable herbage as October or November burns, with March burns being intermediate [102]. Using a backfire following a rain and into a cool wind blowing three to ten m.p.h. has been suggested for use after December 1 [72].

Prescribed burns on wiregrass range at intervals of three to four years were effective in keeping common gallberry and saw-palmetto in check, but they survived by sprouting from the rootstalks [72]. Burning on March 7 followed by two pounds a.e. of 2,4,5-T on August 7 gave the best combination of kill [26]. This combination killed 84 percent of the gallberry, reduced gallberry foliage by 75 to 85 percent, and increased grass production by 300 percent.

Burning in late winter increased the protein content of wiregrass from 6 to 10-12 percent and the phosphorus content from 0.08 to 0.12-0.14 percent in the first two months of growth [72]. However, late winter burning had no advantage in nutritive value of forage beyond early summer. Later winter burning for spring and summer grazing in Georgia increased weaning weights by an average of twenty-six pounds and cow weights in the summer by twenty-eight pounds [151]. On Florida wiregrass range, burning increased cow and calf gains by eighty-seven and thirty-seven pounds, respectively, through the spring and summer [77].

Southeastern cane type. The cane forage type in southeastern United States is characterized by switch cane (*Arundinaria tecta*) frequently associated with a forest overstory of pond pine (*Pinus serotina*). Switch cane provides good grazing and thrives in a fire-maintained community, but stagnates under total fire protection [83, 84]. Prescribed burning renovates decadent cane stands, restores vig-

FIG. 64. *Prescribed burning is an important range improvement practice on pine-wiregrass ranges in southeastern U.S. A, burning with a backfire to minimize damage to pines; B, grazing follows burning to obtain advantage of nutritious, palatable regrowth; C, cattle grazing coarse herbage, mostly pineland threeawn and broom-sedge bluestem, on range protected from fire until pine seedlings are ten feet tall.* (U.S. Forest Service photos provided by Clifford E. Lewis)

243

or and replenishes production, reduces wildfire hazard, and improves accessibility of forage. Where grazing is a major objective, winter burning at intervals of about ten years is recommended. Maximum forage production and increased nutrient content last for two to four years after burning, but good production can apparently be maintained for an additional six to eight years. Deferring grazing during the first part of the growing season after burning is suggested.

Literature Cited

1. Ahlgren, I. F., and C. E. Ahlgren. 1950. Ecological effects of forest fires. Bot. Rev. 26(4):483–533.
2. Aldous, A. E. 1934. Effect of burning on Kansas bluestem pastures. Kansas Agric. Expt. Sta. Tech. Bul. 38.
3. Anderson, Kling L. 1965. Time of burning as it affects soil moisture in an ordinary upland bluestem prairie in the Flint Hills. J. Range Mgt. 18(6):311–316.
4. Anderson, Kling L., Ed F. Smith, and Clenton Owensby. 1970. Burning bluestem range. J. Range Mgt. 23(2):81–92.
5. Arnold, Joseph F. 1963. Uses of fire in the management of Arizona watersheds. Proc. Tall Timbers Fire Ecology Conf. 1963:99–111.
6. Arnold, Joseph F., Donald A. Jameson, and Elbert H. Reid. 1964. The pinyon-juniper type of Arizona: effects of grazing, fire, and tree control. USDA, Agric. Res. Serv. Prod. Res. Rept. 84.
7. Aschmann, Homer. 1977. Aboriginal use of fire. USDA, Gen. Tech. Rep. WO-3, pp. 132–141.
8. Barney, Milo A., and Neil C. Frischknecht. 1974. Vegetation changes following fire in the pinyon-juniper type of west-central Utah. J. Range Mgt. 27(2):91–96.
9. Beaufait, William R. 1965. Characteristics of backfires and headfires in a pine needle fuel bed. USDA, For. Serv. Res. Note INT-39.
10. Bellusci, Albert V. 1973. Thermal brush control. USDA, For. Serv. Equipment Development Center, Missoula, Mont.
11. Bentley, Jay R. 1967. Conversion of chaparral areas to grassland. USDA Agric. Handbook 328.
12. Biswell, H. H. 1958. Prescribed burning in Georgia and California compared. J. Range Mgt. 11(6):293–297.
13. Biswell, H. H. 1967. Forest fire in perspective. Proc. Tall Timbers Fire Ecology Conference 1967:43–63.
14. Biswell, H. H., R. D. Taber, D. W. Hedrick, and A. M. Schultz. 1952. Management of chamise brushlands for game in the north coast region of California. California Fish and Game 38(4):453–484.
15. Biswell, Harold H. 1977. Prescribed fire as a management tool. USDA, Gen. Tech. Rep. WO-3, pp. 151–162.
16. Biswell, Harold H., Harry R. Kallander, Roy Komarek, Richard J. Vogl, and Harold Weaver. 1973. Ponderosa fire management. Tall Timbers Res. Sta. Misc. Pub. 2.
17. Blaisdell, James P. 1950. Effects of controlled burning on bitterbrush on the upper Snake River Plains. USDA, Intermountain Forest & Range Expt. Sta. Res. Paper 20.
18. Blaisdell, James P. 1953. Ecological effects of planned burning of sagebrush-grass range on the upper Snake River Plains. USDA Tech. Bul. 1075.
19. Blaisdell, James P., and Walter F. Mueggler. 1956. Sprouting of bitterbrush (Purshia tridentata) following burning or top removal. Ecology 37(2):365–370.
20. Box, Thadis W., Jeff Powell, and D. Lynn Drawe. 1967. Influence of fire on south Texas chaparral communities. Ecology 48(6):955–961.

21. Box, Thadis W., and Richard S. White. 1969. Fall and winter burning of south Texas brush ranges. J. Range Mgt. 22(6):373–376.
22. Bragg, Thomas B., and Lloyd C. Hulbert. 1976. Woody plant invasion of unburned Kansas bluestem prairie. J. Range Mgt. 29(1):19–24.
23. Britton, Carlton M., and Henry A. Wright. 1971. Correlation of weather and fuel variables to mesquite damage by fire. J. Range Mgt. 24(2):136–141.
24. Brown, Harry E. 1958. Gambel oak in west-central Colorado. Ecology 39(2):317–327.
25. Burma, George D. 1968. Converting California brushland to grass. ASRM, Abstract of Papers, 21st Annual Meeting, pp. 24–25.
26. Burton, Glen W., and Ralph H. Hughes. 1961. Effects of burning and 2,4,5-T on gallberry and saw-palmetto. J. For. 59(7):497–501.
27. Buttery, Robert F. 1958. How to crush chamise brush for prescribed burning. USDA, California Forest & Range Expt. Sta. Mimeo.
28. Buttery, R. F., J. R. Bentley, and T. R. Plumb, Jr. 1959. Season of burning affects follow-up chemical control of sprouting chamise. USDA, Pacific Southwest Forest & Range Expt. Sta. Res. Note 154.
29. Cable, Dwight R. 1957. Recovery of chaparral following burning and seeding in central Arizona. USDA, Rocky Mtn. Forest & Range Expt. Sta. Res. Note 28.
30. Cable, Dwight R. 1965. Damage to mesquite, Lehmann lovegrass, and black grama by a hot June fire. J. Range Mgt. 18(6):326–329.
31. Cable, Dwight R. 1967. Fire effects on semidesert grasses and shrubs. J. Range Mgt. 20(3):170–176.
32. Cable, Dwight R. 1973. Fire effects in southwestern semidesert grass-shrub communities. Proc. Annual Tall Timbers Fire Ecology Conf. 12:109–127.
33. Campbell, Robert S., E. A. Epps, Jr., C. C. Moreland, J. L. Farr, and Frances Bonner. 1954. Nutritive values of native plants on forest range in central Louisiana. Louisiana Agric. Expt. Sta. Bul. 488.
34. Christensen, Earl M. 1964. Changes in composition of a *Bromus tectorum-Sporobolus cryptandrus-Aristida longiseta* community following fire. Utah Academy of Sci. Proc. 41(1):53–57.
35. Clarke, S. E., E. W. Tisdale, and N. A. Skoglund. 1943. The effects of climate and grazing on shortgrass prairie vegetation. Canada Dept. Agric. Tech. Bul. 46.
36. Conrad, C. Eugene, and Charles E. Poulton. 1966. Effect of a wildfire on Idaho fescue and bluebunch wheatgrass. J. Range Mgt. 19(3):138–141.
37. Cook, C. W. 1958. Sagebrush eradication and broadcast seeding. Utah Agric. Expt. Sta. Bul. 404.
38. Cook, C. Wayne. 1966. Development and use of foothill ranges in Utah. Utah Agric. Expt. Sta. Bul. 461.
39. Cooper, C. F. 1961. Controlled burning and watershed condition in the White Mountains of Arizona J. For. 59(6):438–442.
40. Cotner, Melvin L. 1963. Controlling pinyon-juniper on southwestern rangelands. Arizona Agric. Expt. Sta. Report 210.
41. Countryman, Clive M., and Donald R. Cornelius. 1957. Some effects of fire on a perennial range type. J. Range Mgt. 10(1):39–41.
42. Darrow, Robert A., and Wayne G. McCully. 1959. Brush control and range improvement in the post oak-blackjack oak area of Texas. Texas Agric. Expt. Sta. Bul. 942.
43. Daubenmire, R. 1968. Ecology of fire in grasslands. *In* Advances in Ecological Research 5:209–266.
44. DeBano, L. F., P. H. Dunn, and C. E. Conrad. 1977. Fire's effect on physical and chemical properties of chaparral soils. USDA, Gen. Tech. Rep. WO-3, pp. 65–74.
45. Dix, R. L. 1960. The effects of burning on the mulch structure and species composition of grasslands in western North Dakota. Ecology 41(1):49–56.
46. Duvall, V. L. 1962. Burning and grazing increase herbage on slender bluestem range. J. Range Mgt. 15(1):14–16.

47. Duvall, Vinson L. 1971. Personal correspondence.
48. Duvall, Vinson L., and Norwin E. Linnartz. 1967. Influences of grazing and fire on vegetation and soil of longleaf pine-bluestem range. J. Range Mgt. 20(4):241–247.
49. Duvall, V. L., and L. B. Whitaker. 1964. Rotation burning: a forage management system for longleaf pine-bluestem ranges. J. Range Mgt. 17(6):322–326.
50. Dwyer, Don D. 1967. Fertilization and burning of blue grama grass. J. Ani. Sci. 26(4):934.
51. Dwyer, Don D., and Rex D. Pieper. 1967. Fire effects on blue grama-pinyon-juniper rangeland in New Mexico. J. Range Mgt. 20(6):359–362.
52. Ehrenreich, John H. 1959. Effect of burning and clipping on growth of native prairie in Iowa. J. Range Mgt. 12(3):133–137.
53. Ehrenreich, John H., and John M. Aikman. 1963. An ecological study of the effect of certain management practices on native prairie in Iowa. Ecol. Monogr. 33(2):113–130.
54. Ehrenreich, John H., and Robert F. Buttery. 1960. Increasing forage on Ozark wooded range. USDA, Central States Forest Expt. Sta. Tech. Paper 177.
55. Ehrenreich, John H., and John S. Crosby. 1960. Forage production on sprayed and burned areas in the Missouri Ozarks. J. Range Mgt. 13(2):68–70.
56. Fenner, R. L., R. K. Arnold, and C. C. Buck. 1955. Area ignition for brush burning. USDA, California Forest & Range Expt. Sta. Tech. Paper 10.
57. Ferguson, E. R. 1957. Stem kill and sprouting following prescribed fires in a pine-hardwood stand in Texas. J. For. 55(6):426–429.
58. Fowells, H. A., and R. E. Stephenson. 1934. Effect of burning on forest soils. Soil Sci. 38(3):175–181.
59. Fuller, W. H., Stanton Shannon, and P. S. Burgess. 1955. Effect of burning on certain forest soils of northern Arizona. Forest Sci. 1(1):44–50.
60. Gaines, Edward M., Harry R. Kallander, and Joe A. Wagner. 1958. Controlled burning in southwestern ponderosa pine: results from the Blue Mountain plots, Fort Apache Indian Reservation. J. For. 56(5):323–327.
61. Gibbens, R. P., and A. M. Schultz. 1962. Manipulation of shrub form and browse production in game range improvement. California Fish and Game 48(1):49–64.
62. Glendening, G. E., C. P. Pase, and P. Ingebo. 1961. Preliminary hydrologic effects of wildfire in chaparral. Fifth Annual Arizona Watershed Symposium, Sept. 21, 1961.
63. Glendening, George E., and H. A. Paulsen, Jr. 1955. Reproduction and establishment of velvet mesquite as related to invasion of semidesert grasslands. USDA Tech. Bul. 1127.
64. Graves, James E., and Wilfred E. McMurphy. 1969. Burning and fertilization for range improvement in central Oklahoma. J. Range Mgt. 22(3):165–169.
65. Green, Lisle R. 1977. Fuelbreaks and other fuel modification for wildland fire control. USDA Agric. Handbook 499.
66. Green, Lisle R., and Harry E. Schimke. 1971. Guides for fuel-breaks in the Sierra Nevada mixed-conifer type. USDA, For. Serv., Pacific Southwest For. and Range Expt. Sta., Berkeley, Calif.
67. Grelen, H. E., and E. A. Epps, Jr. 1967. Herbage responses to fire and litter removal on southern bluestem range. J. Range Mgt. 20(6):403–404.
68. Grelen, H. E., and E. A. Epps, Jr. 1967. Season of burning affects herbage quality and yield on pine-bluestem range. J. Range Mgt. 20(1):31–33.
69. Grelen, Harold E. 1976. Responses of herbage, pines, and hardwoods to early and delayed burning in a young slash pine plantation. J. Range Mgt. 29(4):301–303.
70. Halls, Lowell K. 1970. Growing deer food amidst southern timber. J. Range Mgt. 23(3):213–215.
71. Halls, L. K. R. H. Hughes, and F. A. Peevy. 1960. Grazed firebreaks in southern forests. USDA Agric. Info. Bul. 226.
72. Halls, L. K., R. H. Hughes, R. S. Rummell, and B. L. Southwell. 1964. Forage and cattle management in longleaf-slash pine forests. USDA Farmers' Bul. 2199.

73. Halls, L. K., B. L. Southwell, and F. E. Knox. 1952. Burning and grazing in coastal plain forests. Georgia Agric. Expt. Sta. Bul. 51.
74. Hanks, R. J., and Kling L. Anderson. 1957. Pasture burning and moisture conservation. J. Soil & Water Cons. 12(5):228–229.
75. Hervey, Donald F. 1949. Reaction of a California annual-plant community to fire. J. Range Mgt. 2(3):116–121.
76. Heyward, Frank, and R. M. Barnette. 1934. Effect of frequent fires on chemical composition of forest soils in the longleaf pine region. Florida Agric. Expt. Sta. Bul. 265.
77. Hilmon, J. B., and R. H. Hughes. 1965. Fire and forage in the wiregrass type. J. Range Mgt. 18(5):251–254.
78. Hilmon, J. B., and C. E. Lewis. 1962. Effect of burning on south Florida range. USDA, Southeastern Forest Expt. Sta. Paper 146.
79. Holtz, S. T., and J. D. Dodd. 1970. Vegetation changes following mechanical, chemical, prescribed burns and combination treatments. ASRM, Abstract of Papers, 23rd Annual Meeting, p. 75.
80. Hopkins, Harold, F. W. Albertson, and Andrew Riegel. 1948. Some effects of burning upon a prairie in west-central Kansas. Trans. Kansas Acad. of Sci. 51(1):131–141.
81. Horton, K. W., and J. E. Hopkins. 1965. Influence of fire on aspen suckering. Canada Dept. of Agric. Pub. 1095.
82. Howard, W. E., R. L. Fenner, and H. E. Childs, Jr. 1959. Wildlife survival in brush burns. J. Range Mgt. 12(5):230–234.
83. Hughes, Ralph H. 1957. Response of cane to burning in the North Carolina coastal plain. North Carolina Agric. Expt. Sta. Bul. 402.
84. Hughes, Ralph H. 1966. Fire ecology of canebreaks. Proc. Tall Timbers Fire Ecology Conf. 1966:149–158.
85. Humphrey, R. R. 1949. Fire as a means of controlling velvet mesquite, burroweed, and cholla on southern Arizona ranges. J. Range Mgt. 2(4):175–182.
86. Humphrey, Robert R. 1962. Range ecology. The Ronald Press Co., New York City.
87. Humphrey, Robert R. 1963. The role of fire in the desert and desert grassland areas of Arizona. Proc. Tall Timbers Fire Ecology Conf. 1963:45–61.
88. Humphrey, R. R., and A. C. Everson. 1951. Effect of fire on a mixed grass-shrub range in southern Arizona. J. Range Mgt. 4(4):264–266.
89. Isaac, Lee A., and Howard G. Hopkins. 1937. The forest soil of the Douglas fir region, and changes wrought upon it by logging and slash burning. Ecology 18(2):264–279.
90. Jameson, Donald A. 1962. Effects of burning on a galleta-black grama range invaded by juniper. Ecology 43(4):760–763.
91. Jameson, Donald A. 1966. Juniper control by individual tree burning. USDA, For. Serv. Res. Note RM-71.
92. Johnson, Donald E., Hashim A. M. Mukhtar, Raymond Mapston, and R. R. Humphrey. 1962. The mortality of oak-juniper woodland species following a wild fire. J. Range Mgt. 15(4):201–205.
93. Kay, Burgess L. 1960. Effect of fire on seeded forage species. J. Range Mgt. 13(1):31–33.
94. Klemmedson, James O., and Justin G. Smith. 1964. Cheatgrass (*Bromus tectorum* L.). Bot. Rev. 30(2):226–262.
95. Klett, W. Ellis, Dale Hollingsworth, and Joseph L. Schuster. 1971. Increasing utilization of weeping lovegrass by burning. J. Range Mgt. 24(1):22–24.
96. Krammes, Jay S. 1960. Erosion from mountain side slopes after fire in southern Califonia. USDA, Pacific Southwest Forest & Range Expt. Sta. Res. Note 171.
97. Kucera, C. L., and John H. Ehrenreich. 1962. Some effects of annual burning on central Missouri prairie. Ecology 43(2):334–336.
98. Launchbaugh, J. L. 1964. Effects of early spring burning on yields of native vegetation. J. Range Mgt. 17(1):5–6.
99. Lay, Daniel W. 1957. Browse quality and the effects of prescribed burning in southern pine forests. J. For. 55(5):342–347.

100. Lay, Daniel W. 1967. Browse palatability and the effects of prescribed burning in southern pine forests. J. Forestry 65(11):826–828.
101. Leege, Thomas A. 1968. Prescribed burning for elk in northern Idaho. Proc. Tall Timbers Fire Ecology Conf. 1968:235–253.
102. Lewis, Clifford E. 1964. Forage response to month of burning. USDA, For. Serv. Res. Note SE-35.
103. Lewis, Clifford E., and Thomas J. Harshbarger. 1976. Shrub and herbaceous vegetation after 20 years of prescribed burning in the South Carolina coastal plain. J. Range Mgt. 29(1):13–18.
104. Lillie, D. T., George E. Glendening, and C. P. Pase. 1964. Sprout growth of shrub live oak as influenced by season of burning and chemical treatments. J. Range Mgt. 17(2):69–72.
105. Lillywhite, Harvey B. 1977. Animal responses to fire and fuel management in chaparral. USDA, Gen. Tech. Rep. WO-3, pp. 368–373.
106. Lodge, Robert W. 1960. Effects of burning, cultivating, and mowing on the yield and consumption of crested wheatgrass. J. Range Mgt. 13(1):318–321.
107. Love, R. M., L. J. Berry, J. E. Street, and V. P. Osterli. 1960. Planned range improvement programs are beneficial. California Agric. 14(6):2–4.
108. Lowe, Philip O., Peter F. Ffolliott, John H. Dieterich, and David R. Patton. 1978. Determining potential wildlife benefits from wildfire in Arizona ponderosa pine forests. USDA, For. Serv. Gen. Tech. Rep. RM-52.
109. Lutz, H. J. 1956. Ecological effects of forest fires in the interior of Alaska. USDA Tech. Bul. 1133.
110. Lyon, L. Jack. 1966. Initial vegetal development following prescribed burning of Douglas fir in south-central Idaho. USDA, For. Serv. Res. Paper INT-29.
111. Manley, Keith, and Charles F. Walker. 1956. Brush control and reseeding for range improvement in central California. J. Range Mgt. 9(6):278–280.
112. Martin, Robert E., Stuart E. Coleman, and Arlen H. Johnson. 1977. Wetline technique for prescribed burning firelines in rangeland. USDA, For. Serv. Res. Note PNW-292.
113. Martin, S. Clark. 1966. The Santa Rita Experimental Range. USDA, For. Serv. Res. Paper RM-22.
114. Martin, S. Clark, and John S. Crosby. 1955. Burning and grazing on glade range in Missouri. USDA, Central States Forest Expt. Sta. Tech. Paper 147.
115. Mayland, H. F. 1967. Nitrogen availability on fall-burned oak-mountain-mahogany chaparral. J. Range Mgt. 20(1):33–35.
116. McCulloch, Clay Y. 1969. Some effects of wildfire on deer habitat in pinyon-juniper woodland. J. Wildlife Mgt. 33(4):778–784.
117. McIlvain, E. H., and C. G. Armstrong. 1966. A summary of fire and forage research on shinnery oak rangelands. Proc. Tall Timbers Fire Ecology Conf. 1966:127–129.
118. McKell, Cyrus M., Alma M. Wilson, and B. L. Kay. 1962. Effective burning of rangelands infested with medusahead. Weeds 10(2):125–130.
119. McLean, Alastair. 1969. Fire resistance of forest species as influenced by root systems. J. Range Mgt. 22(2):120–122.
120. McMurphy, Wilfred E., and Kling L. Anderson. 1965. Burning Flint Hills range. J. Range Mgt. 18(5):265–269.
121. Mueggler, Walter F., and James P. Blaisdell. 1958. Effects on associated species of burning, rotobeating, spraying, and railing sagebrush. J. Range Mgt. 11(2):61–66.
122. Murphy, A. H., and W. C. Lusk. 1961. Timing medusahead burns to destroy more seed—save good grass. California Agric. 15(11):6–7.
123. National Research Council. 1968. Principles of plant and animal pest control. II. Weed control. National Academy of Sciences Pub. 1597.
124. Nelson, Jack R. 1976. Forest fire and big game in the Pacific Northwest. Proc. Annual Tall Timbers Fire Ecology Conf. 15:85–102.
125. Nord, Eamor C., and Lisle R. Green. 1977. Low-volume and slow-burning vegetation for planting on clearings in California chaparral. USDA, For. Serv. Res. Paper PSW-124.

126. Owensby, Clenton E., and Kling L. Anderson. 1967. Yield responses to time of burning in the Kansas Flint Hills. J. Range Mgt. 20(1):12–16.
127. Pase, Charles P., and Carl Eric Granfelt. 1977. The use of fire on Arizona rangelands. Ariz. Interagency Range Comm. Pub. 4.
128. Patton, David R., and Herman D. Avant. 1970. Fire stimulated aspen sprouting in a spruce-fir forest in New Mexico. USDA, For. Serv. Res. Note RM-159.
129. Pechanec, Joseph F., A. Perry Plummer, Joseph H. Robertson, and A. C. Hull, Jr. 1965. Sagebrush control on rangelands. USDA Agric. Handbook 277.
130. Pechanec, Joseph F., and George Stewart. 1954. Sagebrush burning—good and bad. USDA Farmers' Bul. 1948.
131. Plumb, T. R. 1963. Delayed sprouting of scrub oak after a fire. USDA, For. Serv. Res. Note PSW-1.
132. Plummer, A. Perry. 1970. Personal correspondence.
133. Plummer, A. Perry, Donald R. Christensen, and Stephen B. Monson. 1968. Restoring big-game range in Utah. Utah Div. of Fish & Game Pub. 68-3.
134. Pond, Floyd W., and Dwight R. Cable. 1960. Effect of heat treatment on sprout production of some shrubs of the chaparral in central Arizona. J. Range Mgt. 13(6):313–317.
135. Pond, Floyd W., and Dwight R. Cable. 1962. Recovery of vegetation following wildfire on a chaparral area in Arizona. USDA, For. Serv. Res. Note 72.
136. Reynolds, H. G., and J. W. Bohning. 1956. Effects of burning on a desert grass-shrub range in southern Arizona. Ecology 37(4):770–777.
137. Reynolds, Hudson G., and S. Clark Martin. 1968 (Rev.). Managing grass-shrub cattle ranges in the Southwest. USDA Agric. Handbook 162.
138. Robertson, Joseph H., and H. P. Cords. 1957. Survival of rabbitbrush, Chrysothamnus spp., following chemical, burning, and mechanical treatments. J. Range Mgt. 10(2):83–89.
139. Roby, George A., and Lisle R. Green. 1976. Mechanical methods of chaparral modification. USDA Agric. Handbook 487.
140. Sampson, Arthur W. 1944. Plant succession on burned chaparral lands in northern California. California Agric. Expt. Sta. Bul. 685.
141. Sampson, Arthur W., and L. T. Burcham. 1954. Costs and returns of controlled brush burning for range improvement in northern California. California Div. of Forestry, Range Improvement Study 1.
142. Sampson, A. W., and Clark H. Gleason. 1957. Change in burning terminology. J. Range Mgt. 10(3):104a.
143. Sampson, Arthur W., and Arnold M. Schultz. 1957. Control of brush and undesirable trees. Unasylva 10(1):19–29, 10(3):117–128, 10(4):166–182, 11(1):19–25.
144. Sauer, Carl O. 1950. Grassland climax, fire, and man. J. Range Mgt. 3(1):16–21.
145. Schmutz, Ervin M. 1978. The use of fire on southwestern ranges. Rangeman's J. 5(2):35–36.
146. Sharrow, Steven H., and Henry A. Wright. 1977. Proper burning intervals for tobosagrass in west Texas based on nitrogen dynamics. J. Range Mgt. 30(5):343–346.
147. Silker, T. H. 1955. Prescribed burning for the control of undesirable hardwoods in pine-hardwood stands and slash pine plantations. Texas Forest Serv. Bul. 46.
148. Smith, E. F., and V. A. Young. 1959. The effect of burning on the chemical composition of little bluestem. J. Range Mgt. 12(3):139–140.
149. Smith, E. F., V. A. Young, K. L. Anderson, W. S. Ruliffson, and S. N. Rogers. 1960. The digestibility of forage on burned and non-burned bluestem pasture as determined with grazing animals. J. Ani. Sci. 19(2):388–392.
150. Smith, L. F., R. S. Campbell, and Clyde L. Blout. 1958. Cattle grazing in longleaf pine forests of south Mississippi. USDA, Southern Forest Expt. Sta. Occasional Paper 162.
151. Southwell, Bryon L., and Ralph H. Hughes. 1967. Beef cattle management

practices for burned and unburned pine-wiregrass ranges of Georgia. Georgia Agric. Expt. Sta. Res. Rpt. 14.

152. Stoddart, Laurence A., and Arthur D. Smith. 1955 (Second Ed.). Range management. McGraw-Hill Book Co., Inc., New York City.

153. Texas Agric. Expt. Sta. 1976. Burning and the use of fire in Texas agriculture: considerations and alternatives. Texas Agric. Expt. Sta. Misc. Pub. 1255.

154. Tiedemann, Arthur R., and Ervin M. Schmutz. 1966. Shrub control and reseeding effects on the oak chaparral of Arizona. J. Range Mgt. 19(4):191–195.

155. Trlica, M. J., Jr., and J. L. Shuster. 1969. Effects of fire on grasses of the Texas High Plains. J. Range Mgt. 22(5):329–334.

156. Ueckert, Darrell N., Terry L. Whigham, and Brian M. Spears. 1978. Effect of burning on infiltration, sediment, and other soil properties in a mesquite-tobosagrass community. J. Range Mgt. 31(6):420–425.

157. USDA, Rocky Mtn. Forest and Range Expt. Sta. 1964. Chaparral firesprouts attract deer. 1963 Annual Rpt., p. 49.

158. Vlamis, J., and K. D. Gowans. 1961. Availability of nitrogen, phosphorus, and sulfur after brush burning. J. Range Mgt. 14(1):38–40.

159. Vogl, Richard J. 1965. Effects of spring burning on yields of brush prairie savanna. J. Range Mgt. 18(4):202–205.

160. Wahlenberg, W. G., S. W. Greene, and H. R. Reed. 1939. Effects of fire and cattle grazing on longleaf pine lands as studied at McNeill, Miss. USDA Tech. Bul. 683.

161. Waldrip, William J. 1954. Brush control, grazing management, and revegetation of areas infested with blueberry cedar. Texas Agric. Expt. Sta. Prog. Rpt. 1721.

162. Weaver, Harold. 1957. Effects of prescribed burning in ponderosa pine. J. For. 55(2):133–138.

163. Weaver, Harold. 1967. Fire and its relationship to ponderosa pine. Proc. Tall Timbers Fire Ecology Conf. 1967:127–149.

164. White, Larry D. 1969. Effects of a wildfire on several desert grassland shrub species. J. Range Mgt. 22(4):284–286.

165. Wink, Robert L., and Henry A. Wright. 1973. Effects of fire on an Ashe juniper community. J. Range Mgt. 26(5):326–329.

166. Wright, Henry A. 1969. Effect of spring burning on tobosa grass. J. Range Mgt. 22(6):425–427.

167. Wright, Henry A. 1970. Prescribed burning for west Texas. *In* Proceedings of the Eighth West Texas Range Management Conference, pp. 18–22, Abilene Christian College, Abilene, Texas.

168. Wright, Henry A. 1971. Personal correspondence.

169. Wright, Henry A. 1971. Why squirreltail is more tolerant to burning than needle-and-thread. J. Range Mgt. 24(4):277–284.

170. Wright, Henry A. 1974. Effect of fire on southern mixed prairie grasses. J. Range Mgt. 27(6):417–419.

171. Wright, Henry A. 1978. The effect of fire on vegetation in ponderosa pine forests: a state-of-the-art review. Texas Tech. Univ. Range and Wildl. Info. Ser. 2.

172. Wright, Henry A., Stephen C. Bunting, and Leon F. Neuenschwander. 1976. Effect of fire on honey mesquite. J. Range Mgt. 29(6):467–471.

173. Wright, Henry A., Francis M. Churchill, and W. Clark Stevens. 1976. Effect of prescribed burning on sediment, water yield, and water quality from dozed juniper lands in central Texas. J. Range Mgt. 29(4):294–298.

174. Wright, Henry A., and Kenneth J. Stinson. 1970. Response of mesquite to season of top removal. J. Range Mgt. 23(2):127–128.

175. Young, James A., and Raymond A. Evans. 1978. Population dynamics after wildfires in sagebrush grasslands. J. Range Mgt. 31(4):283–289.

Chapter 7

Planning range seeding

Natural Versus Artificial Revegetation

Some depleted ranges can be restored by improved management alone. *Natural revegetation* is based on checking the current cause or causes of depletion and allowing secondary succession to raise range condition to satisfactory levels (Figure 65). Improved management, particularly of grazing, must be provided to restore vigor and accelerate the spread of the remaining desirable forage plants. Adequate seed production and seedling establishment of the desirable species is important, but vegetative spread through tillers, rhizomes, or stolons is equally important for many forage plants. However, when insufficient desirable forage plants remain, consideration must then be given to *artificial revegetation*—usually involving preparation of a seedbed followed by drilling or broadcasting harvested seed.

Determining whether a range can be restored by natural means or will require artificial revegetation is a matter of judgment. However, the decision should be based on the kinds and amounts of plants remaining, the expected rate of recovery and the cost of the alternative approaches, and the climate. Other considerations should include the supplementary treatments that may be used in speeding up natural restoration, present soil conditions including erosion, and whether the site is adapted to present artificial seeding techniques. Reducing stocking rates, changing season of use, initiation of special grazing systems, or improving distribution of grazing by fencing, additional water development, or other practices may be sufficient for range recovery; or natural restoration may need to be accelerated by weed and brush control or land treatment such as pitting, waterspreading, or fertilization.

Natural revegetation is cheaper than artificial revegetation except where extended periods of nonuse are required. Where only adjustments in grazing management are required, the added expense of a natural seeding program may be little or none. The cost of two to four years of complete protection from grazing will amount only to the costs of nonuse—interest and taxes on the land investment or the cost of alternative sources of carrying capacity. Structural improvements or land treatment other than seedbed preparation and artificial seeding may require moderate investments.

FIG. 65. *Natural revegetation of tanglehead and sideoats grama on right side of fence following deferred grazing and proper stocking rates near Uvalde, Texas.* (Soil Conservation Service photo)

By contrast, artificial seeding is generally expensive. The average cost per acre of twenty range seedings made in western Utah in the late 1950s was $18.51 per acre when both direct and indirect costs were included or $8.92 when only direct costs were considered [61]. The cost of converting chaparral to grassland in California on gentle slopes and medium brush levels was estimated in 1967 to range from $12.00 to $54.00 per acre depending on the combination of treatments used [6]. The cost of making warm-season grass seedings in the

Great Plains area in 1979 could be expected to run between $30.00 and $55.00 per acre. On "go-back" land, the direct cost of plowing and seeding to crested wheatgrass may be as low as $12.00 per acre.

The rate of range recovery under natural revegetation depends upon many factors such as the kind and amount of plants in the residual cover, the presence of an adequate seed source, and soil conditions. In Texas, satisfactory recovery by natural spread with careful grazing management has been reported generally to be successful where at least 10 percent of the vegetation consists of desirable plants, when they are uniformly distributed, and when competing weeds and woody plants are controlled [85]. In the Intermountain Region, artificial seeding rather than natural revegetation has been recommended where there is less than one desirable bunchgrass to every ten square feet or one stem of western wheatgrass or other sod-forming, desirable plants to every fifteen square feet [83]. On foothill sagebrush range, a minimum of one desirable grass plant for each four square feet or a 15 percent ground cover of desirable perennial grasses has been used as an index for successful recovery possibilities without artificial seeding [107]. Government cost sharing on range improvement based upon natural improvement generally requires a range condition of 15 percent or higher based on percent plant composition of desirable perennial species.

Lands that have been denuded by continuous farming may take from twenty-five to seventy-five years or even longer to completely revegetate naturally. The size of the land unit, proximity to natural seed sources, intensity of cultivation, competitiveness of invading undesirable plants, amount of rainfall, extent of grazing since cultivation was discontinued, and extent of soil erosion all affect the rate of recovery on go-back land. However, because of high land investment and taxes, it is not feasible to leave private lands out of production for extended periods of time. In some areas such as in the Southern Great Plains, the go-back or increaser grasses have been found to have moderate productivity and palatability [68]. The current productivity of such lands must be compared with the cost of artificial seeding and the risk of subsequent stand failure.

The potential for natural recovery on closed stands composed primarily of undesirable plants is small. Protecting cheatgrass, medusahead, or tarweed range may be hopeless unless there is a fair remnant of the original cover. Dense stands of brush such as juniper, sagebrush, or mesquite often delay or even prevent a reasonable rate of recovery. The effectiveness of natural revegetation depends upon the nature of the climax. Some deserts were never grasslands; dense climax stands of conifers in the Mountain West or hardwood trees in

253

the Midwest are of minor value for grazing. On many difficult sites—those with low rainfall, shallow soils, or steep terrain—artificial seeding is impractical or impossible by current standards. Although drought, poor soil, rodents, or repeated burning may nullify natural revegetation, artificial seeding under these conditions is apt to be even less successful.

Under less intensive or partial seedbed preparation associated with artificial revegetation, the natural recovery of native forage plants often becomes an important part of the forage stand. Artificial seeding programs such as broadcast seeding and chaining, in fact, often depend upon the partial recovery of desirable species in the resident cover. In seeding trials on clay sites in South Dakota which were nearly devoid of vegetation, restriction of grazing combined with good rains the first growing season was just as effective as various combinations of partial seedbed preparation and grass seeding methods [73]. When artificially seeding native plant mixtures, the natural recovery by secondary succession may be highly desirable. However, when cool-season, introduced grasses are seeded in prepared seedbeds, the return of native warm-season perennial grasses such as sand dropseed in the Great Plains is undesirable.

Deciding to Reseed

Range seeding is an action program that readily appeals to ranchers and land managers. Its potential to economically restore range has sometimes been oversold. Range seeding is not a substitute for good range management but, in fact, requires more intensive management of grazing. The costs of range seeding must be covered by the expected benefits. This requires that range seeding be undertaken normally only as a part of a total ranch or allotment plan.

Seeding is often necessary following brush removal (Figure 66). Where the desirable perennials are present in insufficient amounts to naturally revegetate the area or where the brush removal method kills most or all of the understory plants, seeding is required. In the conversion process of chaparral to grassland in California, the establishment of introduced, long-lived grasses is considered the most difficult step in the conversion process [6]. Thus, the need for seeding must be carefully considered in all noxious plant removal programs. Seeding of intermediate wheatgrass, smooth brome, or crested wheatgrass following burning or mechanical brush removal has materially reduced the regrowth and return of Gambel oak, chokecherry, snowberry, snowbrush (*Ceanothus velutinus*), and bigtooth maple (*Acer grandidentatum*) [81].

FIG. 66. *Following brush control by root plowing, area on left side of fence was seeded to a mixture of sideoats grama, blue grama, and blue panicgrass. Untreated area on right of fence; range site near Sylvester, Texas.* (Soil Conservation Service photo)

Range seeding is costly but can significantly increase forage and livestock production and net income when located on the right sites, when properly done, and when well managed. Successful range seeding requires stepwise planning. Risk of failure is involved in range seeding due to unforeseen conditions, but many failures can be avoided by following proven seeding procedures. A few spectacular successes in ideal moisture years frequently encourage range seedings to be made with inadequate planning or seedbed preparation, only to result in failures in average or below average rainfall years.

Important questions to answer before deciding to seed range or related perennial pasture include:

1. How much additional carrying capacity is needed in the present or proposed grazing operation?
2. How much will the new seeding produce over what the range produces now before improvement?
3. Is additional seasonal grazing needed to balance the yearlong forage supply? What season of the year?
4. What kind of livestock or big game can or will utilize the forage crop?
5. Will the seeding provide special use such as lambing or calving pasture, flushing or breeding pasture, pasture fattening, or winter range for deer?
6. Will seeding provide green forage during periods of natural range plant dormancy?
7. Will seeding reduce the amount of hay and supplements needed for overwintering?

255

8. Will the seeding benefit other areas by allowing stocking rates to be reduced, season of grazing altered, turn-out dates delayed, or special grazing systems initiated?
9. Will seeding allow livestock to be moved off hay meadows earlier in the spring and result in increased hay production?
10. Will the expected future benefits discounted to the present exceed the total costs of seeding? Will it be profitable?
11. Are special considerations such as erosion control, firebreaks, and wildlife habitat important?
12. Where are the acres to be seeded? Are they ideally located for optimum use? Are they scattered?
13. Have adequate seeding and management techniques been worked out for the site? Will these be followed?
14. Is the necessary equipment, manpower, and seed available to make the seeding?
15. What is the risk of seeding failure in the area?
16. What is the risk of soil erosion? Will watershed improvement result?
17. How long will the seeding last? Will reseeding have to be done periodically?
18. What will be the cost of maintenance? This may be high in some brush types.
19. Will new fences around the seeding be required?
20. Is stockwater available on the proposed seeding area or can it be easily developed?
21. When can grazing be started after the seeding is made?
22. How should the forage crop be managed after the stand is established?

Range seeding may allow greater versatility in ranch management. Most profitable use of a seeding may require a shift in overall ranch management. For example, a shift is often desirable to use the new grass for yearling steers rather than for the breeding herd. Or cool-season grass seedings may be used by the breeding herd only during the postcalving, prebreeding period and subsequently by stocker cattle. Each range seeding should be planned to meet the needs of the individual ranch or management unit.

Costs of range seedings should be kept as low as possible since profit will result only when total additional returns exceed total additional costs. However, high quality of work is required to produce good stands. Quality should not be lowered for the sake of lower costs to the extent that unsatisfactory results are obtained. Successful and efficient methods are of major importance. The cost of range seeding can often be materially reduced through government cost sharing or by harvesting seed from grass seedings for one to three years before full grazing. Procedures for partial budgeting and estimating the internal rate of return of a range seeding is given in Table 2 of Chapter 1.

Several advantages of range seeding are difficult to measure in dollars but are nevertheless real. Grass seeding helps to stabilize eroding soils since the fibrous roots of grasses hold soil firmly in place. Reduced silt in streams not only improves conditions for fish but increases the life of reservoirs and reduces the need for cleaning irrigation ditches and stockwater developments. Seeding mill sites, skid roads, and haul roads with forage plants after the logging operation is completed prevents weed growth and topsoil erosion, stabilizes access roads, and helps in reducing overly dense stands of conifer seedlings as well as increasing forage production [84].

Increased grazing capacity. Increased grazing capacity is frequently the major reason, but seldom the only reason, for range seeding. A comparison of site potential to produce forage with the amount of forage currently being produced is the best estimate of the increased carrying capacity that can be expected. Obviously, the forage plants species seeded, seeding methods, stand establishment, and subsequent management of grazing as well as many other factors will materially affect the added carrying capacity realized. Increased carrying capacity, in turn, increases land value.

On fifteen BLM cattle allotments in western Utah, the carrying capacity on seeded range varied from 1.3 to 8.2 acres per a.u.m. with an average of 3.8 [61]. Before seeding, the sagebrush-grass ranges had carrying capacities ranging from 7.9 to 13.9 acres per a.u.m. with an average of 11.3. Spring-fall sagebrush range in northern New Mexico seeded to crested wheatgrass increased forage yield from 50 pounds annually per acre to 600 to 1,400 pounds per acre [79]. This was associated with an increase in grazing capacity from 70 acres required per a.u.m. to 2.5 to 7 acres per a.u.m. Converting dense chaparral to grass on fair to good sites in California has commonly increased grazing capacity by 0.5 to 1 a.u.m./acre [6].

In eastern Wyoming, yearling steers grazed throughout the summer on a combination of dryland pastures seeded to adapted introduced forage species made daily gains similar to steers grazed on native range [57]. However, by the fifth grazing season after seeding, the seeded pastures produced 233 percent more carrying capacity and 188 percent more pounds of beef per acre than native range. In Oklahoma, bermudagrass was found to increase carrying capacity by 400 to 500 percent while reducing per head gains by 10 to 15 percent [66]. On lambing range in northern New Mexico, crested wheatgrass seedings produced 400 percent more forage than adjacent native range even in drought years [28]. Increased net returns were sufficient to repay the cost of seeding in four to seven years. On ponderosa pine range in Colorado the direct costs of seeding were repaid in three years of grazing as a result of increased grazing capacity [20].

A balanced seasonal forage supply on many farms and ranches can be provided only by the use of irrigated pastures and temporary pastures in conjunction with native and seeded range. Exploiting the genetic potential in range livestock for higher production often requires the complemental use of high-producing cultivated pasture [67]. Returns from well-managed cultivated dryland or irrigated pastures on good quality farmland compare favorably with other high-value farm crops [106]. Irrigated pastures capable of carrying 2 to 2.5 cow-calf pairs or 4 to 5 yearling steers per acre during the growing season are common. Irrigated pastures in the Intermountain Region have provided up to 12.5 a.u.m. per acre during a May to September grazing season and produced from 1,320 to 1,650 pounds per acre when grazed by steers [34]. However, these levels are obtained only under ideal soil conditions and intensive management of fertility, irrigation, and grazing. In some areas, such as in the Southern Great Plains, irrigated pastures are less efficient as a result of high consumptive use of water and low forage production resulting from high summer temperatures and hot winds, in spite of ample soil moisture [69].

Cultivated pastures can be highly productive where soil fertility and moisture conditions are optimum, but they require additional costs of maintenance and annual or periodic reestablishment. Dry, steep, rocky, or easily eroded lands are better adapted to permanent perennial forage species. Depending upon geographic location and site conditions, temporary forages capable of high production for summer grazing include sudangrass, sorghum-sudangrass hybrids, annual lespedezas, millet, and soybeans. Spring and fall grazing (or winter grazing in the South) can be provided by common rye, wheat, oats, hairy vetch, or crimson clover. Also, cornstalks and other crop aftermath provide needed fall grazing in some areas.

Selecting Sites for Seeding

Range seedings are generally successful and profitable only on the more favorable sites. Low potential or high hazard makes many sites undesirable for seeding. Sites with high potential but currently in low production offer the most favorable cost-benefit ratio. Range in good condition should not normally be disturbed for artificial seeding. The following guidelines have been given for site selection [26]:

1. Sufficient site potential to insure reasonable chances of success.
2. Enough soil for adequate root development and water storage.
3. Expected precipitation must be sufficient to support the species seeded.
4. Soil sufficiently free of rocks and on gentle enough slopes to allow proper seedbed preparation.
5. Soil reasonably free of toxic materials such as alkali.
6. An area that can be easily incorporated into the overall management of the ranch.
7. Adapted to present-day range seeding equipment.

Areas where range seeding is needed and can generally be expected to be beneficial are:

1. Low quality farmland or old fields being returned to permanent grazing land.
2. Go-back land currently producing forage low in quality or quantity.
3. Open rangelands with few desirable forage plants remaining but with high site potential.
4. Grasslands infested with brush and having little grass left.

FIG. 67. *Russian wildrye (upper center and right) and crested wheatgrass (left) surviving but unproductive even under exclusion from grazing at the Desert Range Station near Milford, Utah, where average annual precipitation is less than eight inches.*

5. Woodlands having noncommercial trees and capable of being developed as grazing lands.
6. Wetlands capable of being drained and converted to productive grazing lands.
7. Critical erosion areas such as active sand dunes, blowouts, gullied areas, or eroding watersheds.
8. Dikes, banks, levees, spillways, waterways, trails, and skid roads which need a protective cover to stabilize them.

Rainfall. Range seeding is generally recommended in the West where the average annual precipitation is eleven inches or more with some success expected in the nine- and ten-inch zones [17, 64, 83, 86, 94]. At the lower levels of precipitation, herbage yields are lower, establishment becomes risky, and management becomes precarious (Figure 67) [17]. Fair stands of crested wheatgrass have been established in the eight-inch precipitation zone under a combination of favorable distribution of precipitation, moderate temperatures, and permeable soils [94]. However, seedings established where annual rainfall is low, often established in above average precipitation periods, are made short-lived and low producing by drought coupled with grazing [39]. On chaparral and mountain brush sites in California foothills, seeding perennials on sites with good soils and receiving less than fifteen inches is not recommended [6]. And, on shallower and stonier soils, a minimum of twenty inches annually is suggested.

Precipitation averages are known approximately for most locations in the United States. However, data on yearly and seasonal variability is less complete. The effectiveness of rainfall in the establishment and maintenance of range seedings is greatly affected by its intensity and seasonal distribution and by associated temperatures and wind velocities. Seasonal precipitation just before or after seeding influences establishment in the West more than annual precipitation. Subirrigation, waterspreading, or supplemental irrigation will greatly lower rainfall requirements.

A system of deep furrow seeding of wheatgrasses and "high water" irrigation has been developed for low rainfall areas where early spring runoff can provide one or two supplemental irrigations in the spring [76]. Seeding is made in late fall so seedlings can emerge and make good growth prior to spring irrigation. Seeding in the bottom of four-inch-deep furrows provides decreased moisture evaporation, and the furrows are used to carry the supplemental water. Intermediate, pubescent, and tall wheatgrass were found well adapted to spring irrigation, beginning growth early and yet surviving through normal summer drought.

In New Mexico, saline water has been used successfully to establish selected range shrubs and grasses [99]. Up to 16,000 p.p.m. of total salts in the water was tolerated by bermudagrass, alkali saca-ton, blue panicgrass, and four-wing saltbush when the soil was well drained and when excess water was applied to leach the salt. This level resulted in some difficulty in establishing, but the seeded plants grew well after passing the seedling stage. Postponing irrigation until after emergence or vegetative propagation of blue panicgrass and bermudagrass was suggested.

Temperature and wind. Temperatures of either extreme are detri-mental to seedling establishment. Hot temperatures and drying winds in combination are particularly damaging to young seedlings under minimum moisture conditions. Cold temperatures in the ab-sence of snow cover are also detrimental to young seedlings and may result in high winter kill. Cold temperatures in early spring delay seed germination, but unusually warm spells in either late fall or early spring followed by very low temperatures germinate the seeds only to have the young seedlings killed. Temperatures of either ex-treme also reduce the growth of most forage species. A survey made in the Great Plains revealed that seedling establishment was highest in the northern areas and least in the southern areas [29], suggesting that other climatic factors were more decisive in their effects than annual precipitation.

Soils. Fairly level sites with deep, fertile, and medium textured soils are best for seeding. Clays and sands generally are more difficult to obtain satisfactory grass stands [29]. Clay soils are more compact, are harder to plow and drill, have reduced water penetration, hold mois-ture at the soil surface where it readily evaporates, often crust over, and are more subject to frost heaving. Sandy soils have rapid mois-ture penetration, low moisture-holding capacity, readily dry at the surface, and are subject to wind erosion. Although sandy soils pro-vide a less favorable habitat for establishing seedlings and growing shallow-rooted plants, deep percolation and storage of water favor es-tablished, deep-rooted plants.

In a survey of sagebrush conversions to grassland in the western United States, conversions were found to be most successful in me-dium textured soils with medium water-holding capacities, inter-mediate in coarse soils, and low in fine soils [94]. On sandy soils, crested wheatgrass was successful and productive when established during relatively abundant moisture periods. Soil depth and stoni-ness materially affect site potential for grass seeding. Range seedings are most successful and productive on sites with soil depth of two feet or more [17] and less successful on soils from one to two feet in depth.

Saline and alkali soils are common in the valley bottoms of arid and semiarid regions. These soils present special problems in selecting species for planting, in preparation of adequate seedbeds, and in the germination, establishment, and maintenance of forage stands. Only species adapted to these sites should be used. Where good drainage and irrigation water low in salt content can be provided, leaching of the soil prior to seedbed preparation, planting in furrows or on the edges rather than on the tops of the beds, and frequent irrigation are helpful [4].

On saline-sodic soils in Nevada, it was concluded that under careful management and irrigation from one-half to two-thirds as much beef could be produced per acre with tall wheatgrass and tall fescue as could be produced on more desirable land [49]. However, under dryland conditions little can be done to remedy saline or alkali conditions except to provide for maximum penetration of rainfall by reducing runoff [7].

High levels of soil salinity reduce the amount of water available to plants in the soil and thereby accentuate physiological drought. High levels of salts are also toxic [7]. Sodic soils are characterized by dissolved organic matter deposited as a black surface on the ground, are deflocculated and take water slowly, are sticky when wet but crust over when dry, and are particularly difficult to work into a good seedbed [78]. The stunted growth of species more tolerant to exchangeable sodium results principally from the adverse physical conditions of the soil, while the less tolerant species are also affected by nutritional factors.

Terrain. Better soil moisture conditions are generally found on bottomlands. Excess drainage reduces soil moisture on steep hillsides. Level to gently sloping lands are preferred for range seeding since ground preparation can be more effectively done with machinery and at a reasonable cost. In the Intermountain Region the cost of seedbed preparation and drilling rises sharply on slopes in excess of 20 percent. Except for watershed protection and where special techniques are used, complete seedbed preparation is generally economically infeasible and hazardous on slopes in excess of 30 percent (a 30-foot rise for every 100 feet on the horizontal) [83]. Higher altitude as well as more northerly latitude reduces the length of growing season and may prevent seedset of less adapted species. On semidesert bunchgrass range in southern Arizona, the best sites for seeding are above 3,500 feet where more mesic conditions and milder temperatures exist [64].

Native plant indicators. Native plant indicators are particularly useful in selecting sites suitable for range seeding (Figure 68). In Nevada sites supporting big sagebrush, bluebunch wheatgrass, Canby bluegrass (*Poa canbyi*), Thurber needlegrass, and lupine are generally suit-

FIG. 68. *Sagebrush range suitable for seeding but currently producing minimum herbaceous vegetation. Good site potential is indicated by the vigor of the big sagebrush.*

able for seeding [2]. Sites supporting low sagebrush, black sagebrush, greasewood, bud sagebrush, or winterfat are less suitable for seeding the commonly used forage species. In a survey of the sagebrush zone in the Intermountain Region, grass production following seeding on big sagebrush sites was three times as productive as on former black sagebrush sites [94]. Grass seedings were generally poor on sites where greasewood or shadscale was mixed with big sagebrush. The degree of grass establishment varied directly with the big sagebrush vigor index. Poor vigor of indicator plants may reflect excessive grazing or recent drought rather than site capabilities.

Range types or range sites are commonly used as the basis of site suitability and seeding recommendations since each type or site represents a gross summary of rainfall, soils, topography, elevation, and native vegetation. In the Intermountain Region, sagebrush-grass, mountain brush, aspen, and mountain grasslands are types generally adapted to seeding [107]. Salt-desert shrub including shadscale and greasewood subtypes, desert grasslands, and blackbrush types are least adapted to intensive seeding programs. Conifer, juniper-pinyon, and alpine grassland are often adapted to seeding but local soil conditions and special problems require particular attention.

263

Direct seedings of native or introduced species in the salt-desert shrub zone have generally failed [9, 41]. The arid climate has been found to be the principal cause of failure; but other factors have included low humidity, high evaporation, and high diurnal temperature fluctuations. Although Russian wildrye and crested wheatgrass have been established in some cases, productivity and tolerance of grazing have been very low. Even when successfully established on winterfat winter sheep range, crested wheatgrass has produced less forage than winterfat and is less desirable for winter grazing [107]. It is probable that future efforts toward seeding in this zone will be directed more toward native shrubs and grasses. It is particularly important that adapted species be seeded on arid and saline sites. Rehabilitation through management alone has often been successful.

In the semidesert grasslands of southern Arizona, mesquite and burroweed generally grow on ranges that previously supported grass, and these areas offer good possibilities for seeding [1]. However, plants such as saguaro cactus (*Carnegiea gigantea*), paloverde (*Cercidium* spp.), bursage (*Franseria deltoidea*), creosotebush, and ironwood (*Olneya tesota*) indicate sites unsuited to seeding. In subhumid regions climatic limitations for range and pasture seedings are less marked, and soil factors combined with climatic data reduce the dependency on native vegetation as indicators of site suitability. However, in the semiarid portions of the Great Plains, native vegetation is an important index of site suitability. Sites capable of growing dryland wheat are generally adapted for grass seeding.

Species Adaptation and Selection

Site adaptations. The adaptability of the plant species to the seeding site often means the difference between success or failure. Plant factors that should be considered in selecting forages for seeding on a particular site include (1) drought tolerance, (2) cold hardiness, (3) tolerance of soil salinity, (4) soil texture adaptation, (5) tolerance of high water tables and wet soil conditions, and (6) ease of establishment and aggressiveness and longevity. These characteristics are rated good, fair, and poor for individual forage species in Table 14, *Forage Plants Commonly Seeded on Range and Other Perennial Pasture.* Species adaptations for range or irrigated pasture seeding by geographical regions within the United States are also indicated. (Range and pasture seeding regions are shown in Figure 69.) Other factors that might be considered in evaluating species adaptation are heat tolerance, shade tolerance, and response to soil fertility. Seeding native species on sites that originally supported good stands of these species

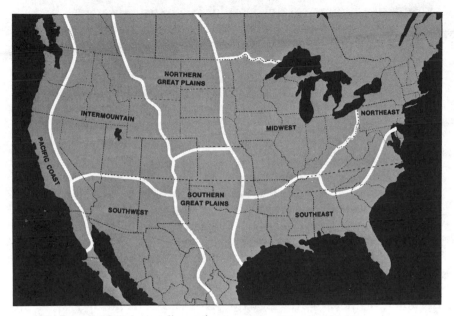

FIG. 69. *Range and pasture seeding regions.*

is one technique used in assuring adaptability. However, certain of the introduced species may be equally or more productive and as well adapted.

The amount of precipitation as indicated by precipitation records and indicator plants is the most important guide to what species can be seeded successfully on a particular site [81]. In the Intermountain Region, crested wheatgrass and Russian wildrye are considered to be the only introduced grasses sufficiently drought resistant for foothill areas receiving twelve inches of annual precipitation or less [107]. A much greater number of species are available for consideration on mesic sites.

An interaction between precipitation and grazing is frequently found on arid and semiarid sites. On an upland clay site in western Kansas, seeded stands of bluestems survived for twenty years under no grazing and made up 90 percent of the total plant populations [88]. However, only small amounts of bluestems survived even under moderate grazing. On this semiarid site, the bluestem mixture on the ungrazed sites yielded no more forage than did seeded stands of blue grama, a shortgrass.

The seasonal distribution as well as the total amount of precipitation should be considered in selecting perennial forages for seeding. In the Southern Great Plains where 70 percent of the precipitation falls during the summer six months, seeding cool-season grasses such

265

as western wheatgrass, Canada wildrye, crested wheatgrass, and intermediate wheatgrass on upland soils is generally unsuccessful [68]. In the Intermountain and Pacific Northwest regions where the precipitation falls primarily in winter and very early spring, warm-season grasses are generally unadapted and only cool-season grasses are recommended [107]. In the Northern Great Plains, the Midwest, and most of the Southeast, both cool-season and warm-season grasses are adapted.

The ability to survive prolonged periods of water inundation or tolerate saturated soil conditions is an important consideration on wet bottomlands, river floodplains, and marshy areas. Reed canarygrass (Figure 70) is so well adapted to seeding on wet sites that it often provides a dense, tough sod able to support livestock and machinery on land previously too boggy for use [110]. The death of plants susceptible to inundation is apparently caused by a lack of oxygen and an accumulation of carbon dioxide rather than a direct effect of the water [87]. It has also been found that grass plants can withstand longer periods of inundation when dormant, semidormant, or during early growth stages while the air and water are relatively cool [87]. Seedlings are less tolerant of flooding than mature plants [83].

FIG. 70. *Reed canarygrass is highly adapted to nonsaline, wet sites but becomes stemmy unless grazed or cut for hay in vegetative stage.*

Soils affected by an excess of basic salts are limited to only a few adapted forage species. Tall wheatgrass is one of but few species that can germinate, establish, and produce at a pH of 10 [12]. Russian wildrye and alkali sacaton are two others adapted to high pH. In evaluating the relative salt tolerance of forage species, it should be noted that this can vary considerably between genetic strains, stages of growth, soil moisture levels, and type of salts in the soil.

Both tall wheatgrass and tall fescue successfully compete with fox-tail barley on wet, salty sites [18, 112]. Forage species such as tall wheatgrass, western wheatgrass, tall fescue, bermudagrass, and bahiagrass will tolerate both moderately high soil salinity and a high water table (Figure 71). However, bermudagrass and bahiagrass will not tolerate the cold winters of the northern half of the United States. Forage species that tolerate moderately high amounts of salinity but are intolerant of prolonged flooding are Russian wildrye, basin wildrye, four-wing saltbush, sweetclover, and birdsfoot trefoil. By contrast forage species such as reed canarygrass, alsike clover, meadow foxtail, and redtop tolerate high water tables but not high levels of salinity.

Other combinations of site adaptations are also of interest. Blue grama, black grama, sand lovegrass, and needle-and-thread will tol-

FIG. 71. *Tall wheatgrass is outstanding in its ability to tolerate both high salinity and a high water table and remain productive.*

erate drought but not high soil salinity. Crested wheatgrass and Russian wildrye are almost unexcelled in reliability to produce early spring grazing under cold, droughty conditions. Other species that are both drought and cold hardy are bulbous bluegrass, buffalograss, blue grama, needle-and-thread, Indian ricegrass, sweetclover, curlleaf mountainmahogany, and many other shrubs. Although rose clover, Lehmann lovegrass, smilograss, and Turkestan bluestem are drought hardy, they have low cold hardiness. Cold hardy species intolerant of drought include Indiangrass, redtop, and timothy.

Adaptation to management. Species selected for range seeding must be adapted not only to the sites selected for seeding but also to the projected use and management of the new forage stands. Considerations relating to management include:

1. Ease of establishment.
2. Stand maintenance.
3. Forage usability by season of year.
4. Season of maximum growth.
5. Tolerance of grazing.

These factors are evaluated in Table 14 for important perennial forage species. Other factors with management implications include:

6. Palatability to different kinds of grazing animals.
7. Stability and amount of forage yield.
8. Nutritive value.
9. Adaptation to use in mixtures.
10. Seed production possibilities for supplemental income.
11. Hay production potential as a supplemental crop.
12. Response to fertilization, weed control, and supplemental water.
13. Ability to regrow following grazing.
14. Length of green period.

From these characteristics can be selected the plant species that best fits the need of the ranch or grazing unit. The kind of grazing animal or animals to be benefited, the seasonal feed requirements, the stage of production, and the level of management the operator plans to provide will determine the needs on the ranch in question.

Palatability is an important characteristic of seeded grasses. A moderately high palatability is generally desirable in increasing forage intake by grazing animals, uniformity of grazing, and alternative grazing uses. Forage species recommended for grazed firebreaks are those resisting fire spread, having a long growing season, remaining green under cold and drought, and being responsive to fertilizer. They must also be highly palatable so they will be kept closely grazed [32]. Because tall wheatgrass has rather low palatability for sheep even when seeded in pure stands but has fair palatability for cattle, uniform grazing is much easier to obtain with cattle. Bermudagrass in Oklahoma was found to have high carrying capacity but to be distinctly less palatable than native grasses by midsummer [66]. It was concluded that bermudagrass was better adapted to a cow-calf operation than a stocker-feeder program where maximum intake and gains are desired.

Palatability of a species can vary considerably. In evaluating the factors that can modify palatability of a forage plant, the following have been noted [89]:

Inherent factors

1. Degree of maturity—new growth is generally more palatable than mature forage.
2. Rate of recovery—regrowth after grazing will be preferred to remaining stubble.
3. Drought resistance—grasses that stay green during dry seasons are preferred over those that dry and mature.
4. Kind of animals—forage species have different palatabilities to different kinds of grazing animals.
5. Local conditions including soil moisture, available sunlight, etc.

Management factors

1. Intensity of grazing—heavier stocking rates increase the relative palatability of species otherwise low in palatability.
2. Other species with which associated—relative palatability is greater when in the presence of species of lower palatability.
3. Proportion in mixtures—species present in large amounts tend to make up a greater portion of the diet.
4. Previous grazing—livestock tend to graze regrowth of plants grazed previously.
5. Nitrogen fertilization often increases palatability of forage grown on nitrogen deficient sites.

Species origin. The relative merits of native versus introduced forage species for range seeding have not been fully agreed upon by range researchers and technicians. Some seeding recommendations are based upon the presumption that only the native species should be considered for permanent seedings. This implies that all introduced species will necessarily prove short-lived, be unable to perpetuate themselves under range conditions, and should be considered only for short-term pasture; but this is untrue.

The advent of highly adapted grasses such as bermudagrass in the Southeast, crested wheatgrass in the Northern Great Plains and In-

FIG. 72. *Crested wheatgrass, an introduced species, seeded near Briggsdale, Colorado, in the 1930s and subsequently grazed by cattle without evidence of stand deterioration.*

termountain Region, and smooth brome in the Midwest and mesic areas of the Intermountain Region has tended to bridge the gap in stand longevity of introduced species. In the sagebrush-grass zone of southern Idaho, twenty- to thirty-year-old stands of crested wheatgrass (*A. desertorum* and *cristatum*) have maintained themselves and even spread despite such adverse factors as heavy use, extremes of temperature and moisture, and disease [46]. When compared to seeded native species, crested wheatgrass apparently withstood drought, cold, and disease just as well, was more tolerant of fire and severe grazing, and spread better. The crested wheatgrass reached maximum production in two to five years before decreasing to stable, long-time production levels [46]. Factors in southern Idaho that favored spread of crested wheatgrass have been lack of competing vegetation, rough soil surface, and soil movement sufficient to provide coverage of seed produced on the site [47].

Crested wheatgrass has maintained stands for over forty years in the Northern Great Plains without showing signs of deterioration (Figure 72) [92]. Looman and Heinrichs [63] concluded that crested wheatgrass was a suitable long-term replacement for native range in dry regions of southern Saskatchewan. An equilibrium had apparently nearly been reached in thirty- to forty-year-old seedings, with invading native grasses seldom making up more than 10 percent of the vegetation, but production still two to three times that of native range. In North Dakota a twenty-eight-year-old stand of crested wheatgrass has produced 89 pounds of gain by yearling steers per acre compared to 104 pounds by a six-year-old stand [91]. Adjacent native range during the same period averaged 42 pounds of gain per acre. Crested wheatgrass seedings on foothill range in central Utah arc productive after over thirty-five years of grazing.

Many smooth brome plantings on western mountain ranges are over sixty years old, and smooth brome appears to be the most widely adapted species on such mesic ranges [81]. Smooth brome or smooth brome-legume seedings on cleared forest land and adjacent broken native grassland in the prairie provinces in Canada eventually reach an equilibrium with the environment and form distinct communities in which some of the composition, mostly under 2 percent, is comprised of native plants [63]. These seedings doubled grazing capacity, were more palatable, and extended the grazing season into the spring and fall.

In eastern and central South Dakota, Russian wildrye has maintained stands for almost thirty years but without apparent reproduction [92]. In this same area smooth brome has maintained itself for fifty to sixty years but tends to become sod-bound and low producing unless fertilized and renovated periodically. Stands of introduced grasses have commonly been invigorated and prolonged by periodic nitrogen fertilization. However, in Saskatchewan the exhaustion of nitrogen by the more productive crested wheatgrass stands has not been a factor in persistence of stands [95a].

The intensity of management that will be provided by the operator is an important factor in deciding what species to seed for perennial pasture; many introduced species require more intensive management than native plant mixtures [68]. On sandy soils in Colorado intermediate wheatgrass was concluded not to provide permanent pasture [22]. One apparent reason was that late spring and summer rains favored invasion of warm-season increaser grasses. Since the presence of alfalfa prolonged the life of the stand in Colorado, it is conceivable that nitrogen fertilization might have been even more effective in maintaining intermediate wheatgrass. On Utah foothill ranges both crested and pubescent wheatgrass were almost closed communities six years after seeding, but neither intermediate nor tall wheatgrass was completely closed even after twenty years [17]. On the more productive sites, it may be practical to replant deteriorated stands of medium lived, high forage yielders such as alfalfa, intermediate wheatgrass, weeping lovegrass, pangolagrass, or perennial ryegrass.

271

The natural distribution of forage species in their native habitat indicates areas of known adaptation. Native forage species are commonly given priority until adapted exotics have been proven adapted [68]. However, Harlan [33] has projected that fewer native species and more introduced species will be planted in the future in the Southern Great Plains and that native range in that area will gradually be replaced by improved materials. He further indicated that most forage plant breeding work in the past had been slanted towards cultivated lands but that more attention will be given to rangelands in the future.

Regional differences in the respective use of native and introduced forage species in permanent and semipermanent pastures are great. In the Northern and Southern Great Plains both native and introduced species are widely used. In the Midwest, Southeast, and Northeast nearly exclusive use is made of introduced perennials. In the Intermountain, Southwest, and Pacific Coast regions some use is made of native grasses and shrubs, but even greater use is being made of introduced grasses and legumes. Introduced species are generally better for spring forage than the native foothill grasses and more palatable and more nutritious over a longer grazing season [17]. But both irrigated and nonirrigated seedings of warm-season, native grasses for summer grazing in the Great Plains have been not only successful but highly productive.

Kinds of plants. Perennial grasses have been much more widely used in range seedings than other kinds of plants. Grasses may not be ideal for all purposes of range seeding but have been the plants for which seed has been commercially available and for which establishment techniques have been developed [6].

Browse seeding has been used principally in the past to increase quality and quantity of forage for big game on winter range. However, palatable native shrubs also show promise for seeding for domestic livestock grazing. Native shrubs interplanted with grasses on semiarid seeded range can improve the quality of forage consumed by livestock, particularly in fall and winter, and can improve the availability of forage during years of drought or deep snows [109]. Many shrubs have great capacity to regrow vigorously following top removal by grazing, fire, or mechanical treatment and produce regrowth of good palatability and nutritive value [70].

Native shrub species showing greatest promise for revegetation in the Intermountain area have included bitterbrush, cliffrose, winterfat, the mountainmahoganies, selected varieties of big sagebrush, four-wing saltbush, and other saltbush species. Bitterbrush has possibly received the greatest attention, being readily established by direct seeding, from rooted stem cuttings, or from nursery and container stock transplants (Figure 73) [104]. *Kochia prostrata,* a promising, long-lived shrub from the arid and semiarid parts of southern Eurasia, has received considerable interest recently for seeding in western U.S. because of its drought resistance, longevity, and productivity [51].

Formidable difficulties in using shrubs for range revegetation in the past have been seed production and collection [70]. Others have included shrub establishment, complexity of managing grazing on multiple-species seedings, and simply a lack of information.

While shrubs are particularly advantageous in providing winter forage, forbs are most beneficial to wildlife and livestock during the spring and summer growing periods. Forbs seldom cure well and generally weather and disintegrate quickly following maturity. Legumes have long been an important constituent of pastures in high rainfall areas where management input is high. However, they have been used sparingly or not at all on rangelands where precipitation is low and seedbed preparation is difficult. Low persistence and bloat hazard have been suggested as the two principal factors limiting the use of legumes in range seedings [36].

Alfalfa is the most important and most widely used forage legume on rangeland. Alfalfa has proven adapted on mesic sites and many semiarid sites receiving fourteen to twenty inches of annual precipitation and on irrigated and subirrigated lands with good drainage. The roots of alfalfa readily reach water tables ten to fifteen feet below the surface. Alfalfa is highly productive and nutritious, tolerates grazing, makes rapid regrowth, and is well adapted to seeding with grasses.

The many varieties of alfalfa vary greatly on such factors as resistance to bacterial wilt and to the spotted alfalfa aphid, winter hardiness, drought hardiness, and rooting habits. Varieties such as Ranger, Vernal, and Ladak are taprooted, grow upright, make rapid regrowth, are high yielding, and have found widespread use for hay production and in irrigated pastures. However, the taprooted alfalfa varieties are susceptible to root cutting by gophers. Rhizomatous, creeping varieties such as Rambler, Travois, and Teton are more resistant to gopher damage, since they are able to regenerate by lateral roots after the taproot has been cut. The creeping rooted varieties also appear to withstand grazing better and maintain stands better under dryland conditions [36]. It appears that longevity and persistence under range conditions will have to be stressed over yield in future alfalfa breeding research, with special attention being given to tolerance or resistance to drought, grazing, wildfire, and insect and rodent pests [36, 113].

FIG. 73. *Bitterbrush, a native woody shrub, has demonstrated its value for browse seeding on both big game and livestock range.*

Various clover species have proven useful in permanent pastures in the eastern half of the United States. They have also been successfully used on subirrigated and other mesic sites, including irrigated land in the West, but lack drought hardiness. Sainfoin (*Onobrychis viciafolia*), cicer milkvetch (*Astragalus cicer*), falcatus milkvetch (*Astragalus falcatus*), and birdsfoot trefoil (*Lotus corniculatus*), all nonbloating legumes, show considerable promise for seeding rangelands in many regions of western U.S. and Canada [36, 81, 101, 113]. However, birdsfoot trefoil appears limited in use for areas where drought is less severe and grazing can be readily controlled.

Extensive trials in the Southern Great Plains led to the conclusion that native legumes offer little promise for range use [55]. Poor seeding habits and difficulties in establishment and maintenance eliminated most native legumes and many introduced exotics. By contrast, yellow sweetclover has naturalized on ranges in South Dakota and Wyoming. In western South Dakota it was found well adapted as a legume for rangelands with heavy clay soils and to be a compatible associate with native vegetation, particularly western wheatgrass [74]. Since sweetclover is a biennial, it must reestablish itself every two years from seed. The hard seeds produced aid this species in surviving bad seed production and low establishment years.

Forbs other than legumes have seldom been used, although some interest has been shown in small burnet (*Sanguisorba minor*), arrowleaf balsamroot, and penstemons in seeding big game ranges [81]. Annual forage species are important in temporary pastures but are seldom included in range seedings. An exception is the seeding of winter annual legumes into annual grass foothill ranges in California, a practice which has nearly doubled forage production, particularly winter forage production [50]. The winter annual legumes used in this area are rose clover, subterranean clover (*Trifolium subterraneum*), and bur clover.

Varieties. Both native and introduced forage species include many varieties or ecotypes. Varietal differences may be large or small, but varietal differences for certain characteristics may be larger than species differences. In alfalfa breeding programs, factors selected in strains for range environments include (1) drought resistance, (2) winter hardiness, (3) resistance to common diseases, (4) persistent grazing survival, and (5) ability to grow and reproduce on the range [93].

Varieties or cultivars must be adapted and serve useful purposes. However, quality and performance are generally secondary to good seed production and good seeding habits in setting user preferences and making the variety successful [33]. Superiority of performance must be very great to make up for any deficiency in reproductive efficiency. There is a continuing need for the breeding and selection of both native and introduced forage plants for extreme range environments. Sufficient genetic variability exists in most species to make selection procedures effective. Some seeding failures with both native and introduced forage species have resulted from planting the wrong ecotypes.

The original source of strains of native species is important in determining their adaptation. For best results seed of warm-season native grasses should originate from local sources or from up to 150 miles south of planting location [14]. Native grass seed should never be used more than 300 miles north or more than 200 miles south of its origin [3]. Also, the seed should be planted in precipitation zones and on soils relatively similar to the origin of the seed. Seed of browse plants should also be obtained from local native sources unless a non-local source has proved superior [19]. Seed collections made in the southern desert shrub region are often unadapted for seeding in the northern desert shrub region [80].

275

Strains from northern sources are generally earlier maturing, lower in forage yields, more cold hardy, and more susceptible to disease than southern strains. On the other hand, southern strains are later maturing and produce better stands and higher yields of forage except where lack of winter hardiness, shortness of season, or limited soil moisture offsets the greater growth potential [14, 105]. With introduced species the place of seed production will have little effect on the adaptability of the seed produced for planting.

Single Species Versus Mixtures

Grass mixtures. Advantages have been claimed for seeding both mixtures and a single grass species on nonirrigated land. These claims are summarized as follows [43, 48, 68, 106]:

Advantages of mixtures
1. Less risk in getting a stand where soil is heterogeneous.
2. Longer green and succulent period.
3. Higher yield.
4. Higher nutritive value.
5. Provide a varied diet.
6. Stand less liable to total loss from adverse climatic conditions or insect damage.
7. Occupancy of land more rapid; better ground cover.
8. Better adapted to variable canopy, soil, terrain, and climatic conditions.
9. More efficient use of the total soil profile.
10. Provides more multiple uses, i.e., livestock and big game.
11. Species can be included when seed is too scarce or limited for full seeding.

Advantages of single species
1. Easier to seed evenly.
2. More uniform grazing because of more uniform palatability.
3. Provides feed at season most needed.
4. Plants have similar growth and regrowth characteristics.
5. More precise site adaptation.
6. More stable plant composition.
7. Cheaper to establish.
8. More rapid establishment of primary species.
9. Only one species may be fully adapted to the site.

In seeding range and other perennial pastures, the trend has been toward single species or simple mixtures rather than complex mixtures. More emphasis is now being given to what is needed rather than a "shotgun" approach. Seeding a single species or a simple mixture of not more than two to five species with similar palatability, season of growth, grazing and drought resistance, and regrowth rate greatly reduces management problems [39, 44, 68]. Added species should be included generally only where added benefits can be expected [103].

In studies on sagebrush-bunchgrass range in Oregon, crested wheatgrass established rapidly and was as effective as mixtures in reducing brush reinvasion [35]. It was concluded that a single species which concentrates growth and competitiveness may control some weedy plants better than a mixture. Including a second grass with crested wheatgrass reduced total herbage production. Herbage production of mixed stands under these semiarid conditions was intermediate between those of pure stands of the respective species seeded alone. Planting various combinations of two-species seed mixtures of native grasses in Kansas did not significantly influence first-year plant numbers when compared with pure species at similar seeding rates [59]. Mixtures of native tall and short grasses seeded on western Kansas uplands reduced annual yield variations somewhat, but the tall grasses seeded in pure stands produced higher average yields but did not reproduce satisfactorily on marginal sites [58]. The taller grasses gave way to lower yielding invading short grasses but more slowly than when seeded in mixtures with them.

The relative palatability of the component species largely determines the future of species seeded in mixtures. The palatability of the various species not only is relative but often varies from season to season. Grazing animals and particularly livestock tend to overgraze the plants in the mixture most palatable at the time of grazing, eventually reducing or killing them out (Figure 74). Wheatgrass species in a mixture seeded on a sagebrush site in Utah produced equal amounts initially [17]. But after a few years of grazing, 46 percent of herbage production was by crested wheatgrass, 34 percent by pubescent wheatgrass, 14 percent by intermediate wheatgrass, and 6 percent by tall wheatgrass. Lower site adaptation and grazing tolerance may also have contributed to the reduction of intermediate and tall wheatgrass.

Pubescent, intermediate, and Fairway crested wheatgrass, but not bluebunch wheatgrass, grew well together in southern Idaho for twenty-five years under irregular, light, spring-fall grazing by sheep [42]. It was projected that the three introduced wheatgrasses in mixture could provide a variety of feed and excellent ground cover under early spring use but that crested wheatgrass would take over the site under late spring and summer use.

Stocking levels geared to the more palatable species wastefully utilize less palatable species. This problem is seldom overcome even under forced rest-rotation grazing. If the less palatable species are productive, suited to livestock production, and capable of replacing the more palatable species as they die out, there may be no advantage in including more palatable species unless they materially increase production in the first few years [21].

Seeding mixtures of grasses for the purpose of having palatable species throughout the grazing season has generally failed since all species in the mixture cannot be maintained [17]. Seeding an early cool-season grass, a late cool-season grass, and a warm-season grass in separate units is suggested. This permits rotating grazing between pastures on the basis of optimum growth and nutrition and when most suited for various phases of livestock production [17, 44].

In Nebraska, cool-season grasses were found to use soil water for a longer period than warm-season grasses [16]. The cool-season grasses depleted soil moisture in the spring that otherwise would have been available for summer growth of warm-season grasses. Thus, seeding cool-season grasses with the warm-season grasses contributed to mid-summer drought suffered by the latter. When planted in separate pastures, each grass can be grazed in its proper season, and a rela-

FIG. 74. *Patchy grazing on a Custer County, Nebraska, seeding resulting from seeding smooth brome (high palatability) and tall wheatgrass (low palatability) in the same mixture.*

278

tively uniform carrying capacity can be maintained throughout a long grazing season each year.

On lowland ranges with fairly even terrain and uniform soil conditions, seeding a single species of grass seems best (Figure 75) [44]. However, in mountainous or other areas having a variety of soil and moisture conditions, simple mixtures are often best [81]. The latter seems particularly true if the area is to be managed extensively, grazed at various times of the year or by different kinds of grazing animals. Or lands to be fenced in with native range in good condition may best be seeded to native species using approximately the same ratio found on the native portion [85]. Definition and singleness of purpose as well as plans for intensive management favor grasses seeded in pure stands or a simple grass-legume mixture.

Grass-legume mixtures. Seeding a legume with range grasses has received renewed interest for increasing production and replacing the need for nitrogen fertilizers. Where both are adapted, a mixture of a grass and a legume generally produces more forage than grass seeded alone [39, 72]. The presence of the legume on low nitrogen sites increases the protein content of the grass component [27] and often the production of the grass component alone if soil moisture is not limiting [8]. The higher yield results from the legume, commonly alfalfa,

FIG. 75. *Crested wheatgrass seeded near Fairfield, Utah, on privately owned range for spring-fall grazing by livestock. Even terrain and uniform soil conditions, ease of management, seasonal feed requirements and high site adaptation favored the seeding of a single species.* (U.S. Forest Service photo)

279

drawing nutrients from deeper depths and supplying nitrogen by bacterial fixation for grass usage [23].

On six-year-old seeded pastures grazed by yearling steers in North Dakota, crested wheatgrass-alfalfa pastures produced 129 pounds of gain per acre annually compared to 104 pounds of gain on crested wheatgrass pastures [91]. In Saskatchewan, adding one pound of alfalfa in dryland seed mixtures (1) increased yields by 55 percent and carrying capacity by 40 percent, (2) increased percent crude protein, (3) increased weight gains when grazed by ewes by 135 percent per head and 215 percent per acre, and (4) reduced consumption per pound of liveweight gain [11]. In eastern Wyoming, adding alfalfa to crested wheatgrass increased carrying capacity for sheep by one-third and produced 50 percent more lamb gain [5].

On sandy soils in northeastern Colorado, alfalfa increased the production of intermediate wheatgrass by 46 percent when seeded in mixture and increased total forage production (alfalfa and intermediate wheatgrass) by 120 percent [22]. The mixture produced 44.4 pounds of yearling cattle gain per acre compared to 25.5 pounds for the grass alone. Adding alfalfa increased average carrying capacity for cattle by 30 percent but by 67 percent in those years when spring drought was not a problem. In Saskatchewan the value of alfalfa in a dryland mixture was greatly diminished in drought periods but still increased yields by 20 to 25 percent [52].

Alfalfa is less commonly recommended for including in warm-season grass mixtures. However, on nonfertilized sandy soils in northeastern Colorado, including alfalfa in the mixture increased total forage production by 93 percent and yield of warm-season grasses by 19 percent [22].

Single grass-single legume mixtures in the West are generally as productive as complex mixtures [48]. Also, after several seasons of use, complex mixtures are apt to become a one grass-one legume combination because of selective grazing and differences in competitive ability of the species used [48]. On mesic sites in the Midwest having several grasses and legumes in addition to alfalfa and smooth brome was of no added advantage [103]. However, on mesic sites and irrigated pasture in the Central Great Plains, adding orchardgrass to smooth brome-alfalfa mixtures has maintained a higher grass yield component during midsummer (Figure 76).

Dryland studies in the Northern Great Plains led to the conclusion that there is little advantage from the use of more than one grass grown in association with alfalfa [53]. Combinations of two persistent grasses inflicted a two-way squeeze on the alfalfa associate. Russian wildrye, streambank wheatgrass (*Agropyron riparium*), and smooth

FIG. 76. *Simple grass-legume mixtures are highly productive on mesic sites and irrigated pastures. Photo of irrigated pastures at the University of Nebraska North Platte Station utilizing electric fences.*

brome were the most competitive grasses. Crested wheatgrass was moderately competitive with the alfalfa. Intermediate wheatgrass and green needlegrass were the least competitive and considered the only species tried that showed promise as a second grass in the mixture.

Seed Quality

Labeling and pure live seed. Grass seed lots often vary widely in quality. Some lots of seed contain inert material, undesirable mixtures, weed seeds, and immature or injured grass seeds that will not grow. The seed label is an important source of information about the seed (Figure 77). Labeling is required by law for all seed, both common (uncertified) seed and certified seed, except that sold by farmers [102]. The label tells the kind or the kind and variety of the pure seed. It also gives the percent of pure seed, hard seed (having impermeable seedcoat or other germination retardant), other crop seeds, weed seeds, and inert matter. Percentage germination (which excludes dormant or hard seed) must be shown, as well as the date of the germination test, which must be not over nine months old in most states. The names and number per ounce or pound of noxious weed seeds are shown. For native harvest seed it is also important to include the origin of the seed, i.e. geographic description, latitude and longitude, vegetation type, and local political subdivision.

Since only live seeds of the desired crop are of value, buying seed on the basis of pure live seed of the desired species or variety assures full value for money spent on seed. This practice also protects producers and handlers of high quality seed against unfair competition from low quality seed. *Pure live seed* (PLS) refers to the portion of a lot of seed that is live seed of the desired kind. Percent PLS is determined by multiplying the percent of pure seed by the percent germination and dividing by 100.

Seed from different sources may be priced by weight on a bulk rate, PLS rate, or both. The price per pound of pure live seed and not the bulk seed price determines which seed lot is the better buy. Two lots of seed can be compared on the basis of PLS as follows:

Source #1 (poor quality)		Source #2 (better quality)	
Purity	50%	Purity	80%
Germination	40%	Germination	50%
Percent PLS	20%	Percent PLS	40%
5 lbs. bulk required to make 1 lb. PLS.		2.5 lbs. bulk required to make 1 lb. PLS.	

Prices are equivalent if source #1 sells for $.40 per pound on a bulk basis, and source #2 sells for $.80 on a bulk basis. Sources #1 and #2 would both sell for $2.00 per pound of PLS according to the following formula:

$$\text{Cost of PLS per lb.} = \frac{\text{bulk seed price per lb.} \times 100}{\text{percent PLS}}$$

Seed certification. Seed certification is a system used to promote production of seed of genetic purity and identity and make the seed available for purchase. It also provides a means of approving and releasing new varieties within a state. Seed is certified in the U.S. on a state basis by an agency or organization authorized by the state legislature such as a crop improvement association, the state department of agriculture, the state extension service, or the agricultural experiment station.

The authorized certification agency regulates the production, harvesting, cleaning, and marketing of each lot of certified seed. It inspects the fields where the seed is produced. All states use a blue tag to identify certified seed and provide a sealing device which prevents a bag from being unofficially opened before reaching the consumer.

A blue tag is the stamp of acceptability from an impartial agency and shows as a minimum the state of certification, the crop and variety name, and identifies the grower by the lot number.

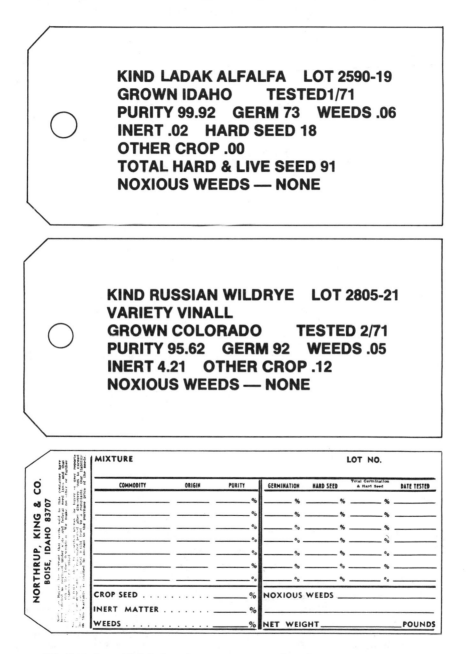

FIG. 77. *Sample seed labels for a legume, a grass, and a mixture.*

FIG. 78. *Certified seed tag (front and back) used in the "two-tag" system of labeling certified seed.*

Certification guarantees the variety or strain, its known varietal potential, that it has been tested under field conditions, and that the variety is recommended for the state in which it is certified [102]. The site on which the certified seed is produced must meet minimum isolation standards to insure continuation of genetic purity. Certified seed must meet minimum requirements as to germination and limitations on weed seeds, other crop seeds, and inert matter. Certified seed must be accompanied by a regular seed label the same as other lots of seed offered for sale showing germination, date tested, weed seeds, and so forth. In contrast to this two-tag system of labeling certified seed (Figure 78), a one-tag system is permitted in some states, which places all of the information about the seed on a single blue card; it is, however, less versatile. Since certified seed is mostly produced under irrigation or on mesic sites, the supply of certified seed tends to be more stable than for native harvest seed.

Production and processing. The production of high quality seed depends upon the use of good seed stock, proper seedbed preparation, the best seeding dates and rates, a wide row spacing and proper planting depth, proper management practices of irrigation, weed control, fertilization, and the control of insects and diseases. The seed must be harvested at the proper stage of maturity by the proper machines and methods. Following harvesting, the seed must be properly dried, packaged, stored, and protected against postharvest damage by insects and disease [3, 102].

Large seeds of most grass species germinate faster, give quicker emergence, and produce more vigorous seedlings than do small seeds. Large seeds are often slightly higher in percent germination also [31]. Cultural practices to obtain maximum development of caryopses as well as large yields are important because of the great importance of seed quality in stand establishment [95]. Good quality seed of grasses such as green needlegrass is difficult to produce because of a lack of uniformity in maturity associated with determinate inflorescences and rapid shattering [54]. Mature seeds of green needlegrass are indicated by a uniform gray color and have higher seed weight, average viability, and quality compared to immature seed recognized by light gray and yellowish green color. Seed color has also been used to evaluate bitterbrush seed since spotted seeds have germinative capacity of only 50 to 82 percent of that of nonspotted seed [24].

Seed debearding and cleaning is often required. The removal of weed seeds and seeds of other plants improves seed purity. The removal of dirt, leaves, stems, and chaff from the seed reduces the bulk of handling and storage, removes moist material that may cause heating in storage, and facilitates flow through seeding equipment. Four-wing saltbush seed dewinged in a hammer mill was found easier to plant at proper depths and germinated more rapidly [97]. Threshed seed of winterfat have been found less subject to wind movement and easier to handle and seed with mechanical equipment than the fluffy whole fruits [96]. However, these treatments are less practical on small seed lots.

Lack of commercial sources of quality seed has greatly limited the use of many productive native plant species. One example is found in central California where the native needlegrasses, purple needlegrass (*Stipa pulchra*) and nodding needlegrass (*Stipa cernua*), have largely been replaced by annual bromegrasses. Although these needlegrasses are valuable for fall and winter grazing, can be readily reestablished by seeding in prepared seedbeds, and can be maintained against encroachment by annual grasses even under grazing, their practical use in range seeding is stymied by lack of commercially available seed [30]. Seed production problems with these needlegrasses have included low seed production, seed shattering, difficulty in harvesting, and the necessity of removing the awns and sharp calluses for effective handling.

The revegetation of roadsides, mining-disturbed areas, and grazing lands in harsh environments has greatly increased the demand for seed of native plants. Both research and the application of knowledge from related fields has been drawn upon to develop efficient techniques for (1) collecting, (2) threshing and cleaning, (3) storing, (4) germinating, and (5) propagating seeds of native plant species [115]. Emphasis has been given to shrub seed in the Intermountain area because of increasing demand.

Seeds of native shrubs in the past have principally been obtained by hand collection from native stands. However, seed production of native stands has varied greatly from year to year, the collection sites have often been of difficult access, and hand collection has provided only small quantities of seeds. However, more use is now being made of browse seed harvesters using vacuum suction for collection, and further improvement in such machines shows much promise in accelerating shrub seed production, particularly along roadsides, seeded plantations, and other readily accessible areas. Four-wing saltbush seed production has been shown to be practical on cultivated agricultural lands [100], thereby offering an alternative to sole reliance on harvesting native stands. Furthermore, such shrub seed orchards under irrigation in Utah have proved profitable not only in producing seed, but also in providing valuable grazing for livestock in winter [108].

Seed dormancy. The inactive condition of seed referred to as *dormancy* (an internal condition of the chemistry or stage of development of a viable seed that prevents its germination even when temperature and moisture are adequate) may be permanent or only temporary. Germination in dormant seeds is often prevented by blocking mechanisms within the seed, such as chemicals that inhibit the growth of the seed embryo, or an impermeable seedcoat.

Freshly harvested seeds of many forage plants have a dormant period during which germination is greatly retarded or completely inhibited. Although this is an advantage in nature by helping plant species avoid complete germination in years of hazardous establishment, the delayed dormancy interferes with artificial seeding success. Winterfat seeds take up to about 10 weeks to complete after ri-

286

pening [98]. Newly harvested seed of some native grass species such as green needlegrass, Indian ricegrass, switchgrass, Indiangrass, and sand lovegrass contain a high percentage of dormant seeds. Keeping such seed in dry storage for an after-ripening period of eighteen months before planting is recommended for increasing germination [14]. When seeding switchgrass and Indiangrass, forty to sixty live seeds of new seed have been required to give stands equivalent to twenty live seeds per square foot of older seed [15]. Unless damaged by improper storage or by insects, maximum germination of most grass seed can be expected from six months to two years of age under proper storage.

Fall or winter seeding of new seed of cool-season grasses has been used with considerable success in breaking new seed dormancy; but this technique is most effective when temperature does not get excessively cold, the seeds in the soil remain moist but not saturated, and the cool-moist period lasts for 30 days without interruption [115]. Prechilling and soaking of seed in the laboratory has given favorable effects similar to overwintering in the soil. Keeping seed in a cool, moist environment for a period of time to simulate overwintering is called *stratification.* Since rapid germination and establishment is important while moisture and temperature are still optimum, cold stratification before spring seeding of species with known new seed dormancy should be considered.

Cold-moist stratification in sand or vermiculite from two weeks to a few months has been an effective and inexpensive method for breaking seed dormancy. Keeping green needlegrass in moist sand at 2° to 4°C for sixty days has proven effective [90]. Stratification for four to six weeks or fall planting has given good germination of bitterbrush seed [77]. Stratifying bitterbrush seed has required as little time as two weeks when kept at temperatures of 0 to 5°C and in a moist environment near field capacity [114]. The germination of new true mountainmahogany (*Cercocarpus montanus*) seed wetted and chilled at 5°C for twenty-two days compared to untreated seed was 24 versus 0 percent three days after seeding, 60 versus 4 percent after seven days, and 62 versus 48 percent after fourteen days [105]. Seedlings from treated seed were more vigorous and freer from mold than untreated seed. Many species including shrubs respond to cold-moist treatment; some require warm-moist treatment or a combination of both.

Although some wetting and drying treatments have increased the germination of crested wheatgrass seed, severe and prolonged wetting-drying treatments have reduced vigor of germinating seeds [65]. The results of stratifying winterfat seed have been inconsistent with generally no advantages from soaking or washing seeds in water [97]. Stratified seed should be dried before planting but may remain somewhat softer than untreated seed. Moist, imbibed seeds may not pass through drilling equipment and must be placed in a seedbed environment conducive to continued germination [113]. While dry seeds of alkali sacaton have withstood temperatures in excess of 110° C without loss of viability, imbibed seeds were reduced in germination by heat treatment at 60°C or more [56].

Cold stratification hastened the emergence of intermediate wheatgrass and mountain rye (*Secale montanum*), aided their survival on dry sites, and made spring planting comparable to fall seeding [25]. Fall planting or storage in a snowbank or refrigerator prior to spring planting enabled a seed crop to be produced the first year. Cold stratification under natural conditions appears to favor range shrubs and forbs becoming established ahead of introduced grasses seeded in the spring [71]. Of the twenty-five native broadleaf species studied in British Columbia, seeds of ten species germinated as well without stratification, five required stratification, and ten germinated better following stratification.

Seed may be scarified to break an impermeable seedcoat and allow absorption of water. Mechanical scarification and sulfuric acid treatment as well as cold stratification have increased the germination of Indian ricegrass [82]. A 70 percent sulfuric acid treatment for sixty minutes was the best treatment for Indian ricegrass seed germinated outdoors, but cold stratification further increased germination. The dual treatment increased germination from only 3 to 54 percent.

Soaking bitterbrush seed for three to five minutes in a 3 percent thiourea solution and then drying has increased germination to 85 percent or more [77]. Thiourea treatment allows spring seeding of bitterbrush and thereby mostly avoiding frost hazard and rodent depredation resulting from fall seeding. The dormancy of curlleaf mountainmahogany (*Cercocarpus ledifolius*) has been effectively broken by a five-minute bath in concentrated sulfuric acid followed by a four-hour immersion in 3 percent thiourea [60]. However, with fourwing saltbush, mechanical and chemical scarification was found ineffective [97]. The optimum length of concentrated sulphuric acid treatment will vary from six seconds to several hours depending upon the plant species being treated [115].

Seed protectants. A combination of Arasan 75 as a fungicide and repellant and endrin as a rodenticide and insecticide plus an adhesive when applied to seed prior to planting has greatly reduced rodent depredation [38]. In one test with bitterbrush, 88 percent of seedspots seeded with untreated seed were disturbed by rodents while only 2

288

percent of those seeded with treated seed were disturbed. Treating the bitterbrush seed with Arasan 75-endrin increased seedling stand ten times. Abnormally colored seed, such as bright yellow, will adequately prevent depredation by birds of planted seeds or poison grain bait. Special fungicides may be needed in some areas for treating grass seed before planting.

Legume inoculation. The successful establishment and production of legumes in range and pasture seedings depends on effective nodulation and nitrogen fixation. In order to assure nodulation, legume seed should be treated with a good commercial inoculant prepared from a strain of *Rhizobia* bacteria specific for the legume being planted. Mixing can easily be accomplished by mixing in a cement mixer or dry feed blender or in a can with a closed lid. Legume seed previously inoculated is now generally available for purchase. When properly inoculated, legumes add nitrogen to the soil and can materially reduce the amount and cost of nitrogen fertilization. Under California dryland conditions, range legumes effectively nodulated were found to fix at least fifty-two pounds of nitrogen per acre in one growing season [37].

Seed pelleting. Although much publicized and extensively tried under range conditions, pelleting seed for range seeding has been unsuccessful and cannot be recommended [45]. Pelleted seed requires a prepared seedbed equally as much as nonpelleted seed [13]. Poor germination and establishment have resulted from broadcasting of pelleted seed by airplane or by hand, and this limitation has not been overcome by increased seeding rates [10, 40]. In addition to sparser stands, the cost of the pelleting and the extra cost of handling the extra bulk have been high [13].

Various substances such as mud, fertilizer, and plastics have been used in making three basic types of seed pellets: (1) coated pellets in which individual seeds receive successive layers of powdered material, (2) extruded pellets made by pressing a pasty seed and soil mixture through holes, and (3) compressed pellets made by running a seed and soil mixture through pressure disks [40]. No pelleting process has shown consistent advantages over using nonpelleted seed on rangelands, and the practice of using pelleted seed has been virtually discontinued. Some pelleting processes have greatly reduced seed germination.

Table 14. Forage Plants Commonly Seeded on Range and Other Perennial Pasture

Common and scientific name	Seeds per lb. (1,000)[4]	Lbs. PLS. per acre seeding rate[7]	Ease of establishment	Stand maintenance	Drought tolerance	Cold hardiness	Salinity tolerance	Soil adaptation Sandy	Silty	Clayey	High water tolerance	Forage usability[5] Early spring	Late spring	Summer	Fall	Winter	Grazing tolerance	Native or introduced	Season of growth[2]
Alfalfa (*Medicago sativa*)	210	4.1	1	1-2	2	1-2	2	1	1	2	2	1-2	1	1	1	3	1-2	I	C-W
Bahiagrass (*Paspalum notatum* and *media*)	166	5.2	1	1-2	2	3	1	1	1	1	1	2	1	2	2	3	1	I	W
Bermudagrass (*Cynodon dactylon*)	1787	1.0	1-2	1	1-2	3	1	1	1	2	1	3	1	1	2	2	1	I	W
Bitterbrush (*Purshia tridentata*)	15	29.0	1-2	1	1-2	1	2	1	1	1	3	1	1	1	1	1	1	N	
Bluegrass, big (*Poa ampla*)	900	1.5	2	1-2	1-2	1	3	1	1	1	1-2	1	2	2	1	2	2	N	C
Bluegrass, bulbous (*Poa bulbosa*)	460	1.9	1	1	1	1	2	2	1	1	3	1	2	3	3	3	1	I	C
Bluegrass, Kentucky (*Poa pratensis*)	2150	.8	1-2	1	2	1	2-3	3	1	1	1-2	1	1	2	1	2	1	I	C
Bluestem, big (*Andropogon gerardi*)	130	6.7	2	1	2	1	2	2	1	1	2	2	1	1	1	2	1-2	N	W
Bluestem, Caucasian (*Bothriochloa caucasica*)	860	1.5	1-2	2	1-2	2	2	2	1	1	2	3	1	1	2	3	1-2	I	W
Bluestem, little (*Schizachyrium scoparium*)	255	3.4	2	1	1-2	1	3	1	1	1	2	3	1	1	2	2-3	2	N	W
Bluestem, sand (*Andropogon hallii*)	113	7.7	2	1	2	1	3	1	2	3	2	2	1	1	1	2	1-2	N	W
Bluestem, yellow (*Bothriochloa ischaemum*)	830	1.5	1	2	1-2	3	2	2	1	1	2	3	1	1	2	3	1-2	I	W

[1]Symbols: 1 = good, 2 = fair, 3 = poor
[2]Symbols: C = cool-season, W = warm-season
[3]The range and pasture regions are shown geographically in Figure 69.
  Symbols: R = range and nonirrigated perennial pasture.
      I = irrigated and subirrigated pasture (Western half of United States only).
[4]Seed weight information compiled from numerous printed sources.
[5]Forage usability by season considers green growth period, palatability, curing, and seasonal grazing tolerance.
[6]Four pure live burs per square foot.

Regional adaptation[s]								Principal cultivars[9]	Special considerations and adaptations
Pacific Coast	Intermountain	Southwest	Northern Great Plains	Southern Great Plains	Midwest	Southeast	Northeast		
RI	RI	RI	RI	RI	R	R	R	Mesa Sirsa, El Unico, Washoe, Ranger, Ladak, Rambler, Teton, Lahontan, Nomad, Travois, Drylander, Vernal, Roamer, Kane	Most widely used legume for range and pasture mixtures. Adapted to irrigated pasture and dryland sites receiving 15″ precip. or more.
					R			Pensacola, Tifhi, Argentine, Wilmington	Keep young by grazing or mowing. Rhizomatous.
RI		RI		RI		R		Midland, Coastal, Suwance, Coastcross, NK 37	Keep young by grazing or mowing and ample fertilization. Named varieties must be grown from sprigs; only common and NK 37 can be seeded.
R	R	R							Principal browse species used in range seeding. Palatable to all grazing species. Some varieties layer, varieties mostly persistent leaved.
R	R	R						Sherman	Seedlings may be pulled up by grazing. Very early growth similar to crested wheatgrass. Seed in pure stands.
R	R							P-4874	Good erosion control; withstands heavy grazing. Spreads by aerial bulblets and swollen stem bases. Low yield; unreliable producer.
					R		R	Delta, Troy	Low production and summer dormancy limit use on range and pasture.
			RI	RI	R			Kaw, Champ, Pawnee	Very palatable and productive on mesic sites. Seeded in warm-season mixtures.
				R					Lower in palatability than King Ranch bluestem but is more winter hardy. Seeded in pure stands generally. An "Old World" bluestem.
			R	R	R	R		Pastura, Blaze, Aldous	Widely used in warm-season grass mixtures on mesic and subhumid sites.
			R	R				Woodward, Champ, Cherry, Elida	Rhizomatous. Very palatable and productive on mesic, sandy soil.
				R				King Ranch, El Kan, Plains, Formosa	Seeded in pure stands; medium palatability. An "Old World" bluestem; also called Turkestan bluestem.

[7]PLS is pure live seed or germinable units; seeding rates based on pure live seeds per square foot drilled into prepared seedbeds as follows: 20 for medium size seeds (65 to 500 thousand per lb.), 10 for large (under 65 thousand), 30 for small (500 to 1,000 thousand), and 40 for very small (over 1,000 thousand).

$$\text{Pounds PLS/A} = \frac{\text{pure live seeds/sq. ft.} \times 43,560}{\text{seed units per lb. of pure seed}}$$

[8]Dewinged seed or shelled seed.

[9]A *cultivar*—derived from *cultivated variety*—is distinguished by any morphological, physiological, cytological, or chemical characters and retains these distinguishing characters when reproduced; equivalent to named varieties; the term cultivar differs from botanical variety, a category below the species written in Latin form (USDA, Soil. Cons. Serv. 1977. *National Plant Materials Handbook.* Washington, D.C.)

Table 14. (Continued)

Common and scientific name	Seeds per lb. (1,000)[4]	Lbs. PLS. per acre seeding rate[7]	Ease of establishment	Stand maintenance	Drought tolerance	Cold hardiness	Salinity tolerance	Soil adaptation Sandy	Silty	Clayey	High water tolerance	Forage usability[5] Early spring	Late spring	Summer	Fall	Winter	Grazing tolerance	Native or introduced	Season of growth[2]
Brome, mountain (*Bromus marginatus*)	70	12.4	1	1-2	2	1	2	3	1	1	2	3	1	1	2	3	1-2	N	C
Brome, smooth (*Bromus inermis*)	145	6.0	1	1	2	1	2	1-2	1	1	2	1-2	1	1	2	3	1	I	C
Buffalograss (*Buchloe dactyloides*)	42	4.1[6]	2	1	1	1	2	3	1	1	1-2	2	2	1	1	1	1	N	W
Buffelgrass (*Canchrus ciliaris*)					2	3		1	1	2		3							W
Burnet, small (*Sanguisorba minor*)	55	7.9	2	3	2	1	3	2	1	2		1	1	2	3	3	2	I	
Canarygrass, reed (*Phalaris arundinacea*)	506	2.6	2-3	1	2	1	2-3	2	1	1	1	1	1	2	2	3	1	N-I	C
Cliffrose (*Cowania mexicana*)	64	6.8	3	1	1	1	2	2	1	1	3	3	3	3	2	1	1	N	
Clover, alsike (*Trifolium hybridum*)	680	1.9	1	2	3	1	2-3	2	1	1	1-2	2	1	1	2	2	1-2	I	
Clover, crimson (*Trifolium incarnatum*)	140	6.2	1	2	3	1	2	1	1	1	3	1	2	2	1	1	1	I	W
Clover, Ladino white (*Trifolium repens*)	850	1.5	1	2	3	2	2-3	2	1	1	1-2	2	1	1	1	2	1	I	
Clover, red (*Trifolium pratense*)	270	3.2	1	2	3	1	3	2	1	1	2	2	1	1	2	2	1-2	I	
Clover, rose (*Trifolium hirtum*)	140	6.2	1	1-2	1	3	2	2	1	1	3	1	1	2	2	1	1	I	
Clover, strawberry (*Trifolium fragiferum*)	295	3.0	1	1-2	3	2	1-2	2	1	1	1	2	1	1	2	2	1	I	
Clover, subterranean (*Trifolium subterraneum*)	65	13.4	2	2	1-2	3	3	2	1	1	2	1	2	2	2	1	1	I	
Dallisgrass (*Paspalum dilatatum*)	220	4.0	2	1-2	2	2	2	2	1	1	1	2	1	1	2	2	1-2	I	W
Dropseed, sand (*Sporobolus cryptandrus*)	5000	.3	1	2	1	1	2	1	1	2	2	3	2	2	2	3	1	N	W

Regional adaptation[3]								Principal cultivars[9]	Special considerations and adaptations
Pacific Coast	Intermountain	Southwest	Northern Great Plains	Southern Great Plains	Midwest	Southeast	Northeast		
R	R							Bromar	Used in native grass mixtures on high mountain sites; used less now than formerly.
RI	RI	RI	RI	I	R	R	R	Lincoln, Southland, Manchar, Achenbach, Lancaster, Homesteader, Carlton, Magna, Baylor, Saratoga, Fischer, Polar	Rhizomatous. Important irrigated pasture grass; adapted to mesic sites on dryland.
			R	R				Mesa	Low production. Seed only in mixtures. Seeded or transplanted by stolons or rhizomes.
				RI				Higgins	Mostly rhizomatous.
R	R								Forb with persistent leaves.
RI	RI		RI		R	R	R	Ioreed, Frontier, Highland, Auburn, Rise, Vantage, Castor	Graze to prevent maturity. Pasture and hay on wet sites. Seeded or spread by sod or culm cuttings.
	R	R							Evergreen shrub palatable to deer; hybridizes with bitterbrush.
RI	RI		I		R			Aurora	Noncreeping. Adapted to cool, moist sites. Commonly used in irrigated pasture mixtures.
R					R	R	R	Dixie, Autauga, Auburn, Talledaga, Chief, Kentucky	Winter annual legume. Readily reseeds itself.
RI	RI	I	I	I	R	R	R	Merit, Pilgrim	Used in pasture mixtures on mesic or irrigated sites. Creeping by stolons.
RI	RI		I		R	R	R	Mammoth, Dollard, Midland, Lakeland, Kenland, Pennscott, Norlac, Kenstar	Short-lived perennial but readily reseeds under mesic conditions. Noncreeping.
R					R			Wilton, Hykon, Kondinin, Olympus	Winter annual. Widely seeded in California on annual grassland and brush burns. Readily reseeds.
RI	I	I						Salina	Creeping by rhizomes; low growing. Best use is on wet, salty sites.
R					R			Tallarook, Mt. Barker, Geraldton	Well adapted for interseeding mesic annual grasslands in Calif. Good winter growth. Winter annual.
RI		I		I	R				Long grazing period in Southeast.
	R	R	R	R					Seeded on dry sites where better forages not adapted.

Table 14. (Continued)

Common and scientific name	Seeds per lb. (1,000)[4]	Lbs. PLS. per acre seeding rate[3]	Ease of establishment	Stand maintenance	Drought tolerance	Cold hardiness	Salinity tolerance	Soil adaptation Sandy	Soil adaptation Silty	Soil adaptation Clayey	High water tolerance	Forage usability Early spring	Forage usability Late spring	Forage usability Summer	Forage usability Fall	Forage usability Winter	Grazing tolerance	Native or introduced	Season of growth[2]
Fescue, hard (*Festuca ovina* var. *duriuscula*)	565	2.3	2	1	2	1	3	2	1	1	3	2	2	2	2	3	1	N	C
Fescue, Idaho (*Festuca idahoensis*)	450	1.9	3	1	2	1	3	2	1	1	3	1	1	1	1	2	2	N	C
Fescue, tall (*Festuca arundinacea*)	227	3.8	1	1	2-3	1-2	1	2	1	1	1-2	2	1	2	2	2	1	I	C
Foxtail, meadow (*Aleopecurus pratensis*)	580	2.2	1-2	1	2-3	1	3	2	1	1	1	2	1	1	1	2	1	I	C
Grama, black (*Bouteloua eriopoda*)	1335	1.3	2-3	1	1	2	3	1	1	3	3	2	1	1	1	1	2	N	W
Grama, blue (*Bouteloua gracilis*)	711	1.8	2	1	1	1	2	2	1	1	3	2	1	1	1	1	1	N	W
Grama, sideoats (*Bouteloua curtipendula*)	191	4.6	1	1	1-2	1	2-3	2	1	1	2	2	2	1	1	2	1	N	W
Hardinggrass (*Phalaris tuberosa* var. *stenoptera*)	350	2.5	2	1	2	2	2	2	1	1	2	1	1	2	2	1-2	1	I	C
Indiangrass (*Sorghastrum nutans*)	170	5.1	2	1	3	1	2	1	1	1-2	1-2	3	1	1	2	2	2	N	W
Johnsongrass (*Sorghum halpense*)	118	7.4	1	1	1-2	2		1	1	1	1	2	1	1	1	2	1-2	I	W
Kleingrass (*Panicum coloratum*)	497	1.8	1-2	1	2	2-3		2	1	1	1	2	1	1	1	2	1-2	I	W
Kochia, prostrate (*Kochia prostrata*)	500	1.7	1	1	1	1	1-2	1	1	1	3							I	
Lovegrass, Boer (*Eragrostis chloromelas*)	2922	.6	2	1	1	2	2	1-2	1	2	3	1	1	1	1	1	1	I	W
Lovegrass, Lehmann (*Eragrostis lehmanniana*)	4245	.4	1	1	1	3	2	1	1	2	3	2	1	2	2	3	1	I	W
Lovegrass, sand (*Eragrostis trichodes*)	1300	1.3	1	2	1	1-2	3	1	2	3	3	2	1	1	2	2	2	N	W

Pacific Coast	Intermountain	Southwest	Northern Great Plains	Southern Great Plains	Midwest	Southeast	Northeast	Principal cultivars[9]	Special considerations and adaptations
R	R		R					Durar	Used mostly in erosion control; generally low palatability; robust form.
R	R								Lack of good seed yields restricts its use in range seeding.
RI	RI	I	RI	I	R	R	R	Ky. 31, Alta, Goar, Kenmont, Kenwell, Fawn	Generally seeded in pure stands, occasionally in irrigated pasture mixes. Winter grazed in South.
RI	RI		I						Creeping foxtail (*A. arundinaceus*) and meadow foxtail are well adapted to mountain meadows. Slightly rhizomatous.
		R						Flagstaff, Sonora, Nogal	Good quality seed is scarce.
			R	R	R			Capitan, Marfa, Lovington	Low yields generally restrict seeding to more droughty portions of the Great Plains. Seeded in warm-season mixtures.
			R	R	R	R		Premier, Butte, Trailway, Coronado, El Reno, Tucson, Vaughan, Uvalde, Pierre, Van Horn	Grows well in mixtures of warm-season grasses. Rhizomatous.
RI		I		R				TAM Wintergreen	Primary species for seeding Calif. coastal and inland zones.
			RI	RI	R	R		Holt, Neb. 54, Cheyenne, Tejas, Llano, Oto	Rhizomatous. Commonly seeded in warm-season mixtures on mesic sites.
		R		R		R			Rhizomatous; prevent from spreading to cultivated lands. HCN potential. Very palatable and productive.
				RI		R		Kleingrass 75	Some varieties are rhizomatous.
	R								
R		R						Catalina	Productive and nutritious.
R		R						A-65, A-68, Kalahari, Cold Hardy	Smaller and less cold tolerant than Boer and weeping lovegrass. Reseeds quickly after fire or other disturbance. Seeded generally in pure stands.
			RI	RI				Neb. 27, Band, Mason	Seed in mixtures. Short lived but readily reseeds itself.

Table 14. (Continued)

Common and scientific name	Seeds per lb. (1,000)[4]	Lbs. PLS. per acre seeding rate[2]	Ease of establishment	Stand maintenance	Drought tolerance	Cold hardiness	Salinity tolerance	Soil adaptation — Sandy	Silty	Clayey	High water tolerance	Forage usability — Early grazing	Late spring	Summer	Fall	Winter	Grazing tolerance	Native or introduced	Season of growth[2]
Lovegrass, weeping (*Eragrostis curvula*)	1500	1.2	1	2	1-2	2	2	1	1	1	2	1	1	2	2	3	2	I	W
Milkvetch, cicer (*Astragalus cicer*)	145	6.0	2	2	2	1-2	2	1	1	1-2	2	2	1	1	2	2	2	I	W
Mountainmahogany, curlleaf (*Cercocarpus ledifolius*)	52	8.4	2	1	1	1	2	2	1	1	3	2	2	2	1	1	1-2	N	
Mountainmahogany, true (*Cercocarpus montanus*)	59	7.4	2	1	2	1	2	2	1	1	3	2	1	1	2	2	1-2	N	
Needlegrass, green (*Stipa viridula*)	181	4.8	2	2	2	1	2	1	1	1	2	1	1	1	1	2	2	N	C
Oatgrass, tall (*Arrhenatherum elatius*)	150	11.6	1	2	2	2	2	1	1	1	2-3	2	1	2	2	3	2	I	C
Orchardgrass (*Dactylis glomerata*)	540	2.4	1	2	2-3	2	2-3	2	1	1	2-3	2	1	1	1-2	2	2	I	C-W
Pangolagrass (*Digitaria decumbens*)			1	2	3	3		1	1	2	1	1	1	2	3	1	I	W
Panicgrass, blue (*Panicum antidotale*)	657	2.0	1-2	1-2	2	3	1-2	2	1	1	2-3	2	1	1	2	3	2	I	W
Redtop (*Agrostis alba*)	4990	.3	2	1	3	1	2	2	1	1	1	1	1	2	2	3	1	I	C
Rhodesgrass (*Chloris gayana*)	2143	.8	2	2		3	1				2							I	W
Ricegrass, Indian (*Oryzopis hymenoides*)	188	4.6	3	1	1	1	2	1	1	2-3	3	1	1	2	2	1	1	N	C
Ryegrass, perennial (*Lolium perenne*)	247	3.5	1	2	3	2	2	2	1	1	2	1	1	2	2	3	1	I	C
Sacaton, alkali (*Sporobolus airoides*)	1750	1.0	2	1	2	1	1	3	2	1	1	3	1	1	2	3	1	N	W

Ratings of plant characteristics[1]

Regional adaptation[3]								Principal cultivars[9]	Special considerations and adaptations
Pacific Coast	Intermountain	Southwest	Northern Great Plains	Southern Great Plains	Midwest	Southeast	Northeast		
		R		R				Ermelo, Morpa, Renner	Palatability low except when young. Seeded mostly in southern Great Plains and in pure stands.
	R		R					Cicar, Lutana, Oxley	Fair to good production on mountain range. Rhizomatous. Erratic in stand establishment. Non-bloating; does not accumulate selenium.
	R								Evergreen. Important winter forage for deer and elk. May grow out of reach of grazing animals.
	R								Deciduous but hybridizes with curlleaf mountainmahogany. Hybrid has persistent leaves.
			R					Green Stipagrass, Lodorm	Seeded in mixtures. Low seed quality; delayed germination.
RI	R	R			R		R	Tualatin	Rapid-developing, short-lived grass adapted to mesic sites. Now infrequently used in new seedings.
RI	RI	I	I	I	R	R	R	Latar, Akaroa, Pomar, Pennlate, Potomac, Sterling, Chinook, Napier, Boone, Pennmead, Clatsop, Nordstern, Palestine	Adapted to irrigated or naturally mesic sites. Develops rapidly and is long lived. Seeded in mixtures. Tolerates shade.
						R			Stoloniferous. Well adapted to tropical and subtropical areas. Established vegetatively by fresh stem and stolon cuttings.
RI		RI		RI				Algerian, A-130	Rhizomatous. Highly productive on good sites.
I	I		I		R		R		Establishes well from broadcasting on wet soils. Widely adapted in mixtures on wet soils.
I			R		R			Bell, Lubbock	High sodium tolerance.
R	R	R						R-2575 (lower percentage of hard seed)	Hard, impermeable seed makes seeding success uncertain.
RI	R			I	R			Linn, Norlea, NK-100	Rapid developing, short-lived perennial. Used as short term pasture mostly.
		R	R	R	R				Merits further study for range seeding on saline lowlands. Seed available from native harvest.

Table 14. (Continued)

Common and scientific name	Seeds per lb. (1,000)[4]	Lbs. PLS. per acre seeding rate[3]	Ease of establishment	Stand maintenance	Drought tolerance	Cold hardiness	Salinity tolerance	Soil adaptation Sandy	Silty	Clayey	High water tolerance	Forage usability[5] Early spring	Late spring	Summer	Fall	Winter	Grazing tolerance	Native or introduced	Season of growth[2]
Sagebrush, big (*Artemisia tridentata*)	2576	.7	1	1	1	1	2	3	1	1	3	3	3	3	2	2	1	N	
Sainfoin (*Onobrychis viciafolia*)	26[8]	16.8	1-2	2	2	2	3				3								
Saltbush, four-wing (*Atriplex canescens*)	60[8]	7.3	2	1	1	1	1	1	1	1	2	2	2	1	1	1	1	N	W
Sandreed, prairie (*Calamovilfa longifolia*)	274	3.2	2	1	1	1	3	1	2	3	3	2	2	1	1	2	1	N	W
Serviceberry (*Amelanchier alnifolia*)	45	9.7	2	1	2	1	3	3	1	1	3	1-2	1-2	2	2	2	2	N	
Smilograss (*Oryzopis mileacea*)	884	1.5	2	1	1	3		1	1	1		2	1	2	2	2	1	I	C
Sweetclover, white (*Melilotus alba*)	260	3.4	1	2-3	1-2	1	1-2	1	1	1	2	2	1	1	2	3	2	I	
Sweetclover, yellow (*Melilotus officinalis*)	260	3.4	1	2	1	1	1-2	1	1	1	2	2	1	1	2	3	2	I	
Switchgrass (*Panicum virgatum*)	389	2.2	1-2	1	2	1	2	1	1	1	1-2	3	1	1	2	3	2	N	W
Timothy (*Phleum pratense*)	1230	1.4	1	2	3	1	2-3	2	1	1	2-3	2	1	1	1	2	2-3	I	C
Trefoil, birdsfoot (*Lotus corniculatus*)	407	2.1	2-3	2	2-3	2-3	1-2	2	1	1	1-2	2	1	1	2	2	2	I	C-W
Vine-mesquite (*Panicum obtusum*)	143	6.1	2	1	2	2	2	2	1	1	1	2	2	2	2	2	1	N	W
Wheatgrass, beardless (*Agropyron inerme*)	142	6.1	2	1-2	1-2	1	2	2	1	1	3	2	1	1	1	2	1-2	N	C
Wheatgrass, bluebunch (*Agropyron spicatum*)	117	7.4	2	2	1-2	1	2	2	1	1	3	2	1	1	1	2	1	N	C
Wheatgrass, Fairway crested (*Agropyron cristatum*)	200	4.4	1	1	1	1	2	2	1	1	2-3	1	1	2	1	3	1	I	C

Ratings of plant characteristics[1]

Regional adaptation[3]								Principal cultivars[9]	Special considerations and adaptations
Pacific Coast	Intermountain	Southwest	Northern Great Plains	Southern Great Plains	Midwest	Southeast	Northeast		
	R	R	R						Seeding limited to critical deer winter range. Considered undesirable on most spring-fall and summer ranges. Palatability varies.
	RI		RI					Melrose, Eski, Remont	Nonbloating legume.
R	R	R	R						Provides outstanding winter forage. Leaves partially evergreen. Use adapted, local strains.
			R						Seeding limited by inadequate seed supplies and low seed quality. Seed common in native grass seed harvest. Rhizomatous.
R	R								Seeding limited to big game range. Seeding success fair.
R									Becomes stemmy with maturity. Adapted to broadcast seeding after fire. Used principally in California.
R	R	R	R	R	R	R	R	Spanish; Evergreen; Cumino; Hubam (an annual variety); Polara	Seed of sweetclover should be scarified. Used for green manure more than forage.
R	R	R	R	R	R	R	R	Madrid; Goldtop; Yukon	More tolerant of drought and competition but has a shorter growth period than white sweetclover. Reseeds better than white sweetclover.
			RI	R	R	R		Neb. 28, Blackwell, Caddo, Greenville, Kanlow, Summer, Pathfinder, Carthage	Rhizomatous. Widely seeded in warm-season grass mixes on mesic sites.
R	R		I		R		R	Climax, Drummond, Essex, Champ, Bounty	Leafy and nutritious as forage but does not tolerate grazing well. Seeded in mixtures.
RI	I	I	I		R		R	Empire, Cascade, Granger, Tana, Viking, Douglas, Maitland	Does not cause bloat. Rhizomatous. Mostly used in irrigated pastures.
		R		R					Stoloniferous. Used principally for erosion control.
R	R		R					Whitmar	
		R							Adaptation and management similar to beardless wheatgrass, but seed less available.
	R	R	R	R				Fairway, Parkway	Stands thicken sooner and spread more than A. desertorum; also leafier and finer stemmed and grazed more uniformly. Seeded alone or with alfalfa.

Table 14. (Continued)

Common and scientific name	Seeds per lb. (1,000)[4]	Lbs. PLS. per acre seeding rate[2]	Ease of establishment	Stand maintenance	Drought tolerance	Cold hardiness	Salinity tolerance	Sandy	Silty	Clayey	High water tolerance	Early spring	Late spring	Summer	Fall	Winter	Grazing tolerance	Native or introduced	Season of growth[2]
Wheatgrass, intermediate (*Agropyron intermedium*)	93	9.4	1	1-2	2	1-2	2-3	1	1	1	2	3	1	1	1	2	2	I	C
Wheatgrass, pubescent (*Agropyron tricophorum*)	90	9.7	1	1-2	1-2	1-2	3	1	1	1	2	2	1	1	1	2	1-2	I	C
Wheatgrass, Siberian (*Agropyron sibiricum*)	206	4.2	1	1	1	1	2	2	1	1	2-3	1	1	2	1	3	1	I	C
Wheatgrass, slender (*Agropyron trachycalum*)	160	5.4	1	2	2	1	1-2	2	1	1	1-2	2	1	1	2	3	1-2	N	C
Wheatgrass, standard crested (*Agropyron desertorum*)	175	5.0	1	1	1	1	2	2	1	1	2-3	1	1	2	1	3	1	I	C
Wheatgrass, tall (*Agropyron elongatum*)	79	11.0	1	1-2	2	1-2	1	3	1	1	1	3	1	1-2	2	2	1-2	I	C
Wheatgrass, western (*Agropyron smithii*)	126	6.9	2	1	1	1	1	3	1	1	1	1-2	1	2	2	2	1	N	C
Wildrye, Altai (*Elymus angustus*)	175	5.0	2	1	1	1	1	2	1	1	2	1	1	2	1	2	1	I	C
Wildrye, basin or giant (*Elymus cinereus*)	95	9.2	2	1	2-3	1	1-2	3	1	1	1-2	2	2	2	2	3	2	N	C
Wildrye, Canada (*Elymus canadensis*)	106	8.2	2	3	2	1	2	1	1	1	2	1	1	2	3	3	2	N	C
Wildrye, Russian (*Elymus junceus*)	175	5.0	2	1	1	1	1	2	1	1	2	1	1	2	1	2	1	I	C
Winterfat (*Ceratoides lanata*)	55	7.9	2	1	1	1	1	1	1	1	3	2	1	1	1	1	1	N	

Ratings of plant characteristics[1] — Soil adaptation — Forage usability[5]

| Regional adaptation[3] | | | | | | | | Principal cultivars[9] | Special considerations and adaptations |
Pacific Coast	Intermountain	Southwest	Northern Great Plains	Southern Great Plains	Midwest	Southeast	Northeast		
RI	RI	R	RI					Slate, Oahe, Greenar, Ree, Amur, Chief, Greenleaf	Productive on mesic sites and under irrigation. Rhizomatous.
R	R	R	R					Topar, Mandan, Luna, Utah 109, Greenleaf, Slate	Similar to intermediate wheatgrass but somewhat more drought tolerant.
R	R	R	R					P-27	Similar to standard crested wheatgrass in adaptation and use but less widely used.
	R	R	R					Primar, Revenue	Short life limits use in range seeding. Seed in mixtures only.
R	R	R	R					Nordan, Summit, Neb. 10, Ruff	Refer to Fairway crested wheatgrass; full stands slightly more productive than Fairway.
RI	RI	RI	R					Neb. 98526, Alkar, Jose, Largo, Orbit	High sodium and salinity tolerance. Seed alone rather than in mixtures.
	R	R	R	R	R			Rosana, Barton, Arriba	Seeded in mixtures or in pure stands. Tolerates alkalinity and silting. Rhizomatous.
	R		R					Prairieland	Similar to Russian wildrye; deep root system.
R	R							P-5797	Vigorous, tall growing bunchgrass. Limited use on bottomlands.
	R		R					Mandan	Lack of stand maintenance and tolerance of grazing has limited its use.
	R		R					Vinall, Sawki, Mayak, Cabree	Seed alone or with alfalfa. Early growth. Very hardy once established. Retains palatability after mature. Provide a weed-free seedbed.
	R	R							Superior palatability, productivity, and adaptability; tolerates grass competition.

Literature Cited

1. Anderson, Darwin, Louis P. Hamilton, Hudson G. Reynolds, and R. R. Humphrey. 1953. Reseeding desert grassland ranges in southern Arizona. Arizona Agric. Expt. Sta. Bul. 249.
2. Artz, John L., J. Boyd Price, Frederick F. Peterson, et al. 1970. Plantings for wildlands and erosion control. Nevada Agric. Ext. Serv. Cir. 108.
3. Atkins, M. D., and James E. Smith. 1967. Grass seed production and harvest in the Great Plains. USDA Farmers' Bul. 2226.
4. Ayers, A. D. 1952. Seed germination as affected by soil moisture and salinity. Agron. J. 44(2):82–84.
5. Barnes, O. K., and A. L. Nelson. 1950. Dryland pastures for the Great Plains. Wyoming Agric. Expt. Sta. Bul. 302.
6. Bentley, Jay R. 1967. Conversion of chaparral areas to grassland. USDA Agric. Handbook 328.
7. Bernstein, Leon. 1958. Salt tolerance of grasses and forage legumes. USDA Agric. Infor. Bul. 194.
8. Bleak, A. T. 1968. Growth and yield of legumes in mixtures with grasses on a mountain range. J. Range Mgt. 21(4):259–261.
9. Bleak, A. T., N. C. Frischknecht, A. Perry Plummer, and R. E. Eckert, Jr. 1965. Problems in artificial and natural revegetation of the arid shadscale vegetation zone of Utah and Nevada. J. Range Mgt. 18(2):59–65.
10. Bleak, A. T., and A. C. Hull, Jr. 1958. Seeding pelleted and unpelleted seed on four range types. J. Range Mgt. 11(1):28–33.
11. Campbell, J. B. 1963. Grass-alfalfa versus grass-alone pastures grazed in a repeated-seasonal pattern. J. Range Mgt. 16(2):78–81.
12. Carter, David L., and H. B. Peterson. 1962. Sodic tolerance of tall wheatgrass. Agron. J. 54(5):382–384.
13. Chadwick, Howard W., George T. Turner, H. W. Springfield, and Elbert H. Reid. 1969. An evaluation of seeding rangeland with pellets. USDA, For. Serv. Res. Paper RM-45.
14. Conard, E. C. 1962. How to establish new pastures. Nebraska Agric. Ext. Serv. Campaign Cir. 165.
15. Conard, E. C., and Raymond G. Sall. 1967. Seed dormancy, age of seed, and rate of planting in relation to stand establishment of switchgrass and Indiangrass. ASRM, Abstract of Papers, 20th Annual Meeting, p. 17.
16. Conard, E. C., and Vern E. Youngman. 1965. Soil moisture conditions under pastures of cool-season and warm-season grasses. J. Range Mgt. 18(2):74–78.
17. Cook, C. Wayne. 1966. Development and use of foothill ranges in Utah. Utah Agric. Expt. Sta. Bul. 461.
18. Cords, H. P. 1960. Factors affecting the competitive ability of foxtail barley. Weeds 8(4):636–644.
19. Cunningham, Richard A. 1975. Provisional tree and shrub seed zones for the Great Plains. USDA, For. Serv. Res. Paper RM-150.
20. Currie, Pat O. 1969. Use seeded range in your management. J. Range Mgt. 22(6):432–434.
21. Currie, Pat O., and Dwight R. Smith. 1970. Response of seeded ranges to different grazing intensities in the ponderosa pine zone of Colorado. USDA (Forest Service) Prod. Res. Rpt. 112.
22. Dahl, B. E., A. C. Everson, J. J. Norris, and A. H. Denham. 1967. Grass-alfalfa mixtures for grazing in eastern Colorado. Colorado Agric. Expt. Sta. Bul. 529-S.
23. Derscheid, Lyle A., Raymond A. Moore, and J. K. Lewis. 1966. A pasture system for you. South Dakota Agric. Ext. Serv. Facts Sheet 307.
24. Ferguson, Robert B. 1967. Relative germination of spotted and nonspotted bitterbrush seed. J. Range Mgt. 20(5):330–331.
25. Frischknecht, Neil C. 1959. Effects of presowing vernalization on survival and development of several grasses. J. Range Mgt. 12(6):280–286.
26. Gates, Dillard H. 1967. Range seedings—success or failure. Oregon Agric. Ext. Serv. Facts Sheet 120.

27. Gomm, F. B. 1964. A comparison of two sweetclover strains and Ladak alfalfa alone and in mixtures with crested wheatgrass for range and dryland seeding. J. Range Mgt. 17(1):19–23.
28. Gray, James R., and H. W. Springfield. 1962. Economics of lambing on crested wheatgrass in northcentral New Mexico. New Mexico Agric. Expt. Sta. Bul. 461.
29. Great Plains Agric. Council. 1966. A stand establishment survey of grass plantings in the Great Plains. Nebraska Agric. Expt. Sta., Great Plains Council Rpt. 23.
30. Green, Lisle R., and Jay R. Bentley. 1957. Seeding and grazing trials of stipa on foothill ranges. California Forest & Range Expt. Sta. Res. Note 128.
31. Green, Norman E., and Richard M. Hansen. 1969. Relationship of seed weight to germination of six grasses. J. Range Mgt. 22(2):133–134.
32. Halls, L. K., R. H. Hughes, and F. A. Peavy. 1960. Grazed firebreaks in southern forests. USDA Agric. Info. Bul. 226.
33. Harlan, J. R. 1960. Breeding superior forage plants for the Great Plains. J. Range Mgt. 13(2):86–89.
34. Harris, Lorin E., Milo L. Dew, and George Q. Bateman. 1958. Irrigated pastures—a way to maintain beef production. Utah Farm and Home Science. 19(3):76–77, 80.
35. Hedrick, D. W., D. N. Hyder, and F. A. Sneva. 1964. Overstory-understory grass seedings on sagebrush-bunchgrass range. Oregon Agric. Expt. Sta. Tech. Bul. 80.
36. Heinrichs, D. H. 1975. Potentials of legumes for rangelands. *In* Robert S. Campbell and Carlton H. Herbel. Improved range plants. Soc. Range Mgt. Range Symposium Series 1, pp. 50–61.
37. Holland, A. A., J. E. Street, and W. A. Williams. 1969. Range legume inoculation and nitrogen fixation by root-nodula bacteria. California Agric. Expt. Sta. Bul. 842.
38. Holmgren, Ralph C., and Joseph V. Basile. 1959. Improving southern Idaho deer winter ranges by artificial revegetation. Idaho Dept. of Fish and Game Wildlife Bul. 3.
39. Hughes, M. D., Maurice E. Heath, and Darrel S. Metcalfe. 1962 (Second Ed.). Forages. Iowa State University Press, Ames, Iowa.
40. Hull, A. C., Jr. 1959. Pellet seeding of wheatgrass on southern Idaho rangelands. J. Range Mgt. 12(4):155–163.
41. Hull, A. C., Jr. 1963. Seeding salt-desert shrub ranges in western Wyoming. J. Range Mgt. 16(5):253–258.
42. Hull, A. C., Jr. 1971. Grass mixtures for seeding sagebrush lands. J. Range Mgt. 24(2):150–152.
43. Hull, A. C., Jr., D. F. Hervey, Clyde W. Doran, and W. J. McGinnies. 1958. Seeding Colorado rangelands. Colorado Agric. Expt. Sta. Bul. 498-S.
44. Hull, A. C., Jr., and Ralph C. Holmgren. 1964. Seeding southern Idaho rangelands. USDA, For. Serv. Res. Paper INT-10.
45. Hull, A. C., Jr., Ralph C. Holmgren, W. H. Berry, and Joe A. Wagner. 1963. Pellet seeding on western rangelands. USDA Misc. Pub. 922.
46. Hull, A. C., Jr., and G. J. Klomp. 1966. Longevity of crested wheatgrass in the sagebrush-grass type in southern Idaho. J. Range Mgt. 19(1):5–11.
47. Hull, A. C., Jr., and G. J. Klomp. 1967. Thickening and spread of crested wheatgrass stands on southern Idaho ranges. J. Range Mgt. 20(4):222–227.
48. Idaho Agric. Ext. Serv. 1961. Idaho forage crop handbook. Idaho Agric. Ext. Serv. Bul. 363.
49. Jensen, E. H., R. O. Gifford, H. P. Cords, and V. R. Bohman. 1965. Forage production on irrigated saline-sodic soils. Nevada Agric. Expt. Sta. Bul. B-4.
50. Kay, Burgess L. 1969. Hardinggrass and annual legume production in the Sierra foothills. J. Range Mgt. 22(3):174–177.
51. Keller, Wesley, and A. T. Bleak. 1974. *Kochia prostrata:* a shrub for western ranges? Utah Sci. 35(1):24–25.

52. Kilcher, M. R., and D. H. Heinrichs. 1965. Performance of some grass-alfalfa mixtures in southwestern Saskatchewan during drought years. Canadian J. Plant Sci. 46(2):177–184.
53. Kilcher, M. R., K. W. Clark, and D. H. Heinrichs. 1966. Dryland grass-alfalfa mixture yields and influence of associates on one another. Canadian J. Plant Sci. 46(3):279–284.
54. Kinch, Raymond C., and Loren E. Wiesner. 1963. Seed quality in green needlegrass. J. Range Mgt. 16(4):187–190.
55. Kneebone, William R. 1959. An evaluation of legumes for western Oklahoma rangelands. Oklahoma Agric. Expt. Sta. Bul. 539.
56. Knipe, O. D. 1974. Effect of heat treatment on germination of alkali sacaton. USDA, For. Serv. Res. Note RM-268.
57. Lang, Robert, and Leland Landers. 1960. Beef production and grazing capacity from a combination of seeded pastures versus native range. Wyoming Agric. Expt. Sta. Bul. 370.
58. Launchbaugh, J. L. 1971. Upland seeded pastures compared for grazing steers at Hays, Kansas. Kans. Agric. Expt. Sta. Bul. 548.
59. Launchbaugh, J. L., and Clenton E. Owensby. 1970. Seeding rate and first-year stand relationships for six native grasses. J. Range Mgt. 23(6):414–417.
60. Liacos, Leonidas G., and Eamor C. Nord. 1961. Curlleaf cerocarpus seed dormancy yields to acid and thiourea. J. Range Mgt. 14(6):317–320.
61. Lloyd, Russell D., and C. Wayne Cook. 1960. Seeding Utah's ranges—an economic guide. Utah Agric. Expt. Sta. Bul. 423.
62. Looman, J. 1976. Productivity of permanent bromegrass pastures in the parklands of the prairie provinces. Can. J. Plant Sci. 56(4):829–835.
63. Looman, J., and D. H. Heinrichs. 1973. Stability of crested wheatgrass pastures under long-term pasture use. Can. J. Plant Sci. 53(3):501–506.
64. Martin, S. Clark. 1966. The Santa Rita Experimental Range. USDA, For. Serv. Res. Paper RM-22.
65. Maynard, Michael L., and Dillard H. Gates. 1963. Effects of wetting and drying on germination of crested wheatgrass seed. J. Range Mgt. 16(3):119–121.
66. McCroskey, Jack. 1966. Bermudagrass and beef cattle. *In* Oklahoma Agric. Expt. Sta. Misc. Pub. 78, pp. 67–72.
67. McIlvain, E. H. 1976. Interrelationships in management of native and introduced grasslands. Annals Okla. Acad. Sci. 6:61–74.
68. McIlvain, E. H., and M. C. Shoop. 1960. An agronomic evaluation of regrassing cropland in the Southern Great Plains. Part I. Rangeland seedings in the southern Great Plains. USDA Agric. Res. Serv. Mimeo.
69. McIlvain, E. H., and M. C. Shoop. 1960. An agronomic evaluation of regrassing cropland in the Southern Great Plains. Part II. Dryland pastures for the Southern Great Plains. USDA, Agric. Res. Serv. Mimeo.
70. McKell, Cyrus M. 1975. Shrubs and forbs for improvement of rangelands. *In* Robert S. Campbell and Carlton H. Herbel. Improved range plants. Soc. Range Mgt. Range Symposium Series 1, pp. 62–75.
71. McLean, Alastair. 1967. Germination of forest range species from southern British Columbia. J. Range Mgt. 20(5):321–325.
72. Murphy, William J., Paul H. Bebermeyer, C. Melvin Bradley et al. 1966. Missouri livestock forage manual for beef and dairy cattle. Missouri Agric. Ext. Serv. Manual 67.
73. Nichols, J. T. 1969. Range improvement on deteriorated dense clay wheatgrass range in western South Dakota. South Dakota Agric. Expt. Sta. Bul. 552.
74. Nichols, James T., and James R. Johnson. 1969. Range productivity as influenced by biennial sweetclover in western South Dakota. J. Range Mgt. 22(5):342–347.
75. Nord, Eamor C., Edward R. Schneegas, and Hatch Graham. 1967. Bitterbrush seed collecting—by machine or by hand? J. Range Mgt. 20(2):99–103.
76. Ogden, Phil R., and Darrell H. Matthews. 1959. Spring pastures from deep-

furrow seeding of wheatgrasses and "high water" irrigation. Utah Farm & Home Sci. 20(2):38–39, 54.

77. Pearson, Bennett O. 1957. Bitterbrush and seed dormancy broken with thiourea. J Range Mgt. 10(1):41–42.

78. Pearson, George A. 1960. Tolerance of crops to exchangeable sodium. USDA AIB 216.

79. Pingray, H. B., and E. J. Dortignac. 1959. Economic evaluation of seeding crested wheatgrass on northern New Mexico rangeland. New Mexico Agric. Expt. Sta. Bul. 433.

80. Plummer, A. Perry. 1971. Personal communication.

81. Plummer, A. Perry, Donald R. Christensen, and Stephen B. Monson. 1968. Restoring big game range in Utah. Utah Div. of Fish and Game Pub. 68-3.

82. Plummer, A. Perry, and Neil C. Frischknecht. 1952. Increasing field stands of Indian ricegrass. Agron. J. 44(6):285–289.

83. Plummer, A. Perry, A. C. Hull, Jr., George Stewart, and Joseph H. Robertson. 1955. Seeding rangelands in Utah, Nevada, southern Idaho, and western Wyoming. USDA Agric. Handbook 71.

84. Pringle, William L., and Alastair McLean. 1962. Seeding forest ranges in the dry belt of British Columbia. Canada Dept. of Agric. Pub. 1147.

85. Rechenthin, C. A., H. M. Bell, R. J. Pederson, D. B. Polk, and J. E. Smith, Jr. 1965. Grassland restoration. Part III. Re-establishing forage plants. USDA, Soil. Cons. Serv., Temple, Texas.

86. Reynolds, Hudson G., and S. Clark Martin. 1968 (Rev.). Managing grass-shrub cattle ranges in the Southwest. USDA Agric. Handbook 162.

87. Rhoades, Edd D. 1967. Grass survival in flood pool areas. J. Soil & Water Cons. 22(1):19–21.

88. Riegel, D. A., F. W. Albertson, G. W. Tomanek, and Floyd E. Kinsinger. 1963. Effects of grazing and protection on a twenty-year-old seeding. J. Range Mgt. 16(2):60–63.

89. Rogler, George A. 1944. Relative palatability of grasses under cultivation in the Northern Great Plains. J. Amer. Soc. Agron. 36(6):487–496.

90. Rogler, George A. 1960. Relation of seed dormancy of green needlegrass (Stipa viridula Trin.) to age and treatment. Agron. J. 52(8):467–469.

91. Rogler, George A., Russell J. Lorenz, and Herbert M. Schaaf. 1962. Progress with grass. North Dakota Agric. Expt. Sta. Bul. 439.

92. Ross, J. G., S. S. Bullis, and R. A. Moore. 1966. Grass performance in South Dakota. South Dakota Agric. Expt. Sta. Bul. 536.

93. Rumbaugh, M. D., G. Semeniuk, R. Moore, and J. D. Colburn. 1965. Travois—an alfalfa for grazing. South Dakota Agric. Expt. Sta. Bul. 525.

94. Shown, L. M., R. F. Miller, and F. A. Branson. 1969. Sagebrush conversion to grassland as affected by precipitation, soil, and cultural practices. J. Range Mgt. 22(5):303–311.

95. Smika, D. E., and L. C. Newell. 1968. Seed yield and caryopsis weight of side-oats grama as influenced by cultural practices. J. Range Mgt. 21(6):402–404.

95a. Smoliak, S., A. Johnston, and L. E. Lutwick. 1967. Productivity and durability of crested wheatgrass in southeastern Alberta. Canadian J. Plant Sci. 47(5):539–548.

96. Springfield, H. W., 1970. Emergence and survival of winterfat seedlings from four planting depths. USDA, For. Serv. Res. Note RM-162.

97. Springfield, H. W., 1970. Germination and establishment of four wing saltbush in the Southwest. USDA, For. Serv. Res. Paper RM-55.

98. Springfield, H. W. 1972. Winterfat seeds undergo after-ripening. J. Range Mgt. 25(6):479–480.

99. Stewart, A. E. 1967. Establishing vegetative cover with saline water. New Mexico Agric. Expt. Sta. Bul. 513.

100. Stroh, James R., and Ashley A. Thornburg. 1969. Culture and mechanical seed harvest of four wing saltbush grown under irrigation. J. Range Mgt. 22(1):60–62.

101. Townsend, C. E., G. O. Hinze, W. D. Ackerman, and E. E. Remmenga. 1975.

Evaluation of forage legumes for rangelands of the Central Great Plains. Colo. Agric. Expt. Sta. General Series 942.

102. USDA 1961. Seeds. 1961 Yearbook of Agriculture. U.S. Govt. Printing Office, Washington, D.C.

103. USDA, Agric. Res. Serv. 1964. Pasture mixture—simple or complex. Agric. Res. 13(6):11.

104. USDA, For. Serv. 1975. Some important native shrubs of the west. Intermountain For. and Range Expt. Sta., Ogden, Utah.

105. USDA, Rocky Mtn. Forest & Range Expt. Sta. 1963. Germination of mountain mahogany seed speeded by chilling. 1962 Annual Rpt., p. 72.

106. Utah Agric. Expt. Sta. 1970. Pasture planting specifications for Utah. Utah Agric. Expt. Sta. Cir. 153.

107. Vallentine, John F., C. Wayne Cook, and L. A. Stoddart. 1963. Range seeding in Utah. Utah Agric. Ext. Serv. Cir. 307.

108. Van Epps, Gordon A. 1974. Shrub seed production—a potential enterprise. Utah Sci. 35(1):21–23.

109. Van Epps, Gordon A., and C. M. McKell. 1977. Shrubs plus grass for livestock forage: a possibility. Utah Sci. 38(3):75–78.

110. Van Keuren, R. W., V. F. Bruns, and D. D. Suggs. 1960. Forage for grazing and weed control in wet areas of the Columbia Basin. Washington Agric. Expt. Sta. Cir. 374.

111. Warnes, D. D., and L. C. Newell. 1969. Establishment and yield responses of warm-season grass strains to fertilization. J. Range Mgt 22(4):235–240.

112. Wilson, D. B. 1967. Growth of *Hordeum jubatum* under various soil conditions and degrees of plant competition. Canadian J. Plant Sci. 47(4):405–412.

113. Wilton, A. C., R. E. Ries, and L. Hofmann. 1978. The use and improvement of legumes for ranges. N. Dak. Farm Res. 36(1):29–31.

114. Young, James A., and Raymond A. Evans. 1976. Stratification of bitterbrush seeds. J. Range Mgt. 29(5):421–425.

115. Young, James A., Raymond A. Evans, Burgess L. Kay, Richard E. Owen, and Frank L. Jurak. 1978. Collecting, processing, and germinating seeds of western wildland plants. USDA Agric. Reviews and Manuals ARM-W-3.

Chapter 8

Range seeding —establishment and management

Failures in range seeding can be caused by many factors acting either alone or in combination. Recognizing the causes of failures is a major step in preventing such failures. Such factors include:

Germination of seed
1. Poor quality seed (low germination)
2. Seed dormancy
3. Unfavorable temperature
4. Insufficient soil moisture
5. Insufficient soil oxygen
6. High soil salinity
7. Depredation by birds and rodents
8. Insufficient soil coverage

Emergence of seedlings
1. Seeding too deep
2. Soil crusting
3. Desiccation
4. Wind and water erosion
5. Rodent and insect damage
6. Poor quality seed (low vigor, shriveled, damaged)
7. High soil salinity
8. Frost heaving

Seedling establishment
1. Drought
2. Competition from weeds
3. Competition of companion crops

4. Soil infertility
5. Insect, disease, and rodent damage
6. Lack of inoculation of legumes
7. Winterkilling and frost heaving
8. Poor soil drainage and flooding
9. High temperatures
10. Grazed too soon
11. Wind and water erosion

Researchers have attributed difficulties of grass establishment in semiarid regions primarily to (1) insufficient moisture, (2) high temperature, (3) high evaporation rates, (4) wind damage to seedlings, and (5) slow growth during the seedling stage [179]. Hyder et al. [96] considered drought and improper seed coverage as the two most common causes of seeding failures on rangelands. Loss of soil moisture combined with crusting has been a major cause of most grass seeding failures on medium to heavy textured soils in the Southern Great Plains [5]. Here it has been observed that the top half-inch of a silty clay loam soil can decrease in moisture content from field capacity to below the wilting point in less than twenty-four hours.

Frost heaving of germinated seed and young seedlings, particularly of legume species but also of grasses, often accounts for a major death loss. In intermediate wheatgrass and smooth brome seedings on mountain range, the average late fall and early spring death loss from frost heaving was 50 percent followed by an additional 40 percent summer loss from other causes [79]. Frost heaving was observed to continue into the fourth growing season.

Frost heaving results from a buildup of ice crystals near the soil surface. It is more severe on clay and silty soils than sandy soils, and saturated soils heave more readily than dry soils. The formation of ice crystals is favored by alternate freezing and thawing. Strong and well-established plants heave less, and coarse-rooted species are more resistant to heaving. Heaving can be reduced by providing a mulch cover to moderate extremes in temperature at the soil surface and by earlier establishment prior to freezing weather [13].

Browse seedlings are generally even more sensitive to frost, rodents, insects, and drought than are grass seedlings. Spring frost damage has been particularly detrimental with bitterbrush. The growing points of browse seedlings are generally more exposed than in grasses. Seeding browse seed too deep or too shallow, high temperatures accompanying drought during the seedling stage, and premature grazing by wildlife and livestock are also problems of browse as well as grass seedings. Legume seedlings are also more sensitive to frost than grasses.

The ideal seedbed for range seeding is (1) very firm below seeding depth, (2) well pulverized and mellow on top, (3) not cloddy nor puddled, (4) free from live, resident plant competition, (5) free of seed of competitive species, and (6) has moderate amounts of mulch or plant residue on the soil surface. Successful range and pasture seeding requires that much or all of the competition from undesirable resident plants be removed first. The major purpose of seedbed preparation is to reduce or eliminate the existing vegetation and reduce the seed supply of undesirable, competitive species.

A good rule to follow is never to drill or broadcast directly into established stands of perennial plants or highly competitive annuals such as cheatgrass (Figure 79) [142]. No forage species has been found that does not suffer when planted in direct competition with resident vegetation. Even strong competitors such as Fairway crested wheatgrass establish best where the original cover is the lowest [66]. However, on mesic and subirrigated sites direct sod seeding and even broadcasting or feeding mature legume-grass hay containing mature seed has readily established adapted legumes when accompanied by phosphate fertilization [177]. On upland sites, direct seeding into drought-depleted range may be successful if followed by abundant rainfall [72]. But direct seeding into reduced vegetation cover is more reliable on mesic sites than on more arid sites [34, 60].

FIG. 79. *Russian wildrye seeded in a drought year at the University of Nebraska North Platte Station. Grass seedlings in foreground established, but those in background succumbed to cheatgrass competition.*

The nature of the resident plant cover affects its competitiveness. The seriousness of perennial competition has been widely recognized. The success or failure of seeding the more productive species has generally been directly proportional to the degree to which the perennial vegetation was destroyed. Complete brush control on foothill range in the Intermountain area during seedbed preparation has allowed seeded species to reach their full production potential in about five years [26]. When the original control was only 60 to 80 percent, herbicidal control was required in seven to eight years, and reaching full potential was delayed until ten to eleven years after seeding.

FIG. 80. *Maverick Point seeding by BLM in southeastern Utah. Area in foreground chained, windrowed with a dozer, and drilled to crested wheatgrass; area in right center undeveloped and very unproductive.*

However, competition from resident annuals can also be severe. Unless cheatgrass can be controlled by plowing or other treatments, densely infested lands may best be managed as cheatgrass grazing range. In greenhouse trials, 4 to 16 cheatgrass plants per square foot have decreased the growth and survival of crested wheatgrass, while 64 to 256 plants per square foot severely decreased shoot and root growth and greatly increased the mortality of crested wheatgrass seedlings [43]. Shading was involved in early growth stages, but depletion of soil moisture was later the primary factor. In the California annual grasslands, seeds of resident annuals have been found to number 1,500 to 2,000 seeds per square foot and to contribute directly to seeding failures [169]. Bitterbrush in the seedling stage has been found able to compete with broadleaf summer annuals but not with cheatgrass [69]. On saline soil, halogeton has been much more competitive to grass seedlings than has Russian thistle [58]. In the establishment of the frail seedlings of Russian wildrye, stands of cheatgrass, halogeton, peppergrass, and Russian thistle were all competitive but most competitive in combination [36]. Although weeds may offer new grass seedlings some protection from sun and wind, they severely compete for soil moisture in dry years.

Seedbed Preparation

Several methods of seedbed preparation have been successfully used in range seeding. The choice depends upon the kind and amount of vegetation present on the land and also on soil factors such as susceptibility to erosion by water and wind, slope, salinity, stoniness, texture, and depth. Accessibility of rangeland to be seeded, limitations of cost, obstructions, and value of resident vegetation will also dictate the extent and method of seedbed preparation. The available choices are (1) mechanical methods, including clean tillage and summer fallow, (2) preparatory crop method, (3) controlled burning, and (4) herbicidal methods.

Mechanical
Seedbed Preparation

A comparison of mechanical equipment available for seedbed preparation on rangeland is given in Table 15. Areas of suitability and disadvantages of each method are included. Clean seedbeds have been satisfactory in the Southeast, Northeast, and Midwest where wind and water erosion are not hazardous and rainfall is not a severe limitation [146]. In the Northern Great Plains, drilling into fallowed seedbeds has been the most successful method of seeding [57, 167]. However, on sandy soils or erosive soils on slopes, clean tillage has often permitted severe erosion. Blowing soil not only blows out seed and seedlings in some spots and buries it in others but can also shear off young plants by sand blasting (Figure 81). Both weed competition and soil erosion of summer-fallowed land can be reduced by leaving the soil rough-plowed until immediately before seeding. Although rough ground surface reduces wind velocity and thus wind erosion, proper seed placement by drilling is difficult on rough-plowed soils [118].

Table 15. Comparison of Mechanical Equipment Used in Seedbed Preparation on Rangeland

Equipment	Areas suitable
Pipe harrow	Adapted to rocky and steep sites where more effective machinery is not adapted. Also, scarifies soil sufficient for covering broadcasted seed on burns, abandoned roads, or excavation scars.
Moldboard plow	Widely used on cultivated soils that are rock free, dense, and tight, but seldom on range.
Pitting	Many types available, including pitting disk plow with cut-away or eccentric disks. Adapted to medium-heavy textured soils on deteriorated range, go-back lands, and areas of low precipitation. Pits trap snow and water and protect young seedlings.
Railing	Adaptation very limited for seedbed preparation.
Chaining and cabling	Adapted to large, nonsprouting brush such as juniper and mature big sagebrush; also used on areas too rocky, rough, or steep for other equipment. Low cost per acre. Used on large acreages where total destruction of understory vegetation is undesirable.
Bulldozing	Generally unadapted as sole treatment in seedbed preparation except where used for blading into windrows (Figure 80). Often preceded by chaining or followed by other treatments such as chaining or plowing for final seedbed preparation.
Standard disk plow	Adapted to deep soils free of rocks. Will penetrate hard soils and withstand sticky soils.
Wheatland plow	Adapted to gentle terrain with nearly rock-free, moderately soft soil; frequently used on land previously cultivated and for light tillage in second time over. Commercially available. Plowing depths readily adjusted on even terrain.
Brushland plow	Adapted to brush up to 2″ in diameter, hard soil, uneven terrain, rocks, and stumps with low breakage. More effective than wheatland plow on uneven, rocky sites. High kill with single treatment.
Offset disk	Heavy models adapted to most areas not too rocky. Commerically available. Adapted to hard, dry, heavy soils. Cutaway disks do a better job of mulching the vegetation. High kill with once-over treatment.
Root plow	Adapted to deep, rock-free soils with brush too large or too dense for other equipment. Leaves mulch and debris on soil surface to reduce evaporation and erosion, but this may interfere with drilling. Undercutting alone may not sufficiently reduce herbaceous plant competition.

	Disadvantages
	Low kill of large shrubs, sprouting or willowy plants, and most herbaceous plants; may not clean itself; brings many rocks to the surface; and must be disassembled before transporting.
	Slow; high power per width of cut; high cost; ineffective on hard, sticky or rocky soil; picks up roots and trash; tolerates only very small brush.
	Does not reduce dense, perennial sod, rhizomatous plants, or brush sufficiently for good seedling establishment. Must often be preceded by plowing.
	Ineffective on herbaceous plants and limber or sprouting shrub species; insufficient soil scarification for adequate coverage of previously broadcasted seed except on sandy soil.
	Low kill of sprouting brush and understory species. Seeding limited to broadcasting prior to final chaining. Special equipment needed for loading and transporting chain. Link modifications aid in soil scarification by chaining.
	Leaves seedbed very rough; costly in complete land treatment; maximum soil disturbance; generally requires additional treatments such as piling brush, plowing, or subsequent chaining of cleared areas.
	High cost per acre; slow; loosens soil too much; high breakage on range; narrow cut and high draft.
	High breakage on rough, rocky, brushy sites; often plows too shallow, second time over generally required.
	High initial cost; difficult to transport; heavy draft; may leave seedbed too soft; not commercially available.
	Difficult to regulate plowing depth, particularly on sandy or loose soils; difficult to transport; seedbeds often too loose without packing by other equipment; leaves dead furrows that may induce water erosion; drafty.
	Not adapted to shallow, rocky, rough, steep, or muddy sites. High initial cost and drafty. Rhizomes of sprouting species must be worked to the surface to prevent sprouting. Seedbed generally too soft for successful seeding until after loose, open soil has been compacted.

In the ten-to-fifteen-inch precipitation zone in the Central Great Plains, the risk of failure in range seedings has been high. Here the establishment of stubble for erosion control has been avoided where possible because of the moisture needed for the cover crop. Fall plowing leaving the soil surface rough, winter fallowing, and planting grass into a clean seedbed in the spring following a light cultivation and smoothing operation has been suggested [11, 126]. The spring cultivation has generally removed the cheatgrass for the rest of the growing season.

In a 9-inch precipitation zone in central Idaho, mechanical fallow and fall drilling was more effective in grass establishment than drilling immediately following fall cultivation or early or late fall burning [62]. For fall seeding perennial wheatgrass on cheatgrass ranges in southern Idaho, summer fallow and shallow drilling was more effective than fall cultivation and drilling or than direct deep furrow drilling in the fall [107]. Summer fallow has been an effective method of controlling weedy, fire-tolerant perennials and of seedbed preparation in California brushlands [35]. Disking after the first fall rains—light disking when competition is not likely to be severe—to destroy any germinating volunteers followed by clover seeding has been recommended in the California annual grasslands [181].

Strip cropping methods of grass seeding have been adapted with partial success as an alternative to clean fallow [11]. Strips thirty feet

FIG. 81. *A clean seedbed is subject to severe wind erosion in some areas such as in western Nebraska, shown above.*

wide between untreated strips of similar width have been mechanically summer fallowed and seeded to grass when stored moisture was available. When used in the Central Great Plains, this method has (1) provided reasonably good wind-erosion control, (2) controlled competing vegetation, (3) conserved soil moisture, and (4) permitted successful establishment of crested wheatgrass and Russian wildrye. Although this further prolongs the return of the land to production, the intervening strips can be summer fallowed and seeded after the first seedlings have become established.

On medium to heavy soils in arid regions such as the Southwest, seedbed preparation to include pits, basins, or interrupted furrows to catch and hold moisture has been the most effective method for achieving good stands [147]. Eccentric disks and other types of pitting machines have been successfully used. Seedings made in the basins have the advantage of extra moisture for germination and establishment. Refer to *Chapter IX* for additional details of pitting and seeding.

A complete revegetation program for mountain meadows with relatively deep soils has been extended over three years [155]. The first year the land is plowed, worked down, rough leveled, and seeded to an annual hay crop. The second year the land is replowed, leveled, and seeded to an annual hay crop. The third year the land is replowed, releveled, and seeded to a grass-legume mixture. On shallow meadow soils where it is undesirable to plow, improvement has been limited to tearing up the sod by chiseling or tandem disking, limiting leveling to topping hummocks, and drilling or broadcasting seed into the tilled sod.

Loose soil problems. A major problem in mechanical seedbed preparation has been loose, soft soil which interferes with proper placement of seed and has poor moisture-holding capacity. Firm seedbeds hold moisture near the surface, help control depth of seeding, and provide anchorage for seedling roots. Some natural packing results from allowing soil to remain undisturbed a few weeks before planting, particularly when accompanied by rainfall. However, soil firming may be required to retain the moisture content in the surface two inches above the wilting coefficient for sufficient time to allow seedlings to establish [91].

In Oregon range studies, rolling has been used to firm seedbeds both before and after seeding [91, 97]. Rolling before drilling improved seed placement and gave better seedling emergence. However, rolling after drilling was generally detrimental. Rolling after broadcasting to cover the seed and firm the soil is effective on freshly plowed seedbeds where compaction above the seed is not excessive.

But rolling can be very detrimental under wet conditions and may be only slightly beneficial on extremely dry soils. Small diameter rollers often skid rather than roll, increase soil density, and may reduce water intake. Seedbeds compacted and smoothed by rolling are more subject to wind and water erosion. However, rolling generally reduces the risk of seeding failures on light textured soils if adequate soil moisture is present at the time of treatment.

Flexible cultipackers have advantages over flat rollers. Flexible cultipackers are able to adapt better to uneven terrain, give greater packing action below the seed placement zone, and leave a more mellow soil surface [97]. Drag harrows and rod weeders are also useful for smoothing and compacting where surface trash is not excessive. Using depth bands on the drill disks only partially alleviates the problems of seed placement on excessively soft seedbeds, and cultipacking allows the depth bands to be more effective on loose soils [124]. Cultipackers or other packing equipment may be pulled directly behind the plow to circumvent the need of a separate operation, but this may reset some of the resident plants under moist soil conditions. Rolling or cultipacking can be expected to be most beneficial when moisture conditions are favorable at planting time by allowing young seedlings to emerge on the existing moisture. Further drill modifications adapted to loose seedbeds are discussed under drilling.

Relatively weed-free seedbeds have been considered a necessity for grass establishment in west central Texas [168]. However, it was found necessary to compact the loose seedbeds left by rootplowing for successful establishment. The most effective combination was rootplowing, broadcast seeding, and roller chopping, resulting in 78 percent of the stands being good and excellent compared to only 25 percent from rootplowing and seeding only.

The prospect of stopping wind erosion by packing small ridges in sandy soils that cannot be controlled by deep plowing alone appears feasible [89, 90, 118]. Plowing and microridge rolling eliminates competing vegetation and has a moderating effect on soil temperature and moisture in the furrows. It appears that a sandy loam soil should contain 9 to 12 percent moisture when packed in order to obtain a surface condition greatly resistant to wind erosion.

Using Nurse Crops

Nurse crops, also synonymous with companion crops, have been commonly used in establishing improved pastures in humid areas and in irrigated pastures. Under these conditions seeding a nurse crop at or near the time when the perennial forage species are sown has had the following advantages: (1) reduction of wind and water erosion, (2) reduction of weed competition including grassy weeds, (3) protection of forage seedlings from wind and severe temperature, and (4) providing forage prior to the development of the perennial species [98].

Nurse crops such as oats or barley, which readily yield to the perennial forage species as they become established, have most commonly been used. Common rye has generally been too competitive to make a good nurse crop with wheat somewhat intermediate. Further suggestions for effective use of a nurse crop include (1) reducing seeding rates of oats or barley to seven to ten pounds per acre, (2) drilling crosswise to direction of seeding perennial species or in alternate rows, (3) frequent and light irrigations on irrigated land until the perennial species are well established, and (4) harvesting the companion crop early [35, 98]. When a nurse crop is used, competition from it must be controlled or establishment of the perennial species may be delayed.

On rangelands where a moisture shortage is likely to develop during the establishment period or on soils where the soil fertility is low, a nurse crop should not be used. When range seedings made with a nurse crop succeed, it is generally in spite of rather than as a result of the nurse crop. In Utah dryland studies, all intensities of grain seeding made in conjunction with grass seeding decreased grass establishment because of competition for moisture [166]. Results from using nurse crops in other semiarid regions have been negative except in unusually wet years [37]. When seeded equally in mixture with sideoats grama and Caddo switchgrass, *Sorghum almum,* a weak perennial resembling sudangrass, was highly competitive with surviving native grasses as well as with the seedlings of sideoats and switchgrass [120].

Preparatory
Crop Method

Plowing followed by planting a residue-producing crop during the growing season before seeding perennial forages and direct seeding into the residue without further seedbed preparation is known as the *preparatory crop method* of seedbed preparation (Figure 82). Advantages of the preparatory crop method are (1) control of both wind and water erosion, (2) reduction of evaporation from around the seeds and the establishing plants, (3) smothering out many weeds that germinate, including grassy weeds such as cheatgrass and the bristlegrasses (*Setaria* spp.), and (4) reducing or preventing a new crop of weed seeds [23]. Protecting the young seedlings from sand blasting and

FIG. 82. *Warm-season grass mixture established in Smith County, Kansas, by drilling into stubble remaining from a forage sorghum crop. The inverted V-shaped area is a drill skip revegetated by seed produced by plants of the original seeding.*

from exposure of seedling roots, protecting new seedlings from extreme temperature fluctuations, and catching snow during the winter are other advantages.

The preparatory crop method has been the most successful method of seedbed preparation in the Southern Great Plains [57]. It has proven effective under dryland conditions where wind and water erosion are hazards or where drying of the soil surface and crusting following rains can be expected [146]. In north central Texas, seeding in the dead litter of a preparatory crop has given 88 percent success while seeding on a clean tilled seedbed has given only 67 percent success [57]. Because of its adaptation to difficult conditions, the preparatory crop method has also been used in the Northern Great Plains and the Intermountain Region.

The preparatory crop method has been less successfully used in areas of the Great Plains receiving less than seventeen inches of precipitation [11] and in the Intermountain Region in areas receiving less than about fourteen inches of average annual precipitation. In a ten-inch precipitation zone in Montana, tilled or standing stubble has suppressed grass-seedling growth in some cases [167]. Although a phytotoxin in the stubble was suspected, reduced available nitrogen, lower soil temperature, and shading were considered possible factors.

318

Although providing somewhat poorer stands than summer fallow culture, adequate stands were also produced under the stubble cultures.

Species most successfully used as a preparatory crop have been sudangrass, grain and forage sorghums, and millet. It is important that the preparatory crop be seeded in rows spaced not more than twenty inches apart [146]. It is also important that the preparatory crop not be allowed to set seed, since volunteer plants may severely compete with the seeded grasses the following growing season. This can be prevented by seeding later in the growing season (late June or early July in the Central and Southern Great Plains) and by grazing or mowing.

In converting native range to seeded range, it has been a common practice in some areas to produce one or two crops of dryland wheat or barley before seeding to grass. Following harvesting the grain in the summer, grass seed is drilled in the fall or the following spring without further seedbed preparation. This approach has been used to reduce the net cost of conversion since crop production is often high for a year or so on virgin cropland. However, the success of seeding grass in stubble left from cereal crops may be limited by excessive volunteering. In dry years this volunteer may so compete with the germinating grass seedlings that the stand fails.

The most effective stubble appears to be the minimum amount consistent with stability of the soil surface [130]. In seeding operations, it is important that the seed is placed in contact with mineral soil. Harvesting the sorghum to a height of twelve to eighteen inches generally leaves sufficient stubble to protect the seedbed and to prevent rapid drying of the soil surface, prevents excessive mulch which is detrimental to seedling establishment, and provides some cash income [110, 130]. Grazing should be used to harvest the excess production only if it can be regulated to prevent locally excessive use.

Residues of grain sorghum, forage sorghum, and sudangrass have been equally effective in grass establishment. Higher amounts of preparatory crop residue have been needed for spring than fall or midwinter seedings. It has been noted that grain sorghum residue was least affected by wind movement, while lighter residue not tied down blew away in high winds [140]. Where mulch is excessive or too coarse to drill through, it may be necessary to chop it up with a cornstalk cutter prior to seeding. Plowing and seeding to annual forages over two or three consecutive years prior to planting perennial forages provide a means of controlling rhizomatous plants such as saltgrass [141, 142].

Killed waste brush was found useful in range seeding in southern New Mexico [173]. Natural range brush applied over seeded areas reduced daily maximum summer soil temperatures from 148° to 108° F and held moisture in the surface soil up to eight times as long as on unshaded soil. The brush mulch also absorbed the force of rain impact and improved moisture penetration. To meet the needs of this dry habitat, a machine has been developed that uproots the brush, works and levels the soil, scoops shallow basins, sows and covers the seed, and then drops the brush back on the ground surface, all in one operation.

Seedbed Preparation
by Burning

The effective use of fire in controlling undesirable plants has been discussed in *Chapter VI*. On sites where the resident plant species are readily killed by fire, burning can be an effective seedbed preparation since it generally leaves the seedbed firm but mellow. However, where sprouting shrubs or forbs occur in the resident stand, the reduction of perennial competition by fire is often insufficient. Dense cheatgrass sites are adapted to seedbed preparation by fire only if competitive perennials can be established quickly.

Hot fires that consume all leaves, twigs, and small stems of shrubs and all ground litter and thereby leave a white ash over mineral soil are most effective in seedbed preparation (Figure 83) [12]. Burning is an effective seedbed preparation only where there is sufficient fuel to give a uniform burn. Spotty burns make very ineffective seedbeds. However, on big sagebrush sites where burning has been successful

FIG. 83. *High kill of resident plant cover and relatively clean seedbed resulting from fire on brushlands near Monrovia, California. Drilling with a rangeland drill is suggested.*

and complete, burning as a seedbed preparation has been comparable to plowing [88]. Clover seeding after controlled burns has also been effective on California ranges [181].

On forest, slash, or dense brush sites, the ashes left by a complete burn may be adequate for direct broadcasting of forage plant seeds following the burn. However, the ash litter is often washed away, leaving a hard surface on which very few seedlings become established [142]. On most other sites including sagebrush sites the ashes left by burns are not deep enough for broadcasting unless followed by mechanical treatments such as chaining or pipe harrowing. Drilling or some type of seed coverage following broadcasting greatly increases seeding effectiveness on most burn sites and is recommended.

Forage plants should be seeded as soon after burns as expected precipitation will allow. The delay of a year on big sagebrush sites [68] or in chaparral [12] often permits the native plants to reestablish and thereby reduce the effectiveness of the seeding. Drilling crested or intermediate wheatgrass in the first fall but not the second fall after summer wildfire allowed the seeded grasses to establish before downy brome had an opportunity to dominate the site and reduced the encroachment by green rabbitbrush (*C. viscidiflorus*) in Nevada [48]. In addition to the use of broadcast burning in seedbed preparation, fire is also often required prior to seeding to clean up slash piles, brush windrows, and even dense accumulations of mulch and trash left by water flow.

Chemical Seedbed Preparation

Herbicides show promise for chemical seedbed preparation. Seeding is made shortly after spraying or after a fallow period maintained by herbicides. These techniques have been made feasible by the development of effective herbicides, improved application methods, and the rangeland drill. Chemical seedbed preparation and direct seeding into the killed mulch without further soil treatment is effective if the herbicide (1) controls a broad spectrum of undesirable plants, (2) dissipates rapidly after weed control is accomplished, and (3) is broken down or leached away by the time seeded species germinate or is not toxic to seedlings of the seeded species [39]. Chemical fallow during the previous growing season has been more successful on low-rainfall sites than spring herbicide treatment and seeding [185].

Seedbed preparation by chemical means has the following advantages when compared with mechanical methods:

1. Leaves a firm seedbed for better grass establishment.
2. Good erosion control since the mulch and litter are left in place.
3. Can be used on land too rocky, steep, erosive, or wet for mechanical treatment.
4. Does not invert the soil profile, which is undesirable on shallow, poorly drained, or poorly structured soil [53].
5. Provides a means of selective plant kill when desirable native forage plants are present.
6. Averts most soil crusting and reduces frost heaving [101].
7. Conserves soil moisture and nitrogen similar to mechanical fallow when used as chemical fallow [39].
8. Improves moisture penetration and retention as a result of mulch cover on the ground.
9. Allows spraying, drill seeding, and fertilization in a single operation while climatic conditions are still optimum [101].
10. Protects grass seedlings by means of the standing brush killed by herbicides [131].
11. Permits seeding an entire field with erosive soil at one time.
12. May be less costly than mechanical preparation.

However, disadvantages of chemical seedbed preparation include the following:
1. No single chemical is yet available which completely kills all resident plants and also dissipates rapidly afterward [53].
2. The dead mulch and litter may be excessive or otherwise hinder seeding.
3. Control of resident cover is greatly affected by drought.
4. May have competition during the seedling year from uncontrolled resident vegetation.
5. May kill insufficient weed seeds resident in the soil unless used as chemical fallow during a growing season.
6. Tillage following chemical application may bring undamaged weed seeds to the surface.
7. Standing dead shrubs may restrict grazing.
8. Hazard of killing desirable broadleaf herbs and shrubs.

2,4-D. Spring plowing and drilling is often infeasible on tarweed flats at higher elevation because of wet roads, uneven drying, and a short planting season. The best method of preparing high elevation tarweed infested ranges for seeding has been chemical fallow [84, 142]. The recommended steps are to spray in the spring with 1.5 to 2.0 pounds per acre of 2,4-D, drill in the fall without further seedbed preparation, and respray the following spring when necessary. Spraying 0.5 to 1.0 pounds per acre of 2,4-D in the spring when grass seedlings are in the two-to-four-leaf stage and tarweed is in the six-to-eight-leaf stage has killed 90 and 95 percent of the tarweed, respectively, without damage to the grass seedlings [106]. Spraying seedlings of intermediate and slender wheatgrass and smooth brome in the one-to-two-leaf stage with 0.5 pounds of 2,4-D per acre doubled the survival of grass seedlings, did not damage grass seedlings, and killed more tarweed than in later grass-leaf stages [81].

Aerial spraying with 2,4-D and drilling with a rangeland drill have been effective for establishing additional perennial grasses on sagebrush-grass and forb-grass sites with a fair understory of perennial grasses [144]. This approach has been particularly effective when the undesirable forbs and shrubs are readily killed by 2,4-D and annual grasses have not become a problem. A second herbicide application may be required in the spring of the establishment year if sprouting shrubs such as rabbitbrush are present or a large number of sagebrush seedlings develop [134]. On sandy soil where plowing is considered hazardous, an effective combination of methods used has been to rotobeat the big sagebrush and rabbitbrush one year, spray the rabbitbrush sprouts with 2,4-D the following year, and drill grass seed in the spring of the third year [65].

Paraquat. Paraquat application and spring seeding has been an effective method of establishing perennial grasses on annual grass sites. Paraquat is very quick acting, leaves no soil residues, and permits planting of perennial grasses immediately after spraying [53, 137]. Paraquat sprayed in the spring at rates as low as 0.5 pounds per acre has given adequate and consistent control of cheatgrass for establishment of perennial grasses [45]. Spring application of paraquat is more effective for cheatgrass control than fall application in the Intermountain Region, since germination of cheatgrass seed is only partially complete by the time of fall seeding [138]. Spring spraying but not fall spraying has also controlled Sandberg bluegrass. However, on California annual grasslands, fall spraying and grass seeding is recommended [100].

Paraquat has been more effective when sprayed before drilling than afterward [45]. Plants covered by soil escape the herbicide, and paraquat kills perennial grass seeds not completely covered by drilling. Paraquat control of cheatgrass has been greatly increased by the use of surfactants [44]. Paraquat at rates up to two pounds has discolored but not killed medusahead [185]. Paraquat does not kill most broad-leaved weeds and gives no permanent control of most perennial grasses [53]. Where broad-leaved weeds and shrubs are growing with cheatgrass, a hormone herbicide such as 2,4-D should be combined with the paraquat application [137].

Spraying paraquat in bands and drilling down the center of each band has been used successfully to establish Hardinggrass and subclover in California [100, 101]. In only five of sixteen tests over a five-year period was full spray coverage superior to a six-inch band. And in only one of sixteen tests was full spray coverage better than the twelve-inch spray band. The twelve-inch band allowed that a six-inch strip be chemically treated on each side of the seeding row and an untreated twelve-inch strip between treated bands. The paraquat band treatment was essential for Hardinggrass establishment but not subterranean clover. The cost of band tilling with paraquat at rates of 0.25 and 0.5 pounds per acre is comparable to or cheaper than cultivation for seedbed preparation.

Dalapon. Dalapon also shows promise in seedbed preparation. Dalapon at two to six pounds per acre has given excellent control of medusahead where competitive plants were quickly provided by seeding [185]. An effective combination of treatments for establishing wheatgrasses in California medusahead stands has been to burn in late spring, spray with three pounds of dalapon early the following spring, spray with one pound of 2,4-D per acre later in mid spring, and seeding to wheatgrass in the fall [170]. This combination allows two successive crops of medusahead to be killed before perennial grass seeding. Another promising combination with medusahead has been tillage one year followed by dalapon treatment of medusahead the following year prior to wheatgrass seeding. Dalapon gives only fair control of cheatgrass [45, 49].

Dalapon is slower acting than paraquat, and the residual remains longer in the soil. Grass seeding should be delayed for at least six weeks following dalapon application [53]. Dalapon gives some control of rhizomatous grasses but is ineffective on broad-leaved weeds [53]. Dalapon at six pounds per acre has reduced red threeawn and sand dropseed, pioneer grasses on sandy loam go-back lands, by 80 to 100 percent [92]. In a related study of chemical fallow with ten pounds of dalapon per acre, establishment and yield of crested wheatgrass were similar to mechanical fallow [49]. The addition of 2,4-D to the dalapon further controlled annual forbs. Dalapon at fifteen pounds per acre has also been used effectively in chemical scalping prior to browse seeding [17, 163].

Atrazine. Atrazine applied at one pound per acre in the fall has shown promise for cheatgrass control and chemical fallow [39, 185]. However, since atrazine is not selective between cheatgrass and seeded perennial grasses, at least one year should be allowed for dissipation prior to grass seeding. In an eight-to-twelve-inch precipitation zone in Nevada, atrazine was applied in the fall of one year and perennial wheatgrasses seeded the following fall for spring germination [40]. When applied at one pound per acre, atrazine dissipation during the interim period was inadequate to prevent damage to wheatgrass seedlings in the field. However, the use of four-inch-deep furrows aided in the removal of contaminated soil from the seedling environment [184].

In the Nevada studies atrazine at 0.5 pounds per acre was equal to the 1.0-pound rate in many instances in controlling cheatgrass, but the results of the lower rate were erratic [40]. It was concluded the 0.5-pound rate could be made more consistent by (1) removing the ground litter before spraying to reduce interception of atrazine or (2) fall application just prior to expected precipitation.

In related studies in Nevada, atrazine applied in a quantity of either .6 or 1.2 lb./a. in the fall generally gave sufficient control of cheatgrass and tumble mustard (*Sisymbrium latissimum*) to enable good stands of crested, intermediate, pubescent, and Siberian wheatgrass from seeding 12 months later [38]. When atrazine is used for chemical fallow, adequate broadleaf control may require spring application of 2,4-D [137]. A viable alternative to rangeland plowing for conversion of degraded downy brome-infested big sagebrush areas in Nevada was concluded to be a chemical seedbed preparation system integrating 2,4-D or picloram spraying for brush control and atrazine fallow for downy brome control [47].

Other herbicides. Other herbicides that show some promise for chemical fallow or preemergence weed control in range and other pasture seedings are propazine [132, 183], amino triazole [53, 92], disodium methanearsonate (DSMA), and siduron and picloram in combination [184]. Four pounds per acre of glyphosate has effectively killed saltgrass in meadows, permitting these areas to be seeded to desirable forage species [174].

Seeding Practices

Season of Planting

Range seedings should generally be made immediately prior to the period of longest favorable growing conditions. A favorable growing period of forty to sixty days will generally assure successful stand establishment, but favorable growing conditions in semiarid areas may be limited to thirty days or less. The seasonal distribution of precipitation is a major factor to be considered; but temperature must also be considered. In areas where winter temperatures are low but the soil remains bare of snow for considerable periods of time, plants must be well established prior to the winter period or seeded late enough that no germination occurs until early spring. In mountainous areas of northerly latitudes where a snow pack can be expected to last through the winter, fall seeding just prior to snowfall is recommended. Cool-season grasses readily germinate under the snow when temperatures are slightly above freezing, are protected by the snow pack, and establish under favorable spring weather [78].

Where winter killing of seedlings is high because of frost heaving or cold, desiccating winds, spring or late summer seeding may be required. However, overwintering in the soil effectively breaks the natural dormancy of seed of many native species and stimulates more rapid spring growth [141]. Late fall or winter seeding reduces the loss of seed to seed-collecting rodents because many of these animals hibernate [141].

Spring seeding in many areas is handicapped by a short period of optimum conditions interspersed between periods of too wet and too dry conditions. Wet soils may prevent seeding until late in the spring, and this becomes serious in areas where hot, dry weather quickly follows. The short optimal spring seeding period does not allow as efficient use of labor and machinery as fall seeding. Seeding should be made reasonably soon after seedbed preparation and plant control operations. If left unseeded for one growing season, the sites may be reoccupied by competitive resident species before the seeded species can establish.

Different species have different temperature and moisture tolerances in the seedling stage [152]. Also, the tolerance of moisture stress in the seedling stage may not be correlated with drought hardiness when mature. Beardless wheatgrass and Russian wildrye in contrast to smooth brome and intermediate wheatgrass have rather specific requirements for temperature and moisture at germination but are more drought hardy once established [122]. Increased moisture stress delays germination, rate of germination, and total germination, but temperature extremes further interfere with germination and establishment. Cool-season grasses under moisture stress have germinated better at 20° C than either 10° C or 30° C [122]. Higher levels of moisture stress can be tolerated during germination and establishment if the temperature is favorable. In semiarid areas, low seedbed temperatures combined with low moisture levels is a major factor in the failure of range seedings [42].

Pacific Coast. Along the Pacific Coast 75 to 85 percent of the total precipitation comes during the winter six months. Because of the typically hot, dry late spring and summer, seeding just prior to the beginning of the fall rains is recommended. If undertaken later, the job may be delayed by storms, and there is the risk of damage from cold winter weather [12]. October is considered optimum for seeding

326

clovers in the California grasslands with midwinter seedings being subjected to cold and frost heaving [181].

Fall planting of bitterbrush seed usually gives three to four weeks earlier emergence than from spring planting, and fall planting has generally been superior on drier sites or in very dry years [73]. However, spring plantings are advantageous where precipitation is adequate and well distributed during the critical spring growing season and where the soil retains moisture well.

Intermountain. In the Intermountain Region 45 to 65 percent of the precipitation comes in the winter months. Dry summers and moist to wet winters are typical. In the valleys and foothills, late fall seeding is the most practical. The seeding of grasses, shrubs, and forbs should be done late enough so that cold weather prevents germination until the following spring [54, 62, 79, 141, 178]. Early fall is satisfactory only if soil moisture is sufficient to assure rapid germination and establishment. Very early spring seeding is effective providing the seedbed was prepared the previous fall and seeding is done within a few days after it is possible to get on the ground [178]. In a Utah foothills study, about 80 percent more fall-planted seedlings were alive at the end of the second growing season than spring-planted seedlings, but there were no differences in production three years after seeding [28].

Irrigated plantings should be made in early spring or in late summer [175]. Legumes must be fully established before killing frosts. In the mountains where summer rains are dependable, both spring and summer seedings have been nearly as effective as fall seedings [178]. However, snowdrifts and differential drying of the soil may interfere with spring seeding. On subalpine ranges, seedings must be made in late fall for spring germination or early enough in the summer that establishment precedes late summer and early fall frost.

Southwest. In the Southwest, temperatures are mild in the winter and hot in the summer except at high elevations. From 50 to 70 percent of the precipitation comes during the summer months, with the peak generally in mid- and late summer. Seeding is recommended in May or June just prior to the beginning of the summer rains in the semidesert grasslands [3, 119, 147]. In northern sections and at medium elevations, September andd October are usually best; July and August are less dependable but produce satisfactory stands in most years; but May and June are poor risks [159]. Four-wing saltbush in New Mexico is more successfully established in spring and midsummer plantings than in fall plantings [164]. On mountainous, pinyon-juniper range in northern Arizona, fall planting was recommended for planting cool-season grasses and summer planting for four-wing saltbush and most warm-season grasses [113].

Great Plains. In the Northern Great Plains, cool-season grasses or grass-legume mixtures are best seeded in late summer for fall establishment if moisture conditions are favorable or irrigation water can be used [23, 153]. If late summer moisture conditions are unfavorable, late fall or winter seeding of cool-season grasses prior to soil freezing is suggested with early spring seeding running a close second. However, grass-legume seedings are preferred in early spring over late fall. Seedings made from late fall to early spring require that seedbeds be well stabilized against wind erosion. Otherwise the seed and seedlings may be blown out of the soil. Fall seeding has been superior to spring seeding when the resident cover has been incompletely killed [72].

In southern Saskatchewan, alfalfa, intermediate wheatgrass and Russian wildrye established best from spring seeding with some success from late fall seeding [102]. Late fall seeding for spring emergence in northern Colorado was significantly better for Russian wildrye than earlier seeding dates, because with earlier dates soil moisture deficiencies did not allow sufficient fall establishment prior to severe cold [11]. In other studies in this area, early spring (April 18) seeding was more reliable in establishing crested or pubescent wheatgrass or Russian wildrye than later spring seedings (May 20 to June 8) [128]. Early spring seedings were concluded to be as good as late fall seedings, except when spring rains or snowstorms upset and delayed the planting schedule.

Warm-season native grasses in the Great Plains are best seeded in mid to late spring depending on the latitude and species involved [23, 153]. Seedlings of warm-season grasses are more apt to be damaged by late fall and early spring frost, and this favors spring over fall seeding. In Nebraska studies, all stands of switchgrass and Indiangrass seeded in the fall were thinner and slower to reach grazing readiness than those from spring seeding [24]. Late spring and early summer seedings of cool-season grasses should be avoided, and summer seedings of warm-season grasses are seldom successful [24].

On the Texas High Plains, seeding sideoats grama in May gave better stands with less competition than seeding in March, April, or June [149]. June seeding was successful in good rainfall summers. However, in the extreme southern part of the Great Plains where frost damage is minimal, seedings may also be made in early fall [9]. Seeding from June through August is not recommended for grass seeding in the Southern Great Plains because of high temperatures and frequent droughts [110]. Cool-season grasses are normally seeded in early spring or fall.

Eastern United States. The eastern half of the United States has a subhumid to humid climate, precipitation is generally well distributed throughout the growing season, and droughts are minor. However, temperatures range from very cold during winter in the northern regions to very hot during summer in the southern regions.

Pasture seedings in the Midwest consist principally of introduced cool-season grasses and legumes and are generally made in the spring or late summer. The grasses usually do best when seeded in late summer, while spring seeding favors legumes [4]. In the Northeast, early spring seedings are generally best while late summer seedings are also successful in the southern part [75].

In the lower portion of the Southeast, warm-season grasses and legumes are predominantly seeded, and plantings should be made in late winter or early spring [61]. In the upper portion of the Southeast, introduced cool-season grasses and legumes play the major role; and seedings are preferably made in the late summer or fall with slightly less dependable results coming from late winter and early spring seedings [75].

Seeding Rates

Seeding rates should provide adequate seed for a good stand while preventing waste of seed resulting from an excessively high seeding rate. When seeded at normal seeding rates by drilling on Utah foothill range, less than 10 percent of the viable wheatgrass seeds produced seedlings and less than 50 percent of these survived beyond the second growing season, i.e. 5 percent establishment [28]. Increasing seeding rates can be expected to increase the number of seedlings emerging and establishing, particularly in favorable years, but higher seeding rates generally reduce the number of plants established per 100 viable seeds planted [28, 111, 127]. (See Table 16.)

Table 16. Effect of Seeding Rates on First Year Response of Native Grasses in Central Kansas*

	Pure live seeds planted/sq. ft.			
	4	12	36	108
First year plants established per sq. ft. (no.)	.37	.64	1.34	2.80
Average percent first year establishment (%)	9.3	5.3	3.7	2.6

*Source [111]

329

Table 17. The Effect of Seeding Rates for Fairway Crested Wheatgrass on Plant Numbers and Yield in Idaho*

Seeding rate (lbs./A)	Number seedlings per square foot in first year	Number plants per square foot in fourth year	% Mortality (first to fourth year)	Tenth year yields (lbs./A)
1	0.7	0.7	0	690
2	1.9	1.0	47	685
4	3.5	1.0	71	705
6	5.0	1.3	74	770
8	7.5	1.4	81	800
10	9.5	1.5	84	715
20	19.0	1.9	90	520
40	37.5	2.0	95	700

*Source: [86].

Increased rates of seeding are suggested for poor seedbeds. And when broadcasting, more seed should be used to give comparable stands with drilling [26, 142]. The higher seeding rate with broadcasting is required to compensate for uneven distribution of depth of seeding. Some of the broadcasted seed remains on the surface where germination and establishment are reduced and depredation by animals is increased. Attempts to mechanically cover the seed after broadcasting is partly effective, but some of the seed is buried to depths that restrict emergence.

The effects of seeding rates on crested wheatgrass, a long-lived and aggressive cool-season grass, have been widely studied. When seeding rates in Idaho were varied from two to twenty-four pounds per acre, all rates produced similar amounts of herbage by the sixth year, but plant numbers differed greatly during the first three years [135]. Similar results from related trials in Idaho are given in Table 17. In New Mexico, seeding rates of two to six pounds per acre with crested wheatgrass gave essentially the same herbage yields, size of plants, and even numbers of plants per foot of row by the fifth through eighth year after seeding [160]. However, light rates have required comparatively long periods of protection for complete stands to develop, and moderate rates are best to produce a full usable stand within a reasonable time [23, 135].

In southern Idaho, higher rates of crested wheatgrass controlled weedy plants and invading brush better than lower rates [88]. Heavy rates up to twenty-four pounds per acre did not cause stand failure from excessive competition between seedlings as might be expected [135]. However, bitterbrush plants with over 2,200 plants per acre by maturity have competed critically for soil moisture and space, have been delayed in development, and have been less vigorous [74]. Based on native grass seedings in Kansas, it was concluded that seeding rates adequate to give rapid ground cover were desirable and that sparse seedling stands often do not develop adequate ground cover until after tillering [111]. It was also noted that poor initial stands of native grasses may never become dominant because of weed and undesirable grass competition.

On thin stands of seeded grasses, plant distribution is irregular, robust and relatively unpalatable plants tend to develop, and subsequent grazing is uneven [86]. Wheatgrasses in thick stands are more readily eaten than when in thin stands, particularly at later stages of maturity [23]. This has been attributed to the fewer seedstalks and finer leafage produced by grasses growing in dense stands [28]. Planting in fairly thick stands and in rows not over twelve inches wide has been recommended with wheatgrasses for maintaining an increased proportion of vegetative shoots and thus palatability [95]. Seeding rates somewhat above average appear justified if there is danger of invasion by noxious plants, on eroding watersheds where it is necessary to establish a cover quickly, or on favorable sites where potential production is greater [142].

PLS basis. Range seedings are now commonly done on the basis of number of pure live seeds per square foot, or, less frequently, pure live seeds per linear foot of row. Although rates of ten to forty seeds per square foot have been used for seeding grass seeds of medium size, seeding rates based on twenty pure live seeds per square foot have become somewhat standardized in the U.S. for seeding grasses on ordinary upland range sites [20, 129, 146]. Hull [83] has recommended increasing seeding rates by 50 percent for species having 0.5 to 1.0 million seeds per pound and by 100 percent for those having over 1.0 million.

In addition to increasing seeding rates on sites having unusually high production potential, or when broadcasting rather than drilling is used, seeding rates should be adjusted for individual species and seedbed conditions where research and local experience have shown this to be desirable. Higher rates may be desirable for species or strains found to have low seedling vigor, high seedling mortality, and low spreading rates associated with reduced tillering, rhizome extension, or seed production [111]. Rates as low as six pure live seeds per square foot may be adequate for some species which spread rapidly or tiller extensively or when seed price is unusually expensive [75].

For the forage plants commonly seeded on rangeland, Table 14 lists the number of seeds per pound, which is indirectly related to seed size but is a direct function of seed weight. Also given in column

Table 18. Computing Bulk Seeding Rates of Mixtures on a PLS Basis*

Species	Percent desired in the mixture	Lbs. of PLS per acre for single species seeding*	Lbs. of PLS needed per acre	% PLS of available bulk seed	Lbs. of available bulk seed needed per acre
Little bluestem	40	3.4	1.36	40	3.40
Sideoats grama	25	4.6	1.15	70	1.64
Blue grama	20	1.8	.36	50	.72
Sand lovegrass	15	1.3	.20	75	.26
	100%				6.02

*Rates from Table 14.

three of Table 14 are approximate full seeding rates for drill seeding single species on ordinary upland sites. These seeding rates should be increased by about 50 percent when drill seeding is being done on bottomland sites and by 100 percent on irrigated and subirrigated sites. Also, seeding rates should be increased by 75 percent (50 to 100 percent depending upon conditions being met) when broadcast rather than drill seeding is to be used.

Table 18 shows the procedure for computing bulk seeding rates of mixtures on a PLS basis. It is intended primarily for grass seed mixtures but can be extended to forb and browse components. However, browse seeding rates have not been well established. Also, it is suggested that alfalfa be added to the calculated mixture on the basis of the discussion in the next section.

Alfalfa seeding rates. On the basis of twenty live seeds per square foot, a full seeding of alfalfa is four pounds per acre, and this is commonly used for dryland hay production. However, when seeded in grass-legume mixtures for irrigated pastures or on other mesic sites, two pounds is considered adequate [175]. For irrigated pastures in Colorado, a 10:2 ratio of grass to large-seed legume or 10:1 ratio for grass to small-seed legume has been recommended [165]. On dryland pasture when seeded in mixture with crested wheatgrass or Russian wildrye, one pound is considered adequate, with little or no advantage from higher rates [104]. In intermediate wheatgrass-alfalfa mixtures for irrigated pasture, one-half to one-pound rate compared to higher rates will provide a better legume-grass balance where alfalfa is well adapted [2]. One pound of alfalfa seed per acre has been suggested for interseeding into range or permanent pasture in South Dakota [154].

Drill seeding provides a means of distributing and covering the seed in a single operation. Providing the seedbed has been properly prepared and is not excessively soft and fluffy, drilling provides the best method of obtaining uniform distribution and depth of seeding. The Range Seeding Equipment Committee [144] advises, "Wherever drills can be used, they are the most satisfactory implements for securing a successful seeding." Satisfactory stands may be obtained by drilling directly into burns, into open stands of less competitive annuals, and into cropland stubble without additional seedbed preparation. However, intensive seedbed preparation should generally precede drilling.

Although there are many types of drills and furrow openers, drilling compared to broadcasting generally favors infiltration and moisture storage, offers some wind protection, is satisfactory on moderately trashy mulch surfaces, produces a more unifom stand of seeded plants, and results in stands reaching full production sooner. In one Idaho range seeding study, each dollar spent on drilling produced three to seven times as much grass as when spent on broadcast seeding when seedings were made into seedbeds prepared by plowing or burning [77]. Drilling at six pounds per acre was superior to broadcasting at twelve pounds per acre. Each dollar spent for plowing or burning followed by drilling produced five to forty times as much grass as airplane broadcasting of pelleted seed into unprepared seedbeds. However, care in covering the seed following broadcasting compares more favorably with drilling.

Planting depth. The ideal seeding depth for grass seeds is about one-quarter inch for small seed such as lovegrasses, one-half inch for average sized seed such as crested wheatgrass, and one inch for very large seed such as tall wheatgrass and smooth brome [79, 85, 147, 178]. Medium and large seeds may be planted somewhat deeper on sandy than on heavy clay soils. A rule of thumb used in Texas has been not to place seed deeper in the soil than seven times the seed diameter [143]. Planting depths here suggested apply to seedings made in broad, deep furrows as well as in shallow or no furrows. Seed placement in range seedings is made in the bottom of the furrows. Placement of seed on the ridges between the furrows or on particularly the south face of the microridges has given reduced grass seedling emergence and survival [80].

Assure that all of the seeds are covered but not planted too deeply. Seeding too deeply delays emergence and reduces total emergence. Seeding too shallowly increases desiccation, depredation, and faulty rooting of seedlings. Larger seeds of the same species are similarly handicapped in emerging from excessive depths [114, 151]. Only a few species such as Indian ricegrass [105] and switchgrass [172] are capable of emerging from depths of two to three inches in sandy soil or where crusting does not occur between planting and emergence.

Regulating the depth of seeding is also important in browse seeding. New Mexico research suggests winterfat should be seeded at a shallow depth, preferably about one-sixteenth inch deep [162]. No emergence was obtained below one-half inch. The optimum seeding depth for bitterbrush in Idaho was one-half inch with deeper depths delaying date of emergence and with no emergence from depths over two inches [10]. Seeding depths of one-quarter to one-half inch are considered the most practical depths for browse seeding [141], or slightly deeper when several seeds are seeded per seed spot.

Row width. Row widths of 10 to 14 inches are most commonly used in range seedings. Row width recommendations for seeding crested wheatgrass in western Canada have been 18 to 24 inches in semiarid areas but 12 or 14 inches in moister areas. However, numerous research reports indicate that total herbage production is not generally affected by row widths between six and eighteen inches after full establishment [11, 23, 82, 127, 160]. A trend of higher yields with narrow spacing immediately after establishment to no difference or higher yields from wider spacing in later years has been reported [123, 158].

Narrower rows appear to better control weeds such as cheatgrass, to produce leafier plants with finer stems, to give better and more rapid soil stabilization, to result in less ridging from buildup at crowns, and to be better adapted where trampling is apt to be excessive [28, 103, 123]. Row spacings of twelve to fourteen inches may be better than narrower rows for plant establishment and survival where there is a premium on moisture and when deep-furrow drills are used [28]. However, on low rainfall sites in northeastern Colorado, wheatgrass seedings made at thirty-inch row spacings produced lower yields than from narrower spacings and were invaded by other species [127]. It has been suggested that the ideal combination of intensity and spacing is that which results in equal distribution among plants in all directions. Wide rows up to forty inches are used in irrigated grass seed production [9].

Alternate row seeding. Seeding of grasses and legumes or browse plants in alternate rows increases the chances of successful seedling establishment of both as a result of decreased competition [59, 98]. This also assists in the establishment and maintenance of grass and legumes in the desired proportion (Figure 84). Slow establishing species such as birdsfoot trefoil generally respond favorably to being seeded in separate rows from vigorous grasses, particularly where soil moisture or fertility are limiting factors.

Seeding each species of mixtures in alternate rows may not greatly affect total forage yields but maintains their original composition better than when species are mixed in each row [55, 133]. Separation of browse seed and herbaceous plant seeds by planting in alternate rows has reduced the vulnerability of browse seedlings to competition for moisture [141]. Alternate row seeding requires that each planter has its own hopper or a section of a divided seedbox. When seeds of different size are being used, alternate row seeding complicates seeding rates with most drills.

Seeds per spot. Studies with bitterbrush have revealed that a lone seedling emerging from a given spot or hill is less likely to survive than a seedling in a group, possibly as a result of mutual protection from heat or in breaking through the soil crust during emergence. From this it has been suggested that several browse seeds be planted at each spot [60]. Placing up to eight seeds per spot had no advantage over a single seed when tried with pubescent wheatgrass, crested wheatgrass, and Russian wildrye [125]. It was concluded that placing eight grass seeds per spot would require the seeding rate be doubled for an equivalent number of seedlings established per acre.

FIG. 84. *Alternate row seeding of alfalfa and crested wheatgrass on sandy soil near Seneca, Nebraska, after eighteen years of spring-fall grazing.*

Ideal drill characteristics. The following characteristics are considered ideal for drills used in range seeding [91, 144]:

1. Ability to encounter adverse terrain, rocks, and brush with minimum breakage and maintenance.
2. Separate seed boxes for small seeds and for large and fluffy seeds.
3. Agitator in seedbox to prevent seed and trash bridging over the seeder openings.
4. Precise metering of seed—force feed usually better than gravity feed except for very fluffy seeds.
5. Baffles to maintain distribution of seed in seedbox on steep slopes.
6. Disk openers equipped with band-type depth regulators.
7. Flexible equipment to allow individual planters to adjust to irregular seedbed.
8. Mechanism for rapid and accurate setting of seeding rates.

At moderate moisture levels, soil compaction with low pressure by press wheels has increased emergence [124, 172]. Wide packer wheels (three-inch) are suggested since narrow (two-inch) wheels, particularly under heavy pressure, bury the seed too deeply [124]. Press wheels are less effective on sandy or dry soils and may contribute to crusting on wet, fine textured soils. Care should be exercised in using drags that the seed is not gouged from the ground or covered too deeply. Depth regulators are generally required to restrict the planting depth on most range seedbeds. However, firm seedbeds are still desired since depth regulators are only partially effective on very loose seedbeds. Although adjustable depth regulators are available for many models of drills, nonadjustable bands fixed at one inch in depth are generally used in range seeding and without apparent difficulty.

Small seeds often cannot be accurately dispensed in conventional farm drills unless diluted with some other material. Rice hulls treated with a methyl cellulose sticker are an effective diluent and generally considered better for this purpose than sand, sawdust, wheat shorts, or other materials [112]. Since neither rice hulls nor methyl cellulose deteriorate, this diluent can be mixed with the seed in advance of seeding whenever convenient. Vermiculite plaster aggregate can also be used to dilute small, free-flowing grass seeds or even small legume seeds after screening out larger particles from the aggregate.

Disk furrow openers. Disk openers are the most common kind of furrow openers used in range seeding. Site conditions under which the drill will most commonly be used determine the type of disk openers that should be used. Double-disk openers make no furrow and are best adapted to well-prepared seedbeds. Breakage from rocks and brush may be high, and the double disks do pick up sticks and clods. Single-disk openers are desirable on firm, litter-covered ground, where a definite cutting action is needed to open the furrow [144]. However, on soft seedbeds the double-disk openers equipped with depth bands are more effective in controlling seeding depth than the single-disk type, hoe type, or lister type [11].

Single-disk openers include the following types: standard, semi-deep furrow, and deep furrow. The standard disk opener leaves practically no furrow and is well adapted to clean, firm seedbeds. The semideep-furrow disk opener is possibly the most satisfactory for all-around range use. Disks are larger and more concave than the standard disk and leave a 1.5- to 2-inch-deep furrow. These furrows help protect against frost damage and trap additional rain and snow.

The single-disk, deep-furrow opener leaves a furrow 2 to 4 inches deep which appears better adapted to dry and weedy sites. Although deep-furrow openers give some weed killing action [8], the amount of weed control is generally inadequate unless preceded by plowing or spraying [45]. Use of a deep-furrow drill with a twelve-to-sixteen-inch spacing or a deep, wide furrow drill with eighteen-to-twenty-inch spacing has been suggested as a means of obtaining superior seedling stands and greater productivity on dry sites [7, 39]. Deep furrows have increased soil moisture during the critical spring germination period in dry years by an average of 50 percent and up to 100 percent [121]. The deep furrows reduce wind velocity and thus soil moisture evaporation, allow seedlings to utilize deeper moisture, reduce maximum temperatures around the seedlings, and generally increase seedling establishment on dry sites or during dry years [121, 46].

Problems of deep furrows include excessive soil sloughing and blowing on unstable soils, drowning seedlings under wet conditions, and burying seedlings by soil washing in torrential rains. It was concluded from a northern Arizona study that wide, shallow furrows with gently sloping sides are the best shape to reduce smothering of young seedlings by excessive soil sloughing [113].

Conventional grain drill. Commercially available grain drills used in farming operations, with some modifications, are adapted for range seeding on relatively smooth, brush-free sites. Row spacings vary from six to twelve inches in width, and drills are available in an assortment of disk openers, types of seedboxes, planting rate controls, and agitation and seed metering equipment. Although well adapted to good seedbeds, the conventional drill is not well adapted to brushy, trashy, rocky, or rough ground. Breakage is often high on rocky or rough ground or where brush snags are common. It is adapted for use following complete burns where terrain and soil factors permit. Depth bands are highly desirable (Figure 85).

Rangeland drill. The rangeland drill (Figure 86) is better adapted to rough, rocky terrain or hard soil than conventional grain drills. It has stronger and heavier construction and special modifications to reduce maintenance and breakage. Each planter unit is hinged separately for independent action. The seeder arms are solidly plated on the underside to free brush and serve as a skid over rocks. A brush guard extending the width of the drill and a protective shield enclosing the gear train further reduce breakage. Although the rangeland drill withstands small and brittle brush stems left after burning or spraying, heavy brush stems will cause costly damage and delays and should be broken down or removed before drilling [12]. A special drill-arm assembly has permitted the rangeland drill to be used for

FIG. 85. **A,** *Grass drill (rear view) equipped with double-disk openers and,* B, *depth bands;* C, *gravity-flow seedbox (front view) with agitator for large, chaffy seed and fluted seedbox for small seed.* (Photos by Miller Seed Co., Lincoln, Nebraska)

deep furrow drilling [8]. This assembly has permitted an adjustable disk angle and has utilized added weight and the removal of depth bands.

Press seeder. The best seedling establishment with a conventional drill on loose, fluffy soil has generally been in the wheel tracks of the tractor [97]. The press seeder developed by the Oregon Agricultural Experiment Station and the Acadia Development Center of the United States Forest Service is a twelve-row seeder which includes the benefits of wheel-track planting in its basic design (Figure 87) [15]. The twelve press wheels serve three purposes: (1) support the weight of the machine, (2) transmit this weight into soil firming, and (3) act as furrow openers. Each wheel is thirty-two inches in diameter, has a face of six inches, and is individually suspended to follow the contour on the ground and operate over rocks and brush. The V-shaped rib on the face of each wheel makes a seed furrow one inch deep in the bottom of the wheel track, and conventional grain drill drag links are used to cover the seed. The furrows are spaced twelve inches apart. The press seeder is not commercially available and must be custom built. Construction plans are available through the Range Seeding Equipment Committee.

FIG. 86. *The rangeland drill is adapted for seeding rough, trashy terrain. Inset shows each planter is hinged separately, is equipped with a depth regulator band, and is provided with a skid under the planter arm.*

The press seeder was developed for use on loose, light soils commonly associated with big sagebrush in Oregon. On such sites, wheel-track planting by the press seeder has consistently produced twice as many seedlings as conventional drilling [15, 91]. Firming the soil under the seed and covering the seed with loose soil retains moisture better and provides an improved seed-soil relationship. However, on firm seedbeds a conventional drill will produce stands equal to or better than the press seeder. On crusted or hard seedbeds or where the soil is rocky, is covered by brush or dense litter, or is wet, the press seeder is ineffective.

Grass seeder-packer. The grass seeder-packer, also called culti-packer-seeder or Brillion drill, is basically a tandem roller-pulverizer with a seeder box mounted above and between (Figure 88). Both rollers consist of a series of metal wheels, each of which has a shallow groover around the perimeter. The first roller firms the seedbed and makes shallow grooves into which the seed drops. The second roller than splits the ridges left by the front roller, covers the seeds, and presses soil around them.

The grass seeder-packer is well adapted where packing is essential but only where the ground has been well prepared, where the soil surface is smooth, and where rocks, brush, and other debris are absent [144]. Seed is uniformly distributed and planted about one-half inch deep. It is not widely adapted to rangeland use and is ineffective on uneven terrain, rocky or brushy surfaces, or where the soil packs excessively. It has been the most consistently successful

FIG. 87. *Press seeder developed in Oregon for seeding loose, fluffy seedbeds.* (Oregon Agricultural Experiment Station photo by Dean E. Booster)

method of grass establishment in the semidesert grasslands of the Southwest when preceded by a pitting operation to conserve moisture [3]. It has also been used to seed disturbance sites such as contour furrows, ripping areas, backslopes of dams, and road cuts.

Sod seeder. Sod seeders, also referred to as no-till planters, are designed to penetrate sod or other untilled sites, open narrow slits in the mineral soil, and deposit seed at a uniform depth in the slits [180]. Also, application of fertilizer and herbicides to control the competing vegetation is often accomplished in the same operation. Various devices such as chisels and disks are employed to make the slits while otherwise minimizing surface disturbance, sometimes preceded by individually spring-loaded ripple coulters.

Sod seeding involves direct seeding of a legume or grass into a site without prior mechanical seedbed preparation. It shows promise as a means of restoring old pastures, meadows, and hay fields, but is mostly limited to moist or wet areas where soil moisture is not a major limiting factor. On such sites sod seeding may have the advantages of conserving fuel energy, conserving nitrogen by introducing legumes, circumventing erosion in complete seedbed preparation, and reducing equipment needs. However, its adaptation is primarily to arable lands rather than rangelands but can be used on moderately stony soils and uneven topography.

Aridland seeder. This complex machine, developed in New Mexico, works behind a standard root plow and additionally picks up brush, forms basin pits and firms the soil, plants seed with press seeder

FIG. 88. *Grass seeder-packer, also called cultipacker-seeder, drops seeds between two roller-pulverizers. It is well adapted where soil packing is essential but requires a relatively smooth, clean seedbed.* (Soil Conservation Service photo)

units, and replaces the brush over the planted area as protective debris [1, 67]. Its use has resulted in about 50 percent of plots seeded in semiarid and arid southern New Mexico being successful. In trials made in creosotebush and tarbush areas, failures were primarily associated with erosive, gravelly, and sandy loam sites, with heavy loam soils where crusting occurred, and on unusually droughty sites. However, even on droughty sites there was generally a good stand of seeded species under the brush cover, particularly when the brush covered a low place where water was concentrated in the basin.

Drill calibration. After presetting for desired seeding rates, all drills require calibration to assure desired seeding rates are achieved. Seed size, weight, mixture, and dryness as well as degree of processing and amount of trash present greatly affect seeding rates. Calibration in the shop can be performed by jacking up the drive wheel of the drill and turning it for a set number of revolutions. The seed can be caught in pans or plastic bags tied to the seeders or on a canvas sheet. Since the vibration of planter boxes and seed tubes in the field affect rate of seeding, it is preferable also to calibrate the drill in the field under typical use conditions including travel speed.

The following formula can be used in calibrating a drill in the field:

$$\text{Lbs. per acre of bulk seed planted} = \frac{43{,}560 \times \text{lbs. seed collected}}{\text{Strip width in ft.} \times \text{strip length in ft.}}$$

In the field, the strip length is measured on the ground. In the shop the strip length should be estimated as follows:

$$\text{Strip length (ft.)} = 1.1 \, (\text{no. of revolutions} \times \text{wheel circumference (ft.)})$$

The factor 1.1 allows for 10 percent wheel slippage in the field not realized in the shop.

Broadcast Seeding

Broadcasting should be used on rangelands only where there is an assurance that a majority of the seeds will be covered [142, 178]. For this reason broadcasting on snow or on thin portions of fully used pastures generally fails. Seeds broadcasted without adequate covering root poorly and rapidly dry out when the soil surface dries [138]. Broadcasting into sparse, unconsolidated litter where surface moisture is often more favorable than on bare soil is preferable where no seed covering operation follows [138]. However, cultipacking may be required to place the seed in contact with the mineral soil.

342

Broadcast seeding may be the only means of seeding on rough terrain where rocks, steep slopes, and trees make drilling impractical. However, lack of better methods of seeding does not necessarily make broadcasting a recommended practice. Attempts to reseed without preparation or follow-up after broadcasting have commonly been failures. Although broadcasting by airplane or helicopter often costs $.50 to $.75 per acre less than drilling, this cost advantage is generally offset by the 50 to 75 percent more seed required with broadcasting.

Under some conditions, broadcasting may be equal or superior to drilling. Broadcasting is adapted to the following conditions:

1. On naturally loose soil where normal sloughing and settling will cover the seed, especially during an overwintering period.
2. Following rough plowing or disking.
3. Ahead of mechanical treatment that gives minimum but sufficient soil disturbance for coverage.
4. Shortly after burning in deep ashes, as in forest burns but not sagebrush burns.
5. Under aspen prior to leaf fall—has given good results.
6. Under Gambel oak, bigtooth maple, chokecherry, and serviceberry prior to leaf fall—has given fair results [141].

Many types of broadcasters are available for use in seeding rangelands. Aerial broadcasting by airplane is rapid and advantageous on large areas, but height and swath is somewhat difficult to control in rough terrain. The helicopter can get closer to the ground, fly at

FIG. 89. *Helicopter being loaded for aerial seeding of rangeland on the Uinta National Forest, Utah. Seed was aerially broadcast between chaining operations.*

slower speeds, and is easier to land (Figure 89). Helicopters are more adapted to small seeding areas and are generally associated with more uniform rate and distribution of seed than by airplane. Aerial application commonly utilizes a Venturi flume mounted under the fuselage. Seed dropped into the flume is spread and carried away through the slipstream as it exits from the flume. Even mixtures can be sown by helicopter with good distribution, providing the wind is not over 10 mph or turbulent and shifting. Light crosswinds change the swath location slightly but the effect is similar for each species [41]. Effective flagging is critical in aerial broadcast seeding. Costs of aerial seeding in California in 1973 were $2.00 to $2.50 per acre for rotary-wing and $1.00 to $2.50 per acre for fixed-wing aircraft [150].

Broadcasters may be mounted directly on the tractor or the tillage implement or may be trailed behind. Power source varies from power takeoff, gasoline, electric, to traction. Broadcast seeding should follow such operations as deep plowing, disking, brush stacking, and root plowing, but precede shallower treatments such as railing, chaining, and pipe harrowing. Broadcasting between two successive chainings has given satisfactory results. Other seed-covering operations which are effective when following broadcasting are cultipacking, rolling with a sheepsfoot roller, dragging brush across the area,

FIG. 90. *Crawler tractor equipped with broadcast seeder. Seed is scattered by air blast coming from the exhaust system.* (Arcadia Equipment Co. photo)

or driving livestock over the seeded area. On very soft seedbeds furrow openers may be taken off standard drills and the seed allowed to fall to the soil surface.

Broadcast seeding is also useful for seeding around range improvements such as stockwater dams, gully plugs, and diversion ditches where disturbance has resulted from construction. Areas disturbed in construction or logging should be broadcast seeded while the soil is still loose. Hand-driven "windmill" broadcasters are well adapted for this purpose as well as for seeding small areas of aspen, spotty timber burns, logged areas, and skidroads at the rate of ½ to 2 acres per hour. Broadcasters have been developed which scatter seed through the air blast created by the exhaust exit of the tractor (Figure 90). A seed dribbler has also been designed to drop seed onto the track pads of crawler tractors. After the seed is carried forward and drops to the ground in front of the track, it is pressed into the soil as the tractor passes over [141]. Seed dribblers are adapted for use with various types of seed, but have been particularly useful in establishing shrubs such as bitterbrush, four-wing saltbush, and cliffrose with low per-acre seeding rates.

Species or strains favored to germinate and establish from broadcast seeding are those which germinate rapidly during brief periods

FIG. 91. *Lehmann lovegrass readily establishes from broadcast seeding into prepared seedbeds as in southern Arizona, shown here.* (Soil Conservation Service photo)

345

of favorable moisture and which can germinate at low temperatures and under relatively dry conditions [138]. Bulbous bluegrass, the lovegrasses (Figure 91), and dropseeds are particularly adapted to broadcasting. Because of the difficulty of regulating drilling depths for small seeding species on soft ground, broadcasting of lovegrass seed has been as effective as drilling in southern Arizona [99]. In western Texas, intensive seedbed preparation and drilling established about twice as many grass seedlings as broadcasting and roller chopping (1.5 versus 0.8 plants per square foot), but the latter combination was considered more practical and more economical for conditions existing in the area [16].

Evaluation of Seeded Stands

Seeded stands should be evaluated to determine the success of seeding. This evaluation should be made as early as possible but not before the planted seeds have had a full chance to germinate and establish. Evaluation of irrigated pasture mixtures and rapidly establishing cool-season grasses can generally be evaluated by the end of the first growing season. No appreciable numbers of viable seed have been found to remain after the first growing season in the wheatgrass species [28] or in Lehmann lovegrass [99]. However, it is generally best to judge the seeding success of most native grasses, and particularly warm-season grasses, during the second growing season because of slower establishment. Seeding failures should be reseeded before complete seedbed preparation is again required. However, many fair stands develop by means of secondary seeding when given extra time and protection.

Several techniques for stand rating have been proposed, but all compare the stand in relation to the potential under existing soil and climatic conditions. The Great Plains Agricultural Council evaluated grass stands in the Great Plains on the basis of *good* (more than 1 seedling established per square foot), *fair* (0.5 to 1 per square foot), and *poor* (less than 0.5 per square foot) [57]. On seeded foothill range in the Intermountain Region in the eleven-to-thirteen-inch precipitation zone, stands were evaluated based on number of seeded plants per square foot as follows: *excellent* (0.75 or more), *good* (0.5 to 0.75), *fair* (0.25 to 0.5), and *poor* (less than 0.25) [28]. In northeastern Colorado, seedings have been considered successful when one or more plants were established per two feet of row by two years after seeding [11]. A good stand of bitterbrush in the Intermountain Region has been considered to contain 800 to 1,000 plants per acre [51]. However, density methods do not measure heterogeneity, are difficult to use on mature stands, and are time consuming [93].

A rapid percentage-stocked method has been used in which the percent of one-square-foot plots containing one or more seeded plants is determined. This method emphasizes heterogeneity of stand which is measured as frequency. A success-rating scale suitable for the ten-to-twelve-inch average annual precipitation zone in Oregon has been suggested as follows [93]:

Success rating	Percentage stocked
Excellent	50% or more
Good	40 to 50%
Fair	25 to 40%
Poor	10 to 25%
Failure	9% or less

A third system of rating seeding success includes a series of observations and measurement [76]. Stands are given a numerical rating ranging from 0 to 10. Characteristics of the stand included in the rating include (1) number of seeded plants per square foot, (2) distribution or homogeneity of seeded plants, (3) vigor of seeded plants, (4) average height of seeded plants, and (5) stage of plant development. This intensive evaluation system is particularly adapted to research plots.

Managing
Seeded Stands

Management of seeded range greatly affects the longevity of the stand and the actual returns and benefits received. After making sizeable investments in range improvement by seeding, it is essential to follow good rules of management. Failure to manage properly even an excellent stand can quickly liquidate the investment made in the seeding.

Assisting Establishment

Grazing of seeded stands should not begin until after the plants have become established. Seeded stands on range should not be grazed until after the second full growing season following seeding [147, 178]. If dry years prevail during establishment of range seedings, an additional year or more may be required to allow seedlings to become firmly rooted. Plants are not sufficiently established if they can be pulled out of the ground by hand. On mesic sites, light grazing toward the end of the first growing season may be tolerated.

347

On irrigated pasture established in late summer, grazing can be begun with care during the middle of the following growing season.

Where big game grazing pressure is heavy, this should be reduced to proper levels by heavier animal kills prior to seeding. Also, much overgrazing by big game and rodents can be prevented by seeding larger areas at a time to prevent damaging concentrations. From survival studies of young bitterbrush in Idaho it was concluded that drastic reduction in deer numbers is unnecessary in a revegetation program if moderate grazing is the rule and where heavy utilization is present only during the most severe winters [50]. Full protection for fifteen years compared to light to moderate winter use by deer increased bitterbrush stands only slightly but doubled plant size.

Weed control during establishment may be required and may be beneficial in average to dry years. In wet years when competition for moisture is low, weed control may not be beneficial. In record drought years, no level of weed control may result in success. Herbicide application rather than mowing is suggested on pure grass plantings because of greater selectivity and less damage to forage plants becoming established. Weeds should be prevented from reaching full size and using up the soil moisture needed by the forage seedlings. Native grass seedings are often weedy during the first year or more, only to have the seeded species become dominant the following year without weed control beyond seedbed preparation [18]. However, even highly competitive species such as crested wheatgrass may materially benefit from 2,4-D application during establishment [25]. New seedings should be protected from damage from rabbits, rodents, and insects where the seriousness of the situation warrants.

Distributing Grazing

Livestock often concentrate their grazing on a seeded stand and ignore adjacent available native forage. This is particularly true when the seeded stand is green and succulent and native plants are dormant and dry or are found on steeper terrain. Fencing will usually be required to prevent grazing during the establishment period. Otherwise, it may be necessary to remove livestock from the adjacent range to prevent grazing on the seeding during the establishment period. However, this temporary nonuse may materially improve vigor and stand on the native range as well. Even after establishment, seeded areas should preferably be grazed separately from the native range unless stocking rates are based on the seeded areas. Grazing seeded range intermixed with native range may not be critical under fall or winter grazing since grazing damage to the seeded grasses after becoming firmly rooted is usually slight when grass is dormant [129].

Grazing damage to seedlings has been apparent in grazed pastures even under very light stocking rates where grazing is not well distributed. The utilization of crested wheatgrass plants growing side by side has been observed to range from 15 to 80 percent for cattle and 5 to 90 percent for sheep (Figure 92) [26]. Such grazing patterns not only damage the plants repeatedly grazed but waste the forage produced by the ungrazed "wolf" plants. Wolf plants in crested wheatgrass have been effectively controlled by heavy grazing every second or third year. Stocking at the rate of 0.5 to 1 acre per cow for a few weeks in early spring after the new growth has reached about three inches has been suggested on Utah foothill range [26]. Although 80 percent utilization annually removed the wolf plants, crested wheatgrass stands deteriorated to about 30 percent of their potential after only seven years of such heavy grazing. Light grazing of crested wheatgrass has encouraged the development of wolf plant more than moderate use [33, 156].

Haying while guarding against severely close mowing is one means of harvesting early growth for beneficial use while preventing harm-

FIG. 92. *Spot grazing, as shown with crested wheatgrass in the picture, can be reduced by occasional forced grazing, haying, or mowing. Burning and nitrogen fertilization have also been useful in some situations.*

ful grazing patterns developing before full establishment. In Nevada, clipping, crushing, or dragging removed the coarse, old growth from established crested wheatgrass plants [6]. Burning completely controlled wolf plants for two growing seasons. Fertilizing with 80 N/A or less reduced wolf plants while 160 to 240 N/A controlled the wolf plants completely.

Maximum carrying capacity can be realized from seeded stands without harmful effects only if every effort is made to insure uniform grazing and prevention of livestock concentration. This may require additional water development or water hauling, cross fencing, or even herding, or a combination of all three. Salt and supplement should be moved away from the water. Differential grazing also suggests that seeded and cleared and seeded but uncleared pastures as well as adjacent pastures seeded to different species should be fenced and grazed separately [26, 178]. Although fencing costs are high, they may be reduced somewhat by using temporary electric fences or by locating permanent fences that will be useful in later years for grazing management [178].

Intensity of Grazing

Grazing crested wheatgrass to 65 percent utilization or two-inch stubble height by the end of the growing season has been rather widely recommended [33, 52]. The tolerance of crested wheatgrass to heavy spring grazing results from early root growth activity, early accumulation of leaf tissue, and significant accumulation of carbohydrates in the underground parts by the time the plants are six inches tall [94]. A four-inch stubble height is generally recommended for tall and pubescent wheatgrass, big bluegrass, and Russian wildrye. Smooth brome will tolerate grazing to a three-inch stubble height, while intermediate wheatgrass being less tolerant to grazing should be grazed on dryland no closer than five inches. On most range sites, at least 40 percent of the growth must be left at the end of the growing season to keep the forage plants vigorous and productive. On dry sites and with native species, half or more of the current growth should be left [178]. But on mesic or irrigated sites 60 percent or more of the herbage can be removed.

Spring grazing of crested wheatgrass in Utah under a moderate rate (65 percent utilization) compared to the heavy rate (80 percent utilization) increased daily cattle weight gains as follows: yearlings, 0.27 pounds (2.55 versus 2.28); pregnant cows, 0.7 pounds (3.1 versus 2.4); lactating cows, 0.92 pounds (2.72 versus 1.80); and calves, 0.12 pounds (1.82 versus 1.70) [52]. Light grazing was beneficial over moderate grazing for weight gains of pregnant and lactating cows only. Cattle gains per acre over an eleven-year period were 43.4 pounds for medium, 39.7 for heavy (greatest in early years), and 36.8 for light. Medium but not heavy spring utilization left sufficient grass for fall grazing.

Spring and fall grazing of crested wheatgrass at different rates with sheep were compared in central Utah [14]. No difference was found in lamb gains from light use (59 percent) and moderate use (71 percent), but heavy grazing (88 percent) reduced lamb gains. Gains in lamb weight per acre favored heavy grazing over the seven-year period as follows: 35.4 pounds for heavy, 32.0 pounds for moderate, and 28.2 pounds for light. However, this occurred in the face of declining vigor of crested wheatgrass under heavy grazing. In the seventh year of the study, lamb gain per acre under the heavy grazing dropped considerably lower than under the lighter treatments.

The quality of the herbage ingested decreases generally if the grazing pressure causes a low availability of herbage, thus reducing the animal's opportunity for selective grazing [19]. Under heavy grazing, the grazing animal ingests a larger portion of the whole plant, lower in digestibility than the immature herbage, than if there was an opportunity for selective grazing. Some selectivity under either continuous or rotational grazing must be allowed if higher animal gains are achieved.

Variation in annual production is a problem on seeded range as well as native range. Compared to the three-year average of 549 pounds per acre, the annual production of crested wheatgrass in northern New Mexico varied from 53 to 1,150 and 443 pounds in consecutive years [170]. In the Intermountain Region during a twelve-year period, annual crested wheatgrass production varied from a low of 425 pounds to a high of 925 pounds per acre [52]. Yearly forage production of crested wheatgrass at Manitou, Colorado, varied from a low of 668 pounds to 2,457 pounds over a ten-year period [32]. Peak productivity often is reached during the first and second year of use before decreasing somewhat to long-time productivity levels [87, 129]. This may mislead into stocking seeded range too heavily.

Grazing Systems

Rotational grazing is generally used in preference to continuous grazing on irrigated pasture. Under rotational grazing four or five pastures of approximately equal size are grazed in succession. Each pasture is commonly grazed from five to seven days and then allowed a regrowth period of twenty-one to twenty-eight days before being regrazed. Each pasture is generally irrigated just prior to and just after the grazing period and once or twice additionally during the regrowth period.

Rotational grazing on irrigated pasture results in more uniform grazing, reduces selectivity between species and plant parts, is generally required to maintain alfalfa in the stand, allows slightly heavier grazing without damage to the forage plants, and reduces bloat when legumes are in the mixture. Avoidance by livestock of mature ungrazed plants may be a special problem when grazing is continuous and poorly controlled. However, rotational grazing seldom increases livestock production per head, and forced utilization of low quality forage may reduce the production of high-producing animals [171].

On seeded dryland range, rotation grazing has often failed to show enough advantage over continuous grazing to justify the extra expense of fencing and water development. When seeded range is grazed for short seasonal periods only, if the terrain allows uniform distribution of grazing, and the forage plants are not easily defoliated under close grazing, controlled continuous grazing may give as good results as rotational grazing. However, rotation grazing on seeded range can benefit highly productive grasses and legumes in low vigor because of heavy, continuous grazing.

On seeded, nonirrigated pastures in the Midwest, two weeks of grazing followed by a rest period at least four weeks is recommended [136]. However, in semiarid areas pastures grazed under rotation are generally grazed not over twice during the grazing season. On foothill crested wheatgrass range in the Intermountain Region, a three-division rotation in which each area was grazed twice each spring generally reduced gains per head slightly but tended to give highest gains per acre (46 vs. 39 lb./A) and slightly higher grass production per acre [52]. When rotation systems are used on seeded range, guard against prolonged concentration of livestock in a small rotation pasture during drought or when root resources are low [21].

Deferred rotation combines the advantage of deferred grazing along with rotation grazing. Preventing a pasture from being grazed at the same part of the growing season each year is important in maintaining plant vigor. On crested wheatgrass range, delaying the start of grazing has improved maximum basal area and grass yields [52]. Shortening the grazing season at the end of the spring encouraged maximum plant numbers. Occasional deferment along with moderate grazing is generally more effective in maintaining or improving the vigor of seeded forages than is light grazing or underuse alone. However, deferment is often enabled only by the addition of rotation to the grazing system.

352

Heavy grazing of crested wheatgrass in alternate years in Nevada studies produced downward trend at two of three locations studied [148]. Moderate grazing each year was concluded better than alternate year rest after heavy use. One year's complete rest did not compensate for overgrazing in the previous year. This suggests that rest-rotation grazing should be used with caution on semiarid ranges, and particularly when seeded species being grazed such as intermediate wheatgrass are not tolerant of close grazing. However, forage production and plant vigor of heavily grazed crested wheatgrass in Idaho has been restored by letting it rest a year or two, deferring grazing, and/or alternating the timing of grazing during the growing season [156].

Two-Crop Versus One-Crop Management

Grazing crested wheatgrass from boot stage to anthesis has been referred to as one-crop management [94, 156]. This method of grazing management (1) permits maximum root growth, (2) harvests maximum amounts of dry matter, crude protein, and phosphorus, but (3) has generally given no late spring regrowth on Intermountain foothill ranges. Two-crop management involves grazing from six-inch leaf height to boot stage and again in late summer or fall after the crested wheatgrass is fully cured.

Two-crop management has been noted to (1) depress root growth slightly (2) harvest a maximum of early feed, (3) reduce total production somewhat, (4) permit high storage concentration of carbohydrates during the summer, and (5) provide late summer or fall forage that is more palatable and nutritious than if not topped early in the spring [94]. Close grazing just prior to the boot stage favors late summer and fall regrowth. Crested wheatgrass grazed by cattle in the fall is more palatable when previously grazed in early spring because of an increase in the ratio of vegetative shoots to reproductive shoots.

For maximum length of spring grazing season or for spring-fall grazing, a rotation combining two-crop and one-crop systems and using three pastures has been suggested [94]. Under this method two two-crop pastures would be grazed in early spring, one one-crop pasture would be grazed in mid spring, and the two two-crop pastures would be grazed again in late spring or in the fall, followed by a recombination of pastures the following year. In a commercial ranching enterprise grazing yearling heifers, the combination of one-crop and two-crop systems maintained good daily gains over an extended grazing season, gave a forage to beef ratio of 10:1, kept forage reasonably nutritious, and prevented the development of wolf plants [65]. However, little second-crop herbage was produced in very dry years.

Spring-fall grazing of crested wheatgrass over a ten-year period at Manitou, Colorado, produced more forage, provided more days of grazing, and produced higher weight gains per acre than either spring or fall grazing alone [32]. On foothill range in central Utah, early spring grazing of crested wheatgrass allowed regrowth prior to fall grazing [63]. Late spring grazing ending in late June was generally followed by insufficient soil moisture for regrowth, and the remaining forage was less valuable for fall grazing. Pastures moderately grazed in early spring and again in the fall provided slightly additional per-head gains in the fall than when fall grazed only during years of no fall regrowth. However, heavy spring utilization generally left insufficient grass for fall grazing and reduced regrowth.

Complementary Grazing Systems

Seeded range as well as temporary pasture complement native range by extending the green growing period, providing special use grazing, increasing productivity, serving as a buffer in reducing fluctuations in grazing capacity, and providing greater versatility. Since native range is of maximum value for a relatively short period of time over much of the West, grazing systems using a combination of native range and seeded forages but in different pasture units appear most realistic [115].

When based 20 percent on crested wheatgrass pastures, 50 percent on native range, and 30 percent on Russian wildrye pastures, the acreage required per animal unit for a 7.5-month grazing season in the Northern Great Plains required only 11.4 acres compared to 24.8 acres when only native range was used [115]. At Manitou, Colorado, adding seeded ranges for spring and fall grazing to a combination of native range and meadow increased the adequate protein intake period for range cows from eight to ten months [117]. Providing a forage that will meet the nutritional requirements for lactating animals on spring ranges has been deemed more desirable than feeding sufficient supplement to meet nutritional deficiencies on native spring ranges [27]. Early growing seeded pastures can also fill a critical need when used for spring grazing to permit deferment of native range.

Crested wheatgrass has been widely recognized for its ability to produce early spring grazing. In central Utah, cows and calves gained an average of 1.73 and 2.02 pounds per day respectively on seeded foothill range but only 1.02 and 1.37 pounds per day on sagebrush-grass range during a five-week spring grazing period [26]. However, it has also been found that crested wheatgrass could be satisfactorily grazed from April to December [63]. At the end of a three-month summer grazing season, calf weights from cows grazing seeded foothill crested wheatgrass were similar to those from cows grazing mountain forest range, but the cows summered on crested wheatgrass weighed 20 pounds less. In other areas, cows have been wintered satisfactorily on crested wheatgrass when provided a protein supplement.

On foothill range at Manitou, Colorado, grazing cow-calf pairs on seeded range in the spring and fall compared to grazing yearlong on native bunchgrass range and meadow increased weaning weights by 33 pounds (451 versus 418 pounds) and gross income per calf by $8.95 [31]. The seeded ranges grazed at Manitou included Russian wildrye, April 15 to May 15; crested wheatgrass, May 15 to June 15; and Sherman big bluegrass, October 15 to December 15. In Wyoming a combination of seeded pastures (crested wheatgrass for spring, intermediate wheatgrass for summer, and Russian wildrye for fall) gave daily gains similar to native range when grazed with yearling cattle but had grazing capacity and gains per acre from two to three times that from the native range [109]. In Canada ewes were reported able to rotate themselves under a free-choice system, going first to crested wheatgrass, then to native range, and lastly to Russian wildrye while gaining two pounds more per acre (18.5 versus 16.5 pounds) than when under a forced rotation system [157].

Seeded Pasture for Breeding Herds

Seeded, high-producing pastures make effective breeding pastures. Their use with artificial insemination has been equally beneficial to use under natural service since concentrating the cows has greatly assisted in checking for heat, thereby reducing labor required in heat checking and accelerating rebreeding. Early growing, cool-season pastures favor rapid calf gains, and fall regrowth grazed by cow-calf pairs can materially increase weaning weights [31]. Good lambing range limits sheep production in many range areas. In New Mexico, using crested wheatgrass for forty-five days in the spring for lambing not only was a convenience but increased lamb crops by 4.5 to 7.0 percent while reducing death losses by 1 to 3 percent [56].

Early spring pasture provided by introduced grasses such as crested wheatgrass and Russian wildrye or grass-legume mixtures furnishes succulent, nutritious forage well ahead of native range and when needed most by range cows. After coming through the winter on maintenance levels of nutrition, spring-calving range cows are subjected to the stress of calving and providing milk for the calf. Research and rancher experience indicates that grazing cool-season spring pasture from calving until breeding provides all or most of the following benefits [176]:

1. Earlier return to estrus following calving—a major benefit.
2. Calving is concentrated in the fore part of the calving season, thus increasing average age and weight at weaning on a predetermined calendar date.
3. Length of breeding season needed is reduced to 45 to 60 days.
4. Percent calf crop is increased at birth and weaning.
5. Milk flow is stimulated and daily gains of calves increased during the spring.
6. Health and stress resistance of cows and calves is improved.
7. Needed supplements and hay are reduced by shortening the overwintering period.

Early spring pastures at the Fort Robinson Beef Cattle Research Station in Nebraska have helped materially in bringing breeding females, particularly heifers, back into heat after calving and in getting them settled [182]. In one trial with three-year-old cows suckling calves, 71 percent of those placed on crested wheatgrass after calving mated during the first twenty-one days of the breeding season while only 43 percent of those on native range were serviced. By eighty days after calving in another trial, 83 percent of the heifers grazing crested wheatgrass had come into heat as compared to only 60 percent of those grazing native grasses. The results for a third trial are shown in Table 19.

Providing spring calving range cows with cool-season pasture from April 23 to June 3 at Miles City, Montana, increased the beef weaned per cow by 46 pounds (395 lb. vs. 349 lb.) when compared to similar cows grazed yearlong on native range [71]. This 13 percent increase in beef production resulted from a nine percent higher per-

Table 19. Three-year-old Cows in Heat after Spring Calving at the Fort Robinson Beef Cattle Research Station, Nebraska*

Kind of pasture (postcalving)	Days after calving							
	40	50	60	70	80	90	100	110
Crested wheatgrass	14%	32%	43%	64%	77%	79%	86%	92%
Native range	6%	16%	28%	41%	57%	70%	85%	95%

*Source: [182].

356

FIG. 93. *Russian wildrye pasture grazed in the spring by lactating beef cows has greatly improved reproductive performance and calf production at the North Platte Station, Nebraska.*

cent calf crop weaned (90.8 versus 81.7 percent) and increased weaning weights (per head of 10 to 20 pounds). These advantages were based on five years of data and were attributed to the earlier growth, greater forage production, and higher protein content of the crested wheatgrass-alfalfa and Russian wildrye-alfalfa pasturage. In Oregon, crested wheatgrass compared to native range for spring grazing increased the calf crop by 10 percent (95 versus 85 percent) and weaning weights by 40 pounds (440 versus 400 pounds) [64]. The practice of spring grazing on improved pasture also increased the number of calves old enough to be worked before spring turnout.

Grazing cow-calf pairs on grass-legume mixtures in the spring at the North Platte Station in Nebraska (Figure 93) was superior to a combination of hay and supplement and native range from the standpoint of calf production and reproductive performance [22]. Cool-season pasture benefits included eleven pounds heavier weaning weights one year and twenty-three pounds the next. However, improved reproduction was even greater. While 89 percent of the young two-year-old cows on improved pasture had returned to heat by the beginning of their second breeding season, only 55 percent of those in the spring hay and range treatment had returned to heat. Corresponding figures for the following year as threes were 76 versus 44 percent. This indicates that cool-season pasture, whether from dryland or irrigated pasture, is important to cow-calf operations.

Preventing Bloat

When legumes such as alfalfa, Ladino clover, or red clover are included in grass-legume mixtures, frothy bloat becomes a possibility.

However, legumes are commonly included in mixtures on subhumid and mesic sites because of the resulting increase in carrying capacity and nutritive value of the pasturage. A compound called poloxalene, an antifoaming agent, appears to prevent pasture bloat for twelve hours if fed in adequate amounts. Poloxalene can be included in molasses blocks, in a liquid molasses supplement, or in granular form mixed in a mineral or grain supplement. However, the cost of feeding and the need for obtaining correct and regular intake must be considered.

Many management practices are helpful in reducing pasture bloat. Although their effectiveness varies greatly from region to region and from pasture to pasture, they are generally not 100 percent effective. Practices useful in reducing bloat include:

1. Seed legumes in pasture mixtures at rates not to exceed one to two pounds per acre.
2. Manage pastures for no more than 40 percent legume production in the mixture.
3. Delay grazing until the legume reaches the late bud to early bloom stage of maturity or until alfalfa reaches the height of twelve to fourteen inches.
4. Graze under a rotation system where a restricted area is grazed for a short period, thereby reducing selectivity.
5. Do not turn hungry livestock into a grass-legume pasture; feed dry roughage first.
6. Keep some dry roughage available at all times; or mow a strip three to four days before grazing.
7. Turn livestock into a new pasture during the heat of the day rather than in early morning.
8. Keep salt and water available at all times.
9. Check livestock frequently; remove chronic bloaters.

Literature Cited

1. Abernathy, George H., and Carlton H. Herbel. 1973. Brush eradicating, basin pitting, and seeding machine for arid to semiarid rangeland. J. Range Mgt. 26(3):189–192.
2. Allred, Keith R. 1966. Up-grading irrigated pastures—use alfalfa-intermediate wheatgrass where water is limited. Utah Farm & Home Science 27(2):47–51.
3. Anderson, Darwin, Louis P. Hamilton, Hudson G. Reynolds, and R. R. Humphrey. 1953. Reseeding desert grassland ranges in southern Arizona. Arizona Agric. Expt. Sta. Bul. 249.
4. Andrew, F. W., G. R. Carlisle, L. R. Fryman, D. F. Wilken, and W. D. Pardee. 1964. Illinois forage handbook. Illinois Agric. Ext. Serv. Cir. 895.
5. Army, J. T., and E. B. Hudspeth, Jr. 1959. Better grass establishment with plastic covers. Texas Agric. Progress 5(4):20, 22–23.
6. Artz, John L., and E. Irving Hackett. 1971. Wolf plants in crested wheatgrass seedings. ASRM, Abstract of Papers, 24th Annual Meeting, p. 17.
7. Artz, John L., J. Boyd Price, Frederick F. Peterson, et al. 1970. Plantings for wildlands and erosion control. Nevada Agric. Ext. Serv. Cir. 108.
8. Asher, Jerry E., and Richard E. Eckert, Jr. 1973. Development, testing, and evaluation of the deep furrow drill arm assembly for the rangeland drill. J. Range Mgt. 26(5):377–378.
9. Atkins, M. D., and James E. Smith. 1967. Grass seed production and harvest in the Great Plains. USDA Farmers' Bul. 2226.
10. Basile, Joseph V., and Ralph C. Holmgren. 1957. Seeding-depth trials with bitterbrush (*Purshia tridentata*) in Idaho. USDA, For. Serv. Res. Paper 54.
11. Bement, R. E., R. D. Barmington, A. C. Everson, L. O. Hylton, Jr., and E. E. Remmenga. 1965. Seeding of abandoned croplands in the central Great Plains. J. Range Mgt. 18(2):53–59.
12. Bentley, Jay R. 1967. Conversion of chaparral areas to grassland. USDA Agric. Handbook 328.
13. Biswell, H. H., A. M. Schultz, D. W. Hedrick, and J. I. Mallory. 1953. Frost heaving of grass and brush seedlings on burned chamise brushlands in California. J. Range Mgt. 6(3):172–180.
14. Bleak, A. T., and A. P. Plummer. 1954. Grazing crested wheatgrass by sheep. J. Range Mgt. 7(2):63–67.
15. Booster, Dean E. 1961. The Oregon press seeder. Oregon Agric. Expt. Sta. Cir. of Info. 605.
16. Brock, J. H., C. E. Fisher, E. D. Robison, et al. 1971. Seedbed preparation, seeding methods, and their effect on the establishment of grasses following brush removal. ASRM, Abstract of Papers, 24th Annual Meeting, p. 31.
17. Brown, Ellsworth Reade, and Charles F. Martinsen. 1959. Browse planting for big game. Washington State Game Dept. Biol. Bul. 12.
18. Bryan, G. G., and W. E. McMurphy. 1968. Competition and fertilization as influences on grass seedlings. J. Range Mgt. 21(2):98–101.
19. Bryant, H. T., et al. 1970. Effect of grazing management on animal and area output. J. Ani. Sci. 30(1):153–158.
20. Burzlaff, D. F., and J. C. Swinbank. 1965 (Rev.). Pure live seed—method for determining requirements for grass seedings. Nebraska Agric. Ext. Serv. Cir. 65–135.
21. Campbell, J. B. 1963. Grass-alfalfa versus grass-alone pastures grazed in a repeated-seasonal pattern. J. Range Mgt. 16(2):78–81.
22. Clanton, D. C., J. T. Nichols, and B. R. Somerhalder. 1971. Young cows on irrigated pastures. *In* 1971 Nebraska beef cattle report, pp. 16–19. Nebraska Agric. Ext. Serv. Cir. EC 71–218.
23. Conard, E. C. 1962. How to establish new pastures. Nebraska Agric. Ext. Serv. Campaign Cir. 165.
24. Conard, E. C., and Raymond G. Sall. 1967. Seed dormancy, age of seed, and rate of planting in relation to stand establishment of switchgrass and Indiangrass. ASRM, Abstract of Papers, 20th Annual Meeting, p. 17.

25. Cook, C. Wayne. 1965. Grass seedling response to halogeton competition. J. Range Mgt. 18(6):317–321.
26. Cook, C. Wayne. 1966. Development and use of foothill ranges in Utah. Utah Agric. Expt. Sta. Bul. 461.
27. Cook, C. Wayne, and L. A. Stoddart. 1961. Nutrient intake and livestock responses on seeded foothill ranges. J. Anim. Sci. 20(1):36–41.
28. Cook, C. Wayne, L. A. Stoddart, and Phillip L. Sims. 1967. Effects of season, spacing, and intensity of seeding on the development of foothill range grass stands. Utah Agric. Expt. Sta. Bul. 467.
29. Cope, Gene E., and Frank C. Petr. 1976. Bloat control of animals grazing on alfalfa pasture. Texas Agric. Ext. Leaflet 1496.
30. Currie, Pat O. 1967. Seeding Sherman big bluegrass. J. Range Mgt. 20(3):133–136.
31. Currie, Pat O. 1969. Use seeded ranges in your management. J. Range Mgt. 22(6):432–434.
32. Currie, Pat O. 1970. Influence of spring, fall, and spring-fall grazing on crested wheatgrass range. J. Range Mgt. 23(2):103–108.
33. Currie, Pat O., and Dwight R. Smith. 1970. Response of seeded ranges to different grazing intensities in the ponderosa pine zone of Colorado. USDA (Forest Service) Prod. Res. Rpt. 112.
34. Derscheid, Lyle A., Raymond A. Moore, and Melvin D. Rumbaugh. 1970. Interseeding for pasture and range improvement. South Dakota Agric. Ext. Serv. Fact Sheet 422.
35. Douglas, Donald S., A. L. Hafenrichter, and K. H. Klages. 1960. Cultural methods and their relation to establishment of native and exotic grasses in range seedings. J. Range Mgt. 13(2):53–57.
36. Drawe, D. Lynn, and I. G. Palmblad. 1977. Competition between Russian wildrye seedlings and four common weeds. J. Range Mgt. 30(3):223–226.
37. Dudley, D. I., and Ethan C. Holt. 1963. Establishment of warm-season grasses on the Grand Prairie. Texas Agric. Expt. Sta. Misc. Pub. 672.
38. Eckert, Richard E., Jr., Jerry E. Asher, M. Dale Christensen, and Raymond A. Evans. 1974. Evaluation of the atrazine-fallow technique for weed control and seedling establishment. J. Range Mgt. 27(4):288–292.
39. Eckert, Richard E., Jr., and Raymond A. Evans. 1967. A chemical-fallow technique for control of downy brome and establishment of perennial grasses on rangeland. J. Range Mgt. 20(1):35–41.
40. Eckert, Richard E., Jr., Gerald J. Klomp, Raymond A. Evans, and James A. Young. 1971. Establishment of perennial wheatgrasses in relation to atrazine residue in the seedbed. J. Range Mgt. 25(3):219–224.
41. Edmunson, George C., Lisle R. Green, and Jay R. Bentley. 1961. Seed distribution in the swath from helicopter sowing. USDA, Pacific Southwest Forest & Range Expt. Sta. Res. Note 190.
42. Ellern, Sigmund J., and Naphtali H. Tadmor. 1967. Germination of range plant seeds at alternating temperatures. J. Range Mgt. 20(2):72–77.
43. Evans, Raymond A. 1961. Effects of different densities of downy brome (*Bromus tectorum*) on growth and survival of crested wheatgrass (*Agropyron desertorum*) in the greenhouse. Weeds 9(2):216–223.
44. Evans, Raymond A., and Richard E. Eckert, Jr. 1965. Paraquat-surfactant combinations for control of downy brome. Weeds 13(2):150–151.
45. Evans, Raymond A., Richard E. Eckert, Jr., and Burgess L. Kay. 1967. Wheatgrass establishment with paraquat and tillage on downy brome ranges. Weeds 15(1):50–55.
46. Evans, Raymond A., H. Richard Holbo, Richard E. Eckert, Jr., and James A. Young. 1970. Functional environment of downy brome communities in relation to weed control and revegetation. Weed Science 18(1):154–162.
47. Evans, Raymond A., and James A. Young. 1977. Weed control-revegetation systems for sagebrush-downy brome rangelands. J. Range Mgt. 30(5):331–336.
48. Evans, Raymond, and James A. Young. 1978. Effectiveness of rehabilitation

practices following wildfire in a degraded big sagebrush-downy brome community. J. Range Mgt. 31(3):185–188.

49. Everson, A. C., D. N. Hyder, H. R. Gardner, and R. E. Bement. 1969. Chemical versus mechanical fallow of abandoned croplands. Weed Sci. 17(4):548–551.

50. Ferguson, Robert B. 1968. Survival and growth of young bitterbrush browsed by deer. J. Wildl. Mgt. 32(4):769–772.

51. Ferguson, Robert B., and Joseph V. Basile. 1967. Effect of seedling numbers on bitterbrush survival. J. Range Mgt. 20(6):380–382.

52. Frischknecht, Neil C., and Lorin E. Harris. 1968. Grazing intensities and systems on crested wheatgrass in central Utah: response of vegetation and cattle. USDA Tech. Bul. 1388.

53. Fryer, J. D., and S. A. Evans (Editors). 1968 (Fifth Edition). Weed control handbook. Volume II. Recommendations. Blackwell Scientific Publications, Oxford and Edinburg.

54. Gates, Dillard H. 1967. Range seedings—success or failure. Oregon Agric. Ext. Serv. Fact Sheet 120.

55. Gomm, F. B. 1964. A comparison of two sweetclover strains and Ladak alfalfa alone and in mixtures with crested wheatgrass for range and dryland seeding. J. Range Mgt. 17(1):19–23.

56. Gray, James R., and H. W. Springfield. 1962. Economics of lambing on crested wheatgrass in northcentral New Mexico. New Mexico Agric. Expt. Sta. Bul. 461.

57. Great Plains Agric. Council. 1966. A stand establishment survey of grass plantings in the Great Plains. Nebraska Agric. Expt. Sta., Great Plains Council Rpt. 23.

58. Haas, Robert H., Howard L. Morton, and Paul J. Torell. 1962. Influence of soil salinity and 2,4-D treatments on establishment of desert wheatgrass and control of halogeton and other annual weeds. J. Range Mgt. 15(4):205–210.

59. Hafenrichter, A. L., John L. Schwendiman, Harold L. Harris, Robert S. MacLauchlan, and Harold D. Miller. 1968. Grasses and legumes for soil conservation in the Pacific Northwest and Great Basin states. USDA Agric. Handbook 339.

60. Halls, L. K., G. W. Burton, and B. L. Southwell. 1957. Some results of seeding and fertilization to improve southern forest range. USDA, Southeastern Forest Expt. Sta. Paper 78.

61. Halls, L. K., R. H. Hughes, and F. A. Peevy. 1960. Grazed firebreaks in southern forests. USDA Agric. Info. Bul. 226.

62. Harris, Harold L., A. E. Slinkard, and A. L. Hafenrichter. 1972. Establishment and production of grasses under semiarid conditions in the Intermountain West. Ida. Agric. Expt. Sta. Bul. 532.

63. Harris, Lorin E., Neil C. Frischknecht, and Earl M. Sudweeks. 1968. Seasonal grazing of crested wheatgrass by cattle. J. Range Mgt. 21(4):221–225.

64. Hedrick, D. W. 1967. Managing crested wheatgrass for early spring use. J. Range Mgt. 20(1):53–54.

65. Hedrick, D. W., W. M. Moser, A. L. Steninger, and R. A. Long. 1969. Animal performance on crested wheatgrass pastures during May and June, Fort Rock, Oregon. J. Range Mgt. 22(4):277–280.

66. Heinrichs, D. H., and J. L. Bolton. 1950. Studies on the competition of crested wheatgrass with perennial native species. Sci. Agric. 30(4):428–443.

67. Herbel, Carlton H., George H. Abernathy, Clyde C. Yarbrough, and David K. Gardner. 1973. Rootplowing and seeding arid rangelands in the Southwest. J. Range Mgt. 26(3):193–197.

68. Hervey, D. F. 1958. Improving sagebrush ranges for grazing use. Colorado Agric. Expt. Sta. Prog. Rpt. General Series 688.

69. Holmgren, Ralph C. 1956. Competition between annuals and young bitterbrush (Purshia tridentata) in Idaho. Ecology 37(2):370–377.

70. Holmgren, Ralph C., and Joseph V. Basile. 1959. Improving southern Idaho deer winter ranges by artificial revegetation. Idaho Dept. of Fish and Game

Wildlife Bul. 3.

71. Houston, W. R., and J. J. Urick. 1972. Improved spring pastures, cow-calf production, and stocking rate carryover in the Northern Great Plains. USDA Tech. Bul. 1451.

72. Houston, Walter R. 1957. Seeding crested wheatgrass on drought depleted range. J. Range Mgt. 10(3):131–134.

73. Hubbard, Richard L., and H. Reed Sanderson. 1961. When to plant bitterbrush—spring or fall? USDA, Pacific Southwest Forest & Range Expt. Sta. Tech. Paper 64.

74. Hubbard, Richard L., Pinhas Zusman, and H. Reed Sanderson. 1962. Bitterbrush stocking and minimum spacing with crested wheatgrass. California Fish & Game 48(3):203–208.

75. Hughes, M. D., Maurice E. Heath, and Darrel S. Metcalfe. 1962 (Second Ed.). Forages. Iowa State University Press, Ames, Iowa.

76. Hull, A. C., Jr. 1954. Rating seeded stands on experimental range plots. J. Range Mgt. 7(3):122–124.

77. Hull, A. C., Jr. 1959. Pellet seeding of wheatgrass on southern Idaho rangelands. J. Range Mgt. 12(4):155–163.

78. Hull, A. C., Jr. 1960. Winter germination of intermediate wheatgrass on mountain lands. J. Range Mgt. 13(5):257–260.

79. Hull, A. C., Jr. 1966. Emergence and survival of intermediate wheatgrass and smooth brome seeded on a mountain range. J. Range Mgt. 19(5):279–283.

80. Hull, A. C., Jr. 1970. Grass seedling emergence and survival from furrows. J. Range Mgt. 23(6):421–424.

81. Hull, A. C., Jr. 1971. Spraying tarweed infestations on ranges newly seeded to grass. J. Range Mgt. 24(2):145–147.

82. Hull, A. C., Jr. 1972. Seeding rates and row spacings for rangelands in southeastern Idaho and northern Utah. J. Range Mgt. 25(1):50–53.

83. Hull, A. C., Jr. 1974. Species for seeding arid rangeland in southern Idaho. J. Range Mgt. 27(3):216–218.

84. Hull, A. C., Jr., and Hallie Cox. 1968. Spraying and seeding high elevation tarweed rangelands. J. Range Mgt. 21(3):140–144.

85. Hull, A. C., Jr., D. F. Hervey, Clyde W. Doran, and W. J. McGinnies. 1958. Seeding Colorado rangelands. Colorado Agric. Expt. Sta. Bul. 498-S.

86. Hull, A. C., Jr., and Ralph C. Holmgren. 1964. Seeding southern Idaho rangelands. USDA, For. Serv. Res. Paper INT-10.

87. Hull, A. C., Jr., and G. J. Klomp. 1966. Longevity of crested wheatgrass in the sagebrush-grass type in southern Idaho. J. Range Mgt. 19(1):5–11.

88. Hull, A. C., Jr., and G. J. Klomp. 1967. Thickening and spread of crested wheatgrass stands on southern Idaho ranges. J. Range Mgt. 20(4):222–227.

89. Hyder, D. N., and R. E. Bement. 1969. A micro-ridge roller for seedbed modification. J. Range Mgt. 22(1):54–56.

90. Hyder, D. N., and R. E. Bement. 1970. Soil physical conditions after plowing and packing of ridges. J. Range Mgt. 23(4):289–292.

91. Hyder, D. N., D. E. Booster, F. A. Sneva, W. A. Sawyer, and J. B. Rodgers. 1961. Wheeltrack planting on sagebrush-bunchgrass range. J. Range Mgt. 14(4):220–224.

92. Hyder, D. N., and A. C. Everson. 1968. Chemical fallow of abandoned croplands on the shortgrass plains. Weed Sci. 16(4):531–533.

93. Hyder, Donald N., and Forrest A. Sneva. 1954. A method for rating the success of range seeding. J. Range Mgt. 7(2):89–90.

94. Hyder, D. N., and Forrest A. Sneva. 1963. Morphological and physiological factors affecting the grazing management of crested wheatgrass. Crop Sci. 3(3):267–271.

95. Hyder, D. N., and Forrest A. Sneva. 1963. Studies of six grasses seeded on sagebrush-bunchgrass range—yield, palatability, carbohydrate accumulation, developmental morphology. Oregon Agric. Expt. Sta. Tech. Bul. 71.

96. Hyder, Donald N., Forrest Sneva, and Clee S. Cooper. 1955. Methods for planting crested wheatgrass. In 1955 Field Day Report, Squaw Butte-Harney Range and Livestock Expt. Sta., Burns, Oregon, pp. 19–20.

97. Hyder, Donald N., Forrest A. Sneva, and W. A. Sawyer. 1955. Soil firming may improve range seeding operations. J. Range Mgt. 8(4):159-163.

98. Idaho Agric. Ext. Serv. 1961. Idaho forage crop handbook. Idaho Agric. Ext. Serv. Bul. 363.

99. Jordan, Gilbert L., and Michael L. Maynard. 1970. The San Simon Watershed: revegetation. Prog. Agric. in Arizona 22(6):4-7.

100. Kay, Burgess L. 1966. Paraquat for range seeding without cultivation. California Agric. 20(10):2-4.

101. Kay, Burgess L., and Richard E. Owen. 1970. Paraquat for range seeding in cismontane California. Weed Science 18(2):238-244.

102. Kilcher, M. R. 1961. Fall seeding versus spring seeding in the establishment of five grasses and one alfalfa in southern Saskatchewan. J. Range Mgt. 14(6):320-322.

103. Kilcher, M. R. 1961. Row spacing affects yields of forage grasses in the brown soil zone of Saskatchewan. Canada Dept. Agric. Pub. 100.

104. Kilcher, M. R., and D. H. Heinrichs. 1968. Rates of seeding Rambler alfalfa with dryland pasture grasses. J. Range Mgt. 21(4):248-249.

105. Kinsinger, Floyd E. 1962. The relationship between depth of planting and maximum foliage height of seedlings of Indian ricegrass. J. Range Mgt. 15(1):10-13.

106. Klomp, G. J., and A. C. Hull, Jr. 1968. Effects of 2,4-D on emergence and seedling growth of range grasses. J. Range Mgt. 21(2):67-70.

107. Klomp, G. J., and A. C. Hull, Jr. 1972. Methods for seeding three perennial wheatgrasses on cheatgrass ranges in southern Idaho. J. Range Mgt. 25(4):266-268.

108. Lang, Robert. 1962. Range seeding and pitting study in the Teton National Forest. Wyoming Agric. Expt. Sta. Mimeo. Cir. 173.

109. Lang, Robert, and Leland Landers. 1960. Beef production and grazing capacity from a combination of seeded pastures versus native range. Wyoming Agric. Expt. Sta. Bul. 370.

110. Launchbaugh, J. L., and Kling L. Anderson. 1963. Grass reseeding investigations at Hays and Manhattan, Kansas. Kansas Agric. Expt. Sta. Tech. Bul. 128.

111. Launchbaugh, J. L., and Clenton E. Owensby. 1970. Seeding rate and first-year stand relationships for six native grasses. J. Range Mgt. 23(6):414-417.

112. Lavin, Fred, and F. B. Gomm. 1968. Stabilizing small seed dilution mixtures. J. Range Mgt. 21(5):328-330.

113. Lavin, Fred, F. B. Gomm, and T. N. Johnsen, Jr. 1973. Cultural, seasonal, and site effects on pinyon-juniper rangeland plantings. J. Range Mgt. 26(4):279-285.

114. Lawrence, T. 1957. Emergence of intermediate wheatgrass lines from five depths of seeding. Canadian J. Plant Sci. 37(3):215-219.

115. Lodge, Robert W. 1970. Complementary grazing systems for the Northern Great Plains. J. Range Mgt. 23(4):268-271.

116. Lodge, Robert W., Sylvester Smoliak, and Alexander Johnston. 1972. Managing crested wheatgrass pastures. Canada Dept. Agric. Pub. 1473.

117. Malechek, John C. 1966. Cattle diets on native and seeded ranges in the ponderosa pine zone of Colorado. USDA, For. Serv. Res. Note RM-77.

118. Marlatt, W. E., and D. N. Hyder. 1970. Soil ridging for reduction of wind erosion from grass seedbeds. J. Range Mgt. 23(3):170-174.

119. Martin, S. Clark. 1966. The Santa Rita Experimental Range. USDA, For. Serv. Res. Paper RM-22.

120. Mathis, Gary W., Merwyn M. Kothmann, and William J. Waldrip. 1971. Influence of root plowing and seeding on composition and forage production on native grasses. J. Range Mgt. 24(1):43-47.

121. McGinnies, William J. 1959. The relationship of furrow depth to moisture content of soil and to seedling establishment on a range soil. Agron. J. 51(1):13-14.

122. McGinnies, William J. 1960. Effects of moisture stress and temperature on germination of six range grasses. Agron. J. 52(3):159-162.

123. McGinnies, William J. 1960. Effects of planting dates, seeding rates, and row spacings on range seeding results in western Colorado. J. Range Mgt. 13(1):37–39.

124. McGinnies, William J. 1962. Effect of seedbed firming on the establishment of crested wheatgrass seedlings. J. Range Mgt. 15(4):230–234.

125. McGinnies, William J. 1966. Effects of spot seeding on establishment of three range grasses. Agron. J. 58(6):612–614.

126. McGinnies, William J. 1968. Effect of post-emergence weed control on grass establishment in north-central Colorado. J. Range Mgt. 21(3):126–128.

127. McGinnies, William J. 1970. Effects of seeding rate and row spacing on establishment and yield of crested wheatgrass. Agron. J. 62(3):417–421.

128. McGinnies, William J. 1973. Effects of date and depth of planting on the establishment of three range grasses. Agron. J. 65(1):120–123.

129. McIlvain, E. H., and M. C. Shoop. 1960. An agronomic evaluation of regrassing cropland in the Southern Great Plains. Part I. Rangeland seedings in the Southern Great Plains. USDA, Agric. Res. Serv. Mimeo.

130. McIlvain, E. H., and M. C. Shoop. 1960. An agronomic evaluation of regrassing cropland in the southern Great Plains. Part II. Dryland pasture for the Southern Great Plains. USDA, Agric. Res. Serv. Mimeo.

131. McKell, Cyrus M., J. R. Goodin, and Cameron Duncan. 1969. Chaparral manipulation affects soil moisture depletion patterns and seedling establishment. J. Range Mgt. 22(3):159–165.

132. McMurphy, Wilfred E. 1969. Pre-emergence herbicides for seeding range grasses. J. Range Mgt. 22(6):427–429.

133. McWilliams, Jesse L. 1955. Effects of some cultural practices on grass production at Mandan, North Dakota. USDA Tech. Bul. 1097.

134. Mohan, Joseph M., and W. F. Currier. 1962. Range rehabilitation by spray and drill. ASRM, Abstract of Papers, 15th Annual Meeting, pp. 28–31.

135. Mueggler, Walter F., and James P. Blaisdell. 1955. Effect of seeding rate upon establishment and yield of crested wheatgrass. J. Range Mgt. 8(2):74–76.

136. Murphy, William J., Paul H. Bebermeyer, C. Melvin Bradley, et. al. 1966. Missouri livestock forage manual for beef and dairy cattle. Missouri Agric. Ext. Serv. Manual 67.

137. National Research Council. 1968. Principles of plant and animal pest control. II. Weed control. National Academy of Sciences Pub. 1597.

138. Nelson, Jack R., A. M. Wilson, and Carl J. Goebel. 1970. Factors influencing broadcast seeding on bunchgrass range. J. Range Mgt. 23(3):163–170.

139. Nichols, James T., and James R. Johnson. 1969. Range productivity as influenced by biennial sweetclover in western South Dakota. J. Range Mgt. 22(5):342–347.

140. Owensby, Clenton E., and Kling L. Anderson. 1965. Reseeding "go-back" land in the Flint Hills of Kansas. J. Range Mgt. 18(4):224–225.

141. Plummer, A. Perry, Donald R. Christensen, and Stephen B. Monson. 1968. Restoring big game range in Utah. Utah Div. of Fish and Game Pub. 68-3.

142. Plummer, A. Perry, A. C. Hull Jr., George Stewart, and Joseph H. Robertson. 1955. Seeding rangelands in Utah, Nevada, and southern Idaho, and western Wyoming. USDA Agric. Handbook 71.

143. Ragsdale, B. J., R. V. Miller, Jr., G. O. Hoffman, and J. D. Rodgers. 1970. Keys to profitable range management in Texas. Texas Agric. Ext. Misc. Pub. 965.

144. Range Seeding Equipment Comm. 1965 (Rev.). Handbook of range seeding equipment. USDA and USDI.

145. Rauzi, Frank, Robert L. Lang, and C. F. Becker. 1962. Mechanical treatments on shortgrass rangeland. Wyoming Agric. Expt. Sta. Bul. 396.

146. Rechenthin, C. A., H. M. Bell, R. J. Pederson, D. B. Polk, and J. E. Smith, Jr. 1965. Grassland restoration. Part III. Re-establishing forage plants. USDA, Soil. Cons. Serv., Temple, Texas.

147. Reynolds, Hudson G., and S. Clark Martin. 1968 (Rev.). Managing grass-shrub cattle ranges in the Southwest. USDA Agric. Handbook 162.

148. Robertson, J. H., D. L. Neal, K. R. McAdams, and P. T. Tueller. 1970. Changes in crested wheatgrass ranges under different grazing treatments. J. Range Mgt. 23(1):27–34.

149. Robertson, Truman E., Jr., and Thadis W. Box. 1969. Interseeding sideoats grama on the Texas High Plains. J. Range Mgt. 22(4):243–245.

150. Roby, George A., and Lisle R. Green. 1976. Mechanical methods of chaparral modification. USDA Agric. Handbook 487.

151. Rogler, George A. 1954. Seed size and seedling vigor in crested wheatgrass. Agron. J. 46(5):216–220.

152. Rosenquist, David W., and Dillard H. Gates. 1961. Responses of four grasses at different stages of growth to various temperature regimes. J. Range Mgt. 14(4):198–202.

153. Ross, J. G., S. S. Bullis, and R. A. Moore. 1966. Grass performance in South Dakota. South Dakota Agric. Expt. Sta. Bul. 536.

154. Rumbaugh, M. D., G. Semeniuk, R. Moore, and J. D. Colburn. 1965. Travois—an alfalfa for grazing. South Dakota Agric. Expt. Sta. Bul. 525.

155. Seamands, Wesley J. 1966. Increase production from Wyoming mountain meadows. Wyoming Agric. Ext. Bul. 441.

156. Sharp, Lee A. 1970. Suggested management programs for grazing crested wheatgrass. Idaho Forest, Wildl., and Range Expt. Sta. Bul. 4.

157. Smoliak, S. 1968. Grazing studies on native range, crested wheatgrass, and Russian wildrye pastures. J. Range Mgt. 21(1):47–50.

158. Sneva, Forrest A., and Larry R. Rittenhouse. 1976. Crested wheat production: impacts on fertility, row spacing, and stand age. Ore. Agric. Expt. Sta. Tech. Bul. 135.

159. Springfield, H. W. 1956. Relation of time of planting to establishment of wheatgrasses in northern New Mexico. USDA, Rocky Mtn. Forest & Range Expt. Sta. Res. Note 24.

160. Springfield, H. W. 1965. Rate and spacing in seeding crested wheatgrass in New Mexico. USDA, For. Serv. Res. Note RM-42.

161. Springfield, H. W. 1966. Effects of three years' grazing at different intensities on crested wheatgrass lambing range in northern New Mexico. USDA, For. Serv. Res. Note RM-65.

162. Springfield, H. W. 1970. Emergence and survival of winterfat seedlings from four planting depths. USDA, For. Serv. Res. Note RM-162.

163. Springfield, H. W. 1970. Germination and establishment of four-wing saltbush in the Southwest. USDA, For. Serv. Res. Paper RM-55.

164. Springfield, H. W., and R. M. Housley, Jr. 1952. Chamiza for reseeding New Mexico rangelands. USDA, Southwest Forest & Range Expt. Res. Sta. Note 122.

165. Stewart, William G. 1973. Irrigated pastures for Colorado. Colo. Agric. Ext. Bul. 469A.

166. Stoddart, L. A. 1946. Rye nurse crops in range seeding. Ecology 27(1):61–64.

167. Stroh, James R., and Vernon P. Sundberg. 1971. Emergence of grass seedlings under crop residue cover. J. Range Mgt. 24(3):226–230.

168. Stuth, Jerry W., and Bill E. Dahl. 1974. Evaluation of rangeland seedings following mechanical brush control in Texas. J. Range Mgt. 27(2):146–149.

169. Sumner, D. C., and R. Merton Love. 1961. Seedling competition from resident range cover often cause of seeding failures. California Agric. 15(2):6.

170. Torell, Paul J., and Lambert C. Erickson. 1967. Reseeding medusahead-infested ranges. Idaho Agric. Expt. Sta. Bul. 489.

171. USDA, Agric. Res. Service. 1960. Utilizing forage from improved pastures. USDA, ARS 22–53.

172. USDA, Agric. Res. Serv. 1961. Improving grass stands on the Southern Great Plains. Agric. Res. 10(3):5.

173. USDA, Agric. Res. Serv. 1970. Waste brush helps grass grow. Agric. Res. 18(7):14.

174. USDA, Agric. Res. Serv. 1975. Renovating saltgrass meadows. Agric. Res. 23(10):7.

175. Utah Agric. Expt. Sta. 1970. Pasture planting specifications for Utah. Utah Agric. Expt. Sta. Cir. 153.
176. Vallentine, John F. 1970. Early pasture important to cow-calf operations. Utah Farmer 90(9):12–13.
177. Vallentine, John F., and Donald F. Burzlaff. 1968. Nebraska handbook of range management. Nebraska Agric. Ext. Cir. 68–131.
178. Vallentine, John F., C. Wayne Cook, and L. A. Stoddart. 1963. Range seeding in Utah. Utah Agric. Ext. Serv. Cir. 307.
179. Welch, N. H., Earl Burnett, and E. B. Hudspeth. 1962. Effect of fertilizer on seedling emergence and growth of several grass species. J. Range Mgt. 15(2):94–98.
180. Welty, Leon. 1977. Sod seeding with minimum tillage. Now 13(4):10–11.
181. Williams, Wm. A., R. Merton Love, and Lester J. Berry. 1957. Production of range clovers. California Agric. Expt. Sta. Cir. 458.
182. Wiltbank, James N. 1964. Reasons for poor reproductive performance. *In* Fort Robinson Beef Cattle Research Station Field Day Report, April 30, 1964.
183. Wise, A. F. 1966. Chemical fallow still costs too much. Crops and Soils 18(1):15.
184. Young, James A., and Raymond A. Evans. 1970. Weed control in wheatgrass seedbeds with siduron and picloram. Weed Science 18(5):546–549.
185. Young, James A., Raymond A. Evans, and Richard E. Eckert, Jr. 1969. Wheatgrass establishment with tillage and herbicides in a mesic medusahead community. J. Range Mgt. 22(3):151–155.

Chapter 9
Special range seeding and treatment techniques

In addition to drilling and broadcast seeding, emphasized in Chapter 8, special revegetation practices such as hay mulch seeding and transplanting may be practical and even necessary for some plant species and under some range situations. Many site stabilization, development, and revegetation techniques utilized in reclaiming sites devastated by mining or road construction differ greatly from those employed on normal grazing lands.

Mechanical land treatments such as contour furrowing, terracing, pitting, and waterspreading are water conserving practices adapted to certain range sites. Expected benefits include increasing the quality and quantity of forage from resident species through more efficient use of rainfall, but simultaneous control of erosion is frequently a major consideration. Contour furrowing, terracing, and pitting prevent water from running off the land; ripping, chiseling, and rotary subsoiling fracture restrictive soil layers and improve infiltration; and waterspreading intercepts runoff water and diverts it to dry soils. When combined with artificial revegetation, these treatments provide a means of establishing additional forage plants following partial to complete seedbed preparation.

Special range treatment practices such as contour furrowing, pitting, and waterspreading are adapted to semiarid regions of the West and in arid regions receiving at least eight inches of average annual precipitation. Terracing has been widely used in restoring Intermountain range watersheds at higher elevations. However, in New Mexico a variety of range treatments including widely spaced terraces, small diversion dams, contour channels, brush dams, and closely spaced contour earthen structures were found ineffective on sandy soil [76]. The low response was attributed to instability of the soil, low water retention in the topsoil, and rodent damage. In humid regions special land treatments have been mostly unsuccessful [11], and complete seedbed preparation prior to artificial seeding has been the standard recommendation.

Interseeding

Interseeding (scalping) provides another means of introducing or reintroducing desirable plant species to a range and may be advantageous over complete seedbed preparation where (1) erosion hazard is

FIG. 94. *Three-row range interseeder provided with lister openers, separate seed boxes for large and small seeds, depth-gauge wheel, packer and drive wheel.* (Miller Seed Co., Lincoln, Nebraska, photo)

high, (2) the preparation of a complete seedbed is impractical, or (3) the purpose is to modify rather than replace the present plant stand. Interseeding consists of scalping a furrow to remove perennial and annual competition and seeding adapted, vigorously competitive species in the center of the furrow. Subsequent management must encourage establishment in the row and spreading into the areas between the rows. Interseeding is a compromise between slow natural reseeding and the relatively quick establishment expected from complete seedbed preparation. Complete renovation and seeding is more rapid and probably more economical and more effective on the more productive, less erosive sites.

Interseeders are now available commercially or can be built locally (Figure 94). An interseeder must have a furrow opener that removes a strip of sod from each row, a seeding unit for each row, adequate control of furrow depth and seeding depth within the furrow, packer wheels, and seed hoppers or a seedbox. Various types of furrow openers have been used including double-wing moldboards or listers, single-wing moldboards, sweeps with wings or shields attached above, heavy disks, inclined disks, or rototillers. In eastern

FIG. 95. *Range interseeding near Haynes, North Dakota, using fourteen-inch tilled strips. Note seedlings establishing in the row and the vigor of the native western wheatgrass between the furrows.* (Soil Conservation Service photo)

Montana the firm seedbed of lister furrows was found to be better suited to the establishment of seeded grasses than was the loose seed-bed of rotary-tilled strips [84]. The latter had the advantage of not leaving a rough surface but needed some method of firming used in combination with it. Seeding in the cleaned strip with a double-disk opener has worked well on sandy soils and soft seedbeds, but a stiff-shank opener has given better results on hard, dry clay or clay loam soils [21]. The best establishment of Russian wildrye on shortgrass range in southeastern Alberta was set in tilled strips 24 inches wide or wider [70].

The width of sod strip that should be removed depends on the vig-or of the existing sod, the soil moisture, and the forage species being interseeded. Wider channels are needed in the more competitive sods and on drier sites where competition for moisture is high. Furrow widths of four to six inches have been adequate for alfalfa and for cool-season grasses interseeded into overgrazed sod [18]. However, tilled strips for native grasses should generally be twelve to sixteen inches wide (Figure 95).

A seeder developed by the SCS for use in the Great Plains makes furrows eight to ten inches wide, two to three inches deep, with straight sides and flat bottoms, and with centers of rows spaced forty inches apart [67]. Shallower furrows did not adequately remove the competition, and deeper furrows allowed wind and water to cover the seed too deeply. Narrower row spacings to encourage more rapid cover by the seeded grasses generally gave excessive ridging between furrows or sacrificed furrow width. Stand establishment success ra-tios with the SCS interseeder were similar to complete seedings [67].

Interseeding has given the greatest success and has most com-monly been used on sandy soils but shows promise on silty soils as well. On clayey sites, problems of equipment operation and soil crusting have been encountered. On arid sites, moisture competition may be too great for interseeding, and complete summer fallow may be required [6]. Making furrows on the contour will catch and hold some additional moisture and thus enhance the chances of seedling establishment. Interseeding is not adapted to rough, rocky, or very steep sites or where heavy brush is found. Interseeding in the South-ern Great Plains has been reported to establish significantly more grass seedlings in buffalograss sod than in taller grasses [21]. Opti-mum planting dates with interseeding have been similar to those used for full seedings. Seeding rates of one-third to one-half of full seedings have been typical. Row widths of 3 to 3.5 feet have general-ly been used on upland sites with narrower widths on mesic sites.

A primary purpose of interseeding has been to reestablish in rows native grasses of higher successional rank than the residual plants be-

FIG. 96. *Interseeding used on sandy soil in Nebraska to reestablish sand bluestem, switchgrass, Indiangrass, and sand lovegrass into native range.*

tween the rows and raise range condition to fair or good in three to five years (Figure 96) [67]. The interseeding of introduced species into native range has met with variable success. In South Dakota, alfalfa and sweetclover have been successfully interseeded into native range [18]. Establishment of 3,000 to 5,000 alfalfa plants per acre is the suggested goal for interseeding in South Dakota [66]. On heavy-clay sites in western South Dakota, sweetclover was readily established by various interseeding and direct sod seeding methods [55]. In some areas of western South Dakota and eastern Wyoming, sweetclover, although a biennial, has become naturalized in shortgrass sites and has increased forage production by up to 140 percent.

On mountain range in Wyoming, interseeding successfully established Nordan crested wheatgrass and Manchar smooth brome [46]. Russian wildrye and pubescent wheatgrass were fair in stand ratings, but interseedings of slender wheatgrass, intermediate wheatgrass, and beardless wheatgrass were considered failures. Several high-producing grasses have been successfully introduced into deteriorated native pastures in central and eastern South Dakota [42].

Interseeding of crested wheatgrass and alfalfa on shortgrass range of eastern Wyoming has given successful establishment [60]. However, the alfalfa was later killed by two years of drought. Crested wheatgrass maintained better stands under grazing on bottom land than on upland sites. By contrast, sod seeding in the same area established crested wheatgrass only in depressions and did not increase production after the third year [63].

In later studies on shortgrass range in southeastern Wyoming, introduced wheatgrasses plus a legume were seeded in 18-inch tilled

371

strips (with 22-inch non-tilled strips between) and in 30-inch tilled strips (with 60-inch non-tilled strips between) and compared with solid seedings [62]. Nordan crested wheatgrass was found to be superior to pubescent wheatgrass, and alfalfa much superior to sainfoin or cicer milkvetch. Plants in the narrow-interseeded treatment never attained as much vigor as in the wide-interseeded or solid-seeded treatment. Both interseeding treatments stimulated the growth of western wheatgrass in the non-tilled strips. Russian wildrye interseeded into native shortgrass sod in Wyoming did not successfully compete with the native species in drought years, particularly in furrows under eighteen inches wide [64].

Experience to date suggests that satisfactory establishment and maintenance of introduced grasses on semiarid upland range sites generally require greater suppression of the native plant community than that provided by interseeding; but this practice shows much promise on subhumid sites.

Periods of grazing protection are essential following interseeding. Since the new seedlings are more palatable and are grazed in preference to the resident vegetation between the rows [70], several years of closely regulated grazing may be required. Some winter grazing may be possible during the establishment period, but grazing should be based on the requirements of the seeded species until fully established. Interseeding can be considered successful only if the seeded species permanently establish, and particularly if they begin to replace the resident cover between the rows. Rhizomatous species established in rows by interseeding have been observed to rapidly invade into the unscalped strips when favored by light tillage. Later tillage of the native strips left in interseeding in southeastern Alberta increased the growth of Russian wildrye, but total production remained greater on plots completely tilled initially [70].

Browse interseeding. Interseeding has been successfully used in seeding browse plants on big game range in the Intermountain region [58]. Use of interseeding has been recommended for (1) removal of cheatgrass or tarweed in strips sufficient to establish shrubs and perennial herbs and (2) establishment of shrubs and forbs in perennial grass stands. Furrows twelve to twenty-four inches wide are preferred for seeding shrubs, but six- to twelve-inch widths have been satisfactory for competitive herbaceous perennials. Release from competition during the critical first year of shrub seedling growth is necessary for effective interseeding.

In a Utah study comparing scalp widths for shrub establishment, the 24-inch scalps proved superior to all narrower scalps in terms of the survival of all the shrubs included in the study [32]. When seeded at approximately 24 seeds per linear foot of row, the number of plants surviving after five years in the 4-inch versus 24-inch scalps was as follows: four-wing saltbush, .02 vs. .15; bitterbrush, .2 vs. 1.56; and cliffrose, .05 vs. .59. However, forage production per plant seemed to have no significant relationship to scalp width. Even wider cleared strips were found by Van Epps and McKell [79] to be optimum for establishing four-wing saltbush in semiarid areas—40 inches for direct seeding and 60 inches for transplanting shrub seedlings.

Hand scalping small plots not less than 2.5 feet square and at least one inch deep with a wide hoe has sufficiently reduced competition to allow shrub seedlings to become established when seeded in three or four spots on the scalped area [34]. Machines for scalping or rototilling and browse seeding in such interrupted plots have also been developed.

Contour Furrowing

Contour furrows, when provided with dams at frequent intervals, have been used effectively in the Northern and Southern Great Plains and in portions of the Intermountain West to control moderate amounts of runoff and improve infiltration for increased forage production. Even when damming is used to interrupt the furrows, placing the furrows on an exact contour is suggested; but this may be less critical than when no dams are used.

Furrowing practices have varied considerably between areas. Small furrows four to six inches across and three to four inches deep, commonly referred to as grooves (Figure 97), have been used on shortgrass and midgrass range. Although this system has been initially effective in improving forage production, the longevity of the furrows has been limited to a few years depending upon soil stability. Wide, shallow furrowing and planting seed in one operation is essentially interseeding, as discussed in the previous section.

Various equipment—including listers, blade tips, and specially designed furrowers—has been used in furrowing. An improved furrower, referred to as the Model B contour furrower and developed in cooperation with the Range Seeding Equipment Committee, has received considerable use in the West, particularly on federal ranges [59]. This implement has been designed to make two furrows five feet apart, with each furrow ranging from eighteen to thirty inches wide and 10 inches deep at the maximum. Furrow and ridge portions represent about 40 to 60 percent of the treated area. Each furrow opener consists of two disks that throw the soil in opposite directions; it is preceded by a ripper adjustable to a depth of twelve inches below the furrow depth. Behind each furrowing unit is a four-paddle dammer adjustable for placing check dams in the furrow at predetermined intervals. The Model B contour furrower is also equipped to broadcast seed slightly behind the dammer. Costs of operation have been about $20 per acre on moderate terrain [83].

Contour furrowing on shortgrass range in Wyoming has given favorable results when the furrows were spaced two feet apart and made four to eight inches deep [2, 3]. However, furrows spaced five to ten feet apart were ineffective. Contour furrowing of native grasslands in central Nebraska reduced runoff by 84 to 94 percent and protected lower lying lands from the effects of siltation and runoff [80]. In another Nebraska study, contour furrowing of range in poor condition, where grazing was uncontrolled, reduced runoff and conserved an average of 1.2 inches of additional moisture annually but did not produce improved stands of perennial grasses [19].

Contour furrowing with the Model B furrower on panspot range sites in southeastern Montana increased herbage production 165 percent and available soil water 107 percent over an 8-year period [83]. Increases in productivity came primarily from thickspike wheatgrass and western wheatgrass, benefitting from the increased soil water coming from overwinter recharge and reduced summer runoff. Benefits were less on saline upland sites and were considered not economical. Associated studies in this area [52] showed added initial water storage in the soil of about one inch but also indicated a decrease

FIG. 97. *Furrowing shortgrass range in western Nebraska.*

over time due to natural weathering, intrafurrow dam failure, and furrow breaching. The average effective life of the furrows was set at 25 years, and it was suggested that leaving intrafurrow dams of undisturbed or compacted soil material was necessary. Branson [11] found additional moisture storage of up to two inches immediately after treatment, with this lever stabilizing over time at about one-half inch.

Furrowing in shortgrass sod in the Southern Great Plains has increased moisture depth and amount of available moisture, increased grass production associated with deeper root development, and prolonged the green growth period in drought periods [47]. However, continued range recovery and increased production has been dependent upon furrows being of sufficient size and lasting ability [15]. On relatively stable soils in the southwest, furrows six to ten inches wide and made with blade tips on graders or with plows or listers have been recommended for the semidesert grassland [1].

A survey made in the Southwest and Intermountain regions indicated that furrowing and pitting were generally successful on Nutall saltbush sites but not on winterfat or black grama sites or on blue grama sites with coarse soils [11]. In Montana contour furrowing on Nutall saltbush sites and seeding to crested wheatgrass increased forage yields by 118 percent from 200 pounds of Nutall saltbush to 500 to 700 pounds of total herbage per acre [10]. However, crested wheatgrass appeared unable to spread from the furrows. Similar results were obtained on Nutall saltbush sites in the Big Horn basin of Wyoming [27]. The furrows had lost 25–30 percent of their capacity after 15 years, with an associated loss of herbage productivity, but added benefits were expected to last for another 20 years. However, on arid saltbush communities in eastern Utah, contour furrows provided initial control of runoff and sedimentation but were short lived and did not increase herbage production [30, 31].

Terracing or Trenching

Terracing or trenching cultivated land is a widely used practice. Graded terraces are designed to intercept and divert runoff at nonerosive velocities to safe outlets while level terraces are designed to impound and conserve rainfall in low-rainfall areas. Even expensive bench terraces are sometimes placed on highly productive croplands as a means of making steep sites available for cultivation by reducing the land slope and controlling erosion [7].

Terraces and trenches as applied to rangeland differ from furrows in being 1.5 to 3 feet deep and of equivalent width. Terraces are de-

FIG. 98. *Equipment for installing trenches;* A, *construction and bank sloping with bull-dozer,* B, *trenching with a double-disk trencher.* (Photos courtesy of U.S. Forest Service)

Contour trenching on the Uinta National Forest, Utah, for watershed protection; C, *trenches being installed in 1960;* D, *same site in 1965 after trenching and seeding.* (Photos courtesy of U.S. Forest Service)

signed primarily for flood control and the reduction of runoff and sedimentation (Figure 98). Since terraces are generally ineffective or at least uneconomical in increasing herbage production, their use on rangelands is considered impractical except as a watershed treatment practice on critical areas [11]. Terraces, as well as contour furrows, place restraints on vehicular traffic.

On shallow soils or in humid areas, terracing frequently reduces herbage production. On shallow, eroded class VII land in eastern Oklahoma, large terraces constructed prior to perennial grass seeding were detrimental to both water conservation and grass production [17]. Although the terraces reduced runoff during the year of seeding and during three years of grass establishment, they caused greater runoff subsequently. Also, after full establishment the unterraced land produced 3,911 pounds of grass per acre compared to 1,203 pounds per acre on the terraced land. The detrimental effects of terracing were attributed to putting most of the remaining topsoil into the ridges, exposing more of the compact, unweathered subsoil between the terraces, actually increasing the land slope from the top of the ridge to the bottom of the furrow of the next terrace below, and maintaining an environment too wet in the furrows and too dry on the ridges.

Restoration of mountain watersheds deteriorated to the point of low ground cover and extensive gully systems may require contour trenching, seeding, and exclusion from grazing. Although such practices are very expensive when intensively applied, they have been successfully used on many potential flood producing sites on high elevation watersheds in the West subject to torrential rainfall. The required size and frequency of the ditchlike trenches partitioned by dams at intervals of twenty to thirty feet depend on such factors as depth and character of the soil, degree of slope, and the amount and intensity of precipitation expected.

Terraces and trenches require the use of large, sturdy equipment for their construction. Such equipment must be able to operate in compacted soil, remove soil from the trench and move it laterally downslope into the ridge, function on up to moderately steep slopes, and operate at a reasonably fast rate of speed. Hula dozers and motor patrols have been commonly used. Tractor mounted contour trenchers have been developed in cooperation with the Range Seeding Equipment Committee [59]. These trenchers consist of one or two double-disk units that can be attached directly to crawler tractors. They work effectively on side slopes up to 45 percent and can be raised or lowered by means of a hydraulic lift.

Pitting on rangelands consists of forming small basins or pits in the soil. This range treatment first came into prominence in the 1930s and since then has been commonly used in the Great Plains and the Southwest. The major objectives of pitting in the Great Plains have been to catch and hold rain and runoff water, to store moisture in and around the pits for plant use, and encourage more productive mixtures of native grasses. However, in the Southwest from western Texas to southern Arizona an equally important objective has been moisture conservation to permit grass seeding and establishment [4].

Pitting equipment (Figure 99). Modified disk plows have been the most commonly used type of rangeland pitters. The modifications have generally consisted of eccentric disks or deeply notched or cutaway disks on a standard disk plow, wheatland plow, brushland plow, or even a tandem disk frame. Still another innovation has been a tractor-mounted disk with an eccentric furrow wheel [82]. Basins made by the disk pitters are three to five feet long, eight to twelve inches wide, and four to eight inches deep.

Pits can also be made with lister bottoms by using an arrangement for raising and lowering the bottoms. Pits made by listers have a flat, relatively wide bottom, are deeper, increase moisture more at the one-to-four-foot zone, but cost more than pits made with disk pitters.

The rotary-drum pitter, sometimes called the Calkins pitter, consists of a series of spike teeth mounted in rows on a cylindrical drum. The fifteen-inch, curved spike teeth punch holes about fourteen inches deep, narrow at the bottom but wider at the top. The rotary-drum pitter makes about 5,000 pits per acre with pits spaced nearly three feet apart [4, 40]. Another implement using spike-tooth openers is the rotary pitter or subsoiler. This implement consists of two spiked wheels with four eighteen-inch spikes each. The spike-tooth pitters have lacked sufficient pit stability for long-range benefits, have given insufficient surface disturbance for seed establishment, and have been a potential livestock hazard for a time after treatment [40, 82]. Although some success has been met in breaking through the subsoil hardpan and temporarily increasing percolation and retention, many pits made with spike teeth have filled in within a year.

Land imprinters, consisting of steel cylinders with angle irons welded to them imprint a variety of geometric patterns on the ground surface and result in infiltration increases and reduction of runoff and evaporation while routing more rainwater to the plant roots [75]. They have been used successfully as primary implements

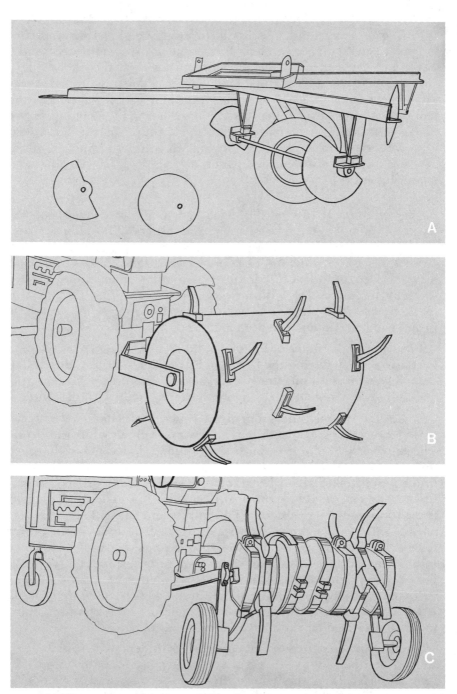

FIG. 99. *Three types of pitting equipment used on rangeland:* A, *disk pitters using notched or eccentric disks;* B, *rotary-drum or Calkins pitter;* C, *rotary pitter or subsoiler.*

on near-barren land areas not otherwise arable or as secondary implements on loose, plowed soil.

Response to pitting. In the Northern Great Plains, pitting has been effective on shortgrass range in fair to good condition [2, 3, 4, 60, 61]. Spacing pits closely and removing about one-third of the vegetation has increased forage production by 30 to 50 percent and even up to 100 percent. The increased productivity has been attributed to better water relations, more plant food available for the reduced plant numbers, and a significant change in vegetation composition toward midgrasses. Although western wheatgrass in mixture with other grasses greatly increases after pitting, pure stands of western wheatgrass have shown little benefit from pitting or furrowing [3]. Also, on poor condition range the response to pitting has been unsatisfactory and weed problems have been accentuated.

New pits made by disk pitters can store 0.3 to 0.6 inch of water besides trapping additional snow in the winter [4]. Pitting often doubles absorption rates thereby reducing runoff during torrential rains. Rows spaced two to three feet apart have been most beneficial, while pitting rows exceeding five feet in spacing have been mostly ineffective in moisture retention and increased forage production [3].

In the Southern Great Plains, pitting has been more effective on upland sites than on lowland sites such as old lake beds and tobosa flats. Pitting on uplands has increased moisture penetration and stimulated plant growth [4, 20, 74]. Some natural seeding has been evident, particularly from sideoats grama. In the Southwest where precipitation comes largely in the form of summer thunderstorms, pitting without seeding has given variable results [4]. In the Big Horn Mountains of Wyoming pitting with an eccentric one-way disk has increased water intake, reduced runoff, improved vigor of native grasses and grasslike plants, and increased forage production by 32 to 68 percent on deep soil sites [45].

The effectiveness of pits declines as they fill in with dirt. On medium textured soils in the Northern Great Plains, pits have been effective for up to fifteen years [60]. In the Southwest, pits may last only three to five years [4]. However, the benefits of changes in plant composition often last much beyond the effective water-holding capacity of the pits [11]. The rapid development of cover following pitting extends the life of pits. However, washing or blowing in of silt reduces the longevity of the pits. The life of pits is shorter on sandier soils than on finer textured soils and shorter where the vegetation is sparse than in dense grass stands.

Pitting has been ineffective not only on sandy soils [4, 84] but also on dense clay sites where extremely loose soil resulting from freezing

381

FIG. 100. *Semidesert grassland site in southern Arizona treated with a disk pitter. Right hand portion of pitted area has also been seeded with a grass-seeder packer or culti-packer-seeder.* (Soil Conservation Service photo)

and thawing encourages silting in and eroding away of the pits [54]. Pitting just before the main growing season not only favors the growth and establishment of the resident forage species but also reduces soil movement and increases longevity of the pits [4]. The optimum season of treatment corresponds closely to optimum periods for artificial seeding.

Pitting and seeding (Figure 100). Many areas in the Southwest must be seeded before beneficial use can be made of water retained by the pits. Also, range seedings made in this area without pitting have a high rate of failure in average and below average rainfall years. Where the resident cover is relatively sparse and where much of the precipitation falls as summer thunderstorms, pitting as a combination of seedbed preparation and water conservation provides more positive establishment of range seedings [4].

Pitting and seeding of grasses into otherwise unprepared seedbeds has generally been unsuccessful in the Northern Great Plains, occasionally successful in the Southern Great Plains, and more successful in the southwest than elsewhere [4]. This practice has generally resulted in failure with attempts to establish grasses and legumes into shortgrass range. Although pitting and seeding sideoats grama in the

382

Southern Great Plains has shown some success in favorable years where the resident cover has not been too competitive, this procedure has often been unsuccessful [82].

Pitting and seeding without additional seedbed preparation is successful only on those sites where the resident cover is relatively sparse and provides minimum competition to the seeded plants. Since the pitting operation should provide some plant control, spike-tooth pitting is inadequate and only disk pitters are suggested [82]. In southwest Texas, drilling after pitting alone on depleted sites was more effective than drilling into a prepared seedbed without pitting [74]. However, seedbed preparation by pitting alone in the southern Great Plains has been less effective than pitting following previous plant control methods [82]. Pitting alone in southern Arizona gave inadequate shrub control for successful seeding [43]. Pitting preceded by chaining for seedbed preparation showed some promise in above average rainfall years, but only disk plowing or root plowing gave consistently high shrub control.

Broad, shallow pits or basins about five by five to eight feet and up to six inches deep at the deepest point have proven longer lasting in the Southwest than conventional pits [68, 69]. Conventional, narrow pits made with disk pitters have often given good initial establishment of seeded grasses but subsequently lost their effectiveness in retaining and storing water after a year or so. This loss of effectiveness has been due largely to rapid filling in of pits with soil following summer storms and to excessive competition from resident plants. In southern Arizona, herbage production of seeded buffelgrass (*Cenchrus ciliaris*) over a four-year period on the basined areas averaged 2.5 times as high as on conventionally pitted areas and 5 times as high as on the unpitted areas [68]. These intermediate-size basins also gave twice as much production from Boer lovegrass, Lehmann lovegrass, and kleingrass as conventional pits, in addition to superior plant establishment and stand maintenance [69].

It has been observed that seeds planted in the bottom of conventional pits are occasionally drowned as seedlings. To give better adaptation to rainfall and provide longer lasting pits, basin-forming machines have been developed to form sloping fan-shaped basins which allow seed to be placed at various depths below the original soil surface [28]. The basins are placed with the straight, vertical side on the downward slope and the fan-shaped side on the upper side. One machine showing promise is a blister roller used for pressing basins and seeding into loose soil following root plowing. Shallow water-retention basins for seeding have also been constructed with a bulldozer [1, 28]. The resulting broad, shallow pits have served the dual purposes of scraping away the cover and seed of the resident plants and providing long-lasting basins. Seeding by broadcasting or drilling should follow rather than precede the pitting operation [74].

Limitations. Soils on which water intake is naturally high, such as sands, should not be pitted. On sandy soils pitting will not materially increase water penetration, the pits will be short lived, and the disturbance may reduce cover and increase wind erosion. Pitting is also not adapted to very rocky soil, brushy sites such as big sagebrush areas, many desert-shrub types, ranges in excellent range condition, or on weedy sites in low condition unless artificial seeding is effective [3, 4]. Pitting is best suited to gently sloping to slightly rolling sites and generally should not be made on slopes greater than 8 percent [4]. Pits made with disks or lister pitters should be made on the contour or at least across the general slope. Although soils of medium and fine texture have generally given the greatest benefits, salinity and pH of the soil have apparently affected results only when extreme [11]. Since pitting causes rough conditions for wheeled vehicles, it is often desirable to skip trails and roadways.

Ripping

The purpose of ripping or deep chiseling is to break or shatter compacted layers that inhibit root development and moisture penetration. Ripping depth varies from ten to thirty-six inches depending upon the depth of the restrictive layer. Ripping should be performed on the contour and when the soil is dry [1]. The practice is not adapted on steep slopes or where the soil is shallow over bedrock.

Ripping on rangeland is expensive and varies between $12 to $20 per acre (1979). Thus, only the more responsive sites should be considered. Ripping has not been effective generally on rangeland in the Northern Great Plains [4]. However, on heavy clay in South Dakota, ripping twelve to fourteen inches deep and at six-foot spacings opened up compacted soils to water infiltration and resulted in a 173 percent increase in the number of western wheatgrass plants and a 444 percent increase in grass production [53].

In the Southwest and Intermountain regions, ripping has generally been effective only when the operation created a lasting furrow [11]. Ripping treatments that did not effectively modify the soil surface failed to increase forage production. On medium and fine soils furrows made with a construction ripper with wide blades lasted up to twenty-four years and increased forage production by 160 percent. Ripped furrows in southern Arizona that left a ridge as high as an average lister furrow had a projected longevity of fifteen years [12].

Adequate seedbed preparation should eliminate any compacted layers formed under cultivation. Seedling emergence is not affected, but hardpan depresses the vigor of established forage plants by limiting root penetration and the area of soil from which water and nutrients can be extracted. Any water below such hardpans is also essentially unavailable to grass roots. Areas having hardpans but having otherwise high forage production potential may justify the use of soil-ripping equipment. Such areas may include bottomlands, sites being developed for irrigated pasture, and waterspreading areas with restrictive layers.

Hay Mulch Seeding

The hay mulching method of seeding, because of cost of hay and hand labor required, is generally too expensive except for treatment of difficult areas such as gullies, dams, spillways, waterways, dunes, and sand blowouts [81]. This requires harvesting hay when the seed is nearly ripe and before it shatters. The hay is then spread by hand, a manure spreader, or a blower-type spreader. At least one ton per acre is generally required and two tons is even better on blowouts or other erosive sites. Anchoring mulch on areas subject to wind or water erosion is suggested. A disk, or tiller, or trampling by livestock will help work the hay into sand or other loose soil. The hay not only provides seed but holds the soil in place while new plants are becoming established.

On sandy soil, no seedbed preparation prior to spreading is generally required. On heavier textured soils, shallow tillage should precede the hay spreading. Mowing at an optimum stage of maturity, raking and stacking immediately, and spreading after at least one winter in the stack has proved the best method of handling the seed hay [81]. Normally hay should be spread in late winter or early spring. The treated area should be protected from grazing, trampling, and excessive compaction until the seeded species have established. The hay mulching method has long been used with buffalograss in the Great Plains but can also be used with most other grass species which produce viable seed. This method requires low cash expenditures and a drill is not needed, but the labor cost is high.

Hay mulch seeding is a technique commonly used to heal sand blowouts by stopping sand movement and establishing a permanent cover of plants. Vegetation must be reestablished on the blowouts before this rangeland can be managed effectively. Steps recommended for stabilizing both the blowouts and the sand deposition areas include [23, 25, 48, 78]:

1. Fencing to keep livestock from trampling and grazing new vegetation on the blowout areas. Permit grazing only after reestablishment of perennial cover and then under close control.
2. Shaping the sharp edges of the blowout into a gradual slope. The sharp embankments give the wind its swirling action and interfere with mulching and seeding equipment.
3. Mulching with straw, hay, brush, or sorghum fodder to stop damage from blowing soil and hold moisture while grasses are becoming established.

In many cases an application of prairie hay mulch containing a large number of desirable seeds is the only source of seed needed. Any type of mulch can be used if underseeded or later drilled into the firmed seedbed. Perennial sand-binding grasses recommended for the Central and Northern Great Plains include prairie sandreed, sand lovegrass, western wheatgrass, sand bluestem, and intermediate wheatgrass, most of which are more effective because they are rhizomatous. Temporary crops such as rye and vetch may be seeded at the time of mulching. However, the mulch itself plays the primary role in stabilizing the site [48]. Since blow sand is often low in fertility, the use of strawy manure or chemical fertilizer high in nitrogen and phosphorus can be expected to accelerate plant establishment.

Vegetative Plantings

Vegetative plantings have not been widely used on rangelands in the past but do offer alternative revegetation methods for special situations. The method of planting sprigs (rhizome sections) is used for establishing bermudagrass [65]. Since Coastal bermudagrass produces no seed, this or other vegetative plantings must be used. Sod planting can be used for bermudagrass and buffalograss but is very costly. Stolon planting can be used for stoloniferous grasses but is less effective because the runners often dry out too quickly.

Spreading sod pieces or freshly cut, well-jointed culm segments on moist soil with a manure spreader is an effective method of establishing reed canarygrass [77]. Covering these pieces with a light disking or trampling them into the mud is beneficial to them also. This method of establishment is often superior to seeding where wet soils prevent adequate seedbed preparation. When worked into mud or moist soil, pieces of root or mature plant cuts of reed canarygrass with a joint establish readily.

Because of widespread interest in revegetating big game winter range and disturbed areas with useful shrubs, and because of the problems encountered in their establishment by seeding, much attention is presently being given to techniques for establishing browse plants by vegetative plantings. Bitterbrush plants can be propagated from stem cuttings four to six inches long when cut with a heel of older wood [56]. Moistening the cuttings with water and dipping them in 0.1 percent indol-3-butyric acid in talc before planting is suggested. About two-thirds of cuts treated in this way have rooted when planted in the field.

Stem cuttings offer a means of vegetatively propagating many other native shrubs as well, including those that are difficult to grow from seed and without loss of genetic integrity [24]. Stem cuttings have reduced cost and time involved even more than containerized planting materials. However, not all shrub species have responded well to stem cuttings.

Transplantings of wildland shrub seedlings and even forb seedlings can also be successfully made in the early spring while the ground is still moist [58]. However, transplanting is a laborious and relatively expensive procedure of germinating seed and growing seedlings in a nursery and then transporting into the field as bareroot or containerized materials [86]. If the seeds exhibit high germination, the containers can be seeded directly. If not, the seedlings must be started in flats, potted into containers, and later transplanted into the field.

Containerized seedlings can substantially reduce the time required to produce high quality shrub seedlings, as has been the case with conifer tree seedlings [16]. In contrast with the 1½ to 2 years required to develop bare-root shrub seedlings sufficiently for outplanting, direct-seeded containerized seedlings of nearly all shrub and forbs species are large enough and hardy enough for field planting at 12 weeks of age and some at 6 weeks [26]. Most shrubs are best handled by germinating seeds in the fall, obtaining optimum growth and hardiness in the greenhouse during the winter, and transplanting in the spring [86].

Container-grown shrub transplanting offers many advantages over direct field seeding but also has the following advantages over bareroot transplanting [26]:

(1) Seedlings can be started at any time in the greenhouse.
(2) Seedlings can be subjected to tailored growing conditions for rapid development and adequate hardiness.
(3) Seed is used more efficiently (the survival rate is possibly twice that of bare-root transplants and 20 to 40 times that of direct field seedings).
(4) Roots suffer less transplanting shock.
(5) Diseases are easier to treat.

(6) No lifting or storage problems exist.

(7) Planting season is extended in some cases.

(8) Fewer outdoor problems exist from rodents, birds, and grazing animals.

(9) Seedling dormancy is not required at planting.

(10) Field survival rate is higher.

However, bare-root stock have the advantages of being easier to pack and ship, not requiring greenhouse facilities for their production, often being more cold hardy, and being perhaps as effectively established as containerized stock on moist sites, although not under severely dry conditions.

Both plantable and non-plantable containers have been used successfully with wildland shrub and forb transplants. Plantable containers (including tubelings) made of decomposable peat, paper, or plastic net are more adapted to mechanized, rapid-planting procedures, but equipment may be less available locally. Whether bare-root stock or containerized seedlings are mechanically or hand planted, the steps for transplanting into the field are (1) opening the soil, (2) inserting the seedling, and (3) closing and firming the soil [86].

Reclaiming Disturbed Sites

Range managers are frequently called upon to reclaim drastically disturbed sites. Such sites include shaft mine spoils, strip-mined areas, dredging spoils, and sand and gravel pits. Still other problem areas include highway and railroad right-of-ways; reservoir, subdivision, and industrial construction sites; and flood debris areas, mud slides, erosion channels, earth slides, sites disturbed by logging, and excessively gullied areas.

The objectives of restoration and the projected future use of the area must be determined for each disturbed site before the plan of operation is prepared and initial steps are taken. Stabilization from further wind, water, and gravity erosion is often a primary objective. Purifying water and streams, restoring wildlife habitat, enhancing aesthetics, or merely improving public relations may be major considerations. Meeting environmental regulatory standards may provide the first objective to be met. Also, a decision must be reached as to whether restoration or improvement in agricultural production is to be provided for.

Many of the same techniques used in improving normal rangelands can also be used in restoring disturbed lands. Species recommended for reclaiming disturbed sites in the West include primarily those typically recommended for dryland seedings but may include additional native or introduced species that provide minimum forage but rapid and lasting ground cover. Reestablishment of the native vegetation that existed before the site was disrupted, with all its inherent complexities, is often an unrealistic if not impossible objective.

Reclamation of greatly disturbed areas often requires techniques peculiar to disturbed areas or special adaptations of widely used techniques. Sources of information on reclaiming disturbed sites are only gradually becoming available. Two important sources pertaining to the reclamation of disturbed arid lands in the western United States are Wright [85] and Thames [73]; two sources for high-altitude disturbed lands are Berg et al. [8] and Zuck and Brown [87]. A useful handbook on equipment for reclaiming strip-mined land has been published by the U.S. Forest Service [13].

Since the availability of soil moisture is commonly the most limiting factor in the reclamation of disturbed arid lands, emphasis should be given to increasing available soil moisture by impeding runoff, increasing infiltration, and reducing evaporation [33]. This is primarily achieved by gradient reduction, cultivation, mulching, topsoiling, buffering, surface manipulation, compaction relief, supplemental water provision, and special seeding techniques. Although continued irrigation is seldom feasible for arid land plantings, the use of sprinkler, drip, or trickle irrigation may be important or even required in initial plant establishment.

The following is a checklist of procedures for reclaiming bare, disturbed sites in arid parts of the West, but it holds considerable application to other areas also.

(1) *Shaping and grading.* Gradients must be reduced to the point of physical stability and permit the use of other equipment. Problems of alignment of spoils, final contour, compaction, settling problems, and stream flow control should be anticipated and be met. Little flexibility may be permitted in grade and contour along highway right-of-ways. Toxic materials should be buried.

(2) *Topsoiling.* Where feasible to do so, topsoil should be segregated from subsoil during the excavation and redistributed over the area after final shaping. Topsoil includes organic matter, better structure and often texture, and fertility that will reduce erosion and accelerate new plant growth. Selectively saving the topsoil, usually including the A and B soil horizons down to 1½ to 6 feet in depth, can generally be done with

389

little or no additional cost. Such topsoil may also be a major source of useful plant seeds.

(3) *Cultivation.* The cultivation of soil aids infiltration, breaks up capillary action, reduces evaporation, conserves moisture available to plant roots, and provides an improved seedbed.

(4) *Surface mulching and organic matter incorporation.* Most spoils are low in or devoid of organic matter. Surface mulching is often needed to impede runoff and erosion, increase available soil moisture, lower soil temperatures, reduce evaporation, and conserve moisture available to plant roots. Organic matter incorporated in the soil will not only improve physical conditions and water-holding capacity but may provide a source of plant nutrients.

(5) *Surface manipulation.* Treatments of the surface area include deep chiseling, offset-listering, gouging, waffling, and basin forming for erosion control and water retention.

(6) *Fertilization.* Nutrients required for plant growth, including both micro- and macronutrients, may be very deficient in subsoils and mine wastes and must be added by mineral or organic fertilizers. Lime will be required on acid sites to improve the pH balance. Soil amendments are sometimes effective in counteracting toxic minerals in mining spoils, but deep burial of the toxic spoils is often more practical.

(7) *Special techniques for plant establishment.* Methods used to prepare the area and aid plant establishment include providing temporary fiber or plastic ground covers or temporary, non-competitive plant cover; using vegetative plantings; making some provision for supplemental soil moisture such as condensation traps, water harvesting, or plant or artificial snow barriers; and hydromulching or hydroseeding.

The direct application of mulch or other ground covers may be justified in revegetating critical, difficult sites. In central Texas, grasses emerged at a much lower moisture level when a mulch covered the seed area but was less beneficial with cool temperatures than with hot temperatures during the germination period [50]. Based on precipitation records it was concluded that a mulch would have been helpful in 31 percent of the years at Big Springs and 36 percent of the years at Seminole, Texas. Straw mulches in South Dakota were only partially successful in roadside grass plantings [36]. Although they did increase water infiltration rate and decreased runoff on steep sites, grass establishment was reduced when cereal seedlings developed from seeds contained in the straw.

In establishment studies with four-wing saltbush, straw maintained cooler soil temperatures and gave best seedling establishment of the mulches used [71]. It was suggested that seeding be done when the soil was still moist and the mulch then applied over the seeded rows. During hot weather, light-colored mulches such as straw and white petroleum resin were effective in reducing afternoon temperatures. It was suggested that dark-colored mulches such as Soil Guard, black asphalt, and black polyethylene might be used to increase soil temperatures for seeding during cooler seasons.

On an arid coal spoil bank, the use of jute netting or straw mulch held in place by wire netting increased grass seedling establishment by six times—by even more when the two methods were used in combination [41]. On high-altitude, barren sites in northern Idaho, native hay held in place by chicken wire but not sawdust, evergreen boughs, or asphalt emulsions increased native grass establishment [29]. Wood chips depressed rather than improved grass establishment on scab ridges in the Pacific Northwest [44]. Rapid-setting asphalt mulches have hastened emergence and development of some grasses, maintained moisture in the topsoil for a longer period, and increased soil temperature; but hail and high-intensity rains tended to break up the asphalt mulch [5].

Waterspreading

Waterspreading on rangeland is the practice of diverting runoff water from stream channels or courses and distributing the water over nearby flood plains or valley floors. Here the velocity of the water is slowed and it is allowed to infiltrate into the soil. Waterspreading has two main functions: (1) increasing forage production by spreading of floodwater and its storage in the soil profile and (2) reducing gully erosion and downstream flooding [51]. Range floodwater spreading systems are constructed so that operation is mostly automatic whenever sudden, torrential storms result in flood flows. Since the water supply is uncertain and intermittent and usually of short duration, its use in normal irrigation systems is too inefficient. The state water rights laws should be checked before spreader systems are constructed. The spreading of floodwaters for agricultural purposes is an old practice in the West, since the early Indians in the Colorado Plateau Region used it extensively in crop production [7].

Area adaptation. Stream channels which are dry most of the time but which flow for short periods following heavy rains or snow melt generally provide the water supply for waterspreading. The watershed area above the spreading site should provide at least one flooding per year for satisfactory forage production, and additional floodings each year are advantageous [49]. A rate of flow large enough for satisfactory spreading is necessary but not so large as to be unmanageable. A prolonged period of flow is desirable to permit greater water penetration. Where water is scarce and the area suitable for spreading is greater than the water supply, applications of three to six inches per acre when available can be expected to produce more forage than heavier applications on smaller areas [72]. However, where a flow can be expected only once or twice a year, higher application rates up to twelve inches may not be too much.

A runoff basin having a rainfall less than nine inches of average annual precipitation generally gives small forage yield responses on the spreading area and does not justify the construction of a water-spreading system [49]. However, the gentle slopes where the water is actually spread often receive only about half the amount of precipitation received by the runoff basin. It is important that there be a proper ratio between the size of the runoff basin and the spreading area. A ratio of 10:1 has been considered correct on the average in the Four Corners area of the Colorado Plateau [39] but has varied from 5:1 to as much as 25:1 over the West. The shape, drainage patterns, terrain, soils, and vegetational cover on the watershed and the frequency and intensity of storms as well as the average annual precipitation must be considered. Available rainfall and runoff records for the area should be carefully studied in planning the spreading system.

The quality as well as the quantity of water should be considered. Runoff from badlands and other sparsely vegetated areas may carry too high a silt load for spreading. Frequent and heavy deposits of sediment may interfere with the operation of the spreading system and harm the soil in the spreading area. It has been suggested that sedimentation be largely restricted to the upper 20 percent of the spreading area [39]. Runoff basin rehabilitation, upstream desiltation basins, using larger spreading areas, or facilities for selecting only low silt flows or partial diversions are means employed to reduce the silt problem. If harmful salts in toxic amounts are suspected in the water, this should be checked.

A broad, smooth, gently sloping plain without gullies or channels is an ideal waterspreading site. Grades under 1 percent are desirable where the sediment load in the water is not great [72]. Where the ponding type of spreading is used, and particularly on soils of low infiltration capacity, slopes should not exceed 2 percent. Where wild flooding is used, and particularly on soils which take water rapidly, slopes up to 5 percent can be tolerated [72]. Although deep, fertile, loam soils are ideal, most soil types are suitable for waterspreading except gravels and deep sands, shallow soils over bedrock, and soils

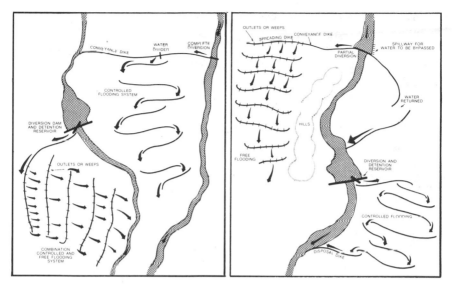

FIG. 101. *Diagrams of waterspreading systems adapted to two different terrains.* (Diagrams adapted from Stokes et al. [72])

high in soluble salts. Unless a satisfactory remnant of native forage plants is scattered evenly over the proposed spreading area, artificial seeding of the spreading area with adapted, productive forage species should be considered.

The spreading system. The best waterspreaders are not built to a pattern but are built to fit the terrain (Figure 101). Since no two diversion and spreading areas are exactly alike, the systems built to fit two different areas will never be alike in all details. The waterspreader may consist of a single furrow or ditch leading from a channel to an area wetted in low flows or it may be a very complex combination of dams and dikes. The complete and more complex systems have up to five rather distinct parts or functions [72]:

1. Reservoir for temporary storage of floodwater.
2. Diversion structures for removing water from the channel or reservoir.
3. Conveyance structures for taking the water to the spreading site.
4. Control structures on the spreading area.
5. Disposal structures for removing excess water without erosion damage.

Although many small waterspreading systems do not have a detention dam and consist only of a dike across the stream bed to divert the water, including a detention dam as part of the system has distinct advantages: (1) makes flood peaks more manageable, (2) in-

393

creases the flooding time from flash storms, (3) provides initial sediment storage, and (4) furnishes a measure of flood control [72]. Since the waterspreading system is built around the detention reservoir when one is included, it should be built for lasting use. Construction details should consider removing loose material from the dam site, excavating a core trench down to solid foundation material, using fill soil that will make an impervious embankment, applying fill material in layers and firmly packing, and providing an adequate spillway to handle excess flow.

Where there is a flat, gentle plain adjacent on both sides of the flood channel and where the banks of the channel are high enough to prevent the water from draining back into the channel, no structures other than the detention dam or diversion dike may be required. However, such ready-made spreading areas are often not available, and the diversion must be extended to convey the water away from the channel vicinity to a suitable spreading area. Free flooding consists of allowing the water to seek its own course and flow unregulated across the flooding area after it has been carried to the point of release.

Controlled flooding, wherein the water flow is further regulated as it flows across the flooding area, spreads the water more uniformly over the spreading area and controls the velocity so that more of the water can soak in. Even distribution of the water prevents excessive and prolonged ponding, benefits a larger area, and produces more forage per unit of water [49]. Efficient distribution of water is accomplished by various combinations of spreading dikes, respreading dikes, contour furrows, and small gully plugs and check dams. Earthen dikes and fitted-rock spreaders have proven more effective and to require less maintenance than loose-rock spreaders, brush and rock spreaders, and rock-rubble spreaders [57]. However, continued maintenance is mandatory for continued efficiency of any waterspreading system.

In cold climates contour furrows about five inches deep were of little value in holding and spreading water because they filled with ice and snow [37]. However, contour spreader dikes about two feet high were quite effective. Spreader dikes and even the main conveyance dike are generally provided with percolators or weeps made of loose rock or other porous materials. Excessive ponding is often prevalent where no drainage through the dikes is provided and results in bared areas, water weeds, and wasted water. On areas where there is overflow from the spreading area, means must be provided to return the overflow to the channel with a minimum amount of erosion.

394

Soil response. The most profound and often deleterious effect of waterspreading on the soil has been the change in soil texture due to sedimentation. This change has varied with the distance from the diversion dike. In New Mexico studies the soil became more sandy over the first 1,500 feet [38]. Beyond this point, the texture either remained the same or became heavier as a result of the deposition. Rapid deposition of more than five inches of sediment greatly damages most grasses except western wheatgrass [39].

The primary purpose of waterspreading—increased soil moisture—is invariably achieved on areas receiving extra water. Although deep sands or soils with gravel subsoils often have low water retention, total retention of moisture in the A and B horizons has had greater influence than soil texture per se on forage yield [49]. It has been suggested that a minimum amount of fine textured sediment may be beneficial on sandy portions of the spreading area. Other effects of waterspreading on sandy soils have included (1) reduced organic matter and nitrogen content in areas of coarse deposits, (2) generally little effect on bacterial activity, total soluble salts, and pH, and (3) raising the ground level and smoothing the ground surface where sedimentation was considerable [39].

Vegetation response. The major response of vegetation to waterspreading has been a general increase in herbage production. Increases have generally ranges from 300 to 1,000 percent on areas

Table 20. Economic Returns from Three Waterspreading Systems in Montana, 1951*

Factors evaluated	Alzada	Klintworth	Hale
Original construction	1944	1948	—
Spreading area, acres	900	195	580
Initial cost	$3,300	$ 982	$3,087
Other expenses**	$1,906	$ 238	563
Total cost to 1951	$5,206	$1,220	$3,650
Grazing capacity prior to construction, A.U.†	10	3	10
Grazing capacity in 1951, A.U.	47	15	65
Cost per acre	$ 5.79	$ 6.37	$ 6.30
Cost per A.U. of increased carrying capacity	$ 141	$ 102	$ 66

*Source: [22].
**This includes cost of ripping, reseeding, and maintenance.
†An animal unit of yearlong carrying capacity (A.U.) is equivalent to 12 AUMs per year.

flooded once or more each year. This has often been equivalent to increased forage production and carrying capacity of 150 to 500 percent over the entire spreading system [35, 37, 39, 51]. However, the effects of waterspreading on botanical composition has often been equally great [9, 35, 39, 49, 51].

Although greatly modified by geographical location and various climatic and soil factors, grasses responding very favorably to waterspreading include western wheatgrass, bermudagrass, streambank wheatgrass, tobosa, vine-mesquite, alkali sacaton, tall wheatgrass, sand bluestem, prairie sandreed, switchgrass, Indiangrass, and foxtail barley. Grasses showing reduced favorable response are crested wheatgrass, Rothrock grama (*Bouteloua rothrockii*), Lehmann lovegrass, blue panicgrass, slender wheatgrass, smooth brome, sideoats grama, little bluestem, and Indian ricegrass. Big sagebrush, pricklypear, blue grama, galleta (*Hilaria jamesii*), Sandberg's bluegrass, sand dropseed, and red threeawn appear poorly adapted to sediment deposition and flooding and thus decrease markedly under waterspreading. Except where introduced grasses are seeded and successfully established, the same grasses that predominate in naturally flooded areas also predominate in the spreading areas [49].

Estimates of economic returns from waterspreaders built by the Bureau of Land Management near Alzada, Montana, are given in Table 20 [22]. The total cost per acre over the spreading area ranged from $5.79 to $6.37. Increased annual carrying capacity ranged from 370 to 550 percent. Costs per animal unit of additional yearlong carrying capacity were $141, $102, and $66. The cost of purchasing additional native rangeland in the area in 1951 was estimated at $250 to $300 per animal unit of carrying capacity.

At Miles City, Montana, the $.29 per acre annual cost of waterspreading gave added gross returns of $2.25 to $2.50 on the basis of herbage production alone [35]. Other benefits noted included (1) earlier grazing, (2) higher permissible utilization, (3) better response to fertilization and seeding, (4) sediment retardation of channel erosion, (5) restoration of ground water levels, (6) reduction of channel erosion, and (7) use of diverted water for stockwater. Other advantages of waterspreading may include a longer green growing period, hay production, and the advantage of special uses such as calving and lambing or breeding pastures.

Literature Cited

1. Arizona Interagency Range Technical Sub-Committee. 1969. Guide to improvement of Arizona rangeland. Arizona Agric. Ext. Bul. A-58.
2. Barnes, O. K. 1950. Mechanical treatment on Wyoming range land. J. Range Mgt. 3(3):198–203.
3. Barnes, O. K. 1952. Pitting and other treatments on native range. Wyoming Agric. Expt. Sta. Bul. 318.
4. Barnes, Oscar K., Darwin Anderson, and Arnold Heerwagen. 1958. Pitting for range improvement in the Great Plains and the Southwest desert regions. USDA Prod. Res. Rpt. 23.
5. Bement, R. E., D. F. Hervey, A. C. Everson, and L. O. Hylton, Jr. 1961. Use of asphalt-emulsion mulches to hasten grass seedling establishment. J. Range Mgt. 14(2):102–109.
6. Bement, R. E., R. D. Barmington, A. C. Everson, L. O. Hylton, Jr., and E. E. Remmenga. 1965. Seeding of abandoned croplands in the central Great Plains. J. Range Mgt. 18(2):53–59.
7. Bennett, H. H. 1939. Soil conservation. McGraw-Hill Book Company, Inc., New York.
8. Berg, W. A., J. A. Brown, and R. L. Cuany (Ed.). 1974. Proceedings of a workshop on revegetation of high-altitude disturbed lands. Colo. State Univ., Env. Res. Center Infor. Ser. 10.
9. Branson, F. A. 1956. Range forage production changes on a water spreader in southeastern Montana. J. Range Mgt. 9(4):187–191.
10. Branson, F. A., R. F. Miller, and I. S. McQueen. 1962. Effects of contour furrowing, grazing intensities, and soils on infiltration rates, soil moisture, and vegetation near Fort Peck, Montana. J. Range Mgt. 15(3):151 158.
11. Branson, F. A., R. F. Miller, and I. S. McQueen. 1966. Contour furrowing, pitting, and ripping on range lands of the western United States. J. Range Mgt. 19(4):182–190.
12. Brown, Albert L., and A. C. Everson. 1952. Longevity of ripped furrows in southern Arizona desert grassland. J. Range Mgt. 5(6):415–419.
13. Brown, Darrell. 1977. Handbook: equipment for reclaiming strip mined land. USDA, For. Serv. Equip. Dev. Cen., Missoula, Mont.
14. Brown, Ray W., Robert S. Johnston, and Douglas A. Johnson. 1978. Rehabilitation of alpine tundra disturbances. J. Soil and Water Cons. 33(4):154–160.
15. Caird, Ralph W., and J. S. McCorkle. 1946. Contour-furrow studies near Amarillo, Texas. J. For. 44(8):587–592.
16. Colby, Marilyn K., and Gordon D. Lewis. 1973. Economics of containerized conifer seedlings. USDA, For. Serv. Res. Paper RM-108.
17. Cox, Maurice B., Harley A. Daniel, and Harry M. Elwell. 1971. Terraces on grassland. Oklahoma Agric. Expt. Sta. Bul. 373.
18. Derscheid, Lyle A., and Melvin D. Rumbaugh. 1970. Interseeding for pasture and range improvement. South Dakota Agric. Ext. Serv. Fact Sheet 422.
19. Dragoun, F. J., and A. R. Kuhlman. 1968. Effects of pasture management practices on runoff. J. Soil and Water Cons. 23(2):55–57.
20. Dudley, R. F., and E. B. Hudspeth. 1964. Pitting and listing treatments on native shortgrass rangeland. Texas Agric. Expt. Sta. Prog. Rpt. 2313.
21. Dudley, Richard F., Elmer B. Hudspeth, and C. W. Gentt. 1966. The bushland range interseeder. J. Range Mgt. 19(4):227–229.
22. Dudley, T. E., and W. R. Shanahan. 1951. Estimates of economic returns from Bureau of Land Management range waterspreading systems at Alzada, Montana. USDI, Bur. Land Mgt. and Geol. Survey. Mimeo.
23. Eck, H. V., R. F. Dudley, R. H. Ford, and C. W. Gentt, Jr. 1968. Sand dune stabilization along streams in the Southern Great Plains. J. Soil and Water Cons. 23(4):131–134.
24. Everett, Richard L., Richard O. Meeuwig, and Joseph H. Robertson. 1978. Propagation of Nevada shrubs by stem cuttings. J. Range Mgt. 31(6):426–429.
25. Everson, A. C., B. E. Dahl, and A. H. Denham. 1966. Controlling blowouts for

forage production. J. Range Mgt. 19(3):147–148.

26. Ferguson, Robert B., and Stephen B. Monsen. 1974. Research with contain-erized shrubs and forbs in southern Idaho. Great Plains Agric. Counc. Pub. 68, pp. 349–358.

27. Fisser, Herbert G., Michael H. Mackey, and James T. Nichols. 1974. Contour-furrowing and seeding on Nuttall saltbush rangeland in Wyoming. J. Range Mgt. 27(6):459–462.

28. Frost, K. R., and L. Hamilton. 1965. Basin forming and reseeding of rangeland. Trans. Amer. Soc. Agric. Eng. 8(2):202–203, 207.

29. Gates, Dillard H. 1962. Revegetation of a high-altitude, barren slope in north-ern Idaho. J. Range Mgt. 15(6):314–318.

30. Gifford, Gerald F., Valdon B. Hancock, and George B. Coltharp. 1978. Effects of gully plugs and contour furrows on the soil moisture regime in the Cisco Basin, Utah. J. Range Mgt. 31(4):293–295.

31. Gifford, Gerald F., Dee B. Thomas, and George B. Coltharp. 1977. Effects of gully plugs and contour furrows on erosion and sedimentation in Cisco Basin, Utah. J. Range Mgt. 39(4):290–292.

32. Giunta, Bruce C., Donald R. Christensen, and Stephen B. Monsen. 1975. Inter-seeding shrubs in cheatgrass with a browse seeder-scalper. J. Range Mgt. 28(5):398–402.

33. Hodder, Richard L. 1978. Dry land techniques in the semiarid west. In The reclamation of disturbed arid lands. Albuquerque: Univ. New Mex. Press, pp. 217–223.

34. Holmgren, Ralph C., and Joseph V. Basile. 1959. Improving southern Idaho deer winter ranges by artificial revegetation. Idaho Dept. of Fish and Game Wildlife Bul. 3.

35. Houston, Walter R. 1960. Effects of water spreading on range vegetation in eastern Montana. J. Range Mgt. 13(6):289–293.

36. Hovland, Dwight, Dean E. Wesley, and Jordan Thomas. 1966. Establishing vegetative cover to protect roadside soils in South Dakota. South Dakota Agric. Expt. Sta. Bul. 527.

37. Hubbard, William A., and Sylvester Smoliak. 1953. Effect of contour dikes and furrows on short-grass prairie. J. Range Mgt. 6(1):55–62.

38. Hubbell, D. S., and J. L. Gardner. 1944. Some edaphic and ecological effects of water spreading on range lands. Ecology 25(1):27–44.

39. Hubbell, D. S., and J. L. Gardner. 1950. Effects of diverting sediment-laden runoff from arroyos to range and crop lands. USDA Tech. Bul. 1012.

40. Hughes, Eugene E. 1962. Moisture penetration under three types of range pit-ting. Texas Agric. Expt. Sta. Misc. Pub. 589.

41. Jacoby, Pete W., Jr. 1969. Revegetation treatments for stand establishment on coal spoil banks. J. Range Mgt. 22(2):94–97.

42. Johnson, Clarence E., Edwin A. Dowding, and Paul M. Wheeldon. 1972. A ma-chine for pasture interseeding. S. Dak. Agric. Expt. Sta. Cir. 206.

43. Jordan, Gilbert L., and Michael L. Maynard. 1970. The San Simon watershed: revegetation. Prog. Agric. in Arizona 22(6):4–7.

44. Klomp, Gerald J. 1968. The use of woodchips and nitrogen fertilizer in seeding scab ranges. J. Range Mgt. 21(1):31–36.

45. Lang, Robert L. 1958. Range-pitting trials in the Big Horn Mountains of Wyoming. Wyoming Agric. Expt. Sta. Bul. 357.

46. Lang, Robert. 1962. Range seeding and pitting study in the Teton National Forest. Wyoming Agric. Expt. Sta. Mimeo. Cir. 173.

47. Langley, B. C., and C. E. Fisher. 1939. Some effects of contour listing on native grass pastures. J. Amer. Soc. Agron. 31(11):972–981.

48. Malakouti, M. J., D. T. Lewis, and J. Stubbendieck. 1978. Effect of grasses and soil properties on wind erosion in sand blowouts. J. Range Mgt. 31(6):417–420.

49. Miller, R. F., I. S. McQueen, F. A. Branson, L. M. Shown, and Wm. Buller. 1969. An evaluation of range flood-water spreaders. J. Range Mgt. 22(4):246–257.

50. Moldenhauer, William C. 1959. Establishment of grasses on sandy soil of the southern High Plains of Texas using a mulch and simulated moisture levels. Agron. J. 51(1):39–41.

51. Monson, O. W., and J. R. Quesenberry. 1958. Putting flood waters to work on rangelands. Montana Agric. Expt. Sta. Bul. 453.

52. Neff, Earl L. 1973. Water storage capacity of contour furrows in Montana. J. Range Mgt. 26(4):298–301.

53. Nichols, J. T. 1966. Effect of ripping on western wheatgrass range. Livestock Field Day, South Dakota Agric. Expt. Sta., Newell Field Sta., May 10, 1966, pp. 31–32.

54. Nichols, J. T. 1969. Range improvement on deteriorated dense clay wheatgrass range in western South Dakota. South Dakota Agric. Expt. Sta. Bul. 552.

55. Nichols, James T., and James R. Johnson. 1969. Range productivity as influenced by biennial sweetclover in western South Dakota. J. Range Mgt. 22(5):342–347.

56. Nord, Eamor C. 1959. Bitterbrush plants can be propagated from stem cuttings. USDA, Pacific Southwest Forest & Range Expt. Sta. Res. Note 149.

57. Peterson, H. V., and F. A. Branson. 1962. Effects of land treatments on erosion and vegetation on rangelands in parts of Arizona and New Mexico. J. Range Mgt. 15(4):220–226.

58. Plummer, A. Perry, Donald R. Christensen, and Stephen Monson. 1968. Restoring big game range in Utah. Utah Div. of Fish and Game Pub. 68-3.

59. Range Seeding Equipment Comm. 1965 (Rev.) Handbook of range seeding equipment. USDA and USDI.

60. Rauzi, Frank. 1968. Pitting and interseeding native shortgrass rangeland. Wyoming Agric. Expt. Sta. Res. J. 17.

61. Rauzi, Frank, and R. L. Lang. 1956. Improving shortgrass range by pitting. Wyoming Agric. Expt. Sta. Bul. 344.

62. Rauzi, Frank, and Robert L. Lang. 1976. Grazing solid-seeded, wide-interseeded, narrow-interseeded and native pastures in southeastern Wyoming. Wyo. Agric. Expt. Sta. Res. J. 109.

63. Rauzi, F., R. L. Lang, and C. G. Becker. 1962. Mechanical treatments on shortgrass range. Wyoming Agric. Expt. Sta. Bul. 396.

64. Rauzi, Frank, R. L. Lang, and C. F. Becker. 1965. Interseeding Russian wildrye—a progress report. Wyoming Agric. Expt. Sta. Mimeo. Cir. 216.

65. Rechenthin, C. A., H. M. Bell, R. J. Pederson, D. B. Polk, and J. E. Smith, Jr. 1965. Grassland restoration. Part III. Re-establishing forage plants. USDA, Soil Cons. Serv., Temple, Texas.

66. Rumbaugh, M. D., G. Semeniuk, R. Moore, and J. D. Colburn. 1965. Travois—an alfalfa for grazing. South Dakota Agric. Expt. Sta. Bul. 525.

67. Schumacher, C. M. 1964. Range interseeding in Nebraska. J. Range Mgt. 17(3):132–137.

68. Slayback, Robert D., and Dwight R. Cable. 1970. Larger pits aid reseeding of semi-desert rangeland. J. Range Mgt. 23(5):333–335.

69. Slayback, Robert D., and Clinton W. Renney. 1972. Intermediate pits reduce gamble in range seeding in the southwest. J. Range Mgt. 25(3):224–227.

70. Smoliak, S., and M. Feldman. 1978. Establishment of Russian wildrye (*Elymus junceus* Fisch.) in strip-tilled *Stipa-Bouteloua* prairie. Proc. Internat. Rangeland Cong. 1:626–628.

71. Springfield, H. W. 1970. Germination and establishment of four-wing saltbush in the Southwest. Rocky Mtn. Forest & Range Expt. Sta. Res. Paper RM-55.

72. Stokes, C. M., Floyd D. Larson, and C. Kenneth Pearse. 1954. Range improvement through waterspreading. Foreign Operation Administration, Washington, D.C.

73. Thames, John L. (Ed.). 1977. Reclamation and use of disturbed land in the southwest. Univ. Ariz. Press, Albuquerque.

74. Thomas, G. W., and V. A. Young. 1956. Range pitting and reseeding trials on the Texas Range Station near Barnhart. Texas Agric. Expt. Sta. Prog. Rpt. 1882.

75. USDA, Agric. Res. Serv. 1977. Tool (imprinter) for water conservation. Agric. Res. 25(11):6–7.
76. Valentine, K. A. 1947. Effect of water-retaining and water-spreading structures in revegetating semi-desert range land. New Mexico Agric. Expt. Sta. Bul. 341.
77. Vallentine, John F. 1967. Nebraska range and pasture grasses. Nebraska Agric. Ext. Serv. Cir. 67-170.
78. Vallentine, John F., and Donald F. Burzlaff. 1968. Nebraska handbook of range management. Nebraska Agric. Ext. Cir. 68-131.
79. Van Epps, Gordon A., and C. M. McKell. 1977. Shrubs plus grass for livestock forage: a possibility. Utah Sci. 38(3):75–78.
80. Wasser, C. H., L. Ellison, and R. E. Wagner. 1957. Soil management on ranges. 1957, USDA Yearbook of Agriculture, pp. 633–642.
81. Wenger, Leon E. 1941. Re-establishing native grasses by the hay method. Kansas Agric. Expt. Sta. Cir. 208.
82. Whitney, R. W., L. O. Roth, D. G. Batchelder, and J. G. Porterfield. 1967. Pasture pitting machines. Oklahoma Agric. Expt. Sta. Bul. 657.
83. Wight, J. Ross, E. L. Neff, and R. J. Soiseth. 1978. Vegetation response to contour furrowing. J. Range Mgt. 31(2):97–101.
84. Wight, J. Ross, and Larry M. White. 1974. Interseeding and pitting on a sandy range site in eastern Montana. J. Range Mgt. 27(3):206–210.
85. Wright, Robert A. (Ed.). 1978. The reclamation of disturbed arid lands. Univ. New Mex. Press, Albuquerque.
86. Young, James A., Raymond A. Evans, Burgess L. Kay, Richard E. Owen, and Frank L. Jurak. 1978. Collecting, processing, and germinating seeds of western wildland plants. USDA Agric. Reviews and Manuals ARM-W-3.
87. Zuck, R. H., and L. F. Brown (Eds.). 1976. Proceedings high altitude revegetation workshop no. 2. Colo. State Univ., Fort Collins, Colo.

FIG. 102. *Mesic sites such as this mountain meadow respond well to nitrogen or nitrogen-phosphorus applications.* (Union Pacific Railroad photo)

Chapter 10

Range
fertilization

Expanded use of fertilizers on range and other pasture with high potential appears assured as research points the way and as demand for meat, wool, and grazing capacity for both domestic livestock and big game increases. The increased use of fertilizers is basic to the trend from extensive to intensive management of soil and forage resources. However, this increased use will require the continued availability of adequate supplies and realistic prices for fertilizers in the future. Nitrogen fertilizers, manufactured primarily from fossil fuels, have greatly increased in price since the early 1970s. It is projected that shortages of fossil fuel will accelerate the use of legumes as nitrogen sources for pasture soils in the future.

Yield responses to fertilizer application are greatest on mesic sites where soil moisture is not a limiting factor and on sites with deep soils of medium texture and good structure but low fertility. Sites in the western U.S. with the highest potential for economical fertilizer use include meadows (Figure 102), irrigated pastures, and seedings of cool-season grass or grass-legume mixtures on sites receiving fifteen inches or more of average annual precipitation. Today, forage production on semiarid rangelands can often be increased 50 to 100 percent and still not be economical. Future changes in economic conditions will primarily determine the use of fertilizers on many such native ranges.

Determining
Soil Deficiencies

Nutrient deficiencies in the soil result in less forage production, modify vegetation composition, and may be the reason for low nutrient content of forages produced thereon. These deficiencies can be

401

corrected by adding fertilizers to provide adequate levels and proper balance of nutrients needed for plant growth. The most profitable use of fertilizers and the greatest increase in yields result from application of the nutrient most limiting in the soil. As higher amounts of this nutrient are applied, increasing amounts of other nutrients may be required for maximum yields.

Soil testing is the most commonly used method of estimating the kinds and amounts of nutrients that should be added to soils through fertilization. Many university and private testing labs are equipped to analyze soil samples for pH (used for estimating lime requirements), nitrogen, phosphorus, potassium, and other macro- and microelements and to provide fertilizer recommendations based on their findings. Other indirect methods of determining the kind and extent of nutrient deficiencies in the soil include (1) examination of plants for morphological and pathological symptoms of nutrient deficiencies, (2) the chemical composition of leaves or other plant tissue, and (3) relating total yields of nutrients in herbage to the potential of a soil to supply nutrients.

In addition to the above indirect measurements of the expected benefits from fertilization, small scale trials (Figure 103) are sugges-

FIG. 103. *Small-scale plots in Nebraska being harvested to measure actual plant response to fertilizer under field conditions. The response of the old seeding of crested wheatgrass on this sandy site is obvious even from the photo.*

402

ted as the best means of measuring actual field response. When made on sites typical of the areas on which the results will be applied, a series of small plots allows various rates and even combinations of fertilizers to be evaluated and the benefits measured. Such field trials are generally desirable even after soil tests have been made since the latter can provide only projected rather than actual plant response. A combination of experience, the results from experiment station research, and field trials greatly assists in applying the results of soil tests.

Some laboratory analyses are more reliable estimates of fertilizer needs than others. Only those methods and procedures that yield results closely correlated with plant responses are useful. In Illinois, soil tests for pH, P_1 (available phosphorus), P_2 (reserve phosphorus), and K, but not total nitrogen, are suggested in forage production [3]. Although tests for organic matter and total nitrogen have been poorly related to forage yields, nitrate-nitrogen levels and nitrification rate tests have been better correlated with forage yields.

Fertilizer Types and Needs

Fertilizer label guarantee. Fertilizer laws in all states in the United States require the following: (1) the registration and accurate labeling of all brands and grades of fertilizers to be sold, (2) guarantee of percent total nitrogen (N), percent available phosphoric acid (P_2O_5), and percent soluble potash (K_2O), and (3) penalties for failure to meet the guarantees. In addition to guarantees of percent composition of N, P_2O_5, and K_2O, each fertilizer tag must guarantee the content of other plant nutrients listed as being present in the fertilizer.

In order to provide greater accuracy, simplicity, and uniformity in reporting soil, plant, and fertilizer analyses, it has been recommended that all plant nutrients be expressed as the elemental form rather than the oxide form [129]. Great confusion has resulted in the past from expressing nitrogen in the elemental form but phosphorus and potassium and most other nutrients in the oxide form. As a result many state laws have been changed to require or permit dual labeling, and the elemental basis of reporting has come into widespread use.

Because the basis of reporting composition percentage of fertilizer components is transitional, care must be taken to find out whether the old or the new system is being used. For example, a listing of 6-12-12 under the old system meant that the fertilizer contained 6 percent N, 12 percent P_2O_5, and 12 percent K_2O. However, under the

Table 21. Average Composition of Fertilizer Materials*

Fertilizer	% N	% P
Nitrogen fertilizers		
Ammonia, anhydrous	82	
Ammonium nitrate	33.5	
Ammonium phosphate sulfate	16	9
Ammonium sulfate	20	
Di-ammonium phosphate	21	22
Mono-ammonium phosphate	11	21
Potassium nitrate	14	
Urea	45	
Sodium nitrate	16	
Phosphorus fertilizers		
(see also under nitrogen fertilizers)		
Calcium metaphosphate		28
Rock phosphate		15
Superphosphate, single		9
Superphosphate, triple		20
Phosphoric acid		24
Mono-potassium phosphate		23
Potassium fertilizers		
(see also under nitrogen and phosphorus fertilizers)		
Potassium chloride (muriate of potash)		
Potassium sulfate		
Organic fertilizers		
Manure, dairy (fresh)	0.7	.13
Manure, poultry (fresh)	1.6	.55
Manure, steer (fresh)	2.0	.24
Sulfur fertilizers		
(see also under nitrogen and phosphorus fertilizers)		
Calcium sulfate (gypsum)		
Magnesium sulfate		
Soil sulfur		
Sulfate potash magnesia		
Liming fertilizers		
Calcium oxide		
Dolomite		
Limestone, ground		
Shell meal		

*Sources of data: California Fertilizer Association [18] and Teuscher et al. [131].
**Compared to 100 basicity for $CaCO_3$.

% K	% P$_2$O$_5$	% K$_2$O	P solublity in water	% S	CaOO$_3$ equivalence** Basicity	Acidity
						147
						60
	20		over 75%	16		88
				24		110
	50		over 75%			75
	48		over 75%	2.6		58
38		46			23	
						71
					28	
	64		slight		neutral	
	33		1% or less		basic	
	20		over 75%	12	neutral	
	46		over 75%	1	neutral	
	54		over 75%			110
29	52	35	over 75%		neutral	
50		60			neutral	
44		53		18	neutral	
.54	.30	.65	50%		slight	
.75	1.25	.9	50%		slight	
1.59	.54	1.92	40%		slight	
				18.6		acidic
				13		acidic
				99		acidic
21.5		26		18		acidic
					178	
					110	
					95	
					95	

new system of labeling this fertilizer would be listed as 6-5-10, i.e., 6 percent total nitrogen, 5 percent available phosphorus, and 10 percent soluble potassium. The different forms of phosphorus and potassium are readily converted to the other form as follows:

(1) $P_2O_5 \times .44 = P$ (3) $P \times 2.29 = P_2O_5$

(2) $K_2O \times .83 = K$ (4) $K \times 1.20 = K_2O$

The average composition of common fertilizers is given in Table 21 for N, P, K, P_2O_5, K_2O, and S. Note that, although classified by type of fertilizer, several commercial fertilizers supply significant amounts of more than one plant nutrient.

Nitrogen. Nitrogen is the major element used in range and pasture fertilization in the West and throughout much of the United States. In the Great Plains, Rocky Mountain Region, Intermountain Region, and Southwest, only nitrogen and sometimes phosphorus have generally given beneficial results on range and most improved pastures. Nitrogen fertilizer rates vary from a low of 20 or 30 pounds per acre (N/A) annually on semiarid sites to 150 to 200 N/A on irrigated grass pastures. Soils developed under low rainfall are characteristically low in nitrogen even in their virgin state.

Nitrogen reserves in the soil are principally in the form of proteins, amino acids, and amides, and are incorporated in humus and mulch. In this form the nitrogen is nearly unavailable for plant use. By the process of ammonification, bacterial action breaks down the organic matter into ammonium compounds. In this form the nitrogen is held by the organic matter and colloidal particles, its movement in soil moisture is minimized, and it is slightly available to plants. By the process of nitrification, bacterial action oxidizes the ammonium salts to nitrites and then to nitrates.

Nitrate-nitrogen is soluble, moves readily with soil moisture, and is quickly available to plants. This is the most usable form in the soil for plant use. The conversion rate of ammonium compounds to nitrates in the soil depends on several factors:

1. The presence of adequate ammonium levels.
2. Soil pH of 5.5 to 10.0 with the optimum around 8.5.
3. Adequate soil oxygen.
4. Soil temperature between 34° F and 104° F with the optimum about 86° F.

Even soils high in organic matter and total nitrogen are frequently low in available nitrogen, particularly early in the spring. Nitrate-nitrogen accumulates in the soil during spring, summer, and fall, but high winter precipitation greatly reduces spring nitrate-nitrogen levels [39]. Natural nitrification rate in early spring when soil temperatures are low may be too slow to promote early cool-season growth [114]. Under these conditions, early spring application of readily available nitrogen could be expected to give yield reponse above what fall soil tests might predict. On Oregon high deserts, it was concluded that depletion of soil nitrate was more apt to restrict grass production prior to June 1 than lack of moisture [127].

In South Dakota it is recommended that introduced cool-season grasses receive annual applications of 40 N/A by fertilization in the western part of the state to 70 N/A in the eastern part, regardless of the soil test [1]. By contrast, it was recommended that phosphorus fertilizers be applied only when soil tests showed a deficiency. Although nitrogen fertilizers high in nitrate-nitrogen provide nitrogen that can be quickly used by plants, the source of nitrogen did not affect forage yields or stands of crested wheatgrass in North Dakota when applied annually as ammonium nitrate, ammonium sulfate, calcium nitrate, or urea [102]. Nitrogen fertilizers commonly used on range and pasture are ammonium nitrate, ammonium sulfate, and ammonium phosphates. When applied at low to medium rates in both semiarid as well as more humid regions, urea shows promise as an alternative nitrogen source, provided volitization losses are not excessive [101].

Summer fallowing promotes both soil moisture retention and nitrification. In seedbed preparation comparisons in Nevada, fall NO_3-N levels in the top six inches were as follows [39]:

Atrazine fallow—43 pounds/A
Mechanical fallow—27 pounds/A
No fallow—5 pounds/A

The atrazine treatment had the advantage of a full year's fallow beginning the previous fall. Poor weed control in the summer greatly reduced NO_3-N accumulation in the mechanical fallow. Although nitrogen is added to soil by lightning and legume fixation, losses of N from erosion and leaching can be high.

Nitrogen fertilizers including ammonium sulfate, ammonium phosphate, ammonium phosphate sulfate, mono-ammonium phosphate, di-ammonium phosphate, and urea are also commonly used as fire retardants [18]. These ammonium compounds cannot cause nitrate poisoning in livestock directly but must first enter the soil, be converted to nitrates, and then be absorbed and accumulated by plants [30]. Although nitrate poisoning of livestock on range is rare, poisoning in special hazard cases can be prevented by deferred grazing for three to four weeks until the plants can convert excess nitrates into protein.

Phosphorus. On soils deficient in phosphorus or where legumes are an important component of the forage stand, phosphorus fertilization of pasture becomes important. Range soils in the Northern Great Plains and westward contain fairly large amounts of phosphorus. Very low phosphorus levels are found along the Gulf and Atlantic coasts. Intensive pasture management in the Southern Great Plains, in the Midwest, the Southeast, and New England generally requires phosphorus fertilization.

The availability of soil phosphorus for plant use depends primarily on its degree of water solubility. Available phosphorus, the portion of phosphorus soluble in water plus the additional portion soluble in citrate solution, is an estimate of the total amount of phosphorus readily available for use by plants. However, since even the citrate-insoluble phosphorus represents a phosphorus reserve in the soil, both total phosphorus and available phosphorus content of soils and fertilizers supply useful information. Much of the soluble phosphorus applied in fertilizers is converted in the soil to insoluble forms.

Phosphorus availability is also affected by soil conditions, particularly pH. Its availability is highest at 6.0 to 7.0 pH. Maintaining a pH above 6.5 is beneficial in maximizing phosphorus availability on acid soils. Phosphate in fertilizers such as the superphosphates, most ammonium phosphates, and phosophoric acid are adequately water soluble to be highly useful to plants. Others such as rock phosphate and colloidal phosphate are nearly insoluble. At pH 6.5 or higher, rock phosphate is considered to be insoluble [3]. Bone meal, basic slag, and fused phosphate are only slightly soluble. Most other phosphorus fertilizers have intermediate solubility levels.

Potassium and calcium. Over most of the western range, potassium and calcium in the soil are adequate for forage production. However, over much of the Midwest, the Northeast, the Southeast, and the northern Pacific Coast, both potassium and calcium are generally required in addition to nitrogen and phosphorus for satisfactory forage production. The greatest deficiencies occur in the Southeast along the Gulf and Atlantic coasts.

Lime in strict chemical terminology refers to calcium oxide, but the term is now applied to all limestone-derived materials applied to neutralize acid soils. In addition to the plant-nutrient value of calcium and/or magnesium provided, liming is used to raise the pH of acid soils to 6.0 or higher. Raising the pH to near neutral increases nitrification rate, increases nitrogen fixation, increases organic matter content of the soil, and increases phosphorus availability. The effectiveness of various liming chemicals in terms of $CaCO_3$ equivalence is shown in Table 21. The relative basicity or acidity is also shown for the other fertilizers.

Sulfur and other plant nutrients. Sulfur fertilization has given profound benefits on many California ranges and on primarily legume pastures on sandy soils in Nebraska, in Minnesota, and in the Southeast. Legumes are much more apt to be sulfur deficient than perennial grasses. Common sulfur sources include gypsum (calcium sulfate), soil or elemental sulfur, ammonium sulfate, single superphosphate, and magnesium sulfate. Equivalent amounts of sulfur from elemental sulfur and from gypsum appeared equally effective on California annual grassland in increasing forage yields, sulfur uptake and concentration in the herbage, and the percentage of clover in the stand [70]. Forty pounds of sulfur from fine elemental sulfur supplied needs for two years, but equivalent amounts from gypsum were sufficient for only one year. However, coarse elemental sulfur was not readily available for several years [5]. Gypsum (calcium sulfate) does not raise pH, and thus it is not a substitute for lime. Using sulfur to neutralize high pH soils is impractical under pasture and range conditions.

Trace minerals such as boron, iron, copper, zinc, manganese, and molybdenum are generally adequate in amount and availability for range forage production. Only in local areas of Florida, along the Atlantic and Gulf coasts, and on the Pacific Coast are such deficiencies common. In localized areas in the West high pH may render much of the trace minerals unavailable. Where the minor elements are seriously deficient such as in parts of Florida, small amounts applied as fertilizer often give large increases in forage production.

Fertilizer Application

Season of application. The optimum time of year to apply fertilizer is determined by the amount and distribution of rainfall, type of forage plants, and the kind of fertilizers. Fertilizers are applied to rangeland most often in the spring or fall except in the southwestern United States. Summer application of fertilizer results in reduced yield responses, particularly by cool-season grasses. Since phosphorus and potassium move very slowly in the soil, fall application may favor movement into the root zone with winter precipitation. An added benefit of fall fertilizer application is that it coincides with the slack labor season on most ranches.

Nitrogen fertilizer is much more mobile than phosphate fertilizers. In winters of high precipitation, nitrogen applied in the fall may be leached before the spring growing season begins. On mountain meadows in Nevada there appeared to be little difference between spring and fall for N and P fertilizers if the fall application was made after all of the growth and irrigation had stopped and if the soils were not kept wet during the winter by flooding [48]. Similar results were also obtained on Oregon meadows with late fall and early spring application except where spring flooding occurred [22]. However, N and P fertilizers applied in the fall to seeded grasses on mountain rangelands in northeastern Utah and southeastern Idaho apparently were leached away by thirty to forty inches of late fall and winter precipitation [62]. When checked in June following fall application, all of the N applied had been leached from the top forty-two inches of the soil. The use of slow-release nitrogen fertilizers may provide a means of greatly reducing leaching while extending the period of effective nitrogen fertilization. For example, the use of sulfur-coated urea, a slow-release fertilizer, was found useful in meeting these objectives for both minerals when applied to range soils in eastern Oregon and Washington [76].

Nitrogen should not be applied in the fall to lowlands which are subject to flooding in the winter or spring. However, on dry foothill range in the Intermountain Region, fall application has allowed nitrogen to be moved into the root zone for early spring growth [21]. Spring application may be too late to move nitrogen into the root zone, particularly in dry springs, unless applied early.

On upland sites in the Northern Great Plains, spring and fall application of N fertilizer has had similar effects on yield and protein content of seeded, cool-season grasses [1]. Nitrogen fertilization of seeded, cool-season pastures in eastern Nebraska in mid-April promoted leafy, vegetative growth while late fall or very early spring fertilization promoted seedstalk production in early summer followed by dormancy until cool fall weather arrived [143]. Warm-season grasses on productive sites responded well to small amounts of fertilizer (30 N/A) when applied in late May or early June [143]. Earlier or heavier rates tended to favor cool-season grasses over the warm-season grasses. On subirrigated meadows in the Nebraska Sandhills, March-April fertilization with nitrogen is recommended where cool-season grasses predominate, but late May or early June is recommended where warm-season grasses predominate or are to be encouraged [40].

On semidesert grasslands of southern Arizona and New Mexico, the best time to apply fertilizer for optimum grass production and quality is near the beginning of the summer rainy season [130]. Applying fertilizer after soil moisture becomes available is helpful in preventing losses during a dry season. However, application toward

FIG. 104. *Fertilizer application equipment.* A, *gravity flow equipment applying lime in the Southeast* (New Holland Division of Sperry-Rand photo); B, *truck equipped with whirling disk applicator;* C, *sod seeder equipped for fertilizer placement above or below ground* (American Potash Institute photo); D, *banding fertilizer in conjunction with seeding.* (American Potash Institute photo)

411

the end of the rainy season may reduce the magnitude of production increases. On California annual grassland range, September to March applications of N have been compared [66]. Early fall was better than late fall for winter feed production. For spring feed all dates were similar except for March which was too late. However, no significant differences were found between fall, winter, and spring application of manure on California annual grasslands [88].

Method of application (Figure 104). Broadcasting on the soil surface has been the most widely used method of applying granular fertilizers on range and pasturelands. Liquid application of fertilizers or fertilizers plus herbicides or insecticides can be applied with ground rigs at low cost. Noncorrosive fertilizers in liquid form can also be uniformly spread through sprinkler irrigation systems. This method allows the fertilizer to be applied at lower rates and at more frequent intervals, thereby increasing nitrogen utilization. Broadcasting fertilizer by airplane permits treatment of areas inaccessible by ground rigs, but application costs may be excessive with higher application rates. On annual grasslands in California, helicopter application was considered fast and practical on rangeland too steep for ground application [33]. Sixty pounds S/A was applied at a cost of $2.70 to $3.00 per acre (1966) with good control of application rate and fertilizer distribution.

On subirrigated meadows in Nebraska, surface application of phosphorus fertilizer was better than drilling to a depth of three to four inches [91]. Higher yields and higher percent phosphorus in the forage resulted from the surface placement. The drilled phosphorus was used less by the resident grasses than the broadcasted phosphorus but red clover absorbed more phosphorus from the deeper depths than did the grasses.

On native grasslands and seeded crested wheatgrass in North Dakota, subsurface placement of fertilizer showed a slight benefit, but the advantage was found due to the cultivating effect of the operation rather than to the placement of the fertilizer [122]. Disking-in or rotary tilling of phosphorus fertilizer into the soil or band placement had no advantage over broadcast application on Dallisgrass-white clover pasture in south Texas [112].

Anhydrous ammonia and aqueous ammonia rapidly volatize unless incorporated into the soil. This is also true, but to a lesser extent, for urea. Although incorporation into sods on arable lands is possible with special equipment, the problems of its use on range have mostly prevented this method of application. Foliar-applied nitrogen fertilizer has also been effective in increasing herbage yields, crude protein content, and protein yields of range grasses [54]. However, ammo-

412

FIG. 105. *Nitrogen fertility greatly affects grass seed production. On this irrigated site near Oxford, Kansas, high nitrogen rates were applied to sideoats grama on the left with no nitrogen applied on the right.*

nium nitrate has been more effective in this method of application than urea, apparently due to gaseous losses of ammonia from the urea.

Banding is the placement of narrow, continuous bands of fertilizer in the soil. The fertilizer bands are covered by the soil but are not mixed with it. On sites low in fertility greater efficiency or yield from a given rate of fertilizer, particularly phosphorus, results from banding. Banding reduces the contact of the fertilizer with the soil and thereby reduces the proportion that becomes fixed on the soil colloids [147]. It also places the fertilizer beyond the reach of many surface weeds. However, established pastures are apt to remove more fertilizer nutrients from the soil when broadcast than when placed in bands below the surface [112].

Growth Responses to Fertilizer

Grass vigor. Nitrogen fertilizer has increased the vigor of grasses, particularly cool-season grasses, through increased basal area per plant, increased height and number of seedstalks per plant, increased seed production (Figure 105), increased size of stems and leaves, and greater number and weight of roots [21]. The increased yield resulting from nitrogen fertilization of grasses has been closely correlated with the increased depth and surface area of root growth and the resulting increased subsoil moisture extraction [11]. When nitrogen is

413

applied in a good rainfall year, the increased root growth may aid drought resistance and general hardiness in subsequent drought years. Nitrogen or nitrogen-phosphorus fertilizers have also reduced the pull-up problem of Sherman big bluegrass by stimulating root growth [136].

"Sod-binding" in rhizomatous, cool-season grasses is characterized by low vigor and production associated with nitrogen deficiency. Nitrogen fertilization is much more effective than mechanically tearing up old sod in preventing and correcting sod-bound conditions [1]. Since sod-binding is largely a symptom of nitrogen deficiency, annual application of nitrogen on mesic, productive sites generally is necessary to maintain vigor with smooth brome, crested and pubescent wheatgrass, and redtop. Less frequent application is required on more arid sites. Nitrogen fertilization has been the most effective technique for renovating old, deteriorated stands of crested wheatgrass, particularly when combined with a rest period [21]. Nitrogen or nitrogen-phosphorus fertilization in California developed strong, vigorous roots on annual range plants allowing them to extract moisture from deeper levels [135]. The decline of annual ryegrass seeded after a few years into brush burns in California was concluded to result primarily from the accumulation of annual ryegrass mulch immobilizing the available soil N released by the burn [97].

In contrast with nitrogen fertilization, phosphorus fertilizer generally does not significantly affect yield of grass herbage or plant vigor unless phosphorus levels in the soil are quite low or phosphorus is applied along with high rates of nitrogen. Energy fixation in mixed prairie ecosystems in the Northern Great Plains has been greatly increased by nitrogen addition alone, but nitrogen and phosphorus in combination have been required to obtain maximum energy fixation aboveground [144]. Nitrogen rates of 100 N/A, 300 N/A, and 900 N/A increased energy fixation by 150, 210, and 210 percent, respectively; but these rates plus 200 P/A increased energy fixation by 170, 300, and 330 percent, respectively.

Nitrogen fertilizer was beneficial on warm-season tallgrasses in Nebraska when annual applications were limited to 30 to 40 N/A and application was delayed until May [141]. When nitrogen was applied after the seeded warm-season stands had become dominant, it increased vigor and rhizome length and increased ability to compete with weeds. Erosion was reduced and production increased. Greenhouse trials in New Mexico demonstrated that nitrogen fertilizer increased seedhead production of blue grama by 375 percent but did not affect root production [35]. The nitrogen treatment did not significantly increase shoot production when plants were clipped every ten days to a one-inch height, but did significantly increase shoot production when clippings were made at a two-inch height.

Interspecific grass competition. Most cool-season grasses respond favorably to nitrogen fertilizers while some warm-season grasses do not (Figure 106). This differential response has been found on sub-irrigated meadows as well as upland ranges [13]. This has been attributed to cool-season grasses making most of their growth in early spring when moisture is less limiting and when the release of soil nitrogen is slower [1]. Greater benefit from nitrogen fertilization on native range in the Northern Great Plains than the Southern Great Plains appears at least partially to be a result of the higher cool-season grass component found further north [114]. And introduced, cool-season grasses generally have a higher potential of responding to fertilization than do native, cool-season grasses [145].

Nitrogen fertilization of a native shortgrass stand in eastern Wyoming did not change the plant composition [25]. However, a shortgrass-western wheatgrass stand changed to western wheatgrass under nitrogen fertilization. On mixed prairie in North Dakota nitrogen fertilization also decreased blue grama in basal cover while increasing western wheatgrass [45]. It was concluded in Texas studies that blue grama is a relatively inefficient user of both moisture and nitrogen [79].

FIG. 106. *Nitrogen fertilization greatly stimulated seeded crested wheatgrass, a cool-season grass, over invading warm-season native grasses on this infertile sandy site in northern Nebraska. Seventy-five N/A (left) gave a fivefold increase in forage production over the check (right).*

415

Warm-season tallgrasses and midgrasses such as the lovegrasses, switchgrass, bluestems, Indiangrass, and sideoats grama respond more favorably to nitrogen fertilization than shortgrasses such as blue grama and buffalograss [1, 87]. However, the indiscriminate application of nitrogen, even after warm-season tallgrasses have become established, can be serious in increasing competition from cool-season grasses such as Kentucky bluegrass while the warm-season grasses are dormant [96, 141]. Even low N application rates in Nebraska encouraged cool-season invading grasses over seeded warm-season grasses when applied in the spring, but no such encroachment was found when applied in late spring, even with rates up to 80 N/A [107]. However, the importance of avoiding carry-over of added N from one year to another was pointed out. Changes in species composition resulting from nitrogen application are generally gradual, and changes resulting from low to moderate rates can be considerably controlled by timing of fertilizer application and by season and intensity of grazing [145].

The application of up to 250 pounds/A of a 16-20-0 fertilizer to a mixture of cool-season, introduced grasses on upland sites in Saskatchewan did not influence the competitive ability of any grass species [49]. On Alberta native range low nitrogen fertilization rates did not greatly affect vegetation composition, but high rates decreased blue grama, needle-and-thread (*Stipa comata*), and prairie junegrass (*Koeleria cristata*) while increasing western wheatgrass, thickspike wheatgrass, fringed sage, foxtail barley, and annual weeds [65].

On native range in North Dakota heavily grazed for thirty-five years or more, 90 N/A annually for two years improved range condition and productivity more than six years of complete rest [114]. However, other ranges in low condition often fail to respond satisfactorily to fertilizer because of the stimulation of cool-season weeds and weedy grasses. On deteriorated native shortgrass range in eastern Wyoming, increased production from nitrogen application was primarily from annuals and woody shrubs [25]. Nitrogen fertilization of blue grama range in New Mexico has been recommended only on good to excellent condition range with little or no brush [36]. Cool-season annual bromes growing with seeded warm-season grasses in west central Kansas responded more to the nitrogen fertilizer than did the seeded grasses [77]. In Oklahoma it was similarly concluded that range containing much low quality vegetation should not be fertilized since this did not speed up range recovery [61]. This suggests that spraying weeds with herbicides may be required sometimes in order to reap benefits from fertilization [87].

On native range in the twelve-to-thirteen-inch precipitation zone in central Washington, 40 N/A doubled native grass production while increasing cheatgrass production from 19 to 58 percent [98]. However, nitrogen rates of 60 and 80 N/A greatly increased cheatgrass to the extent that perennial grasses were reduced. A single application of 80 N/A increased cheatgrass production from 13 to 82 percent of the total production. This serious retrogression of the range resulted from a severe decrease in Idaho fescue and considerable reduction of bluebunch wheatgrass. Sandberg bluegrass was not greatly affected. In southeastern Oregon nitrogen application at even low rates of 20 and 30 N/A increased cheatgrass on native range in improved condition [125]. On California range, 60 N/A stimulated soft chess, an annual, but not Hardinggrass, an introduced perennial [83]. A rate of 120 N/A stimulated soft chess similarly but Hardinggrass even more.

On bluebunch wheatgrass range in southeastern Oregon, neither ammonium nitrate nor calcium sulfate greatly increased cheatgrass yield [149]. But over a four-year period repeated application of ammonium sulfate sufficient to supply 80 N/A annually depressed blue-

FIG. 107. *Balanced fertilizer applications are important in maintaining desired grass-legume ratios. Legumes are maintained in this irrigated pasture mixture at the North Platte Station, Nebraska, by adequate but not excessive phosphorus rates. Highly soluble, noncorrosive fertilizers can be distributed through the irrigation system.*

417

bunch wheatgrass yield by 50 percent while increasing cheatgrass yield 600 percent. The resulting dominance of the cheatgrass was attributed to the early spring moisture drain by the cheatgrass. A combination of nitrogen and phosphorus in California similarly increased yield of cheatgrass in nine out of eleven years [71].

Fertilization of cheatgrass might be advantageous on lambing range or on a small area of spring range to shorten the winter livestock feeding period as a result of early growth provided by the cheatgrass [149]. However, in California, fertilization of cheatgrass was not a dependable means of producing additional forage in dry years [71]. Also, difficulty was reported in readily utilizing the extra herbage in good years by existing livestock programs. In related studies in California, sulfur fertilization consistently increased the yield of annual grasses on annual grass range deficient in sulfur [138]. It was concluded that sulfur can directly enhance the growth of annual grasses when their need for nitrogen has been satisfied.

Grass-legume competition. Botanical composition greatly affects the response to fertilizer applications. However, the type of fertilizer can also greatly affect the grass-legume ratio in the forage mixture (Figure 107). Heavy rates of nitrogen application on grass-legume mixtures stimulate the grass while depressing the legumes [25]. On the other hand, heavy phosphorus application on mesic sites can so stimulate the legumes that they become dominant over the grasses. Such a rapid shift in the species composition toward legumes can greatly increase the hazard of bloat. However, on nitrogen-deficient sites where soil moisture is not limiting, phosphate fertilization often increases yields of both the grasses and the legumes by stimulating nitrogen fixation by the legumes. Because of nitrogen fixation, mature alfalfa plants growing in pure stands normally do not benefit from application of nitrogen fertilizer [1].

For improved pastures in the Midwest it has been suggested that when the legume stand is 30 percent or more of the mixture, the main objective in fertilizing should be to maintain the legume [3]. After the legume has declined to less than 30 percent of the mixture, it was suggested the objective of fertilization should then be to increase grass yields. On irrigated pastures or subirrigated meadows, almost complete suppression of the legume or grass component respectively can result from high rates of nitrogen or phosphorus [13]. Similar results can be expected on mesic sites on western ranges except where grazing selectivity overrides the fertilizer effect. However, on semiarid ranges the legumes are seldom sufficiently competitive to dominate perennial grasses even when favored by phosphate fertilization.

418

The relationship of annual clovers and annual grasses has been studied extensively on California annual grassland ranges. Sulfur fertilization has resulted in about twice as many legume plants per acre and increased legume plant weight by 500 percent [47]. The sulfur also doubled the number of nodules and increased nodule weight by 800 percent. Although annual grasses have responded directly to sulfur addition [138], the increased nitrogen fixation by the legumes associated with sulfur fertilization further increased grass production [70]. Clovers decreased with incresing levels of nitrogen fertilization while the grasses increased [67]. However, a balance of sulfur and nitrogen fertilization minimized the effects of drought on grass-legume composition.

In the Northeast, potash application has increased the yield and percent composition of legumes in grass-legume stands by 100 percent [63]. However, except in potassium-deficient soils, potash has had little effect on grass-legume mixtures.

Growth season. Nitrogen stimulates earlier growth of particularly cool-season grasses. Nitrogen fertilizer at 25 to 30 N/A applied to crested wheatgrass in Utah advanced spring range readiness by 11 to 13 days, permitting the savings in hay costs alone to pay for the fertilizer [86]. Crested wheatgrass fertilized with nitrogen at Mandan, North Dakota, provided 50 percent more forage available for grazing at the beginning of the spring grazing season than unfertilized crested wheatgrass [115]. On semidesert grasslands in Arizona nitrogen application increased the green feed period from two to six weeks, particularly at higher rates up to 100 N/A [52]. Delaying nitrogen application until near the end of the summer rainy period on Arizona semidesert grassland extended the green feed period into the fall but delayed the start of the green period and reduced yield compared to earlier application [130]. Under more mesic ponderosa pine site conditions in Arizona, fall nitrogen fertilization also increased the amount of green growth during late summer and early fall [78].

On native range in southeastern Alberta, nitrogen addition produced earier green growth by seven to ten days [65]. Higher levels of nitrogen or nitrogen and phosphorus also extended the green growth period later into the fall. Sulfur-nitrogen and nitrogen fertilization of annual grass range in California induced earlier plant growth in the fall and maintained a more rapid growth rate during the winter and early spring when temperatures were minimal [82, 150]. Applying 80 N/A in addition to sulfur on California annual grasslands increased the number of days in the green season by 17 days (113 versus 96 days) compared to 60 S/A alone [20].

419

Although nitrogen often extends the growing season where moisture is high and the grasses are grazed to maintain vegetative growth, the growing season may be terminated earlier where soil moisture is low. On Oregon high deserts receiving an average of ten to eleven inches of rainfall annually, fertilized grasses depleted soil moisture more rapidly and caused an advancement in the growing season and shortened the green feed period [127]. On loamy upland bluestem range in Kansas, nitrogen application reduced soil moisture in dry years with the moisture deficit becoming progressively greater as the season advanced [96]. Under California conditions, nitrogen or nitrogen-phosphorus fertilizer in low rainfall years also reduced late spring growth of annual range plants [135].

Fertilizer and rainfall. Fertilizer application often fails to give expected benefits in drought years. Ample moisture supplies are required to maximize the benefits of fertilizer whether on irrigated pasture, introduced pasture on nonirrigated sites, or native range. On semiarid sites significant response to fertilizer may be given only in above average rainfall years. In the Great Plains area the response to fertilizer is highly correlated with moisture conditions and particularly May precipitation [73, 102]. However, some of the nitrogen applied in dry years and not used may be carried over and used the following year if desirable moisture conditions return [96]. This suggests that lower rates of fertilizer should be considered following a drought year.

Seasonal rainfall (July 1–June 30) in California greatly affected the pounds of beef produced per pound of nitrogen applied to rangeland [82]. When applied at average rates of 67 N/A, maximum yield of 2 pounds of beef per pound N occurred between fifteen and twenty inches seasonal rainfall. Between twelve and thirty inches seasonal rainfall, beef yield of 1.5 pounds per pound N or greater was

Table 22. Efficiency of Water Use on Fertilized Smooth Brome-crested Wheatgrass Meadow, South Dakota*

Nitrogen rates	Hay yields per inch of additional water (lbs./A)
0 N/A	87
40 N/A	201
80 N/A	306
160 N/A	344

*Source of data: [132].

420

obtained. It was concluded that under twelve inches was too droughty and over thirty was impractical, because of leaching, for nitrogen application. It has been recommended that nitrogen application on blue grama rangelands of the Southern Great Plains be limited to areas of at least fourteen inches annually with a well-defined peak during the growing season of the dominant grasses [36].

Although fertilizer benefits may be minimized by drought, increased forage production with increased moisture supply may not be realized without additional fertilizer. Smika et al. [120] found in North Dakota that grasses responded with increased yields to added moisture primarily where adequate amounts of nitrogen were available. They concluded that the effect of nitrogen fertilizer on forage yields may be equivalent to the effect of several inches of additional water. On fertilized smooth brome-crested wheatgrass pasture in South Dakota [132] higher nitrogen application rates increased the efficiency of water use as indicated in Table 22.

When applied to crested wheatgrass stands at Mandan, 80 to 160 N/A resulted in 380 pounds of additional herbage production for each acre-inch of water above five inches [102]. The effects of phosphorus fertilization were small and were relatively independent of water supply. On established stands of smooth brome in eastern Nebraska, forage production at 40 N/A applied annually did not respond to changing quantities of precipitation but at 80 N/A was linearly related to amount of precipitation [19]. On mixed grass-clover annual range in California, application of manure produced the greatest response in wet years but was considered highly valuable in drought years as well [88].

Dense stands of perennial grasses tolerate moderately high levels of nitrogen fertilization without toxic effects except under drought conditions. Under such conditions stimulation of early, rapid growth may cause moisture stress to be reached earlier. On native sandy range in western Nebraska, early spring application of 30 N/A applied at the beginning of what later developed into an extended and severe drought permanently damaged the grass stand [16]. High death losses of particularly warm-season grasses such as prairie sandreed and bluestems and even blue grama occurred, thereby opening up the stand to weed invasions.

On native range in eastern Wyoming nitrogen fertilization stimulated the growth of western wheatgrass, a cool-season species, thereby further reducing moisture available to warm-season species later in the growing season [104]. Nitrogen fertilization of warm-season grasses in dry years at Woodward, Oklahoma, caused drought symptoms in some instances but generally did not reduce yield [87]. When coupled with extremely dry years, high nitrogen application was the principal cause of marked reduction of intermediate wheatgrass in California where cheatgrass was present in sufficient amounts to cause rapid depletion of soil moisture [72]. Selective grazing further increased damage to the one-year-old stand of intermediate wheatgrass. However, on the Oregon high deserts rates up to 40 N/A did not injure seeded stands of crested wheatgrass in any year [127].

Fertilizer and Plant
Nutritive Content

After reviewing the effects of soil fertility upon the yield and nutritive value of forages, Ward [140] concluded that a deficiency of a mineral nutrient reduces forage yield and the concentration of that nutrient in the forage. A second conclusion was that additions of a mineral nutrient at high rates may cause its luxury consumption and result in higher concentration in herbage or in lower composition of other nutrients. A third conclusion was that a balanced mineral nutrient supply, whether at low or high levels, usually results in forage of approximately the same mineral nutrient composition but in different yield.

Nitrogen. An increased level of nitrogen in the soil nearly always increases the nitrogen concentration in the herbage of grasses growing there. Nitrogen application increases crude protein in grasses more at the beginning than at the end of the growing season [21]. However, soil factors, growing conditions, and time and rate of application influence the effect of nitrogen fertilization on protein content [13].

Crude protein levels in crested wheatgrass leaves at the beginning of the grazing season in North Dakota were 14.08, 19.10, and 22.34 percent, respectively, with 0, 40, and 80 N/A applied annually [115]. At the end of the spring grazing season, the crude protein in the remaining vegetation was 9.58, 10.42, and 12.54 percent under 0, 40, and 80 N/A, respectively. In South Dakota, nitrogen fertilization and association with alfalfa were not significantly different in increasing the protein content of various grasses [64]. Nitrogen fertilization of smooth brome in eastern Nebraska increased protein content during the forepart of the growing season but not by late summer (Table 23).

In Texas studies, nitrogen application to a bunchgrass-annual forb community increased crude protein percentage in the herbage from 8.1 percent on the check plots to a maximum of 18.1 percent at 600 N/A [32]. Between the extreme application rates of 0 and 240 N/A,

422

Table 23. Effects of Source and Amount of Nitrogen on Protein Content of Smooth Brome, Nebraska*

Date	No N	50 N/A	Grass grown with legumes
May 6	13.9%	22.7%	19.4%
June 6	7.2%	9.4%	10.1%
August 23	4.4%	4.0%	5.1%

*Source of data: [31].

each ten-pound unit of nitrogen applied to smooth brome in northeastern Nebraska increased crude protein by 0.3 to 0.6 percent [90]. Protein content of brome-crested wheatgrass hay in South Dakota was increased by nitrogen application but was decreased by greater rainfall [132]. In South Dakota, 60 N/A increased the protein content in crested wheatgrass hay from 8.4 to 10.7 percent [1].

On native range in British Columbia, 60 N/A increased the protein content of bluebunch wheatgrass by 59 percent (3.9 versus 6.2 percent) [85]. Although nitrogen has increased the stem-leaf ratio in grasses, total protein content has been increased because of increased protein in both the stems and the leaves [13]. On mountain meadows in Nevada, nitrogen application rates over 100 N/A generally increased protein content of the hay by 2 to 3 percent [48].

June application of 40 N/A to blue grama did not increase protein levels through maturity for grazing the following spring dormancy period in New Mexico [99]. But when applied to warm-season prairie grasses in Texas during June, nitrogen fertilizers did increase crude protein through the winter [28]. On December 29 the crude protein content of the blue grama and buffalograss averaged 7.3, 9.7, 12.0, and 12.7 percent, respectively, when nitrogen application levels were 0, 33, 100, and 300 N/A. Neither N or P fertilizers influenced in vitro dry matter digestibility of warm-season grasses or Kentucky bluegrass in Nebraska [108]. N fertilization had a marked influence on the protein percentage of both, and the increase was linear at maturity; P fertilizer increased P levels in both immature and mature plant tissue. However, in west central Kansas nitrogen application did not consistently increase the crude protein content of warm-season grasses [77].

After reviewing the effects of nitrogen fertilization on the nutritive value of pasture grasses, Blaser [12] concluded that added nitrogen:

1. Generally gives large increases in yields and livestock products per acre.

423

2. Generally did not improve output per animal.
3. Increased both protein content and apparent digestibility.
4. Did not generally alter cellulose or crude fiber content or lignification.
5. Reduces soluble carbohydrates but does not appreciably alter digestible energy levels.

Blaser [12] concluded that energy rather than protein content was generally limiting for livestock production in intensively managed pastures during the growing season.

Nitrogen application and winter burning in combination increased early summer protein levels of weeping lovegrass more than the sum of the two treatments applied individually [74]. Crude protein content was 3.5 percent on the control, 5.0 percent with 44 N/A, and 7.5 percent of the burned plots. However, when the treatments were applied in combination, crude protein content was 10.5 percent.

Both paraquat and atrazine have been shown to increase forage protein levels more than nitrogen fertilizers alone. Paraquat is effective for rapid curing and thereby maintains the higher protein levels resulting from nitrogen fertilization. The results of one study in Oregon is shown in Table 27. Increases in protein digestibility, as well as productivity, was anticipated. Atrazine and simazine cause tolerant plants to increase nitrogen uptake from the soil. On shortgrass range in northeastern Colorado, 1 pound per acre active ingredients of atrazine and 40 N/A increased protein yield for fall and winter grazing by 126 percent in a good rainfall year [55] and by 90 percent during a dry period [56]. The most practical treatment was considered to be 1 pound simazine active ingredients and 20 N/A [56].

Nitrogen fertilization seldom increases phosphorus content of grasses [13, 32, 41]. In North Dakota, fifteen tons of manure was approximately equal to 30 N in raising the protein content of grass but increased the phosphorus content of the hay also [121]. However, when applied to grass-legume mixtures at low to moderate rates, nitrogen often reduces crude protein content of the herbage by reducing the legume component [13, 84]. But high levels of nitrogen application may increase the protein levels even though the legume component is reduced.

Phosphorus. The application of phosphorus fertilizer on grass stands does not increase protein content. However, on grass-legume pastures phosphorus application stimulates the legumes and thereby generally increases the protein content of the herbage [13, 84]. On phosphorus deficient sites phosphorus fertilization can greatly increase the phosphorus content of both grass and legume herbage. On subirrigated meadows in Nebraska, the application of 35 P/A doubled the phosphorus content of grass-legume hay, reduced the consumption of mineral supplements by cattle, and generally provided adequate dietary phosphorus without phosphorus supplements needed [13]. In other Nebraska studies, P fertilizer increased P levels in both immature and mature plant tissue of seeded warm-season grasses and Kentucky bluegrass [107]. Phosphate fertilization of wet meadows in Oregon resulted in an increase of clover content from 1 to 50 percent and increased the nitrogen content of the hay from 6.5 to 12.5 percent [22].

Phosphorus fertilizer applied only once on northeastern Montana range increased phosphorus content in the range herbage for four years [11]. The 50 P/A rate increased phosphorus content of crested wheatgrass from 0.105 to 0.141 percent and increased phosphorus content of native range herbage from 0.097 to 0.120 percent. On shortgrass range in eastern Wyoming, phosphorus fertilization increased the phosphorus content of the grasses slightly but increased the phosphorus content of the total herbage more by encouraging greater amounts of forbs [25]. However, 87 P/A when applied to smooth brome in eastern Nebraska had little effect on yield, crude protein or phosphorus content, indicating that phosphorus was not deficient on this site [19].

When applied to phosphorus deficient range in south Texas, fertilization with triple superphosphate providing from 20 to 160 P/A did not affect protein content of the range herbage made up principally of paspalums, gramas, and bluestems but greatly increased phosphorus content [111]. The recommended rate of 40 P/A every five years supplied adequate phosphorus to meet dietary requirements of range cattle except during severe drought in the Texas studies. On semidesert grasslands in Arizona, fertilization at 80 P/A increased phosphorus content of the herbage from 0.14 to 0.25 percent but had no effect on protein content [41].

Heavy application rates of rock phosphate on native range in south Florida increased phosphorus content in the herbage by 100 to 400 percent [80]. The rock phosphate also included sufficient calcium oxide to materially increase the calcium content of the herbage. Fertilizing with the phosphorus-calcium fertilizer eliminated the need for feeding these minerals in supplements to the grazing cattle. It also increased both herbage yields and palatability. In experiments using coastal bermudagrass and crimson clover in the Southeast [2], increasing rates of N, K, and P fertilization increased the content of these nutrients in the bermudagrass. Increasing rates of K and P but not N fertilization increased the content of these nutrients in the clover.

Other nutrients. Ward [140] concluded that increasing rates of fertilization with K, Ca, S, or Co on sites deficient in these nutrients can be expected to increase their concentration in grasses.

**Fertilizer and
Forage Palatability**

Plants stimulated by added nitrogen are more palatable to livestock than unaffected plants. This effect of nitrogen fertilization on livestock preference has been found to be nearly universal whether the forage was from grasses or forbs, native or introduced species, or weeds rather than recognized forage species [32, 42, 52, 87, 105]. The resulting increased utilization and reduced selectivity have important management implications. Phosphorus, in contrast to nitrogen, generally has had no significant effect on palatability of grasses grown without legumes [21].

Nitrogen application has materially reduced ungrazed "wolf plants" in crested wheatgrass stands and resulted in more uniform grazing of individual plants for a few years [21]. When only a portion of a pasture receives nitrogen fertilization, this portion is subject to increased grazing pressure. This can be an advantage as well as a disadvantage. If overgrazed areas are fertilized, even heavier grazing can be expected. Fertilization can extend areas of excessive use if improperly used. However, application of nitrogen fertilizer on undergrazed portions of a grazing unit shows promise as a means of improving distribution of grazing.

Forage utilization by cattle on Wyoming mountain range was increased by two to five times by application of 67.5 N/A [123]. Cattle tended to graze untreated areas in the immediate vicinity of the treated plots. The increased utilization on the fertilized areas extended into the second growing season. On sloping land in mountainous areas of Utah, utilization on fertilized areas was increased when animals were drifted into the areas [21]. The cattle did not seek out the fertilized areas on the slopes but did utilize them more intensively once there. The use of nitrogen fertilization solely to improve distribution appears uneconomical; however, its use to increase forage production should be coordinated with herding, salting, and water development while increasing forage production [53].

Increased rates of nitrogen fertilization have increased the degree of use. On desert grasslands in southern Arizona 25, 50, and 100 N/A increased utilization of grasses by 3, 4, and 5 times, respectively [52]. However, 44 N/A in Texas did not improve utilization of weeping lovegrass in Texas [74]. On California annual range sulfur fertilization increased the palatability of both green and dry forage and was effective in making herbage more attractive on open slopes and reducing grazing on the swales [5, 9].

Nitrogen fertilization was found in the Black Hills of South Dakota to increase the palatability of forage grasses to deer [123]. In New Mexico browse plants fertilized with urea exhibited greater use by mule deer than unfertilized plants, but the application of ammonium sulfate apparently did not attract the deer [4]. On big game winter range in Utah, nitrogen fertilizer, but not phosphorus fertilizer, increased both the yield and palatability of bitterbrush and sagebrush for elk [8]. However, it was concluded that the palatability effects on shrubs was not sufficient to be used as a means of controlling movements and distribution of big game.

Use of Starter Fertilizer

The use of fertilizer as a stimulus for increased emergence and establishment of forages is common in humid regions. In semiarid regions the use of starter fertilizers has often failed to be advantageous except on irrigated lands and very deficient soils. The benefits of starter fertilizer have been more consistent with legumes than with grasses.

Legumes. Alfalfa, clover, and most other legumes respond to starter fertilizers applied at or near the time of seeding. In the Midwest the use of phosphorus fertilizers in starting legumes is widely recommended, particularly on sites low or deficient in phosphorus [3]. Phosphorus or phosphorus-sulfur fertilization is generally required at the time of seeding range clovers on California annual rangelands [147]. Starter fertilizers of 20 P/A with or without 10 to 15 N/A are commonly used with legumes and sometimes grasses. However, Holland et al. [51] concluded that nitrogenous fertilizers should not be used when establishing legumes in a seeding mixture. He found that well-nodulated legumes from inoculated seed do not need additional nitrogen from the soil and that adding nitrogen gave a competitive advantage to non-legume species. However, nitrogen fertilization has improved legume establishment on sandy soils in Nebraska [17].

The banding of fertilizer near the legume seed has generally been more advantageous than broadcast application on the soil surface [1]. Banding allows the fertilizer to be placed for immediate access by emerging seedlings and out of reach of many competing plants. The use of band-seeder attachments on grain drills allows the phosphorus to be placed 1.5 to 2 inches below the soil surface, places the seed about 1 inch above the fertilizer band, and then covers and presses the seed in one operation [3]. However, effective banding requires an excellent seedbed and ideal moisture conditions. On dry seedbeds the banding may reduce emergence and establishment of grasses—probably also legumes—by added seedbed disturbance [113].

The great value of starter fertilizer in seeding legumes into subirrigated grasslands has been amply demonstrated. Direct seeding of clover accompanied by an application of 20 to 40 P/A is recommended for subirrigated meadows in Nebraska [13]. On Oregon hay meadows, 20 to 30 P/A plus delayed cutting to allow the annual white-tip clover (*Trifolium variegatum*) found in the meadow to set seed, shatter, and spread enabled the legume component in the stand to increase from 1 percent up to as much as 50 percent [22].

Grasses. The results of providing starter fertilizer with grass seeding has been quite variable but frequently of no or little benefit. In Utah neither nitrogen nor phosphorus at either 20 or 40 pounds/A applied at seeding had any beneficial effects on seedling emergence, plant survival, or subsequent production of wheatgrasses [21]. This was true on both foothill and mountain range. In southeastern Idaho, fertilization of new seedings did not increase stand establishment of seeded plants when applied in fall or spring [62]. Fertilizer elements tried in the Idaho studies without apparent benefit included N, P, K, S, Cu, Fe, Mg, Mn, Zn, Bo, and Mo. Fertilizer application on high altitude, barren sites in northern Idaho slightly increased grass emergence but not seedling survival [43].

Warm-season grass establishment in Nebraska was not benefited from planting-time application of a mixed nitrogen-phosphorus fertilizer [141]. The use of fertilizer with spring and summer establishment was usually detrimental in increasing weed competition. In Oklahoma, nitrogen and phosphorus starter fertilizer both alone and in mixture did not increase the stand establishment of weeping lovegrass, switchgrass, and Indiangrass [14]. However, in South Dakota the use of 20 or 40 N/A alone and in combination with 20 P/A increased western wheatgrass establishment [24]. The nitrogen-phosphorus combinations appeared better than nitrogen alone. Starter nitrogen fertilizer has been recommended for grass seeding in South Dakota if weeds can be controlled [1].

In Texas studies starter fertilizers including nitrogen and phosphorus increased seedling growth, improved stands, and enhanced seedhead development of warm-season grasses in some years but not in others [60, 139]. In later studies, fertilizer had no effect on seedling emergence of any of the grass species in any of the years studied [142]. However, nitrogen-phosphorus combination and nitrogen

428

alone, to a lesser extent, increased the growth of grass species with medium and high but not low seedling vigor; but fertilizer generally had no effect on final stands of the grasses. In Oklahoma, grasses with strong seedling vigor such as weeping lovegrass, blue panic, and Caddo switchgrass responded to a mixed starter fertilizer [134], but the stimulation of weeds made the practice questionable. In California, both nitrogen and a mixed nitrogen-phosphorus starter fertilizer failed to increase the early seedling growth of Italian ryegrass (*Lolium multiflorum*), a species known for its high seedling vigor, or smilograss, a species known for low seedling vigor [7].

Interseeding. Starter fertilizer has been of uncertain value in interseeding [29]. Adding 30 N/A to an interseeding of sideoats grama near Amarillo did not increase seedling establishment [98]. Neither interseeding Russian wildrye into shortgrass range nor interseeding western wheatgrass into established crested wheatgrass was generally improved by starter fertilizer [24, 103]. However, phosphorus starter fertilizer should be considered when legumes are being interseeded into established stands of vegetation [93].

Regional Fertilizer Considerations

California annual rangelands. Sulfur fertilization is recommended on California annual plant ranges for open, rolling land if the soil is deficient in this element but not on rocky or brushy sites, steeper slopes, or shallower soils [9]. Sulfur alone or with nitrogen or phosphorus greatly increased production of bur clover, and the effects of application lasted from three to seven years [5]. It was concluded that an effective legume stand generally must be established before sulfur or phosphorus is economical.

In one study elemental sulfur applied at the rate of 60 S/A every two to four years increased animal gains annually by 20.4 to 64.2 pounds per acre during the green growth period and 17.4 to 18.5 pounds per acre during the dry forage season without the use of supplements [137]. However, in this study neither steer gains nor grazing capacity were increased during the winter season. In related studies sulfur application increased the yield of annual legumes by 1,086 pounds per acre (1,186 pounds versus 100 pounds) with above average rainfall [47]. Varietal difference in the response of subterranean clover, rose clover, and annual *Medicago* species to phosphorus and sulfur application were as great as differences between species [69].

When annual grasslands in California were fertilized with nitrogen, the primary benefit was the 80 percent increase in production

429

during the winter season when forage growth was normally slow [67]. Eighty pounds N/A gave near maximum increase in total yields with the annual grasses and filaree (*Erodium botrys*) giving the greatest response. Based on the composite results of fifty-four trials, the addition of about 57 N/A increased carrying capacity for yearling beef cattle from 57 to 86 grazing days/A and livestock gain/A from 74 to 142 pounds [82]. Beef produced/pound of N was 1.17 pounds. On phosphorus and sulfur deficient sites, the addition of these elements resulted in 2.75 and 2.54 pounds of beef/pound of N, respectively. Nitrogen plus phosphorus or sulfur provided relatively greater carryover effects into the second growing season than nitrogen alone.

Subterranean clover establishment was compared with nitrogen fertilization of California annual grasslands [68]. Yields were similar in a moisture deficient year, but the subterranean clover-grass produced much more than the fertilized grass in moisture adequate years and produced forage of better quality. Nitrogen fertilization increased grass production rapidly after application, which suppressed legume growth, while the uptake of nitrogen from the clover was much more uniform throughout the growing season. As a result protein content in the forage was increased by the clover association during the dry season when needed most. While nitrogen application had to be repeated each year, a good stand of clover once established could be maintained for many years by proper grazing and periodic maintenance applications of phosphorus and sulfur [68].

With benefits lasting into the third growing season, the application of a ton of chicken manure to annual grassland range increased forage production up to 1,600 pounds/A [88]. This treatment increased quality and palatability, but higher rates of manure initially decreased the abundance of legumes. Based on first year benefits alone, the manure produced additional carrying capacity at a cost of $1.56 to $2.18 per a.u.m.

Intermountain. On Utah foothill range receiving about twelve inches of precipitation annually, single application of 30 to 40 N/A increased herbage production by 800 to 1,000 pounds over a three-year period [21]. About 60 percent of this increase was produced the first year, 30 percent the second year, and 10 percent the third year. Phosphorus fertilization was ineffective in increasing herbage yield in this study.

On Oregon high desert sites seeded to crested wheatgrass and receiving an average of ten to eleven inches of precipitation annually, 20 N/A was the most efficient application rate with 25 pounds of additional herbage being produced per pound of nitrogen added [127]. This rate of application increased mean spring yield by 64%, gave

430

similar yield whether applied annually or double rates were applied biennially, but did not offset the decreasing yield trend with age of stand [128]. When applied to big bluegrass or Whitmar beardless wheatgrass stands, 20 N/A annually from fall application increased forage production on spring pasture by 80 percent or 230 pounds/A but at a cost of $24.00 per ton [125]. When grazed only after forage maturity, nitrogen application increased herbage production at a cost of $8.00 to $9.50 per ton; this was the only fertilization practice recommended. Nitrogen application had only a minor effect on sagebrush plant numbers.

Crested wheatgrass seeded in Nevada on sites receiving eight to twelve inches of average annual precipitation responded to nitrogen fertilization [38]. However, since the best rate was 60 N/A and this increased grass production at a cost of $60 per ton (1961), fertilization on similar sites was not recommended. No grass response resulted from additions of Cu, Bo, Mg, Fe, S, Zn, and P; Mo and K depressed production.

On native bluebunch wheatgrass range in British Columbia receiving eleven inches of average annual precipitation, added yields from single application of fertilizers lasted from one to four years and produced additional forage at a cost of $6.40 to $98.00 per ton [59, 85]. Sixty N/A increased forage production 68 percent the first year, 35 percent the second year, 14 percent the third year, and 6 percent the fourth year. The heavy rate of 240 N/A increased yields by 73 percent the first year, 58 percent the second year, 92 percent the third year, and 101 percent the fourth year. An average of 507 pounds of herbage/A was produced annually on the check plots. Phosphorus fertilization was ineffective on the native bluebunch wheatgrass sites. However, it should be noted that nitrogen application in Washington damaged native bluebunch wheatgrass sites by stimulating large increases in cheatgrass [98].

When applied to established stands of smooth brome on mountain range in northern Utah, 40 to 80 N/A produced about 1,000 pounds more herbage the first year and 100 pounds more the second year [21]. Nitrogen recovery in the forage yield was 94 percent in an unusually good year, and this compared to 19 to 43 percent from foothill wheatgrass seedings. On native mountain range in the area, 80 N/A increased grass production in dry meadows by 500 pounds while 80 P/A increased grass production only 150 pounds [21]. However, a combination of 80 N/A and 80 P/A increased grass production by 1,000 pounds.

Fall fertilization with nitrogen of intermediate wheatgrass stands seeded in the ponderosa pine zone near Flagstaff, Arizona, increased yields over a period of several years [78]. Although the biggest increase in production came during the first year, the 33 N/A rate significantly increased yields over the first three growing seasons while the 66 N/A and 99 N/A rates were significant over the first four growing seasons. Four-year intermediate wheatgrass yields from 0, 33, 66, and 99 N/A were 3,464, 4,484, 5,754, and 6,175 pounds, respectively. In producing 2,300 pounds of additional herbage over the four-year period with a nitrogen-grass ratio of 1:35, the 66 N/A rate was considered the most advantageous. Phosphorus applied alone or in combination with nitrogen was of no advantage in the study.

On pinegrass range in southern British Columbia, ammonium nitrate at 90 and 180 N/A increased yield, nutritive value, and palatability of the pinegrass [42]. The two-year herbage production was 750 pounds on the check, 1,100 pounds for 90 N/A, and 1,800 pounds for 180 N/A. At a conversion ratio of 1:5 to 1:6, it is obvious that these treatments were not economical. Sulfur additions slightly accentuated the response from nitrogen, but P, K, and micronutrients gave negative responses. Most of the sulfur and all of the nitrogen had been depleted from the upper root zone by the end of the second growing season. Nitrogen application on upland range in the Big Horns of Wyoming gave maximum efficiency at 25 N/A but did not approach economic feasibility [124]. On ponderosa pine, spruce, and aspen sites in Colorado, minor elements gave no response, P and K were indefinite, and N additions increased herbage yields only on some soils [109].

On high altitude mountain hay meadows in Nevada, 80 to 160 N/A increased hay yields by 1.0 to 1.6 tons per acre [48]. The response to phosphorus depended on soil phosphorus levels; on very low phosphorus soils 10 to 30 P/A greatly increased production. On very low phosphorus soils nitrogen application without additional phosphorus was ineffective. On meadows with over 25 percent legumes, little response to nitrogen fertilization was obtained where the clovers had been properly nodulated. However, in mountain meadow studies in Colorado, Willhite [146] concluded that it was more profitable to utilize commercial nitrogen as a source of nitrogen than depend solely on the nitrogen produced by the legumes. In his work 180 N/A applied to a 92-percent-grass–8-percent-legume stand produced 4.3 tons of hay per acre while an unfertilized 36-percent-clover–64-percent-grass stand produced only 2.7 tons annually.

On wet meadows in Oregon, application of N and P but not K, Cu, Bo, Mn, or Zn influenced hay yields [22, 23]. All levels of fertilization from 0 to 60 N/A increased hay yields the first year, but there were no carry-over benefits into the second year. The 60 N/A rate increased yield one-third ton the first year, one-fifth ton the second, and none the third. Based on 1955 prices, $.15 of N (one pound) gave $.35 of additional hay and $.10 of P (one pound) gave $.28 of additional hay. It was recommended, based on this study, that 60 to 80 N/A be applied annually on bluegrass meadows and on rush-sedge-grass meadows. However, it was recommended that fertilizer not be applied to nearly pure baltic rush (*Juncus balticus*) sites, on saline sites, or in deep swales submerged in water over one foot. For mountain meadows in Wyoming, 80 N/A annually was considered the most economical application rate for introduced grasses [118].

A combination fertilizer and herbicide treatment of depleted ponderosa pine-bunchgrass range in Colorado restored forage production to a good level, while grazing systems including protection, alternate year rest, and seasonal spring or fall grazing were ineffective [26]. Spraying alone was effective but required more time for grasses to come back. Fertilizer alone increased ground cover and herbage yield, but the latter included big responses from non-forage species.

Northern Great Plains. Applications of 30 to 50 N/A annually have commonly doubled forage production and resulted in an N-use efficiency of 20 pounds dry matter or an equivalent of 1 pound beef per pound of N when applied to introduced grasses on the better sites in the Northern Great Plains [145].

The effects of added nitrogen from commercial fertilizer and from fixation by alfalfa on productivity of crested wheatgrass have been studied intensively at Mandan, North Dakota [115]. Annual applications of 40 N/A increased forage production by 81 percent, carrying capacity by 72 percent, and beef gains per acre by 68 percent (Table 24). Pasture productivity was increased only slightly more by 80 N/A than 40 N/A. On pastures in which alfalfa made up about one-third of the production, forage production was 45 percent greater than on the unfertilized crested wheatgrass pastures, carrying capacity was 26 percent greater, and beef gains per acre were 34 percent greater. The levels and sources of added nitrogen did not affect gains per head, and daily forage consumption by the yearling steers was approximately twenty-five pounds per head regardless of treatment. The added herbage was mostly above the two-inch height on the plants.

In western South Dakota, hay yields alone from smooth brome-crested wheatgrass did not justify the use of nitrogen, phosphorus, or combination fertilizers [132]. However, the increases in yield of protein from all nitrogen rates, which ranged from 20 to 160 N/A anually, were profitable. Annual application of 80 N/A, the apparent best rate, increased annual hay yields by 129 percent (1,351 versus 594 pounds/A) and annual protein yields by 226 percent (182.8 versus 56.0 pounds/A). With hay valued at $15 per ton and protein at $.13 per pound and nitrogen fertilizer costing $.15 per pound of nitrogen, added annual returns in hay dollars per acre at the 80 N/A annual rate was –$2.32 but was $8.48 in terms of protein dollars. Yield response from a single application of 160 N/A lasted as long as five growing seasons. The recommended nitrogen fertilization rates in South Dakota for seeded, cool-season grass stands are shown in Table 25.

Fertilization of smooth brome pastures in northeastern Nebraska with 60 to 80 N/A annually nearly doubled forage and beef yields per acre [90]. One pound of nitrogen ($.10) frequently gave one pound ($.25) or more of additional beef with yearling steers. Each annual application of 80 N/A yielded an average of 0.6 ton of additional forage. It was concluded that the full benefit of nitrogen fertilization was not reached within the first year but during the years following repeated fertilizer use. At Lincoln, Nebraska, annual application of 80 to 120 N/A was recommended for smooth brome [19]. These rates increased smooth brome production up to 380 percent. In other studies at Lincoln, yearling steer gains were 86 pounds per acre annually on unfertilized smooth brome, 168 pounds per acre on smooth brome fertilized with 50 N/A annually, and 191 pounds per acre on unfertilized smooth brome-alfalfa pasture [31].

In renovating old, low-producing crested wheatgrass pastures in North Dakota, the application of nitrogen fertilizer to undisturbed sod was the most effective of ten treatments tried [81, 116]. Annual applications of 30 and 60 N/A respectively increased crested wheatgrass herbage yield by 2.5 times (1,503 versus 605 pounds) and 3.7 times (2,265 versus 605 pounds), respectively. Tearing up one-third of the sod by mechanical means provided no lasting effects. The application of nitrogen to torn-up sod gave no increase in yields over application to the undisturbed sod. A combination of nitrogen fertilization and weed control gave good increases in production on undisturbed sod and was suggested for stands containing a fair to good component of low vigor plants where moisture was adequate to promote growth. In Montana 50 N/A, harrowing, or plowing and seeding all failed to renovate crested wheatgrass during drought years [57].

Old stands of dryland crested wheatgrass at Archer in eastern Wyoming were made more productive by adding nitrogen, but the practice was economical only in above average rainfall years [119]. The most efficient rate was 66 N/A; this rate increased yields by 88 percent, produced 6.7 pounds of hay per pound of added nitrogen, but on the basis of hay yields produced only one-third of the yield required to pay for the fertilizer. Also in southeastern Wyoming, the continuous annual application of 60 N/A over an eight-year period reduced crested wheatgrass stands by 50 percent and crested wheatgrass-alfalfa stands by 38 percent [10]. Only manure application increased crested wheatgrass yields throughout the eight-year period. Ten tons of manure per acre annually produced an additional half ton of crested wheatgrass herbage each year but only 300 pounds of crested wheatgrass-alfalfa each year.

Nitrogen fertilizer effectively renovated intermediate wheatgrass stands without alfalfa on sandy soil at the Eastern Colorado Range Station, Akron [27]. Fifty pounds N/A increased intermediate wheatgrass production from about 260 to nearly 800 pounds per acre the first year and from 360 up to 500 pounds the second year. However, the nitrogen also nearly doubled the production of sand dropseed, a native, warm-season perennial. It was concluded that early spring application of nitrogen was advantageous to the cool-season, seeded grass only with early spring rains. It was recommended that intermediate wheatgrass-alfalfa not be fertilized with nitrogen since this treatment reduced the alfalfa while increasing the intermediate wheatgrass. Beef production on the fertilized intermediate wheatgrass was similar to that on the unfertilized intermediate-alfalfa but was less profitable. Broadcast application of phosphorus apparently had no influence on forage yield in either the intermediate wheatgrass or wheatgrass-alfalfa pastures.

Fertilization of native range in the Northern Great Plains with nitrogen or nitrogen and phosphorus has commonly increased forage yields from 25 to 100 percent but has generally been uneconomical. With increases in forage production of 25 to 50 percent from 30 N/A, it was concluded unwise to consider fertilization of native range in Nebraska in the fifteen-to-nineteen-inch rainfall belt at 1967 price levels [15, 16]. Total herbage production was increased by N fertilizer on Nebraska Sandhills range, but undesirable broadleaf weed production increased when not controlled by herbicides [92]. Nitrogen fertilization of native rangeland at Archer and Gillette, Wyoming, was concluded not to be economical [104]. In eastern Montana 90 N/A increased perennial grass yields by 50 percent and annuals by 20 percent [58].

On dense clay range sites depleted by drought and overgrazing in South Dakota, nitrogen and 2,4-D in combination increased production and slightly improved range condition when applied individually but were more effective when applied together [94, 95]. Three years' total production of perennial grasses was increased 391 pounds per acre by 2 pounds of 2,4-D, 594 pounds by 120 N/A, but by 1,640 pounds by a combination of both. On shortgrass range in eastern

Table 24. Pasture Productivity of Crested Wheatgrass as Affected by Nitrogen Fertilization and Alfalfa, Mandan, North Dakota*

Annual treatment	Forage production		Carrying capacity	
	Lbs./A	% increase over check	Steer days/A.	% increase over check
Check	1,680	—	38.3	—
40 lb. N/A	3,040	81	65.9	72
80 lb. N/A	3,260	94	71.2	86
Grass-alfalfa mixture	2,440†	45	48.2	26
Average	2,605	—	55.9	—

*Data are ten-year averages (1956–1965). Grazing by yearling steers annually from May 16 to about July 1. Average annual precipitation during the study was 15.4 inches. Source of information: Rogler and Lorenz [115].
**Calculated from livestock gains.
†The mixture averaged 1,800 lbs./acre of grass and 640 lbs. Ladak alfalfa/acre. Alfalfa increased in the grass-alfalfa mixture from 10% in 1956 to 37% in 1965.

Colorado, manuring increased production by 15 to 20 percent but P, N, and K commercial fertilizers did not increase production [75].

On native grassland range at Mandan, North Dakota, annual fertilization has given more favorable response. When moderately grazed range was fertilized annually at 0, 30, and 90 N/A, annual forage production was 656, 1,314, and 2,007 pounds/A, respectively [114]. The increased production was primarily due to the increase in western wheatgrass. Each pound of nitrogen at the 30 N/A rate produced 21.6 pounds of herbage; each pound at the 90 N/A rate produced 15.0 pounds of herbage. In related studies 0, 40, and 80 N/A

Table 25. Nitrogen Fertilization Rates in South Dakota Recommended for Seeded, Cool-season Grass Stands*

Site	Western half (lbs. N/A)	East central (lbs. N/A)	Eastern (lbs. N/A)
Irrigated	160	160	160
Nonirrigated			
Good moisture	50	75	100
Average moisture	40	50	80
Low moisture	0	0	40
Subirrigated or waterspreaders	75	—	—

*Source of data: [148].

436

Yearling gains per head		Beef gains per acre		TDN**	
Lbs./day	% increase over check	Lbs./A	% increase over check	Lbs./A	% increase over check
2.66	—	100.7	—	375	—
2.63	0	168.8	68	652	74
2.53	–5	176.3	75	688	83
2.82	6	135.4	34	505	35
2.66	—	145.3	—	555	—

annually produced 46, 82, and 108 pounds of beef per acre, respectively [116].

The effects of fertilizing subirrigated meadows in the Nebraska Sandhills have been intensively studied [13, 117]. However, the results obtained over the many meadows included in the trials were quite variable depending upon soil factors, vegetation, and height of water table. Combinations of nitrogen and phosphorus gave higher yield increases than either element applied alone. About 90 percent of the meadows responded to various combinations of N and P. Generally, no benefits resulted from applying K or lime either alone or in combination with N or P. Of the meadows fertilized with 80 N/A and 40 P_2O_5/A (about 17.6 P/A) annually, 94 percent of the meadows responded with an average annual increase of one ton per acre.

On Sandhill meadows residual effects of nitrogen fertilization into the second year were found only when application rates were above 80 N/A or when drought reduced first year utilization [13]. However, when legumes were present, the effects of phosphate fertilization were often as great in the second year as in the first and frequently carried ino the third and fourth years. Forty and 80 P_2O_5/A respectively gave a four-year total increase in yield of 1.5 and 2.0 tons, respectively. Eighty P_2O_5/A the first year (or 40 P_2O_5/A the first and third year) plus 40 N/A each spring gave a four-year total increase in yield of three tons per acre. A single application of 160 P_2O_5/A every four years, 80 P_2O_5/A every two years, or 40 P_2O_5/A annually resulted in similar yield increases over the four-year period.

Southern Great Plains. Fully established stands of warm-season grasses are generally benefited by nitrogen application provided the rates are not excessive and application is made in late spring. However, such fertility treatments are not always economical. Nitrogen fertilization of twelve problem sites in southern and eastern Nebraska previously seeded to warm-season native grasses produced vigorous growth which in turn controlled erosion and reduced weed invasion [141]. Application rates of 30 to 40 N/A annually increased the production of late maturing varieties of switchgrass and Indiangrass respectively by approximately 2,000 pounds and 1,800 pounds per acre. Both untimely mowing and untimely fertilization increased cool-season weeds.

Although fertilization of seeded stands of native warm-season grasses in west central Kansas increased yields considerably, the practice was considered uneconomical based on yield response alone at the rates used in the study [77]. Fertilizer rates applied were 80 N/A and 60 P_2O_5/A applied alone and in combination in the first and third years of the four-year study. Four-year yields were increased 1,460 pounds (2,266 versus 806 pounds on the check) by nitrogen alone and 2,701 pounds (3,507 versus 806 pounds) on the nitrogen-phosphorus combination. Phosphorus alone did not increase yields. The phosphorus-nitrogen combination increased yields over nitrogen alone on low organic matter sites but not on high organic matter sites. In this Kansas study, switchgrass was more responsive to nitrogen fertilization than blue grama, sideoats grama, or buffalograss.

On both native and introduced warm-season grasses at Woodward, Oklahoma, the cost of fertilizer exceeded the value of the additional forage produced [87]. Phosphorus was beneficial on the grasses only when applied in conjunction with nitrogen. Little nitrogen carried over into the second year, and neither nitrogen nor phosphorus increased livestock gains per head [87]. On loamy upland bluestem range in eastern Kansas, 50 N/A increased herbage yields by 0.75 to 1.0 tons per acre in normal or above average precipitation years [96]. The nitrogen greatly increased moisture-use efficiency, but almost no nitrogen was carried over in wet years. However, it was concluded that nitrogen fertilization in true prairie should be held in abeyance until some economic means, such as properly timed burning or herbicide application, could be developed to control the shift to cool-season species.

Application of 40 N/A to native hay meadows in east central Oklahoma increased hay yields by 700 pounds while 40 N/A plus 40 P_2O_5/A increased hay yields by 1,460 pounds [89]. Phosphate alone

438

did not increase yields. In related studies, 300 pounds of super-phosphate every third year and 100 pounds of ammonium nitrate each year slightly reduced grass production on native range [61]. This lack of response was associated with stimulation of weed growth. Fertilization of little bluestem in southeastern Texas at 80 pounds of superphosphate per acre increased little bluestem yield by 50 percent [106]. Potassium decreased grass production because of an increase in weeds. The failure of little bluestem to respond to nitrogen application was attributed to droughty conditions during the study period.

On native prairie in excellent condition in north central Oklahoma, neither nitrogen application nor burning alone significantly increased hay yields [44]. However, the burning-100 N/A combination increased hay production by 59 percent (4,921 versus 3,088 pounds on the check) and the burning-50 N/A combination increased hay yields by 45 percent (4,492 versus 3,088 pounds). The burning-nitrogen treatments were also effective in reducing forbs. Big bluestem and Indiangrass were the most efficient users of the

Table 26. Herbage Production and Livestock Production Increased by Phosphate Fertilization in South Texas*

Part A. Herbage response	Amount of superphosphate (lbs./A)	P equivalent (lbs./A)	7-year average yield (lbs./A)	Phosphorus content of herbage (%)
	0	0	2,610	.092
	100	20	3,480	.136
	200	40	3,540	.154
	400	80	4,010	.180
	800	160	3,680	.242
Part B. Livestock response (6-year averages)			No fertilizer	200 lbs. superphosphate per acre (40 P/A)
	No. of cows per section		42	62
	Calf weaning weights (lbs.)		489	551
	Percent calf crop		69	98
	Pounds of calf produced per acre		93	176
	Fall weight of yearlings		802	1,015
	Gross return per acre		$13.91	$19.97

*Source of information: Reynolds et al. [111]. Superphosphate applied in the spring of the first year only and not repeated in subsequent years. Paspalums, gramas, and bluestems were the principal forage species. Average annual precipitation was twenty-five inches during the study period.

Table 27. Chemical Curing of Nitrogen-fertilized Crested Wheatgrass with Paraquat, Oregon

| | Effects over control when fall harvested | | |
	Paraquat only	N only	Paraquat and N
Herbage yield	–23%	+80%	+40%
Crude protein concentration	+100%	0%	+68%
Crude protein yield	+60%	+79%	+148%

Source: [126]. Ammonium nitrate applied at 20 N/A the preceding fall and paraquat applied at 0.2 lb./A active in 10 gal. water in late June.

added nitrogen. In related studies in eastern Oklahoma, each pound of nitrogen produced over 36 pounds of total herbage when applied to burned plots and speeded up range recovery [46].

On south Texas brush range, fertilization and mowing together improved range condition faster and increased grass production more than either mowing or fertilizer used alone [100]. Nutrients provided in the fertilizer treatments (N, NP, P) were used more readily by weedy grasses than climax grasses on unmowed areas. Mowing combined with fertilization reduced competition from invader and increaser plants and allowed more of the fertilizer nutrients to be used by the climax grasses.

On blue grama rangeland in south central New Mexico, annual application of 40 N/A increased herbage production by 53 percent (870 versus 1,328 pounds) and 60 N/A by 85 percent (870 versus 1,608 pounds) [36]. The added nitrogen did not alter forb production, and the increase in production came almost entirely from grasses, mostly blue grama. Preliminary grazing results indicated that annual application of 40 N/A would increase yearling cattle gain from twenty pounds to fifty pounds/A. However, these benefits came from increased grazing capacity rather than increased summer gains per head [37]. Burning of blue grama range in the same area increased blue grama production by 56 percent while 60 N/A plus burning increased blue grama production by 174 percent over the check [34].

In south Texas, phosphate fertilization of grassland was shown to be an effective means not only of increasing herbage yields but also of supplementing livestock with phosphorus and thereby greatly improving livestock production [110, 111]. Although results were decreased in drought years, range forage production was increased 35

440

to 84 percent over a seven-year period by a single application of superphosphate at different rates made in the first year of the study (Table 26). Percent calf crop was increased from 69 to 98 percent, and weaned calf weight per acre was approximately doubled. The effects of phosphorus were found to persist for six years even on sandy soil. Two hundred pounds of superphosphate (40 P/A) every five years was recommended as being the most economical.

Southwest. On semidesert grasslands in southern Arizona, Freeman and Humphrey [41] concluded that fertilization with the resulting small increases in production was not economical based on 1956 prices. The study site was in an area receiving approximately sixteen inches average annual precipitation. The predominant grasses on the site were curly mesquite and sprucetop grama (*Bouteloua chondrosiodes*). The fertilizer rate giving the best return was 100 pounds of ammonium phosphate per acre. This rate produced additional hay yields at a cost of slightly over $20 per ton.

Rates of 100, 200, and 400 pounds of superphosphate, ammonium phosphate, and ammonium nitrate were used in the Arizona study [41]. Only the 400 pound rate of superphosphate materially increased production based on single-year responses. Rates above 100 pounds of ammonium nitrate did not further increase yields. The highest yield increase (548 pounds) was given by 400 pounds of ammonium phosphate. However, in a year receiving summer precipitation five inches above average, Holt and Wilson [52] found that fertilization greatly increased productivity of semidesert grassland where Santa Rita threeawn (*Aristida glabrata*) and Lehmann lovegrass predominated. Fertilization at 25 N/A from ammonium phosphate increased yields by 1,638 pounds.

On tobosa floodplains in southern New Mexico receiving nine inches average annual precipitation but additional run-in water from adjacent mountains and foothills, it was concluded that nitrogen and phosphorus fertilization was uneconomical in all except the best moisture conditions [50]. In a good moisture year when favorable soil moisture conditions existed above a continuous sixty-day growing period, 90 N/A increased production of tobosa by 4,164 pounds per acre. But the benefits of annual treatment were small in two other above-average years. It was concluded that some possibilities exist for 60 to 90 N/A plus 0 to 13 P/A where this can allow longer growing season deferment of adjacent upland pastures by increasing grazing capacity on the floodplains.

In the Southwest, fertilization should be restricted to those areas receiving adequate rainfall or additional water from natural runoff or waterspreading [6]. Present information suggests that nitrogen at rates not exceeding 50 pounds/A are most apt to give best results on native range and seeded grasslands, but the economics of such practices should be carefully analyzed.

Literature Cited

1. Adams, Earl P., and Edward J. Langin. 1966. Fertilizing pasture and hayland. South Dakota Agric. Ext. Serv. Fact Sheet 316.
2. Adams, William E., Matthias Stelly, R. A. McCreery, et al. 1966. Protein, P, and K composition of coastal bermudagrass and crimson clover. J. Range Mgt. 19(5):301–305.
3. Aldrick, S. R., D. W. Graffis, E. L. Knake, et al. 1970. Illinois agronomy handbook. Illinois Agric. Ext. Serv. Cir. 1012.
4. Anderson, Bruce L., Rex D. Pieper, and Volney W. Howard, Jr. 1974. Growth response and deer utilization of fertilized browse. J. Wildl. Mgt. 38(3):525–530.
5. Anonymous. 1968. Range fertilization revival. California Agric. 22(11):12–14.
6. Arizona Interagency Range Technical Sub-Committee. 1969. Guide to improvement of Arizona rangeland. Arizona Agric. Ext. Serv. Bul. A-58.
7. Ayeke, Cyril Alende, and Cyrus M. McKell. 1969. Early seedling growth of Italian ryegrass and smilo as affected by nutrition. J. Range Mgt. 22(1):29–32.
8. Bayoumi, Mohamed A., and Arthur D. Smith. 1976. Response of big game winter range vegetation to fertilization. J. Range Mgt. 29(1):44–48.
9. Bentley, J. R., L. R. Green, and K. A. Wagnon. 1958. Herbage production and grazing capacity on annual-plant range pastures fertilized with sulfur. J. Range Mgt. 11(3):133–140.
10. Birch, Thomas. 1961. The effect of continuous application of nitrogen fertilizer and manure on yield and stands of crested wheatgrass and alfalfa. Wyoming Range Mgt. Issue 147.
11. Black, A. L. 1968. Nitrogen and phosphorus fertilization for production of crested wheatgrass and native grass in northeastern Montana. Agron. J. 60(2):213–216.
12. Blaser, R. E. 1964. Symposium of forage utilization: effects of fertility levels and stage of maturity on forage nutritive value. J. Anim. Sci. 23(1):246–253.
13. Brouse, E. M., and D. F. Burzlaff. 1968. Fertilizers and legumes on sub-irrigated meadows. Nebraska Agric. Expt. Sta. Bul. 501.
14. Bryan, G. G., and W. E. McMurphy. 1968. Competition and fertilization as influences on grass seedlings. J. Range Mgt. 21(2):98–101.
15. Burzlaff, Donald F. 1965. Should I fertilize my rangeland? Nebraska Agric. Expt. Sta. Quarterly 11(4):5–7.
16. Burzlaff, Donald F. 1967. Fertilizers for rangeland. Nebraska Cattleman 23(6):12, 14.
17. Burzlaff, Donald F. 1971. Personal correspondence.
18. California Fertilizer Assoc. 1965 (Fourth Ed.). Western fertilizer handbook. 719 "K" St., Sacramento, California.
19. Colville, W. L., Leon Chesnin, and D. P. McGill. 1963. Effect of precipitation and long term nitrogen fertilization on nitrogen uptake, crude protein content and yield of bromegrass forage. Agron. J. 55(3):215–218.
20. Conrad, C. Eugene, E. J. Woolfolk, and Don A. Duncan. 1966. Fertilization and management implications on California annual-plant range. J. Range Mgt. 19(1):20–26.
21. Cook, C. Wayne. 1965. Plant and livestock responses to fertilized rangelands. Utah Agric. Expt. Sta. Bul. 455.
22. Cooper, Clee S. 1955. More mountain meadow hay with fertilizer. Oregon Agric. Expt. Sta. Bul. 550.

23. Cooper, Clee S., and W. A. Sawyer. 1955. Fertilization of mountain meadows in eastern Oregon. J. Range Mgt. 8(1):20–22.
24. Cosper, H. R., and A. Y. Alsayegh. 1964. Can fertilizer aid in establishing grass on native ranges? South Dakota Farm & Home Res. 15(1):3–5.
25. Cosper, H. R., J. R. Thomas, and A. Y. Alsayegh. 1967. Fertilization and its effect on range improvement in the Northern Great Plains. J. Range Mgt. 20(4):216–222.
26. Currie, Pat O. 1976. Recovery of ponderosa pine-bunchgrass ranges through grazing and herbicide or fertilizer treatments. J. Range Mgt. 29(6):444–448.
27. Dahl, Bill, and J. J. Norris. 1967. Fertilizer as a possible means of renovating intermediate wheatgrass grown in sandy soil. Colorado Agric. Expt. Sta. Prog. Rpt. 206.
28. Dee, Richard F., and Thadis W. Box. 1967. Commercial fertilizers influence crude protein content of four mixed prairie grasses. J. Range Mgt. 20(2):96–99.
29. Derscheid, Lyle A., Raymond A. Moore, and Melvin D. Rumbaugh. 1970. Interseeding for pasture and range improvement. South Dakota Agric. Ext. Serv. Fact Sheet 422.
30. Dodge, Marvin. 1970. Nitrate poisoning, fire retardants, and fertilizers—any connection? J. Range Mgt. 23(4):244–247.
31. Dowe, T. W., E. C. Conard, V. H. Arthaud, and J. M. Matsushima. 1953. Fertilized and unfertilized bromegrass pasture. In 41st Annual Feeders' Day, Nebraska Agric. Expt. Sta., Lincoln.
32. Drawe, D. Lynn, and Thadis W. Box. 1969. High rates of nitrogen fertilization influence coastal prairie range. J. Range Mgt. 22(1):32–36.
33. Duncan, Don A., and Jack N. Reppert. 1966. Helicopter fertilizing of foothill range. USDA, For. Serv. Res. Note PSW-108.
34. Dwyer, Don D. 1967. Fertilization and burning of blue grama grass. J. Anim. Sci. 26(4):934.
35. Dwyer, Don D. 1969. Greenhouse production of blue grama as influenced by clipping and nitrogen fertilization. New Mexico Agric. Expt. Sta. Bul. 549.
36. Dwyer, Don D. 1970. Nitrogen fertilization of blue grama rangeland. In Proc. of the 8th West Texas Range Management Conference, Abilene Christian College, Abilene, Texas.
37. Dwyer, Don D., and Jerry G. Schickendanz. 1971. Vegetation and cattle response to nitrogen-fertilized rangeland in south-central New Mexico. New Mex. Agric. Expt. Sta. Res. Rep. 215.
38. Eckert, Richard E., Jr., A. T. Bleak, and Joseph H. Robertson. 1961. Effect of macro- and micronutrients on the yield of crested wheatgrass. J. Range Mgt. 14(3):149–155.
39. Eckert, Richard E., Jr., Gerald J. Klomp, James A. Young, and Raymond A. Evans. 1970. Nitrate-nitrogen status of fallowed rangeland soils. J. Range Mgt. 23(6):445–447.
40. Ehlers, Paul, Glenn Viehmeyer, Robert Ramig, and E. M. Brouse. 1952. Fertilization and improvement of native subirrigated meadows in Nebraska. Nebraska Agric. Expt. Sta. Cir. 92.
41. Freeman, Barry N., and Robert R. Humphrey. 1956. The effects of nitrates and phosphates upon forage production of a southern Arizona desert grassland range. J. Range Mgt. 9(4):176–180.
42. Freyman, S., and A. L. van Ryswyk. 1969. Effect of fertilizer on pinegrass in southern British Columbia. J. Range Mgt. 22(6):390–395.
43. Gates, Dillard H. 1962. Revegetation of a high-altitude, barren slope in northern Idaho. J. Range Mgt. 15(6):314–318.
44. Gay, Charles W., and D. D. Dwyer. 1965. Effects of one year's nitrogen fertilization on native vegetation under clipping and burning. J. Range Mgt. 18(5):273–277.
45. Goetz, Harold. 1969. Composition and yields of native grassland sites fertilized at different rates of nitrogen. J. Range Mgt. 22(6):384–390.
46. Graves, James E., and Wilfred E. McMurphy. 1969. Burning and fertilization

443

for range improvement in central Oklahoma. J. Range Mgt. 22(3):165–168.

47. Green, Lisle R. 1959. Some effects of sulfur fertilization on nodulation and growth of annual range legumes. USDA, Pacific Southwest Forest & Range Expt. Sta. Res. Note 156.

48. Hackett, Irving, B. Brooks Taylor, Norman Nichols, and E. H. Jensen. 1969. Fertilization of mountain meadows. Nevada Ranch & Home Rev. 4(5):8–11.

49. Heinrichs D. H., and K. W. Clark. 1961. Clipping frequency and fertilizer effects on productivity and longevity of five grasses. Canadian J. Plant Sci. 41(1):97–108.

50. Herbel, Carlton H. 1963. Fertilizing tobosa on flood plains in the semidesert grassland. J. Range Mgt. 16(3):133–138.

51. Holland, A. A., J. E. Street, and W. A. Williams. 1969. Range legume inoculation and nitrogen fixation by root-nodule bacteria. J. Range Mgt. 19(2):71–74.

52. Holt, Gary A., and David G. Wilson. 1961. The effect of commercial fertilization on forage production and utilization on a desert range site. J. Range Mgt. 14(5):252–256.

53. Hooper, Jack F., John P. Workman, Jim B. Grumbles, and C. Wayne Cook. 1969. Improved livestock distribution with fertilizer—a preliminary economic evaluation. J. Range Mgt. 22(2):108–110.

54. Houston, W. R., and D. H. Van der Sluijs. 1973. Foliar-applied urea and ammonium nitrate fertilizers on shortgrass range. J. Range Mgt. 26(5):360–364.

55. Houston, W. R., and D. H. Van der Sluijs. 1973. Increasing crude protein content of forage with atrazine on shortgrass range. USDA Prod. Res. Rep. 153.

56. Houston, W. R., and D. H. Van der Sluijs. 1975. Triazine herbicides combining with nitrogen fertilizer for increasing protein on shortgrass range. J. Range Mgt. 28(5):372–376.

57. Houston, Walter R. 1957. Renovation and fertilization of crested wheatgrass stands in the Northern Great Plains. J. Range Mgt. 10(1):9–11.

58. Houston, Walter R. 1963. Some methods of range improvement. *In* U.S. Range Livestock Expt. Sta. Field Day, May 24, 1963.

59. Hubbard, W. A., and J. L. Mason. 1967. Residual effects of ammonium nitrate and ammonium phosphate on some native ranges of British Columbia. J. Range Mgt. 20(1):1–5.

60. Hudspeth, E. B., Judd Morrow, Earl Burnett, et al. 1959. Seed and fertilizer placement studies on selected native grasses at Big Spring. Texas Agric. Expt. Sta. Prog. Rpt. 2097.

61. Huffine, Wayne W., and W. C. Elder. 1960. Effect of fertilization on native grass pastures in Oklahoma. J. Range Mgt. 13(1):34–36.

62. Hull, A. C., Jr. 1963. Fertilization of seeded grasses on mountainous rangelands in northeastern Utah and southeastearn Idaho. J. Range Mgt. 16(6):306–310.

63. Hunt, O. J., and R. E. Wagner. 1963. Effects of phosphorus and potassium fertilizers on legume composition of seven grass-legume mixtures. Agron. J. 55(1):16–19.

64. Johnson, James R., and James T. Nichols. 1969. Crude protein content of eleven grasses as affected by yearly variation, legume association, and fertilization. Agron. J. 61(1):65–68.

65. Johnston, A., S. Smoliak, A. D. Smith, and L. E. Lutwick. 1967. Improvement of southeastern Alberta range with fertilizers. Canadian J. Plant Sci. 47(6):671–678.

66. Jones, Milton B. 1960. Responses of annual range to urea applied at various dates. J. Range Mgt. 13(4):188–192.

67. Jones, M. B. 1963. Nitrogen fertilization of north coastal grasslands—yield, percent protein, total uptake. California Agric. 17(12):12–14.

68. Jones, Milton B. 1967. Forage and protein production by subclover-grass and nitrogen-fertilized California grasslands. California Agric. 21(10):4–7.

69. Jones, M. B., P. W. Lawler, and J. E. Ruckman. 1970. Differences in annual clover responses to phosphorus and sulfur. Agron. J. 62(4):439–442.

444

70. Jones, M. B., and J. E. Ruckman. 1966. Gypsum and elemental sulfur as fertilizers on annual grassland. Agron. J. 58(4):409–412.
71. Kay, Burgess L. 1966. Fertilization of cheatgrass ranges in California. J. Range Mgt. 19(4):217–220.
72. Kay, Burgess L., and Raymond A. Evans. 1965. Effects of fertilization on a mixed stand of cheatgrass and intermediate wheatgrass. J. Range Mgt. 18(1):7–11.
73. Kilcher, Mark R. 1958. Fertilizer effects on hay production of three cultivated grasses in southern Saskatchewan. J. Range Mgt. 11(5):231–234.
74. Klett, W. Ellis, Dale Hollingsworth, and Joseph L. Schuster. 1971. Increasing utilization of weeping lovegrass by burning. J. Range Mgt. 24(1):22–24.
75. Klipple, G. E., and John L. Retzer. 1959. Response of native vegetation of the Central Great Plains to application of corral manure and commerical fertilizer. J. Range Mgt. 12(5):239–243.
76. Koch, G. O., J. M. Geist, and A. R. Tiedemann. 1975. Response of orchardgrass to sulphur in sulphur-coated urea. The Sulphur Inst. J. 11(3-4):1–5.
77. Launchbaugh, J. L. 1962. Soil fertility investigations and effects of commercial fertilizers on reseeded vegetation in west-central Kansas. J. Range Mgt. 15(1):27–34.
78. Lavin, Fred. 1967. Fall fertilization of intermediate wheatgrass in the southwestern ponderosa pine zone. J. Range Mgt. 20(1):16–21.
79. Lehman, O. R., J. J. Bond, and H. V. Eck. 1968. Forage potential of irrigated blue grama with nitrogen fertilization. J. Range Mgt. 21(2):71–73.
80. Lewis, Clifford E. 1970. Response to chopping and rock phosphate on south Florida ranges. J. Range Mgt. 23(4):276–281.
81. Lorenz, Russell J., and George A. Rogler. 1962. A comparison of methods of renovating old stands of crested wheatgrass. J. Range Mgt. 15(4):215–219.
82. Martin, W. E., and L. J. Berry. 1970. Effects of nitrogenous fertilizers on California range as measured by weight gains of grazing cattle. California Agric. Expt. Sta. Bul. 846.
83. Martin, W. E., Cecil Pierce, and V. P. Osterli. 1964. Differential nitrogen response of annual and perennial grasses. J. Range Mgt. 17(2):67–68.
84. Martin, William E., Victor V. Rendig, Arthur Haig, and Lester J. Berry. 1965. Fertilization of irrigated pasture and forage crops in California. California Agric. Expt. Sta. Bul. 815.
85. Mason, J. L., and J. E. Miltimore. 1969. Yield increases from nitrogen on native range in southern British Columbia. J. Range Mgt. 22(2):128–131.
86. McCormick, Paul W., and John P. Workman. 1975. Early range readiness with nitrogen fertilizer: an economic analysis. J. Range Mgt. 28(3):181–184.
87. McIlvain, E. H. 1961. Summary of range and pasture fertilization studies in the Southern Great Plains. USDA, Agric. Res. Serv., Woodward, Oklahoma. Mimeo.
88. McKell, Cyrus M., Victor W. Brown, Robert H. Adolph, and Cameron Duncan. 1970. Fertilization of annual rangeland with chicken manure. J. Range Mgt. 23(5):336–340.
89. McMurphy, Wilfred E. 1970. Fertilization and deferment of a native hay meadow in north central Oklahoma. Oklahoma Agric. Expt. Sta. Bul. 678.
90. Moline, W. J., E. J. Schwartz, R. Moomaw, et al. 1970. Nitrogen can increase bromegrass pasture profits. Nebraska Quarterly 17(2):17–18.
91. Moore, A. W., E. M. Browse, and H. F. Rhoades. 1968. Influence of phosphorus fertilizer placement on two Nebraska sub-irrigated meadows. J. Range Mgt. 21(2):112–114.
92. Morrow, L. A., and M. K. McCarty. 1976. Effect of weed control on forage production in the Nebraska Sandhills. J. Range Mgt. 29(2):140–143.
93. Musson, A. L., and L. B. Embry. 1957. Does it pay to fertilize native hay? South Dakota Farm and Home Res. 8(3):24–31.
94. Nichols, J. T. 1969. Range improvement practices on deteriorated dense clay wheatgrass range in western South Dakota. South Dakota Agric. Expt. Sta. Bul. 552.

95. Nichols, James T., and Wilfred E. McMurphy. 1969. Range recovery and production as influenced by nitrogen and 2,4-D treatments. J. Range Mgt. 22(2):116–119.
96. Owensby, Clinton E., Robert M. Hyde, and Kling L. Anderson. 1970. Effects of clipping and supplemental nitrogen and water on loamy upland bluestem range. J. Range Mgt. 23(5):341–346.
97. Papanastasis, Vasilios. 1976. Factors involved in the decline of annual ryegrass seeded on burned brushlands in California. J. Range Mgt. 29(3):244–247.
98. Patterson, J. K., and V. E. Youngman. 1960. Can fertilizers effectively increase our rangeland production? J. Range Mgt. 13(5):255–257.
99. Pieper, Rex D., R. Joe Kelsey, and Arnold B. Nelson. 1974. Nutritive quality of nitrogen fertilized and unfertilized blue grama. J. Range Mgt. 27(6):470–472.
100. Powell, Jeff, and Thadis W. Box. 1967. Mechanical control and fertilization as brush management practices affect forage production in south Texas. J. Range Mgt. 20(4):227–236.
101. Power, J. F. 1974. Urea as a nitrogen fertilizer for Great Plains grasslands. J. Range Mgt. 27(2):161–164.
102. Power, J. F., and J. Alessi. 1970. Effects of nitrogen source and phosphorus on crested wheatgrass growth and water use. J. Range Mgt. 23(3):175–178.
103. Rauzi, Frank, R. L. Lang, and C. F. Becker. 1965. Interseeding Russian wildrye—a progress report. Wyoming Agric. Expt. Sta. Mimeo. Cir. 216.
104. Rauzi, Frank, Robert L. Lang, and L. I. Painter. 1968. Effects of nitrogen fertilization on native rangeland. J. Range Mgt. 21(5):287–291.
105. Read, D. W. L. 1969. Residual effects from fertilizer on native range in southwestern Saskatchewan. Canadian J. Soil Sci. 49(2):225–230.
106. Reardon, P. O., and D. L. Huss. 1965. Effects of fertilization on a little bluestem community. J. Range Mgt. 18(5):238–241.
107. Rehm, G. W., R. C. Sorensen, and W. J. Moline. 1976. Time and rate of fertilizer application for seeded warm-season and bluegrass pastures. I. Yield and botanical composition. Agron. J. 68(5):759–764.
108. Rehm, G. W., R. C. Sorensen, and W. J. Moline. 1977. Time and rate of fertilization of seeded warm-season and bluegrass pastures. II. Quality and nutrient content. Agron. J. 69(6):955–961.
109. Retzer, John L. 1954. Fertilization of some range soils in the Rocky Mountains. J. Range Mgt. 7(2):69–72.
110. Reynolds, E. B., J. F. Fudge, J. M. Jones. 1951. Supplying phosphorus to range cattle through the fertilization of range land. Texas Agric. Expt. Sta. Prog. Rpt. 1341.
111. Reynolds, E. B., J. M. Jones, J. F. Fudge, and R. J. Kleberg, Jr. 1953. Methods of supplying phosphorus to range cattle in south Texas. Texas Agric. Expt. Sta. Bul. 773.
112. Riewe, Marvin E., and J. C. Smith. 1955. Effect of fertilizer placement on perennial pastures. Texas Agric. Expt. Sta. Bul. 805.
113. Robertson, Truman E., Jr., and Thadis W. Box. 1969. Interseeding sideoats grama on the Texas High Plains. J. Range Mgt. 22(4):243–245.
114. Rogler, George A., and Russell J. Lorenz. 1957. Nitrogen fertilization of Northern Great Plains rangelands. J. Range Mgt. 10(4):156–160.
115. Rogler, George A., and Russell J. Lorenz. 1969. Pasture productivity of crested wheatgrass as influenced by nitrogen fertilization and alfalfa. USDA Tech. Bul. 1402.
116. Rogler, George A., Russell J. Lorenz, and Herbert M. Schaff. 1962. Progress with grass. North Dakota Agric. Expt. Sta. Bul. 439.
117. Russell, J. S., E. M. Browse, H. F. Rhoades, and D. F. Burzlaff. 1965. Response of sub-irrigated meadow vegetation to application of nitrogen and phosphorus fertilizer. J. Range Mgt. 18(5):242–247.
118. Seamands, Wesley J. 1966. Increase production from Wyoming mountain meadows. Wyoming Agric. Expt. Sta. Bul. 441.
119. Seamands, Wesley J., and Robert L. Lang. 1960. Nitrogen fertilization of

446

crested wheatgrass in southeastern Wyoming. Wyoming Agric. Expt. Sta. Bul. 364.

120. Smika, D. E., H. J. Haas, and J. F. Power. 1965. Effects of moisture and nitrogen fertilizer on growth and water use by native grass. Agron. J. 57(5):483–486.

121. Smika, D. E., H. J. Haas, and G. A. Rogler. 1960. Yield, quality, and fertilizer recovery of crested wheatgrass, bromegrass, and Russian wildrye as influenced by fertilization. J. Range Mgt. 13(5):243–246.

122. Smika, D. E., H. J. Haas, and G. A. Rogler. 1963. Native grass and crested wheatgrass production as influenced by fertilizer placement and weed control. J. Range Mgt. 16(1):5–8.

123. Smith, D. R., and R. L. Lang. 1958. The effect of nitrogenous fertilizers on cattle distribution on mountain range. J. Range Mgt. 11(5):248–249.

124. Smith, Dixie R., and Robert L. Lang. 1962. Nitrogen fertilization of upland range in the Big Horn Mountains. Wyoming Agric. Expt. Sta. Bul. 388.

125. Sneva, F. A. 1963. A summary of range fertilization studies, 1953–1963. Oregon Agric. Expt. Sta. Spec. Rpt. 155.

126. Sneva, Forrest A. 1973. Nitrogen and paraquat saves range forage for fall grazing. J. Range Mgt. 26(4):294–295.

127. Sneva, Forrest A., Donald N. Hyder, and C. S. Cooper. 1958. The influence of ammonium nitrate on the growth and yield of crested wheatgrass on the Oregon high desert. Agron. J. 50(1):40–44.

128. Sneva, Forrest A., and Larry R. Rittenhouse. 1976. Crested wheat production: impacts on fertility, row spacing, and stand age. Ore. Agric. Expt. Sta. Tech. Bul. 135.

129. Soil Science Society of America. 1962. More simple terms for fertilizers. Crops and Soils 14(6):5–7.

130. Stroehlein, J. L., P. R. Ogden, and Bahe Billy. 1968. Time of fertilizer application on desert grasslands. J. Range Mgt. 21(2):86–89.

131. Tuescher, H., R. Adler, and Jerome P. Seaton. 1960. The soil and its fertility. Reinhold Publishing Co., New York City.

132. Thomas, J. R. 1961. Fertilizing bromegrass-crested wheatgrass in western South Dakota. South Dakota Agric. Expt. Sta. Bul. 504.

133. Thomas, J. R., H. R. Cosper, and W. Bever. 1964. Effects of fertilizers on the growth of grass and its use by deer in the Black Hills of South Dakota. Agron. J. 56(2):223–226.

134. Thompson, Lyell F., and C. C. Schaller. 1960. Effects of fertilization and date of planting on establishment of perennial summer grasses in south central Oklahoma. J. Range Mgt. 13(2):70–72.

135. USDA, Agric. Res. Serv. 1961. Fertilized range plants drink deep. Agric. Res. 9(11):10–11.

136. USDA, Rocky Mtn. Forest and Range Expt. Sta. 1969. Forestry research highlights, 1969. Annual Report, USDA, Rocky Mtn. Forest and Range Expt. Sta., p. 10.

137. Wagnon, K. A., J. R. Bentley, and L. R. Green. 1958. Steer gains on annual-plant range pastures fertilized with sulfur. J. Range Mgt. 11(4):177–182.

138. Walker, Charles F., and William A. Williams. 1963. Responses of annual-type range vegetation to sulfur fertilization. J. Range Mgt. 16(2):64–69.

139. Walker, Harvey, E. B. Hudspeth, and Judd Morrow. 1958. Seed and fertilizer placement study on selected native grasses at Lubbock. Texas Agric. Expt. Sta. Prog. Rpt. 2027.

140. Ward, George M. 1959. Effect of soil fertility upon the yield and nutritive value of forages. A review. J. Dairy Sci. 42(2):277–297.

141. Warnes, D. D., and E. C. Newell. 1969. Establishment and yield responses of warm-season grass strains to fertilization. J. Range Mgt. 22(4):235–240.

142. Welch, N. H., Earl Burnett, and E. B. Hudspeth. 1962. Effect of fertilizer on seedling emergence and growth of several grass species. J. Range Mgt. 15(2):94–98.

143. Weldon, M. D. 1966. Tailor your fertilizer program to your own needs. Ne-

braska Agric. Ext. Serv. Agronomy Tips 191.

144. Wight, J. R., and A. L. Black. 1971. Rangeland ecosystems of the Northern Great Plains as affected by high levels of nitrogen and phosphorus fertilization. II. Effect on energy fixation. ASRM, Abstracts of Papers, 24th Annual Meeting, pp. 35–36.

145. Wight, J. Ross. 1976. Range fertilization in the Northern Great Plains. J. Range Mgt. 29(3):180–185.

146. Willhite, Forrest M. 1961. Mountain meadow forage and beef production as affected by commercial nitrogen. J. Anim. Sci. 20(3):665–666.

147. Williams, William A., R. Merton Love, and Lester J. Berry. 1957. Production of range clovers. California Agric. Expt. Sta. Cir. 458.

148. Williamson, Edward J., Edward P. Adams, Lyle A. Derscheid, et al. 1969. Fertilizer for pasture and hayland. South Dakota Agric. Ext. Serv. Fact Sheet 425.

149. Wilson, A. M., G. A. Harris, and D. H. Gates. 1966. Fertilization of mixed cheatgrass-bluebunch wheatgrass stands. J. Range Mgt. 19(3):134–137.

150. Woolfolk, E. J., and D. A. Duncan. 1962. Fertilizers increase range production. J. Range Mgt. 15(1):42–45.

Chapter 11

Rodent and insect control

Nature of Rodent Problems

When rangelands are surveyed for grazing capacity and evaluated for use and recovery, the effects of rodent populations are often ignored or minimized. Rodents and rabbits often impose a greater impact on the plant community than the more conspicuous game or livestock species and can retard or even prevent succession on poor condition ranges [107, 114]. Rodents and rabbits are capable not only of grazing range vegetation closer than cattle but may even dig up root systems for feeding during dry periods [83].

Almost all rangeland is occupied by rodents, but not all rodents are detrimental to the range. Many species are rare or do not graze plants used by livestock or big game. Rodents such as grasshopper mice (*Onychomys* spp.) possibly help rangelands because their diets consist primarily of insects [27]. However, when the food habits of rodents coincide with or overlap with livestock or big game, their grazing competes and thereby lowers the overall grazing capacity for the larger herbivores. The degree of competition depends on the quantity and botanical composition of the forage crop, the season of grazing, the degree of herbage removal, and the population density of the respective rodent species [16]. Competition of rodents with large grazing animals increases when forage supplies are depleted and rodent populations are high.

Newly planted forages are particularly subject to damage and even destruction by rabbits, rodents, and insects [78, 89]. Rabbits can be very destructive because of concentrated, close grazing [75, 78]. Mice, kangaroo rats (*Dipodomys* spp.), and ground squirrels (*Citellus* spp.) can cause similar devastation, particularly through their

449

collection and removal of seeds. Rodents affect browse seedings equally as severely as grass seedings—or often even more severely. Rodents are responsible for more failures in artificial bitterbrush seedings than any other factor, and bitterbrush seeding without adequate rodent control measures is a gamble at best [10]. Rodent problems have been similar with seedings of four-wing saltbush [89]. Where rodent depredations are probable, it is inadvisable to attempt artificial seeding without accompanying rodent control measures [76].

Rodents and range conditions. High rodent populations are commonly associated with low range condition. In southern New Mexico, total rodent numbers were 3.5 times greater on mesquite sandhills (low condition range) than on black grama areas (high condition range) [74]. Climax bunchgrass sites in southern New Mexico had the fewest rodent species, the lowest densities of rodents, and the least fluctuations in rodent numbers [114]. By contrast an annual weed type produced the greatest number of plant species, the greatest number of rodent species, and the highest density of a majority of the rodent species. Range sites in intermediate condition were also intermediate in the number and density of rodent species.

In southern Arizona, jackrabbits and some kangaroo rats were more abundant on ranges in moderately low condition and less abundant on both the high and the extremely low condition sites [48, 63]. Arnold [2] concluded that jackrabbits had little effect on the beginning stages of deterioration of perennial grass ranges in southern Arizona but were a primary cause of deterioration in the final stages. Martin [63] reported that a relatively small rodent population has the ability to consume the entire grass seed crop on poor condition, low rainfall range, prevent plant reproduction, and reduce the resident forage stand. Studies in southern New Mexico also led to the conclusion that rodents including rabbits can greatly curtail vegetation improvement in the deteriorated mesquite sandhills [74, 76].

In Oklahoma, jackrabbits were more abundant on moderately overgrazed areas, while cottontails preferred undisturbed grasslands and sites offering considerable cover [77]. Jackrabbits in Kansas increased on low condition range, and the damage they caused was often aggravated by drought [24]. In California, ground squirrels, jackrabbits, and pocket gophers (*Thomomys* spp.) increased due to intensive livestock grazing in association with reduced herbage production and lowered range condition [37, 91].

Although northern pocket gophers (*Thomomys talpoides*) in western Colorado declined sharply with the replacement of forbs by perennial grasses [98], plains pocket gophers (*Geomys bursarius*) in eastern

450

Colorado thrive on high condition grassland range [69]. The highest gopher populations on one mountain site in Colorado occurred where grazing was excluded [28]; but on another site cattle use in summer and fall had no apparent effect on pocket gopher numbers [104]. On Great Plains grasslands, overgrazing by bison (*Bison bison*) and later by livestock followed by drought apparently caused the reduction of taller grasses and lowered range condition. This, in turn, provided ideal habitat for prairie dogs (*Cynomys* spp.) [58, 75].

Because of their increase on low condition range, animals such as prairie dogs and jackrabbits and many insects have been labeled "animal weeds" [24, 75]. Although Kalmbach [48] concluded that most present-day rodent problems on range have stemmed from earlier abuse of the vegetation, Howard [37] concluded that an increase in certain kinds of rodents is likely to occur whenever land is used and not just when it has been abused by man. For example, planting alfalfa alone or in mixture with grass often causes a manyfold increase in pocket gophers. Hansen and Flinders [26] concluded that the population density of jackrabbits should be considered an expression of the habitat preferences and the available food supply for jackrabbits rather than as an indicator of some previous abuse of the rangeland.

Howard [38] proposed that the most important factor governing the distribution and observed differences, but not the upper limits, in densities of local populations of rodents is the condition of the habitat. He concluded that rodents are self-limiting and that internal stress factors tend to limit rodent populations more than environmental factors such as food, disease, cover, weather, cohabitants, and natural enemies. Howard [37] further concluded that the mere presence or absence of rodent or rabbit species is not a good indicator of range condition.

Rodent Damage and Food Habits

The amount of range herbage consumed or destroyed by rodents can be serious whether caused by a single species or the total rodent population. On the Santa Rita Experimental Range in southern Arizona, rodents and rabbits removed only fifty-six pounds of herbage per acre in one study year, but this amounted to 40 percent of the total annual production that year [63]. A single species, the antelope jackrabbit (*Lepus alleni*), accounted for 65 percent of the total forage loss or thirty-five pounds per acre.

On desert range in southern New Mexico, rodents ate almost four tons of vegetation per section (approximately half of the total herbage production) and approximately two tons of insects annually [114]. Two species of kangaroo rats (*Dipodomys* spp.) and two species of wood rats (*Neotoma* spp.) combined to completely utilize the total food resource without great interspecific competition for any specific food item. In southern Idaho some rodent species were selective in their food habits, but other species were opportunistic in their feeding and subsisted on a variety of foods dictated by availability [44].

Forage disappearance or utilization more aptly describes the impact of rodents on rangeland than forage consumption since they often waste or destroy more forage than they actually eat. On California annual grassland range, actual consumption was less than 10 percent of that destroyed by rodents [16]. In this California study none of the rodent species utilized more than 35 percent of the total herbage production. However, a combination of three rodent species removed 76 percent of the production and greatly competed with livestock during the green forage period.

Jackrabbits. Rabbit populations are often cyclic, and this is particularly true of jackrabbits. High reproductive potential combined with local migrations allows rapid population buildups. When jackrabbit populations are high and green forage is scarce, they can be very damaging, particularly when concentrated into well-vegetated areas [48]. On deteriorated range in New Mexico, jackrabbits removed as much as 99 percent of the perennial grass herbage [76].

The black-tailed jackrabbit (*Lepus californicus*) is distributed throughout the West and probably causes more widespread range damage than any other jackrabbit species. However, severe damage can also be caused by the antelope jackrabbit in the Southwest and by the white-tailed jackrabbit (*Lepus townsendii*) in the Great Plains. Competition for range forage between black-tailed jackrabbits and livestock is less evident and spectacular than the former's heavy damage to crops, but it is probably more general and persistent [107].

Black-tailed jackrabbits on salt-desert shrub range in Utah demonstrated an order of preference for range plants closely resembling that of sheep [13]. They grazed grasses, forbs, and shrubs in the early spring but preferred grasses throughout late spring and summer. Shrubs were preferred in late fall and winter until grass regrowth began. When grazing was for equivalent time periods, 5.8 jackrabbits during the winter and 10.2 jackrabbits during the spring consumed or wasted forage equivalent to one sheep. Jackrabbits commonly clipped the grasses to the soil line and often took the two- and three-year-old growth of shrubs, thus wasting current year's growth and destroying lateral buds [23]. Except for migrations from the valleys to the foothills in hot weather, jackrabbits remained yearlong on a given range in contrast with seasonal use only by sheep. Shrubs most used were Nutall saltbush, winterfat, shadscale, and big sagebrush [13]. Girdling main stems and clipping branches of big sagebrush and big rabbitbrush as observed in some years are beneficial on many sites in the Great Basin but undesirable on some winter ranges.

Widespread girdling of shrubs by black-tailed jackrabbits during winter was also found in northeastern California with the closest and most frequent use on four-wing saltbush and hopsage (*Grayia spinosa*) [66]. Bitterbrush, desert peach (*Prunus andersonii*), and big sagebrush were also grazed but less severely. In southern Nevada, the major forage plants consumed from spring through summer were grasses, particularly *Oryzopsis, Stipa,* and *Bromus,* and herbaceous annuals [32]. From late summer through winter the diets consisted mostly of woody perennials, chiefly *Larrea, Coleogyne, Krameria,* and *Eurotia.*

Estimates made in 1937 at the Santa Rita Experimental Range showed that the antelope and black-tailed jackrabbits, respectively, consumed 175 and 120 pounds of herbage annually [63]. Arnold [2] concluded that 62 black-tailed or 48 antelope jackrabbits ate as much as a mature cow on the basis of total forage consumption. However, if only perennial grasses were considered, it took 260 or 164 black-tailed and antelope jackrabbits, respectively, to be equivalent to one cow. In other Arizona studies, Vorhies and Taylor [106] found jackrabbit densities may reach one per acre and at this density may eat as much as four to ten cattle per section. They concluded that 148 and 30 black-tailed jackrabbits, respectively, ate as much as one cow or one sheep, while 74 and 12 antelope jackrabbits, respectively, ate as much as one cow or one sheep. Stomach content analyses indicated that grass made up 45 percent of the yearly diet of the antelope jackrabbit and 24 percent for the black-tailed jackrabbit.

On sandhill range in eastern Colorado, grasses made up 50 percent of the yearly diet of black-tailed jackrabbits [88]. The greatest competition with cattle occurred in early spring when the highest levels of grasses and sedges were eaten. Forbs made up 30 percent of the yearly diet and were preferred by jackrabbits during the summer and early fall. Shrubs made up 10 percent of the yearly diet, being eaten principally during fall and winter. The major plants in the diet were western and intermediate wheatgrass, alfalfa, sixweeks fescue (*Festuca octoflora*), sand dropseed, pricklypear, and sand sagebrush. The researchers concluded that jackrabbits decreased the longevity of reseeded forages such as intermediate wheatgrass-alfalfa and introduced large amounts of sand dropseed seed into the seeded stands through their droppings. Succulent grasses and forbs were preferred foods in the diet of white-tailed jackrabbits in Colorado during the spring, summer, and early autumn and were the primary foods when available [5]. However, shrubs were the major food items during the winter period.

Studies in Kansas [97] indicated that black-tailed jackrabbits ate primarily grasses and the smaller forbs. However, they ate almost any green vegetation during periods of food shortage. Jackrabbits were dispersed throughout the hills during periods of normal food availability but migrated to brushy draws and cropland when drought and overgrazing depleted the food supply in the hills [9]. These fluctuations in local population density were as much a function of migration as they were of total jackrabbit numbers.

Jackrabbits may thwart efforts to artificially seed rangeland by cropping off seedling grasses as fast as they emerge [107]. This is particularly true in small seeded areas bounded by unimproved range since jackrabbits move from unimproved areas and concentrate on the new seedings. Jackrabbits and even cottontails can severely damage or destroy seedlings of four-wing saltbush or bitterbrush by severe trimming [36, 89]. Unless planted during a low in the population cycle, extensive jackrabbit control may be required to protect range seedings. On old stands of crested wheatgrass, jackrabbits in Utah mostly grazed the outer 1000-foot perimeter of the pastures and utilized only 10 percent of the forage within this area even during a high jackrabbit population period [111].

Pocket gophers. Pocket gophers (*Thomomys* spp., *Geomys* spp., and *Cratogeomys* spp.) are widespread over much of the West, being uncommon or absent only in arid areas or areas with compacted soils. In the mountains, pocket gophers are most common on meadows, grassy openings, and timbered areas thinned by logging. In the Great Plains, they are most abundant on sandy soil and in alfalfa fields. Pocket gophers are burrowing rodents that spend much of their life underground. They develop extensive burrow systems and deposit the excavated soil in a series of mounds above the ground. During the winter, the excavated soil is packed into snow tunnels forming the typical winter casts.

Ellison and Eldous [15] suggested that northern pocket gophers (*Thomomys talpoides*) on subalpine range in Utah may serve a useful function in loosening compacted soils. Ellison [14] concluded that gophers in this area apparently did not destroy enough vegetation to cause accelerated erosion, although gopher diggings in gullies were readily swept away by spring and summer runoff. However, in California the burrows of both pocket gophers and ground squirrels (*Ci-*

tellus spp.) were found to be precursors of gullies [60]. Here subsurface flow enlarged the burrows until they caved in. Subsequently livestock trampled in the edges of the burrows, and overland flow widened them into gullies.

Pocket gophers may deleteriously affect vegetation in several ways [46, 67, 91]:

1. Destroy the root crowns and stem bases of forbs and grasses.
2. Graze aboveground plant parts to a lesser extent.
3. Destroy or weaken plants by undermining and by pulling seedlings and small plants into their tunnels for food and nesting materials.
4. Bury plants while forming mounds and winter casts.
5. Partially seal the soil against water infiltration by the drying of winter casts.
6. Aerate the soil by tunneling and thus aggravate summer drought.

Gopher mounds provide bare areas for establishment of unwanted plants. These disturbed areas frequently are taken over by noxious annuals such as tarweed [46, 67]. Laycock [59] concluded that gopher mounds are microsites capable of perpetuating pioneer species even in climax communities.

A large portion of the pocket gopher's diet comes from roots, root crowns, and regenerative organs such as corms, bulbs, and rhizomes [47]. Studies in mountain meadows in western Colorado revealed that northern pocket gophers ate mostly roots and tubers during the winter but ate large amounts of underground plant parts throughout the year [28]. Stems and leaves made up a considerable part of the diet only during the spring and summer when the plants were growing rapidly and were succulent. Some aboveground foraging was also noted in the snow tunnels.

Perennial forbs are preferred foods of the northern pocket gopher and comprise 93 percent of its diet in western Colorado [110]. The most important species were common dandelion, yarrow, penstemons, lupines, aspen fleabane (*Erigeron speciosus*), hairy goldaster (*Heterotheca villosa*), slenderleaf gilia (*Collomia linearis*), geranium, and mountain dandelion (*Agoseris* spp.) [29, 110]. Significant amounts of grass stems and leaves were consumed only during the summer, although gophers ate and destroyed root crowns of grasses beneath the snow in winter [28]. Pocket gophers generally appear to be less abundant on sheep range than on cattle range and probably compete more with sheep for preferred plants [28].

The preference of northern pocket gophers for forbs has been further demonstrated by spraying gopher-infested sites with 2,4-D. This

455

treatment on a mountain meadow in Idaho reduced gopher mounds by 93 percent and winter casts by 94 percent on the average when compared to unsprayed areas [42]. On Grand Mesa in Colorado spraying 2,4-D reduced the production of perennial forbs by 83 percent, increased grass production by 37 percent, and reduced the gopher population by 87 percent [52]. From an original diet of 82 percent forbs and 18 percent grasses, the remaining gophers changed to a diet of 50 percent forbs and 50 percent grasses. Tests made in conjunction with the Grand Mesa studies [29] and in related studies on Black Mesa [98] showed the sharp decline in gopher population following 2,4-D application was:

1. Caused by an inability to survive on treated areas when the main diet, the perennial forbs, was removed.
2. Not caused by movements from the sprayed areas nor by direct or indirect toxicity of 2,4-D to gophers.

The reduction in forbs associated with application of 2,4-D in western Colorado studies [25, 45], in addition to sharply reducing pocket gopher numbers, had little effect on deer mice (*Peromyscus maniculatus*) but increased the populations of meadow voles (*Peromyscus pennsylvanicus*), and montane voles (*Microtus montanus*). However, as the forb composition returned to pretreatment levels, pocket gophers increased to original numbers or even higher [29, 45].

On sandhill grasslands in northeastern Colorado, the yearly diet of the plains pocket gopher was approximately 64 percent grass although forbs were preferred when they were succulent in spring and summer [27, 69]. The proportion of forbs in the diet ranged from 93 percent in September to 43 percent in March. Although pricklypear, *Astragalus* spp., needle-and-thread, and scarlet globemallow (*Sphaeralcea coccinea*) were the most preferred plants, six of the ten most preferred plants were grasses. From this it was concluded that the plains pocket gopher was adapted to a diet high in grass similar to that of cattle.

Up to 70 percent of the diets of pocket gophers in Nebraska was roots, and grasses constituted the bulk of the diet [92]. Their activities sometimes cut forage production by 50 percent, and the gopher mounds on sandy soils often covered 5 to 25 percent of the soil surface, preventing growth of underlying vegetation and accelerating wind and water erosion. Pocket gophers in this area preferred sites in good to excellent conditions; following decline in range condition from gopher damage, these sites were abandoned for new, undisturbed areas.

Direct gopher control by poisoning on Grand Mesa rapidly increased forb production while grass production varied from no change to moderate increases [101]. However, 2,4-D application indirectly reduced gopher densities but increased grass production more than did the gopher control. Pocket gopher control in Oregon mountain meadows in poor condition increased the grazing capacity for sheep by six times [67]. It was concluded from the Oregon study that a few gophers may hold a poor condition range from improving but, where the range was in fair or better condition, pocket gophers would eat only a small portion of the vegetation. It was further concluded that when forage production was low and gopher populations were high, control was definitely needed.

Pocket gopher damage appears even more serious on artificially seeded range and pasture. In Utah, Julander et al. [46] concluded that establishment and maintenance of introduced grasses on depleted mountain range having dense populations of pocket gophers require direct gopher control. During the first five years following grass seeding, 1,195 pounds of herbage was produced annually where the gophers were controlled compared to 355 pounds where they were uncontrolled. Although excellent stands were initially established on both sites, grass stands and yields declined rapidly on uncontrolled sites. An important portion of the grass loss was caused by gopher damage to the root crowns and stem bases of well-established grass clumps.

In mountain range studies in Oregon old stands of crested wheatgrass were not greatly affected by gopher burrowing and feeding, but the natural reproduction between the drill rows was definitely impaired [21]. It was recommended that narrow drill rows and ample seeding rates be used to offset this loss. Another consideration was to destroy all forbs preferred by gophers during seedbed preparation and until the grass stand had established, thereby making the planting site unattractive to gophers. It was noted that rhizomatous grasses such as pubescent wheatgrass were less damaged than bunchgrasses such as tall oatgrass.

In Utah seeding studies, slender wheatgrass and mountain brome appeared less susceptible to gopher damage than timothy, orchardgrass, tall oatgrass, or smooth brome [46, 47]. The seeded grasses most resistant to pocket gopher damage in the Colorado mountains appeared to be fine-leaved fescues and meadow fescue (*Festuca elatior*) while gophers thrived on smooth brome [28]. Gopher populations are often much higher in grass-legume mixtures than where grass alone is seeded, and mixtures including legumes are more apt to require gopher control [27, 28]. Pocket gophers find alfalfa very attractive and seriously reduce this forage species, particularly the nonrhizomatous varieties. Pocket gophers can be very destructive to

457

browse seedlings such as bitterbrush by severing and girdling the main roots [36].

Prairie dogs. The prairie dog (*Cynomys* spp.) is represented by seven subspecies (sometimes considered as separate species) arbitrarily placed in two groups [58]. The white-tailed group includes the white-tail, Utah, Gunnison, and Zuni subspecies; occurs in the Rocky Mountains and westward but not on the Pacific coast; and its habitat is the foothills, mountain parks, and shrub lands. The black-tailed group includes the black-tail, Arizona, and the Mexican subspecies and occurs in the Great Plains and grasslands of the Southwest.

Damage to range caused by prairie dogs is more obvious than that of any other range rodent. Prairie dogs are gregarious and live in colonies or towns often spreading over large areas. They exert heavy grazing pressure locally on the range, particularly in drought years. Prairie dogs live in burrows with crater-shaped mounds commonly three to six feet in diameter around the openings. From the burrow they forage out into the surrounding area with the intensity of foraging in direct proportion to distance from the burrow. The mounds and the immediate surrounding area are greatly disturbed by foraging and digging for roots, thus favoring the establishment of pioneer forbs which are perpetuated by continued rodent grazing [58]. Prairie dogs were found on shortgrass range to cause vegetation diversity, including the invasion of many weedy species, but were concluded not always to be destructive to rangelands [8].

Prairie dogs are highly competitive with livestock, particularly cattle, and generally eat grass species in the same order of preference [69]. Taylor and Loftfield [94] in their early work on a wheatgrass type in northern Arizona found that the Zuni prairie dog destroyed 69 percent of the wheatgrass, 99 percent of the sand dropseed, or 80 percent of the total potential annual forage production. On a blue grama type the loss of forage production was 80 percent. After evaluating stomach contents of three prairie dog subspecies from Montana to Arizona, Kelso [53] found that valuable forage plants averaged 78 percent of the diet and grass 45 percent of the diet.

Koford [58] concluded that about 25 percent of the diet of the black-tail prairie dog was not livestock feed and that this animal was an opportunistic feeder. But its activities did sharply reduce the midgrasses and tallgrasses and shrubs in favor of the shortgrasses and weeds. Other factors that favored shortgrasses over the taller grasses also favored a buildup in the prairie dog population. Koford also reported that prairie dogs cut down tall plants apparently as a means of removing obstacles to their vision and movement. Osborn and Allan [75] reported that maintaining a good stand of tallgrasses nearly excluded prairie dogs in Oklahoma.

Control by poisoning has been widely applied over large areas of the prairie dog's former habitat and has been highly effective. Koford [58] reported that 1,210,000 acres in Colorado alone were treated with Compound 1080 to kill prairie dogs in 1947. As a result prairie dogs have been eliminated in many areas, and the need of control programs is now limited to local problem areas.

Ground squirrels. Ground squirrels (*Citellus* spp.) are widespread and include many different species. The extent of damage to the range varies from slight to severe depending on species, population density, type of range, and season of the year. In western Canada it was found that the diet of the golden-mantled ground squirrel (*Citellus lateralis*) was very diversified and that fungi and leaves of forbs constituted 87 percent of the diet with seeds, flowers, bulbs, fruit, and even flesh making up the remainder [65]. Although the ground squirrel is a burrowing rodent, most of its food is found on the ground surface with the burrow being used as a den, for safety retreat, and for occasional storage of food [91].

Range damage by ground squirrels has been serious in California. On pasturelands here, ground squirrels (*Citellus beecheyi*) eat considerable quantities of grass and herbage usable by livestock such as filaree, bromegrasses, and fescues [40, 91]. During late winter and spring they feed mostly on green herbage and minor amounts of seed. Competition with livestock is greatest during the winter forage period when range feed is in short supply. Forage production in the spring period is normally adequate for both ground squirrels and livestock. During the summer and fall, ground squirrels eat mainly nonforage plants and seeds.

In Washington it was estimated that 385 ground squirrels ate as much as one cow and that 30 ate as much as one sheep [87]. On pasturelands in California, 450 ground squirrels ate as much grass and herbage as one steer [91]. On annual grass range the common density of 12 ground squirrels per acre reduced the forage crop 35 percent by the end of the green forage season with large amounts being wasted [16]. Each ground squirrel eliminated about 91 pounds of herbage annually. On pasture where ground squirrels were poisoned, seasonal winter gains of heifers were increased by 74 pounds [40]. This was equivalent to additional heifer gains of 4.5 and 2.2 pounds per year respectively during the following two years for each squirrel destroyed.

Kangaroo rats. Kangaroo rats (*Dipodomys* spp.) are most abundant on rangelands in the Southwest where vegetation is sparse. In southern

New Mexico, kangaroo rats were the dominant rodents on all range sites studied except the climax black grama site [114]. The bannertail kangaroo rat (*D. spectabilis*) completely destroyed perennial grass within twenty to fifty feet of its home burrow. These denuding activities prevented the production of livestock feed on ten percent of the range area.

The Merriam kangaroo rat (*D. merriami merriami*) does not appear primarily responsible for range deterioration in the Southwest [81]. Heavy grazing by livestock lowers perennial grass density and results in an increase in number of rats. The main factors favoring kangaroo rat populations were dry climate, light textured soils, and low density of perennial grasses. Ninety-five percent of their diet consisted of seeds from a large number of species, but green vegetation occasionally made up a moderate portion of the diet. Reynolds [81] concluded that on good to excellent condition range that the rats were possibly beneficial to range maintenance and improvement because of their "seed planting" characteristics.

However, the feeding habits of the Merriam kangaroo rat help spread undesirable species such as mesquite and cactus [81, 82]. The positive relation between mesquite spread and kangaroo rat activity is discussed in detail in *Chapter II*. Many of the seeds collected and buried in shallow caches are not consumed and later sprout and result in limited invasions. As mesquite replaces grass, kangaroo rats increase in number, and this in turn further increases the mesquite. Reynolds [81] concluded that control of kangaroo rats is often required before improved grazing practices can result in range improvement.

Merriam kangaroo rats in Arizona readily found and excavated large seeds even when planted to a depth of one inch. Seeds smaller than 500,000 per pound were not taken in significant amounts. Reynolds [81] concluded that range seedings using large seeds should be accompanied by rat control in areas where kangaroo rats are prevalent. The destruction of seed is more severe on seeded sites of limited acreage because of high kangaroo rat concentration.

Mice. Various species of mice frequently interfere with range seeding programs by consuming the seed. Deer mice (*Peromyscus maniculatus*) were primarily responsible for consuming 98 percent of grass seed broadcasted in one study [70]. Deer mice may also prevent successful establishment of browse plants such as bitterbrush unless the seed is treated with a repellent or the mice are poisoned [10, 36].

On fully established crested wheatgrass seedings in Utah deer mice were more numerous where herbage removal was the lightest [18]. Deer mice and pocket mice (*Perognathys* spp.) increased in numbers on the Kaibab plateau following dozing of pinyon-juniper [100]. These increases were attributed to the additional protective cover and a greater variety and abundance of food sources on the bulldozed areas. Clean cultivation has been suggested as a helpful preventive treatment in reducing mice populations before artificial seeding [91].

High mice populations have also been known to damage young forage plant seedlings or even established plants. Meadow voles (*Microtus* spp.) may damage new seedlings of bitterbrush [10, 36] and cut established forage plants to make surface runways [91]. Meadow voles commonly forage on green vegetation in the summer and woody and herbaceous plants in the winter, while harvest mice (*Reithrodontomys* spp.) consume seeds and tops of succulent vegetation [85]. Heavy concentrations of mice or voles (*Microtus* spp.) have caused extensive and severe damage to established seeded stands under a snow cover [17]. This damage consisted of chewing off old plant residues at the crown and eating the current season stem primordia and other vital parts of the crowns, resulting in retarded growth and reduced herbage yields.

Rodent Control Methods

Rodents and rabbits should be artificially controlled when required to halt or prevent serious damage to range and pasturelands, but control practices are often difficult and costly. Control should be directed toward problem species and toward eliminating or minimizing the loss rather than eradicating the species [24]. Effective rodent control depends on a knowledge of the life cycle, reproduction, and food habits of the problem species. Successful rodent control programs frequently require changes in range and pasture management practices or other improvement practices [48].

Natural control. Natural control of rodent populations depends on predation, disease, and food supply. Predators such as the coyote, bobcat, fox, weasel, badger, striped skunk, ferret, hawks, owls, eagles, rattlesnake, and gopher snake depend largely on rodents and rabbits for food. Under certain conditions these natural enemies probably exert considerable control. Sampson [85] recommended that a strong, virile population of rodent eaters, except for the coyote and the bobcat which prey on livestock, should be maintained. However, Howard [38] concluded that the balance of nature, as far as rodents and their predators are concerned, is due to self-limitation of rodents rather than to predation. He further raised the possibility that predators may increase rodent populations in some cases by reducing

461

the amount of internal stresses rather than reducing numbers by direct predation.

Hansen et al. [28] also considered that predators may find sufficient food among the "biological excess of rodents" without reducing the population below carrying capacity for rodents. However, they expressed the probability that predation slows down the rate of rodent buildup while not affecting the final population peak. From their research observations it appeared that coyotes probably exert little control on pocket gophers. Howard [37] reported that coyotes killed only about 7 percent but that rattlesnakes took about 35 percent of the annual increase of ground squirrels.

Based on studies in Kansas, Gier [22] found that rabbits make up 53 percent, other rodents 7 percent, and carrion 27 percent of the October-May diets of coyotes. He concluded the average benefits derived from coyotes in controlling rabbits and rodents exceeded the average losses due to predation on poultry, game birds, and livestock. However, he suggested "killer" coyotes be destroyed and that reduction of numbers in excess of the natural food supply may sometimes be necessary. Although increased predation on lambs has been reported in some areas as herding has been replaced by loose running of sheep under fence, some ranchers contend that lamb losses to coyotes are reduced as long as jackrabbits or other rodents are abundant [107].

Artificial control. Effective control of rodent problems on rangeland generally requires artificial control. The principal methods of artificial control include (1) poisoning, (2) trapping, (3) shooting, and (4) exclusion [91]. Trapping, shooting, and exclusion by fencing are generally practical only on small, highly productive sites or on research plots. Small-mesh wire fences extending twelve inches above the ground to twenty-four inches below the ground, or slightly deeper in lighter soils, have excluded most small rodents [91]. In 1961 the cost of constructing 760 feet of a similar fence to exclude pocket gophers was $627 [51]. Since gophers occasionally came over on the snow or burrowed under the fence, complete control required periodic poisoning around the edge of the exclosure. When the top is extended to thirty-six inches above the ground, the gopher fence will also exclude rabbits [91].

Rodenticides applied in baits or as gas are by far the most effective and widely used method available at present for controlling rodents. Although strychnine has been the most widely used rodenticide on range and pasturelands, other rodenticides rather widely used include zinc phosphide, hydrogen or calcium cyanide, carbon bisulfide, endrin, and various blood anticoagulants [105, 95]. Since all of these rodenticides are poisonous to humans, they must be handled carefully. The following are requirements for use of a rodenticide: (1) effective against target species, (2) safety to man, (3) safety to nontarget species, (4) reasonable cost, and (5) chemical properties that allow formulation and use.

The placement of rodent baits is important and depends upon the species being controlled. Baits are commonly placed in burrows, around rodent mounds, by grid or broadcast for species that forage aboveground, or in cans with punched holes for access. Aerial baiting from aircraft is effective on species that forage rather widely, such as ground squirrels [62].

Jackrabbits. The control of jackrabbits is difficult over large areas. Hunting and shooting is only partially effective. Rabbit drives allow large numbers of jackrabbits to be rounded up and killed when properly planned and enough people participate. Exclusion by fencing is impractical over large acreages. Poisoning is recommended in extreme cases.

Strychnine-treated alfalfa hay or poisoned grain effectively controls jackrabbits when put out in the winter when snow is on the ground [24, 78, 95]. Poisoned lettuce, cabbage or other vegetables are eagerly consumed after the regular herbage loses its succulence but is more difficult to handle than poison grain or grain heads. Strychnine-treated salt is effective in spring and summer when placed in one-inch-deep, one-inch-wide auger holes drilled in short 2 × 4 blocks or wood slabs. One part powdered strychnine to two parts of salt or with an added two parts of alfalfa can be mixed dry and then moistened and pressed into the holes.

Pocket gophers (Figure 108). Pocket gophers are controlled by placing strychnine or zinc phosphide baits directly in the underground burrows. On small areas or as a clean-up practice, probes can be used to locate burrows and the bait placed by hand directly into the runways. This method requires about one tablespoon of poisoned grain for each hole which is then plugged to keep light out of the runway.

On large areas such as meadows and rangeland the most practical bait placement is with the burrow-builder [109]. This machine constructs an artificial burrow at a controlled depth of usually eight to ten inches and distributes bait within the burrow in a single operation. The artificial burrow is formed by a torpedo-shaped shoe about 2.5 inches in diameter and is preceded by a rootcutting coulter. The burrow-builder is constructed either as a wheeled implement or made to attach directly to a three point hitch. The artificial

463

FIG. 108. *Pocket gophers on western ranges:* A, *range damage from excavation and placement of soil in winter casts;* B, *commercial Elston burrow-builder for underground placement of poison bait;* C *and* D, *forest land burrow-builder and its use on wildlands.* (Bureau of Sport Fisheries & Wildlife photos by Victor G. Barnes, Jr.)

464

burrows are placed to intercept the natural runways of pocket gophers and to provide ready access to them.

The burrow-builder can be used on any area where it is possible to drive a tractor and having soil that will form a clean burrow [109]. A moist, medium texture soil forms best. Hard, rocky, and sandy soils are more difficult. A heavy-duty burrow-builder for operating in difficult terrain has been successfully used on forest lands in distributing strychnine bait [3]. Its heavy construction resists breakage from large rocks and roots.

Baits should be applied in the spring or fall when soil moisture is adequate and gopher activity is high [3]. Spring application may be advantageous because gopher numbers are lowest at this season. Artificial burrows are usually spaced at intervals of twenty to twenty-five feet with an application rate of two pounds of bait per acre [109]. About fifty acres can be treated per day. Although 90 to 95 percent control is often obtained, some follow-up treatment with poison bait or trapping is generally required. Complete retreatment is generally required every two to four years depending on size of the treated area [47]. Gopher numbers may build up rapidly after control is discontinued, particularly on small areas surrounded by untreated areas.

Indirect control of pocket gophers by reducing forbs, their preferred foods, with 2,4-D has been previously discussed. However, on sheep range or other areas where the forbs are desired, this means of reducing pocket gopher numbers cannot be used, and direct control by poisoning is required. In Utah studies sheep herds grazing over treated areas frequently caved in the gopher tunnels and disrupted control effectiveness [84]. Control less than 90 percent was not considered effective. Hand baiting is not generally considered an operational tool on rangeland.

Prairie dogs. Strychnine-treated grain is widely used to control prairie dogs but is more effective when green forage is unavailable [24]. One tablespoonful of bait should be scattered around each burrow opening. Late spring and summer is the best time to poison, and one quart of prepared grain is sufficient to treat about forty burrows [55].

Fumigation is a selective and effective way of killing prairie dogs provided the burrows are closed after application [24]. Two tablespoons of liquid carbon bisulfide on a cotton swab or one and one-half tablespoons of calcium cyanide flakes placed into the tunnel are recommended treatments. Calcium cyanide is less effective when the ground is dry. Any reopened or new burrows should be retreated within a week. Fumigation is often used as a follow-up after exposure to poison bait.

Ground squirrels. Grain baits treated with strychnine or zinc phosphide are used to control ground squirrels [62, 91]. Grain bait should be hand baited on cleared ground near the burrows or aerially broadcast in strips. Treat when the squirrels are aboveground and foraging, and particularly when seeds are being collected and stored.

Kangaroo rats and mice. Kangaroo rats are easily controlled with strychnine-treated grain bait hand broadcast near the burrows [91]. Similar treatments are also effective on other seed collecting rodents including mice [78, 91]. Strychnine-treated seeds can be planted around the borders of seedings to prevent rodent movement into the seeded areas [24]. Since the seed is not damaged by strychnine, rows next to the edge can be seeded with treated seed. Such a practice can be used against most seed gathering rodents. Meadow voles are effectively controlled by broadcasting strychnine or zinc phosphide treated grain or alfalfa leaves [91].

Insect Control

Grasshoppers, Mormon crickets, army cutworms, and various other insects overgraze rangelands just as severely as cattle, sheep, goats, and deer [57]. The damage may be particularly severe on newly seeded areas [78]. Ranchers and land managers tend to overlook losses in forage and grazing capacity from insects except in drought years when the effects are obvious and destructive. Although rodents and larger mammalian herbivores have been extensively considered, only limited research has been done on the even more selective grazers among the insects which abound on these same ranges [41, 57].

Although insects compete for nearly all important livestock forage plants, many are beneficial [57]. Insects differ widely in their feeding habits. Many are selective and have food-plant preferences, but many others are opportunists and consume whatever is available. Feeding habits may change as individuals develop or adapt to changes in environmental conditions. Local migrations are common. Thus, control should be directed toward specific species when damage is imminent or occurring.

The development of new insecticides and aerial application techniques has greatly facilitated insect control on rangeland. However, recent and expected restrictions or bans on insecticides make lasting recommendations on forage crops difficult to make. Insecticides commonly used on forage lands are malathion, parathion, methoxychlor, carbaryl (sold as Sevin), diazinon, and naled [35, 96]. However, federal registrations [105], as well as special state registrations and restrictions, should be checked frequently. State departments of agriculture and state extension entomologists are suggested sources of recommendations on insecticide use.

Grasshoppers. Grasshoppers often severely damage range grasses and alfalfa, and their overgrazing is one of the principal reasons for loss of productive grasslands in many western states [103]. Grasshoppers have been and still are the major invertebrate pest of rangelands in western U.S. [33]. More than 100 different grasshopper species are found on range and pastureland in the United States. Some species prefer the same forage plants as do livestock [79]. Competition becomes more severe as damage by grasshoppers increases, and the remaining herbage is of lower nutritive value and palatability. Grasshopper problems are most common on overgrazed and mismanaged grazing lands [50] and recently seeded range.

Comprehensive studies in the Northern Great Plains revealed that most grasshoppers are selective grazers [68]. The degree of selectivity was inherent in each species, although it was considerably affected by the habitat. Most species concentrated on one plant species but often ingested other species in lesser amounts. Some species showed no distinct preferences but were capable of being highly destructive. Some grasshopper species were primary agents in damaging grasslands while others were merely indicators of habitat conditions but did not greatly affect them.

The High Plains grasshopper (*Dissosteira longipennis*) is a migratory species that has seriously damaged range grasses [108]. It is confined to the shortgrass belt, principally the blue grama association. Outbreaks have followed periods of drought and subsided when precipitation was above normal. High populations are capable of removing nearly all vegetation whether also grazed by livestock or not. Severe grasshopper infestations are more apt to occur during seasons when dry weather conditions prevail [50].

Studies on the San Carlos Indian Reservation in Arizona showed that grasshoppers competed with livestock for available forage and were at least partly responsible for the decline of shortgrass stands [71, 72, 73]. Grasshoppers consumed annually an average of 23 percent of the perennial grasses, 30 percent of the annual grasses, and 26 percent of the forbs. Herbage removal generally ranged from 8 to 63 percent, but up to 99 percent of all vegetation was removed locally in the worst drought years. Damage to perennial grasses was greatest where the vegetation was sparse and grasshopper populations were high. Areas of sparse shortgrass dominated by low-growing winter and spring annuals were the preferred habitats of early hatching grasshoppers.

Grasshoppers not only remove great quantities of range forage under certain conditions but also destroy grass seedings. In years when grasshoppers are abundant, seeding following plowing or clean summer fallowing is suggested [103]. More problems are experienced from drilling into heavily infested, unworked stubble than into clean seedbeds. In the Southwest, grasshoppers seriously injure emerging four-wing saltbush plants by eating the cotyledons and stems of many small seedlings [89]. Many seedling stands have been completely destroyed. Grasshoppers injured a higher proportion of four-wing saltbush seedlings than rabbits, but the degree of injury was generally less. High grasshopper populations have also damaged bitterbrush seedlings and even well-established plants [10, 36].

Grasshopper control is advised when the infestation reaches threatening levels. Fifteen grasshoppers per square yard on mature forage stands, or less on establishing stands, are considered economic populations in South Dakota [50]. However, knowledge of the insect species involved and the condition and current growth of the forage plants should be considered as well as total grasshopper numbers [1]. Insecticides should be applied in the late spring and summer after the dominant grasshopper species have hatched but before they start laying eggs again [103]. Baits have largely been replaced by sprays except in areas where total leaf surface is too small to intercept adequate amounts of insecticide. Malathion applied at one pound/A in ultra low volume broadcast spray is currently recommended for grasshopper control on pasture and range [102, 105]. Rates down to 8 oz per acre have also been effective [64]. Carbaryl and methyl parathion have also been effective [103].

Mormon crickets. Mormon crickets (*Anabrus simplex*) are found in the Northern Great Plains, Rocky Mountains, and the Intermountain Region [12]. Breeding grounds for the Mormon cricket are located in and near the mountain ranges of these areas. The greatest damage occurs in the salt-desert shrub and sagebrush-grass range types and on seeded pastures and croplands in these areas.

Mormon crickets are generally more localized than grasshoppers and are more readily controlled. However, because the cricket is gregarious and subject to population explosions, losses of up to 40 percent of total herbage production are not uncommon [12]. Mormon crickets prefer plants with fleshy, succulent leaves such as balsamroot, dandelion, mustards, and young Russian thistle but will take almost any plant when preferred plants are not readily available. A considerable portion of the herbage removed is potential livestock forage. Swain [93] reported that when 45 percent of the herbage was removed by Mormon crickets, it could result in a 74 percent and 98 percent reduction of forage available respectively to cattle and sheep.

Following rapid population buildups Mormon crickets migrate on the ground in large bands [12]. At common travel rates of 0.5 to 1 mile per day, bands may travel 25 to 50 miles in a single season. For effective control, poison baits are spread in front of and across the direction of migration. Either wet or dry baits can be spread by hand, or dry baits can be distributed by aircraft. Oil-on-water or fence barriers have now generally been replaced by bait barriers [12].

Black grass bugs. Black grass bugs, commonly called wheatgrass bugs, and principally *Labops hesperius, Irbisia brachycera* and *pacifica*, have attacked range seedings in epidemic numbers in western U.S. and Canada [7, 43, 57]. Grass bugs remove chlorophyll from healthy plants in early spring. Affected plants are stunted and discolored with large patches on dried yellowish, brownish to whitish foliage. Yield reductions have ranged from minimum to 60 percent [7, 61, 99], and the herbage left standing is generally unattractive to livestock. Seedhead production has also been greatly depressed.

Monocultures of introduced wheatgrasses and Russian wildrye have been more susceptible to black grass bug damage than native vegetation or mixtures of seeded and native plants, resulting in part from the mixtures harboring more predators [31]. Introduced grasses that have been most damaged in pure stands are intermediate wheatgrass, pubescent wheatgrass, crested wheatgrass, Kentucky bluegrass, orchardgrass, smooth brome, and Russian wildrye [35, 43]. Native wheatgrasses and wildryes such as western and slender wheatgrass and giant wildrye and mountain brome have also been damaged, but with little effect on the gramas, galleta, Letterman needlegrass, sheep fescue, and mountain muhly.

Revegetation methods that perpetuate some desirable native grasses and browse species seem to reduce damage to the seeded grasses [31], but such mixtures are not always desirable. Since grass straw provides the principal oviposition material, winter protection, and favorable habitat for survival, cultural practices that reduce the fall-winter carryover of grass herbage reduce black grass bug populations [49, 99]. Recommended practices have included full grazing by fall, burning, and haying. Rest grazing systems have aggravated the problem, and N, but not K and P, fertilization has increased populations of black grass bugs, apparently by increasing palatability. Reduction of early maturing grasses such as Sandberg bluegrass, cheatgrass, and bulbous bluegrass has been helpful. It has also been noted that paraquat, in curing perennial grasses prematurely, has re-

469

duced black grass bug populations through starvation and reducing egg oviposition [49].

Although losses from black grass bugs often do not justify the use of insecticides, and cultural practices can greatly alleviate the problem generally [49, 99], malathion at one pound active per acre by ground rig, half pound ULV malathion by aircraft application, or diazinon at 12 oz active per acre by ground rig has given a high degree of control [11, 35]. Effective control by insecticide may last a few years before reinvasion of black grass bugs occur [30].

Western harvester ant. The western harvester ant (*Pogonomyrmex occidentalis*) is the common reddish colored, mound-building ant of large areas of the West and is a serious pest on rangelands. It creates bare areas of 3 to 35 feet in diameter around each ant hill, maintaining this barren condition on up to 10 percent of the land area on some Intermountain ranges [54, 57]. In addition to the denuded areas, the ants forage up to 100 feet from their nests to collect seeds, which comprise their principal diet [6]. This may reduce perennial grass reseeding and natural range recovery on several times the land area actually denuded.

In northern New Mexico, western harvester ants have destroyed vegetation on up to 15 to 20 percent of some ranges [80]. In southern Idaho studies it was concluded that the harvester ant is a result rather than a cause of poor range condition [86]. Ant clearings occupied 3.5 percent of the area on Nutall saltbush sites in good condition and 5 to 8 percent on depleted sites. Clearings were often surrounded by weeds such as halogeton. However, another study of ants in a saltbush community indicated that the loss of forage from denuded sites was at least partly compensated for by increased production around the perimeter [112].

Miscellaneous insects. Army cutworms (*Corizagrostis auxilaris*) occasionally have eliminated forage plants over large areas [57]. General outbreaks of this noctuid larva have recurrently affected vast acreages of rangelands throughout the Intermountain Region. Cutworm larvae have caused as high as 90 percent seedling mortality of bitterbrush, resulting in even greater damage than that caused by grasshoppers [10]. Cutworms, comprised of larvae of certain moth species, sever seedling bitterbrush stems at or just below ground level but do not damage old, established plants [36].

Tent caterpillars (*Malacosoma* spp.) have killed mature bitterbrush, serviceberry, and true mountainmahogany over considerable acreages [57]. A sagebrush defoliator, the Aroga moth (*Aroga websteri*), damages large areas of big sagebrush [57]. This injury or destruction may be highly beneficial on some ranges but detrimental on winter

470

ranges. The Say stink bug (*Chlorochroa sayi*) in Idaho extensively feeds on ripening bitterbrush seeds by making small punctures and sucking out the contents [4]. Seeds turn black and have no germinative power.

The list of injurious insects on range could be greatly expanded. However, little is known about the extent of damage, food preferences, control practices, and benefits of control for many of them. Information should be obtained from local and state entomologists before control programs are undertaken. An excellent published source on the implications of entomology to range management is *Rangeland Entomology* [34], published by the Society for Range Management. Two publications emphasizing insects that affect important western native shrubs are Furniss [19] and Furniss and Barr [20].

Literature Cited

1. Anderson, Norman L., and John C. Wright. 1952. Grasshopper investigations on Montana rangelands. Montana Agric. Expt. Sta. Bul. 486.
2. Arnold, Joseph F. 1942. Forage consumption and preferences of experimentally fed Arizona and antelope jackrabbits. Arizona Agric. Expt. Sta. Tech. Bul. 98.
3. Barnes, Victor G., Jr., Paul Martin, and Howard P. Tietjen. 1970. Pocket gopher control on Oregon ponderosa pine plantations. J. Forestry 68(7):433–435.
4. Basile, Joseph V., and Robert B. Ferguson. 1964. Say stink bug destroys bitterbrush seed. J. Range Mgt. 17(3):153–154.
5. Bear, G. D., and R. M. Hansen. 1966. Food habits, growth, and reproduction of white-tailed jackrabbits in southern Colorado. Colorado Agric. Expt. Sta. Tech. Bul. 90.
6. Bohart, G. E., and G. F. Knowlton. 1953. Notes of food habits of the western harvester ant. Proc. Ent. Soc. 55(3):151–153.
7. Bohning, J. W., and W. F. Currier. 1967. Does your range have wheatgrass bugs? J. Range Mgt. 20(4):265–267.
8. Bonham, Charles D., and Alton Lerwick. 1976. Vegetation changes induced by prairie dogs on shortgrass range. J. Range Mgt. 29(3):221–225.
9. Bronson, Franklin H., and Otto W. Tiemeier. 1959. The relationship of precipitation and black-tailed jackrabbit populations in Kansas. Ecology 40(2):194–198.
10. Brown, Ellsworth Reade, and Charles F. Martinsen. 1959. Browse planting for big game in the State of Washington. Washington State Game Dept. Biol. Bul. 12.
11. Burkhardt, C. C. 1974. Grass bug control in grass and wheat. Wyo. Agric. Expt. Sta. Res. J. 82.
12. Cowan, F. T., and Claude Wakeland. 1962. Mormon crickets—how to control them. USDA Farmers' Bul. 2081.
13. Currie, Pat O., and D. L. Goodwin. 1966. Consumption of forage by black-tailed jackrabbits on salt desert ranges of Utah. J. Wildl. Mgt. 30(2):304–311.
14. Ellison, Lincoln. 1946. The pocket gopher in relation to soil erosion on mountain ranges. Ecology 27(2):101–114.
15. Ellison, Lincoln, and C. M. Aldous. 1952. Influence of pocket gophers on vegetation of subalpine grassland in central Utah. Ecology 33(2):177–186.
16. Fitch, Henry S., and J. R. Bentley. 1949. Use of California annual-plant forage by range rodents. Ecology 30(3):306–321.

17. Foster, Ronald B. 1965. Effect of heavy winter rodent infestation on perennial forage plants. J. Range Mgt. 18(5):286–287.
18. Frischknecht, Neil C. 1965. Deer mice on crested wheatgrass range. J. Mammal. 46(3):529–530.
19. Furniss, M. M. 1972. A preliminary list of insects and mites that infest some important browse plants of western big game. USDA, For. Serv. Res. Note INT-155.
20. Furniss, Malcolm M., and William F. Barr. 1975. Insects affecting important native shrubs of the western United States. USDA, For. Serv. Gen. Tech. Rep. INT-19.
21. Garrison, George A., and A. W. Moore. 1956. Relation of the Dalles pocket gopher to establishment and maintenance of range grass plantings. J. Range Mgt. 9(4):181–184.
22. Gier, H. T. 1957. Coyotes in Kansas. Kansas Agric. Expt. Sta. Bul. 393.
23. Goodwin, DuWayne L. 1960. Seven jackrabbits equal one ewe. Utah Farm & Home Sci. 21(2):38–39, 51.
24. Halazon, George, and E. H. Herrick. 1956. Controlling damage caused by animals. Kansas Agric. Expt. Sta. Cir. 328.
25. Hansen, R. M. 1961. Effects of herbicide on rodent populations in western Colorado. ASRM, Abstracts of Papers, 14th Annual Meeting, pp. 41–42.
26. Hansen, R. M., and J. T. Flinders. 1969. Food habits of North American hares. Colo. State Univ. Range Sci. Ser. 1.
27. Hansen, Richard M., and Terry A. Vaughan. 1965. Plains pocket gopher studies. Colorado Agric. Expt. Sta. Proj. Rpt. 164.
28. Hansen, Richard M., Terry A. Vaughan, Donald F. Hervey, et al. 1960. Pocket gophers in Colorado. Colorado Agric. Expt. Sta. Bul. 508-S
29. Hansen, R. M., and A. L. Ward. 1966. Some relations of pocket gophers to rangelands on Grand Mesa, Colorado. Colorado Agric. Expt. Sta. Tech. Bul. 88.
30. Haws, B. Austin (Proj. Coord.). 1978. Economic impacts of *Labops hesperius* on the production of high quality range grasses. Utah State Univ., Logan, Utah.
31. Haws, B. Austin, Don D. Dwyer, and Max G. Anderson. 1973. Problems with range grasses? Look for black grass bugs. Utah Sci. 34(1):3–9.
32. Hayden, Page. 1966. Food habits of black-tailed jackrabbits in southern Nevada. J. Mammal. 47(1):42–46.
33. Hewitt, George B. 1977. Review of forage losses caused by rangeland grasshoppers. USDA Misc. Pub. 1348.
34. Hewitt, George B., Ellis W. Huddleston, Robert J. Lavigne, Darrell N. Ueckert, and J. Gordon Watts. 1974. Rangeland entomology. Society for Range Management Range Sci. Ser. 2.
35. Higgins, Kurt M., James E. Bowns, and B. Austin Haws. 1977. The black grass bug (*Labops hesperius* Uhler): its effect on several native and introduced grasses. J. Range Mgt. 30(5):380–384.
36. Holmgren, Ralph C., and Joseph V. Basile. 1959. Improving southern Idaho deer winter ranges by artificial revegetation. Idaho Dept. of Fish and Game Wildlife Bul. 3.
37. Howard, Walter E. 1953. Rodent control on California ranges. J. Range Mgt. 6(6):423–434.
38. Howard, Walter E. 1961. A concept about rodent-predator relationships on rangelands. ASRM, Abstracts of Papers, 14th Annual Meeting, pp. 44–45.
39. Howard, Walter E., and Burgess L. Kay. 1957. Protecting range forage plots from rodents. J. Range Mgt. 10(4):178–180.
40. Howard, W. E., K. A. Wagnon, and J. R. Bentley. 1959. Competition between ground squirrels and cattle for range forage. J. Range Mgt. 12(3):110–115.
41. Huffaker, C. B. 1957. Fundamentals of biological control of weeds. Hilgardia 27(3):101–157.
42. Hull, A. C., Jr. 1971. Effect of spraying with 2,4-D upon abundance of pocket gophers in Franklin Basin, Idaho. J. Range Mgt. 24(3):230–232.
43. Jensen, Frank. 1970. Reseeding and *Labops*. Range Improvement Notes

16(1):6–9. (U.S. Forest Service, Intermountain Region.)
44. Johnson, Donald R. 1961. The food habits of rodents on rangelands of southern Idaho. Ecol. 42(2):407–410.
45. Johnson, Donald R., and Richard M. Hansen. 1969. Effects of range treatment with 2,4-D on rodent populations. J. Wildl. Mgt. 33(1):125–132.
46. Julander, Odell, Jessop B. Low, and Owen W. Morris. 1959. Influence of pocket gophers on seeded mountain ranges in Utah. J. Range Mgt. 12(5):219–224.
47. Julander, Odell, Jessop B. Low, and Owen W. Morris. 1969. Pocket gophers on seeded Utah mountain range. J. Range Mgt. 22(5):325–329.
48. Kalmbach, E. R. 1948. Rodents, rabbits, and grasslands. In Grass, 1948 USDA Yearbook of Agriculture, pp. 248–256.
49. Kamm, J. A., and J. R. Fuxa. 1977. Management practices to manipulate populations of the plant bug (Labops hesperius Uhler). J. Range Mgt. 30(5):385–387.
50. Kantack, B. H., Wayne L. Berndt, and P. A. Jones. 1970. Grasshopper control in South Dakota. South Dakota Agric. Ext. Serv. Fact Sheet 490.
51. Keith, James O. 1961. An efficient and economical pocket gopher exclosure. J. Range Mgt. 14(6):332–334.
52. Keith, James O., Richard M. Hansen, and A. Lorin Ward. 1959. Effect of 2,4-D on abundance and foods of pocket gophers. J. Wildl. Mgt. 23(2):137–145.
53. Kelso, Leon H. 1939. Food habits of prairie dogs. USDA Cir. 529.
54. Knowlton, G. F. 1963 (Revised). Western harvester ant control for Utah. Utah Academy of Sciences, Arts, and Letters 43(2):20–21.
55. Knowlton, G. F., and O. Morris. No Date. Controlling prairie dogs. Utah Agric. Ext. Serv. Leaflet 14.
56. Knowlton, George F., and Reed S. Roberts. 1971. Grass bugs. Utah Extension Entomology Newsletter 64.
57. Knowlton, George F., and Reed S. Roberts. 1971. Range entomology—a subject of major importance. Utah Extension Entomology Newsletter 67.
58. Koford, Carl B. 1958. Prairie dogs, whitefaces, and blue grama. The Wildl. Soc., Wildl. Monogr. 3.
59. Laycock, William A. 1958. The initial pattern of revegetation of pocket gopher mounds. Ecology 39(2):346–351.
60. Longhurst, William M. 1957. A history of squirrel burrow gulley erosion formation in relation to grazing. J. Range Mgt. 10(4):182–184.
61. Malechek, John C., Alan M. Gray, and B. Austin Haws. 1977. Yield and nutritional quality of intermediate wheatgrass infested by black grass bugs at low population densities. J. Range Mgt. 30(2):128–131.
62. Marsh, Rex E. 1968. An aerial method of dispensing ground squirrel bait. J. Range Mgt. 21(6):380–384.
63. Martin, S. Clark. 1966. The Santa Rita Experimental Range. USDA, For. Serv. Res. Paper RM-22.
64. McEwen, Lowell C. 1971. Effects on wildlife of grasshopper control with insecticides. Soc. for Range Mgt., Abstract of Papers, 24th Annual Meeting, pp. 23–24.
65. McKeever, Sturgis. 1964. The biology of the golden mantle ground squirrel (Citellus lateralis). Ecol. Monogr. 34(4):383–401.
66. McKeever, Sturgis, and Richard L. Hubbard. 1960. Use of desert shrubs by jackrabbits in northeastern California. California Fish & Game 46(3):271–277.
67. Moore, A. W., and Elbert H. Reid. 1951. The Dalles pocket gopher and its influence on forage production of Oregon mountain meadows. USDA Cir. 884.
68. Mulkern, Gregory B., Kenneth P. Pruess, Herbert Knutson, et al. 1969. Food habits and preferences of grassland grasshoppers of the North Central Great Plains. North Dakota Agric. Expt. Sta. Bul. 481.
69. Myers, Gary T., and Terry A. Vaughan. 1965. Food habits of the plains pocket gopher in eastern Colorado. J. Mammal. 45(4):588–598.
70. Nelson, Jack R., A. M. Wilson, and Carl J. Goebel. 1970. Factors influencing

broadcast seeding in bunchgrass range. J. Range Mgt. 23(3):163–170.

71. Nerney, N. J. 1958. Grasshopper infestations in relation to range condition. J. Range Mgt. 11(5):247.

72. Nerney, N. J. 1960. Grasshopper damage on short-grass rangeland of the San Carlos Apache Indian Reservation, Arizona. J. Econ. Entom. 53(4):640–646.

73. Nerney, N. J. 1961. Range grasshopper studies on the San Carlos Apache Indian Reservation. ASRM, Abstract of Papers, 14th Annual Meeting, p. 44.

74. Norris, J. J. 1950. Effect of rodents, rabbits, and cattle on two vegetation types in semidesert rangeland. New Mexico Agric. Expt. Sta. Bul. 353.

75. Osborn, Ben, and Philip F. Allan. 1949. Vegetation of an abandoned prairie dog town in tall grass prairie. Ecology 30(3):322–332.

76. Parker, K. W. 1938. Effects of jackrabbits on the rate of recovery of deteriorated rangeland. New Mexico Agric. Expt. Sta. Press Bul. 839.

77. Phillips, Paul. 1936. The distribution of rodents in overgrazed and normal grasslands of central Oklahoma. Ecology 17(4):673–679.

78. Plummer, A. Perry, Donald R. Christensen, and Stephen B. Monsen. 1968. Restoring big-game range in Utah. Utah Div. of Fish & Game Pub. 68-3.

79. Putnam, L. G. 1963. The damage potential of some grasshoppers of the native grasslands of British Columbia. Canadian J. Plant Sci. 42(4):596–601.

80. Race, S. R. 1966. Control of western harvester ants on rangeland. New Mexico Agric. Expt. Sta. Bul. 502.

81. Reynolds, Hudson G. 1958. The ecology of the Merriam kangaroo rat (*Dipodomys merriami* Mearns) on the grazing lands of southern Arizona. Ecol. Monogr. 28(2):111–127.

82. Reynolds, H. G., and G. E. Glendening. 1949. Merriam kangaroo rat a factor in mesquite propagation on southern Arizona rangelands. J. Range Mgt. 2(4):193–197.

83. Reynolds, Hudson G., and S. Clark Martin. 1968 (Rev.). Managing grass-shrub cattle ranges in the Southwest. USDA Agric. Handbook 162.

84. Richens, V. B. 1965. An evaluation of control on the Wasatch pocket gopher. J. Wildl. Mgt. 29(3):413–425.

85. Sampson, Arthur W. 1952. Range management practices and principles. John Wiley & Sons, Inc., New York.

86. Sharp, Lee A., and William F. Barr. 1960. Preliminary investigations of harvester ants on southern Idaho ranges. J. Range Mgt. 13(3):131–134.

87. Shaw, W. T. 1921. Washington's annual losses from ground squirrels. Washington Agric. Ext. Bul. 69.

88. Sparks, Donnie R. 1968. Diet of black-tailed jackrabbits on sandhill rangeland in Colorado. J. Range Mgt. 21(4):203–208.

89. Springfield, H. W. 1970. Germination and establishment of four-wing saltbush in the Southwest. USDA, For. Serv. Res. Paper RM-55.

90. Stoddart, Laurence A., and Arthur D. Smith. 1955. (Second Ed.). Range management. McGraw-Hill Book Co., Inc., New York.

91. Storer, Tracy I., and E. W. Jameson, Jr. 1965. Control of field rodents on California farms. California Agric. Expt. Sta. Cir. 535.

92. Stubbendieck, Ronald Case, Kathie J. Kjar, and Michael A. Foster. 1979. Plains pocket gophers—more than a nuisance. Rangelands 1(1):3–4.

93. Swain, Ralph B. 1943. Nature and extent of Mormon cricket damage to crop and range plants. USDA Tech. Bul. 866.

94. Taylor, W. P., and J. V. G. Loftfield. 1924. Damage to range grasses by the Zuni prairie dog. USDA Bul. 1227.

95. Thomson, W. T. 1976–1977 (Rev.). Agricultural chemicals. Book III. Fumigants, growth regulators, repellents, and rodenticides. Thomson Pub., Fresno, Calif.

96. Thomson, W. T. 1977 (Rev.). Agricultural chemicals. Book I. Insecticides, acaracides, and ovicides. Thomson Pub., Fresno, Calif.

97. Tiemeier, Otto W., M. F. Hansen, Monroe H. Bartel, et al. 1965. The black-tailed jackrabbit in Kansas. Kansas Agric. Expt. Sta. Tech. Bul. 140.

98. Tietjen, Howard P., Curtis H. Halvorson, Paul L. Hegdal, and Ancel M.

Johnson. 1967. 2,4-D herbicide, vegetation, and pocket gopher relationships, Black Mesa, Colorado. Ecology 48(4):634–643.

99. Todd, J. G., and J. A. Kamm. 1974. Biology and impact of a grass bug (*Labops hesperius* Uhler) in Oregon rangeland. J. Range Mgt. 27(6):453–458.

100. Turkowski, Frank J., and Hudson G. Reynolds. 1970. Response of some rodent populations to pinyon-juniper reduction on the Kaibab Plateau, Arizona. The Southwest Naturalist 15(1):23–27.

101. Turner, George T. 1969. Responses of mountain grassland vegetation to gopher control, reduced grazing, and herbicide. J. Range Mgt. 22(6):377–383.

102. USDA, Agric. Res. Serv. 1967. Suggested guide for the use of insecticides to control insects affecting crops, livestock, households, stored products, forests, and forest products. USDA Agric. Handbook 331.

103. USDA, Agric. Res. Serv. 1977 (Rev.). Grasshopper control. USDA Farmers' Bul. 2193.

104. USDA, Rocky Mtn. Forest & Range Expt. Sta. 1969. Forestry research highlights, 1969. Annual Report, USDA, Rocky Mtn. Forest & Range Expt. Sta.

105. U.S. Environmental Protection Agency. 1972. EPA compendium of registered pesticides, 5 vols. Office of Pesticides Programs, U.S. Env. Prot. Agency, Washington, D.C. (Updated by supplements periodically.)

106. Vorhies, C. T., and W. P. Taylor. 1933. The life history and ecology of jackrabbits (*Lepus alleni* and *L. californicus* spp.) in relation to grazing in Arizona. Arizona Agric. Expt. Sta. Tech. Bul. 49.

107. Wagner, Frederic H. 1964. The jackrabbit in its western habitat. Utah Farm and Home Sci. 24(3):64–65, 78–79.

108. Wakefield, Claude. 1958. The High Plains grasshopper—a compilation of facts about its occurrence and control. USDA Tech. Bul. 1167.

109. Ward, A. Lorin, and Richard M. Hansen. 1960. The burrow-builder and its use for control of pocket gophers. USDI, Fish & Wildl. Serv. Special Scientific Report: Wildlife 47.

110. Ward, A. Lorin, and James O. Keith. 1962. Feeding habits of pocket gophers on mountain grasslands, Black Mesa, Colo. Ecology 43(4):744–749.

111. Westoby, Mark, and Frederic H. Wagner. 1973. Use of a crested wheatgrass seeding by black-tailed jackrabbits. J. Range Mgt. 26(5):349–352.

112. Wight, J. Ross, and James T. Nichols. 1966. Effects of harvester ants on production of a saltbush community. J. Range Mgt. 19(2):68–71.

113. Wood, John E. 1965. Response of rodent populations to controls. J. Wildl. Mgt. 29(3):425–438.

114. Wood, John E. 1969. Rodent populations and their impact on desert rangelands. New Mexico Agric. Expt. Sta. Bul. 555.

Chapter 12

Range animal handling facilities

Range Water Developments

Planning water developments. Few ranch investments give annual returns as high as needed stockwater developments [68]. Water developments may allow increased stocking rates by lengthening the season of use, spreading usage more evenly over the range, or opening up more range to grazing. Stocking rates cannot be evaluated only in terms of forage since it must be accompanied by adequate drinking water for the grazing animals. Additional watering places are often required to handle the number of grazing animals the forage supply will carry. Some range areas produce good range feed but are waterless [13]. Maximum livestock gains can be obtained only when both the forage and the water supply are adequate.

There are various types of water developments. These include *natural* water supplies such as lakes, ponds, streams, springs, and seeps. These also include *man-made* developments such as wells, reservoirs, dugouts, sand tanks, and catchment basins. A combination of two or more types of water developments is often more advantageous than one type only. Temporary water sources, although less dependable, are often needed to relieve pressure on permanent or semipermanent water sources.

Stockwater problems arise on the range:

1. When there are too few watering places (Figure 109).
2. When water yield or storage, or both, is inadequate.
3. When water sources are poorly distributed.
4. When water developments are wasteful because of leakage or high evaporation.
5. When there are erosion problems at present facilities.

Plans for developing adequate water for drought years cannot be postponed until drought begins. Ranchers and other range managers must carry out year-to-year programs of developing and maintaining water supplies. During drought periods water shortages are often as severe as forage shortages or even more severe. When water is short, ranchers may be forced to move their stock from the range before the forage is fully grazed. Even more common is a heavy concentration of animals at remaining water sources after the less dependable springs and reservoirs dry out.

On the other hand, care must be taken that additional water development is not used to crowd more livestock onto a fully stocked range. There must be adequate forage to go with the new development. Water development plans must consider the adequacy of water presently on the range, how more can be developed if needed, and what grazing adjustments should be made to better use existing water.

The planning of range water developments must include provisions for future maintenance. Structures should be planned for minimum maintenance. However, plans must be provided for cleaning watering structures; repairing leaks in pipes, troughs, and storage structures; repairing damage caused by grazing animals, rodents,

FIG. 109. *This large blowout in the Sandhills of Holt County, Nebraska, resulted from heavy concentration of livestock around a single watering place.* (Soil Conservation Service photo)

478

floods, or vandals; and repairing moving parts. An application for a water right must be filed and approved in many states before particularly larger range water developments can legally be begun. The respective state laws concerning developing additional stockwater should be checked before initiating work.

Water development and grazing distribution. The location of water developments on rangeland is important in controlling the movement, distribution, and concentration of livestock. Improper distribution of grazing frequently results from improper distribution of watering places. Cattle will graze an area close to water again and again rather than travel long distances to better forage. This results in deterioration of forage resources near the water supply and wastes forage at long distances from water. The excessive travel associated with this inefficient harvesting of forage is harmful to the grazing animals and to the rangeland as well.

More watering places are required on rough than on level or rolling terrain. Cattle should not have to travel more than one-quarter to one-half mile from forage to water in steep, rough country, or more than one mile on level or gently rolling range [86, 87, 88]. However, sheep and horses can travel longer distances. On highly productive range or pasture, it may be realistic to provide at least one watering place per section and possibly two if water can be developed at low cost. Another rule of thumb that has been proposed is that at least one watering facility should be provided for every fifty to sixty animal units for full growing season use.

Drinking water is also an important consideration in big game range development. Water is considered a major controlling factor of game populations in arid areas [92]. For example, dependable, year-round water is an essential part of successful management of bighorn sheep in the Southwest [26]. The development of water increases available bighorn range, lowers disease and predation potential, reduces pressure on other limited watering places, and allows more uniform range utilization. In arid mountainous country in the Southwest a watering place every five miles in favorable habitat was considered minimum for bighorn sheep [26]. Studies in western Utah indicated that water development can encourage better distribution of antelope where natural water sources are limited and particularly during dry seasons or drought years [3].

Along the Missouri River breaks in Montana, deer grazing decreased at distances over one mile from water, elk seldom grazed over a mile from water, and cattle grazing was mostly within three-quarters of a mile from water [48]. Grazing distance from water was more restricted for all three species in summer and fall than in win-

479

ter and spring. Water distribution was concluded to have little effect on mule deer distribution in the central Intermountain Region as long as the feed is green and succulent [39]. However, during periods when forage is dry, particularly in the fall, deer may trail to water and tend to concentrate there. Water may be an important factor in distribution of deer on desert ranges, particularly in drought years.

Recommendations for game range improvement in New Mexico [40] included the following provisions for drinking water: (1) that water be available in all pastures through all seasons for antelope, (2) that at least one watering place per four sections be provided for deer, and (3) that elk and bighorn sheep are not more than one mile from water. Drinking water is also important to upland game birds such as sage grouse, quail, turkey, chukar, and mourning dove.

In some areas only temporary water can be developed. These water sources are less reliable and may be dry when needed most. In such areas or when temporary sources are interspersed with inadequate sources of permanent water, livestock should be grazed in areas near temporary water when the water is available. This will allow nearby ranges to be used more effectively and help relieve congestion around permanent water. A system of rotating access to permanent water sources has been used in the Southwest to enable deferment of grazing on unfenced ranges [52]. This seasonal opening and closing of watering places was effective in reducing grazing intensity near water sources only if the ranges were otherwise properly stocked, the watering places were not too close together but reasonably well distributed, and the pastures were not too small.

Water requirements. Daily water requirements and frequency of watering of range livestock and big game should be considered. Normally, 8 to 10 gallons per day is ample for mature cows, 0.75 to 1 gallon per day for range sheep and goats, and 10 to 12 gallons per day for horses [77, 87]. Lactating cows on dry summer range in Oregon drank 12.4 to 16.7 gallons per day, and 3½-month-old calves consumed about 1.1 gallons per day [75]. In order to meet maximum free choice water consumption and allow some evaporation, the United States Forest Service [86] recommends that water development plans consider 12 to 15 gallons per day for cattle and horses and 1 to 1.5 gallons per day for ewe-lamb pairs. Live weight and animal condition, stage of production, amount of activity, and environmental factors will affect amount of water consumption.

High temperatures, low humidity, high salt or protein content in the diet, dry feeds, and increased feed intake all increase water consumption [97]. Green succulent forage intake decreases water consumption. On desert range in the Intermountain Region ewes were found to drink daily about 1 gallon in the fall, 0.75 gallons in the winter, and 0.50 gallons on spring foothill range [36]. Ewe-lamb pairs drank up to 1 gallon daily on summer range but much less when forage was succulent and morning dews were frequent. On dry range in late winter ewes drank up to 2 gallons daily when the weather was warm, but they seldom required any water when snow was available. Cattle on a salt-meal mix consume an additional one to two pounds of salt per day [12]. For each pound of salt eaten in salt-meal mixes, an additional 5 gallons of water is commonly consumed and should be provided.

Free choice consumption of water without attempts to limit water intake should normally be permitted. Since water is both a nutrient, a medium for metabolic functions in the body, an important tissue constituent, and provides the means of waste disposal, adequate consumption is necessary for animal health and production. Daily animal gains are directly related to the amount and quality of feed consumed each day, but feed consumption can be sharply reduced by inadequate water intake. Restricting water intake by failing to provide adequate water supplies sharply reduces milk flow in lactating females, reduces gains in both weaned and unweaned animals, and may contribute to or even cause death losses if severe [68, 88].

Restricting water intake of 920-pound steers has been studied in Utah under drylot conditions with controlled air temperatures of 30° to 45° F [11]. At the end of twenty-six days, steers with free access to water, 25 percent restriction, and 50 percent restriction weighed 940, 920 and 870 pounds, respectively. All three groups were then allowed free access to water for a second period of twenty-four days. By the end of the second period, both the free access group and the 25 percent restricted group averaged 990 pounds, but the 50 percent restricted group still weighed 18 pounds less. It was concluded that 25 percent reduction appeared only to reduce intestinal fill of water but that the 50 percent reduction reduced actual body weight [11].

Allowing mature cattle on Oregon high desert range consisting of seeded crested wheatgrass and native range to drink every other day or requiring them daily to trail one or two miles to water reduced water intake 25 to 35 percent of that consumed under unlimited access to nearby water [75]. Although dry cows were not adversely affected, it was concluded that watering every other day should not be done if the herd consists of lactating cows with calves. The suckling calf was found to be more susceptible to water stress through reduced milk production and was less likely to travel to water. It was also concluded that adequate water will be consumed for good livestock performance if stockwater is located to assure even utilization of range forage.

Suckling calves without direct access to water in the Oregon studies [75] gained 0.4 lb. less daily over a 60-day period. Yearling heifers watered every 72 hours during July lost weight but survived, but became subject to water intoxication after 96 hours without water. Reduced water intake by pregnant heifers forced to trail considerable distances to water reduced gains but seemed not to affect reproduction. It was suggested that dehydrated cattle should be watched closely for a period exceeding four hours after return to water to prevent water intoxication from overconsumption.

The practice of trailing sheep three to five miles to permanent snowbanks every second to third day during the winter used to be common on desert range in the Intermountain Area [36]. During a forty-day winter trial at the Desert Range Station in western Utah, ewes watered every day gained 3.4 pounds, those watered every second day gained 0.8 pounds, but those watered every third day lost 6.0 pounds [36]. It was also noted that a better percent lamb crop and less abandoned lambs were associated with more frequent watering. Sheep watered daily drank slightly less water on a weekly basis than those watered every other day.

In more recent studies in the same area it was concluded that watering of sheep on alternate days may be an acceptable management practice during periods of cool temperature (below 40° F) or as an emergency measure up to 78° F [14]. It was also concluded that soft, wet snow was as effective as open water for ewes on winter range [10]. However, caution was suggested in relying on frozen or grainy snow or allowing access to snow for only two hours or less per day.

Water consumption has also been related to halogeton poisoning of sheep [37]. Where water intake with range sheep is restricted, sheep generally decrease intake of forage. Following watering after an extended period without water, hungry sheep readily graze whatever is available including halogeton. Most sheep losses from halogeton have been associated with this situation. Toxic oxalates from halogeton are also more rapidly absorbed from a nearly empty rumen [37]. Salt-meal mixes are commonly fed as an indirect method of controlling urinary calculi. The increased water intake associated with higher salt levels appears to aid in reducing the concentration of minerals such as silica in the urinary tract, flushing them away, and thereby reducing the incidence of urinary calculi.

In western Utah, antelope did not drink water even if readily available when forbs were succulent and their moisture content was 75 percent or more [3]. However, they drank up to 3 quarts per day during extremely dry periods. Nichol [58] estimated that the average daily water consumption of mule deer is 1 to 1.5 quarts per day per

100 pounds of live weight in winter and twice that amount in the summer. However, the need for water in the free state for most big game species appears to be at least partially offset by the consumption of succulent forages.

Water quality. Stockwater of good quality is clear and colorless, includes no disease organisms, has no undesirable flavors or odors, and is free of objectional gasses and minerals [17]. If in doubt, water should be tested for disease organisms, algae, dissolved gasses, and the more common mineral salts. Stockwater should be kept as clean and as free of debris and decay as possible. Decomposition of dead fish and other animals may not only make the water objectionable to livestock but toxic as well. Stagnant water, even if nonpoisonous, often decreases water intake by large animals. Providing a flat board to float on the water generally prevents birds from drowning and decomposing. Also, tying a piece of canvas or leather on the side of the tank and hanging into the water will allow rodents that fall in to crawl out and escape [33].

Filamentous green algae (pond scum) and mosses growing on ponds and tanks, although troublesome, will not injure range animals. However, blue-green algae sometimes produces a poisonous "water bloom" of greenish hue which can be deadly. Copper sulfate pentahydrate (commonly shortened to copper sulfate or called blue vitriol or blue copperas) is a blue crystal and the most commonly used algaecide [85]. It is moderately toxic in pure form (LD_{50} 300 mg/kg) but is safe on most fish except trout at 1 p.p.m. in the water. It is safe for drinking water at 2.5 p.p.m. or even up to 7.5 p.p.m. Blue-green algae is killed at 0.1 to 0.5 p.p.m. in the water. Filamentous green algae and most other algae are killed at 0.5 to 1.0 p.p.m. in soft water or 1 to 2 p.p.m. in hard water. Endothal at 0.05 to 0.2 p.p.m. is also an effective algaecide and considered safe for fish and drinking water [85].

One level teaspoon of copper sulfate per 1,500 gallons of water or 1 oz per 8,000 gallons is equivalent to 1 p.p.m. The copper sulfate can be placed in a cloth bag and dragged through the water or dissolved ahead of time in water and sprayed on the water. For a spray mix, dissolve 2 oz of copper sulfate in a pint of water and apply 3 tablespoons of this mixture to each 1,000 gallons of water for 1 p.p.m. [32].

Reduced gains and even death losses may result from range livestock drinking salty water from seeps and dugouts. In South Dakota, levels of soluble salts up to 7,000 p.p.m. caused no apparent harm to livestock, but they drank less of the salty water [19]. However, it was indicated that toxic effects can be expected from concentrations of

10,000 p.p.m., regardless of the type of salts. The mineral salts involved in South Dakota were principally sodium sulfate, sodium chloride, and magnesium sulfate.

Toxic amounts of salts in water from seeps and dugouts come largely from ground water rather than runoff [22]. Groundwater should be analyzed for total salt content in questionable areas before seeps or dugouts are constructed or before shallow wells are dug. Salt concentration increases when drought reduces water flow and evaporation is high. Placing seeps and dugouts so that rapid spring runoff will partially flush them out may have merit. Manually flushing out storage tanks, drinking tanks, and troughs is also suggested to reduce dissolved mineral concentrations due to evaporation. Precautions must be taken to insure that salt water pumped in oil and gas fields is safely disposed of and not allowed to contaminate animal drinking water. Fortunately, livestock do not normally drink harmful amounts of salt water if good water is readily accessible.

Natural Water Supplies

Lakes, ponds, and streams. Natural water supplies such as perennial lakes, ponds, and streams are usually reliable and require low maintenance. These water supplies generally permit direct use by livestock. However, their occurrence is limited or inadequate on

FIG. 110. *Perennial stream of clean water greatly increases the value and usability of this mountain range.* (Union Pacific Railroad photo)

most ranges and their use may be prevented by land ownership or location in deep canyons. The value of ranch lands is greatly increased if there are perennial streams flowing through the center of each range unit (Figure 110). On public grazing lands the use of such water supplies by livestock often must coordinate with water-based recreational use and culinary use of the water.

Natural water sources can often be improved by (1) fencing off bogs and quicksand, (2) building additional access ways into canyons, (3) relocating improperly placed fences, (4) controlling poisonous plants near water sources, and (5) artificially increasing the

FIG. 111. *Spring development. A, muddy spring site greatly in need of development* (Soil Conservation Service photo); B, *collection box and lead-out pipe provided at a spring site.*

water storage or flow. Ephemeral streams fed by snowmelt or rainstorms require a reservoir or other storage facility to prolong water availability.

Springs and seeps. Springs and seeps are formed by ground water emerging naturally through cracks in the rock or porous earth strata. Springs generally emerge on hillsides, and the inflow is in a downward or horizontal direction [28]. A seep, also referred to as a depression spring, occurs where the ground surface dips into the water-conveying stratum or water table, and the inflow rises vertically. There are obviously many gradations between the true spring and the seep.

Springs and seeps can often be developed with little expenditure except labor and generally provide a dependable water supply for range animals. Even a small flow may be worth developing if the spring is strategically located. A spring that yields only one gallon every six minutes should provide enough water for 25 head of cattle, 300 head of antelope, or 240 goats.

Springs which have a steady flow, produce water in excess of maximum drinking water needs, and are situated for easy access and maintenance of water quality may need no further development. However, most springs can be improved by varying degrees of development (Figure 111). Complete spring development consists of locating the true water-bearing outcrop, cleaning out the area where the water emerges, and providing means for collecting and utilizing the outflow. Water can be concentrated by use of a surface ditch, perforated tile, unglazed open-joint tile, or a rock-filled trench [86]. The central collection area made by excavation should be walled up with a metal or concrete box, the latter either prefabricated or constructed at the site. On impervious rock or soil, a concrete or stone curbing can be used to collect the flow. Or an adequate collection basin can sometimes be made by digging toward the spring source and filling with rounded rock.

It is generally better to cover the collecting basin and to pipe the water a short distance to a conveniently located tank or trough rather than allow livestock direct access to the basin. If the site is boggy or if a removable cover is used, the spring should be fenced to prevent trampling and contamination by livestock. Also, the spring should be protected from surface runoff and flash floods. Periodic checks should be made to insure there are no leaks or blockage in the basin, pipe, or trough. The collection basin should be cleaned when needed, and a removable top expedites this job.

Water yield from a spring can often be increased immediately by removing water-loving plants and brush from the immediate vicinity

FIG. 112. *Stockwater wells.* A, *well in Cherry County, Nebraska, with windmill, circular tank, and overflow reservoir* (Soil Conservation Service photo); B, *well in Kane County, Utah, with diesel engine in pump house, storage tank, drinking tank for cattle, and drinking trough for sheep (not shown).*

[4]. Most brush species can be killed with 2,4,5-T or silvex sprayed at the rate of two pounds acid equivalent per acre. Cattails and rushes are most effectively killed with dalapon at the rate of fifteen pounds acid equivalent per acre. Water-loving plants such as cattails and rushes can also be used as plant indicators in locating springs and seeps that do not surface. Rocky Mountain iris (*Iris missouriensis*) in dense stands is considered particularly useful as an indicator plant with falsehellebore being slightly less reliable [24]. However, both may indicate run-in water areas rather than underground water sources.

Seeps are somewhat more costly and more of a gamble to develop than are springs. They are usually found on flat, often boggy terrain where it is difficult or impossible to pipe water away for better access. Direct access by stock may be a hazard on boggy sites where the animals may become bogged down. Using crushed rock or timbers on access points will improve the footing. To develop a seep, excavate an open water source with a bulldozer, back-hoe, or dragline or by dynamiting where the water table is near the surface.

**Man-made
Stockwater Supplies**

Wells. Wells equipped with a windmill are the most common type of water development in the Great Plains and other areas where wind is reliable (Table 112). Although wells are more expensive in initial cost, they normally furnish a dependable year-round water supply. Where the water table is deep and no perennial lakes or streams exist, wells may be the only feasible source of permanent water.

Wells have a distinct advantage in that they can be drilled near the forage supply if adequate ground water is present. They provide a safe place for livestock and game animals to water thoughout the year. In addition to supplying water on the site, wells provide a source of water for piping or hauling to adjacent range. Except for flowing wells, powered pumps are required. Greater use is now made of electric, gasoline, and LP gas engines as a source of power. Even where windmills are provided and are generally reliable, auxiliary power is desirable to meet prolonged calm periods. However, attaching spring counterbalances to the plunger rods will materially reduce wind velocity needed to drive windmills.

Throughout the Midwest and much of the Central and Northern Great Plains, ground water is plentiful. However, in many areas of the Intermountain and Southwest, reliable water sources are found only at deeper depths if at all. Although the yield of most stock wells need not exceed five to ten gallons per minute, drilling wells in high-risk areas should be thoroughly investigated before digging begins [62, 63]. Advice and assistance in selecting well sites can be obtained from experienced well drillers, state geologists, or United States Geological Survey or Soil Conservation Service geologists.

Reconnaissance methods of examination are used by the United States Geological Survey in locating sites with aquifers at reasonable drilling depth and pervious enough to supply the needed water yield [62]. Use of the following evidence in locating promising drilling sites has given 85 to 90 percent success upon drilling:

1. Performance and depth of existing wells in the vicinity and information about the aquifers tapped.
2. Features such as seeps, springs, streams, and phreatophyte growth, with the latter indicating a shallow water table up to fifty to seventy feet deep.

3. Permeability of the aquifer and favorable situation with regard to recharge.

Drilling horizontally is a relatively new method of well development used in the Southwest and mountain states [93, 94]. Success depends upon intercepting water flow or a perched water table by drilling horizontally into a hillside. The resulting "cased spring" has some definite advantages over a vertical well and many other range water supply systems:

1. The water loss is minimized since the flow can be controlled by float valve or turned off when not used.
2. Maintenance is very low.
3. Cost is generally the least expensive method of water development except for natural springs.
4. Greater success ratio than developing conventional springs or vertical well drilling in difficult areas.
5. Provides an excellent source of sanitary water.

The cost of drilling, casing, and providing automatic watering facilities in Arizona has averaged $500 (1969) per horizontal well at one-half to one-fifth the average cost of vertical wells [93]. Site selection for horizontal drilling has been based on such indicators as old springs, seeps, the presence of water-loving plants, and geological formations.

FIG. 113. *Stockwater reservoir in Keya Paha County, Nebraska, with dam site and spillway fenced from livestock.* (Soil Conservation Service photo)

Reservoirs and dugouts. Both reservoirs and dugouts are used to intercept overland flows of water and provide storage while making the water available to livestock and big game. A reservoir is formed by means of an earth fill across a narrow channel or valley, and most of its capacity comes from the natural basin formed behind the earth fill (Figure 113). The dugout, also referred to as a "water hole" or charco, differs from the reservoir in that most of its capacity comes from excavation (Figure 114). The dugout is adapted to flatlands where some overland flow occurs. It is best suited to locations where a comparatively small water supply is sufficient and where the soil is impervious. Diversion ditches leading into the dugout are often required.

Unless fed by perennial streams, seeps, or springs, reservoirs and dugouts generally provide only temporary or seasonal sources of wa-

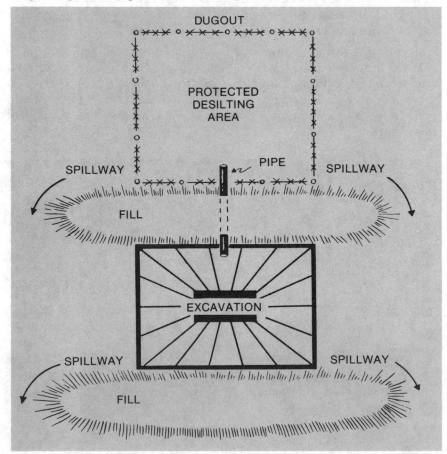

FIG. 114. *Diagram of dugout provided with a desilting area and protection from overland flooding.* (Adapted from University of Wyoming drawing [9])

ter. Both the reservoir and the dugout are relatively inexpensive to construct. They are, however, dependent on surface runoff or seepage, so may be empty when most needed in dry seasons. Proper location and construction in areas receiving moderate amounts of rainfall can materially extend their season of effectiveness. Both require a drainage area of adequate size, a low ratio of surface area to storage capacity, and protection against excessive silting.

Each reservoir must be provided with a stable spillway big enough to carry away excess water. The spillway exit should be three to four feet lower than the top of the embankment to prevent overtopping by waves. Providing desilting areas above the reservoir or dugout or channeling heavy silt-laden flows around the structures is often required for reasonable life expectancy of the development. Fencing off the reservoir and piping the water to a trough located outside the fence is often a desirable practice. Detailed reservoir construction plans are given in several printed sources [8, 9, 28, 86]. Recommended maintenance practices for small earth dams include [8]:

1. Keep spillway clear of debris and barriers and repair eroded parts.
2. Provide a floating log boom close to the dam or rip-rap with stone to prevent wave damage on large reservoirs.
3. Control rodents from burrowing into the earth fill and spillway area.
4. Check for seepage and repair immediately.
5. Remove marginal brush, trees, or other large phreatophytes as needed.

The principal problems associated with reservoirs and dugouts as range water developments are the uncertainty of filling, poor quality of water, and losses through percolation, seepage, and evaporation [41]. Developments should be placed only where there is reasonable expectation of adequate frequency, quantity, and quality of water flow. Reducing the surface area while deepening the reservoir area is a principal means of reducing evaporation. A monomolecular alcohol has been used on open reservoirs with some success in reducing evaporation, and a water saving of up to 25 percent or more has been reported [89].

Because of the prevalence of seepage losses through or under the earth fill or deep percolation through the bottom of the reservoir site, special consideration should be given to these, both in construction and subsequent maintenance. Selecting sites with deep, heavy clay soils where possible, careful preparation of the reservoir site and particularly the earth fill area of the dam, using water-tight material in the embankment and in the reservoir bottom and compacting well the earth fill will prevent most seepage losses. The site should be drained and the vegetation, stumps, debris, and loose rock removed before construction begins. Small patches of gravel or sand in a reservoir or dugout should be covered with a blanket of clay soil four to six inches deep or deeper and crevices and holes filled in and thoroughly compacted [67].

Where the soil contains adequate amounts of silt and clay, rolling with a sheepsfoot roller several times often effectively seals the soil against seepage. Salting livestock in the basin before it fills with water will help reduce seepage losses because the animals will trample and puddle the soil. Where additional treatment is required and the soil contains at least 50 percent of silt and clay, applying chemical dispersing agents helps in puddling and sealing the soil [67]. Sodium chloride (common salt) or sodium polyphosphate should be spread at rates of 0.20 to 0.33 pounds or 0.05 to 0.10 pounds per square foot respectively, disked into the soil, and then compacted. In other trials with calcium-aggregate soils, seepage was more effectively controlled by sodium carbonate than sodium chloride or sodium phosphate when mixed into the top three or five inches of soil in the bottom of the reservoir [66].

Where construction of reservoirs or dugouts on sandy or gravelly soil cannot be avoided, lining the water basin and slightly above the maximum water line with bentonite or other special clays or with flexible, impervious membranes can be considered. Reservoir basins requiring such special treatment should be fenced from livestock and big game to prevent erosion at the waterline and protect the linings, and the water should be piped out [42, 67, 98]. Bentonite clay is a hydrous silicate of sodium composed chiefly of montmorillonite which swells up eight to twenty times its dry bulk when saturated. It is applied at rates of one to three pounds per square foot depending upon the swelling ability of the particular batch [67]. When thoroughly mixed by disking into the top six inches of the soil and then packing with a smooth roller, it is capable of expanding when wet, filling up the soil interstices, and nearly eliminating seepage losses [71, 73]. However, bentonite is only 75 to 80 percent effective when used on sites with widely fluctuating water levels and prolonged dry periods [71].

Playa sediments high in fine clays and sodium also look promising for use as reservoir sealants [70]. Most playa sediments exhibit low swelling (120 to 360 percent) compared to commercial bentonites and should be applied at least two inches deep in the bottom and walls of the reservoir [72]. Burying the playa sediments seal beneath an earth cover and puddling by livestock trampling are additionally helpful.

492

Although plastic and butyl rubber liners have been effectively used on irrigation reservoirs where the sites were well prepared and the linings were covered with soil to a depth of six inches or more [42, 67], their use has not proven practical generally with small stockwater reservoirs [88]. Lining water basins with concrete, soil cement, or asphalt paving is generally too costly to consider.

Sand tanks. A sand tank is made by placing a small dam across a sand wash. To minimize seepage, the dam should be built within a rock-bound channel and bonded to bedrock. The sand tank takes advantage of the fact that water can be stored in the sand trap above the dam.

Twenty-five to 30 percent of the volume of sand is available for water storage [79]. A collection box of loosely joined rocks should be constructed at the base of the upper side of the dam and the water piped from the collection box through the dam to an appropriate watering place below the sand tank. Sand tanks have an advantage over reservoirs in that they greatly reduce evaporation and avoid silting-in following heavy storms.

Rain traps. Water harvesting by rain traps, also variously known as catchment basins, paved drainage basins, and guzzlers, is now being used in low rainfall areas and on coarse-textured or shallow soils where cheaper and more dependable sources of water are not available. A rain trap consists of a sloping, slightly concave, water-tight collecting area and a closed reservoir for storing the collected water (Figure 115) [46]. This water development collects only the precipitation falling directly on the collection area. The catchment or collecting area is ideally located on smooth ground with a 5 to 10 percent slope so that the rain water will collect in the lower corner. The edges of the collecting area are elevated to confine the water falling on the site while preventing run-in. Where sheets of liner materials are used, the outer edges should be buried in trenches on the outside of small dikes. Additional diversion dikes may be needed above the collecting sites to prevent overland flow of water which would carry silt and debris into the rain trap.

Materials used in paving or treating collecting areas of rain traps have included the following [13, 16, 21, 41, 86]:

1. Galvanized iron sheets or corrugated metal roofing.
2. Polyethylene and vinyl plastic sheets.
3. Butyl rubber sheets.
4. Asphalt-treated jute or fiberglass.
5. Heavy weight roofing paper.
6. Concrete about four inches thick or good quality soil cement.
7. Asphaltic road mix rolled about four inches thick.

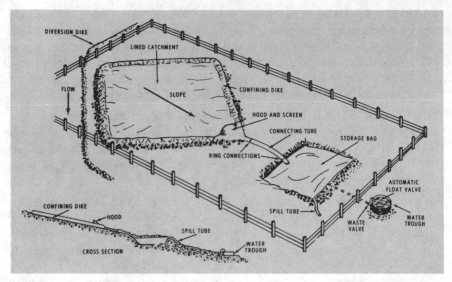

FIG. 115. *Diagram of rain trap installation for collecting and supplying water to range animals.* (Drawing from Lauritzen and Thayer [46])

8. Paved highways.
9. Fine asphaltic emulsion or paraffin wax sprayed on the ground to stabilize the soil and repel water; or a sodium dispersant mixed with the soil.
10. Natural rock outcrops.

Natural rain traps consisting of rock paved slopes with rock tanks at the base occur through the desert areas of the West. Many such rain traps have been used since pioneer times with varying degrees of modification. Lauritzen [46] concluded that butyl rubber 0.03 inch thick was the ideal artificial ground cover since it resists aging from heat, sunlight, and ozone and is also tough, flexible, and resistant to most chemicals. Although asphalt-treated jute and plastic were cheaper, they were short-lived, easily damaged, and readily penetrated by weeds [41]. Collection basins must be tightly fenced to exclude livestock and preferably big game also in order to protect the artificial basin covers.

Rain traps for intercepting and storing water were concluded to be practical on arid and semiarid range and also on farms [46]. However, the cost of the water produced is generally more costly than that from wells or reservoirs but often less costly than water hauling. Each inch of precipitation falling over a 1,000 square foot catchment basin would theoretically yield 623 gallons of water. Using 1960 figures it was estimated that the cost of water yield from a rain trap would be about $3.10 per 1,000 gallons or $.95 per cow month in a twelve-inch precipitation zone [41]. This was based on a ten-year life expectancy of the rain trap, initial cost of a ground cover of $1.00 per square foot, and a seven percent interest charge on investment. Recent (1975) costs of water harvesting have ranged from as low as $0.20 per 1,000 gallons using rock outcroppings up to over $6.00 per 1,000 gallons when long-lasting materials such as concrete are used [21].

The rain trap is best adapted for use in areas of summer rain showers. Since the rain trap is unable to collect water effectively during freezing weather, its use is generally limited to summer except in areas of mild winter climates or when the water is collected and stored in summer and dispensed for winter use. For this reason median seasonal rainfall is a better index of water yield potential than average annual precipitation. If grazed during the seasonally dry periods when no refilling can be expected from precipitation the storage capacity associated with the rain trap must be sufficient to store the full grazing season needs. On yearlong range in the Southwest, a good correlation was found between precipitation, and thus water storage in rain traps, and forage production and grazing capacity [35].

The trick tank is a modification of the rain trap in which the collection basin is elevated and the storage tank is placed directly below [60]. It provides water for livestock and wildlife, but the elevated rain collector also provides shelter for livestock or supplemental feeds. The elevated rain collector is generally smaller than collecting basins placed at ground level but also serves as the top of the water storage structure.

Water collection by rain traps or trick tanks may be greatly increased from collecting snow melt in some cases [86]. Selecting snowdrift areas for locating the catchment area may be necessary in winter precipitation areas. Or, snow fences may be constructed to drift snow over the collecting area if provision is made for its collection as melting takes place. It has also been practical to construct snow basins utilizing snow melt for summer and fall use by birds and animals [98].

Water hauling. Hauling water to livestock in tank trucks may be used in emergencies, may be used to provide a temporary source of water, or used as a continuous practice on rangeland where no other sources of water are available. Since water hauling is time consuming and costly, it should be used only where it is impractical to develop other types of watering facilities. The cost per cow month or sheep month varies greatly depending upon travel distance, condition of equipment, and the amount of water required per head daily.

In the Pacific Northwest the cost of water hauling on rangeland in 1957 averaged $1.00 to $1.50 per cow month [15]. On the West Desert in Utah in 1958, average costs of hauling water to sheep were estimated from $.10 to $.15 per head per month [36].

Where water hauling is required, lightweight, portable troughs shaped for easy stacking allow much greater flexibility and ease in loading and unloading and in placing water where desired (Figure 116). Their use allows daily watering and moving of troughs with the accompanying benefits to the range. In spite of the high cost of water hauling to livestock, the advantages of water hauling can include the following [15, 36, 98]:

1. Places water where forage is available.
2. Reduces livestock trailing and resulting damage to the range.
3. Improves distribution of grazing.
4. Increases livestock performance.
5. Permits grazing at the most appropriate time.
6. Provides a dependable water supply in dry seasons and in emergencies.
7. Provides a means of rotating or deferring grazing where permanent water is unavailable.
8. Permits regular inspection of livestock.
9. Less expensive than regular water developments on some small or infrequently used pastures.

FIG. 116. *Water hauling to sheep on the Desert Range Station near Milford, Utah. The use of lightweight, portable troughs allows sheep to be watered near the forage supply.* (Intermountain Forest & Range Experiment Station photo by Ralph Holmgren)

Piping water. Water may be piped distances of up to two to five miles from central water sources to tanks placed in areas where natural water sources do not occur. This practice allows expanding the effective area serviced by water development and spreads the high cost of deep wells over large land acreages (Figure 117). Flexible plastic pipe made of polyvinyl chloride is particularly well adapted to traversing rough terrain. Delivery pipelines of 2-inch plastic pipe with laterals of 1 to 1½-inch plastic pipe are commonly used.

A dozer blade, subsoiler, or heavy ditcher can be used to make and cover a pipeline trench. When a pipe-laying attachment is added, the trenching and pipe laying can be accomplished in a

FIG. 117. *Piping water. A, tank receiving water from underground plastic pipe—part of a fifty-one-mile pipeline system in Sioux County, Nebraska, providing water to 34,980 acres and eight ranch headquarters* (Soil Conservation Service photo); *B, auxiliary pump at windmill near Scottsbluff, Nebraska, for forcing water through pipeline.*

497

single operation. Care should be taken in laying the plastic pipe that opportunities for air locking and sedimentation are not provided along the line.

Occasional freezing will not harm plastic pipe when only partially filled with water. However, freezing may split plastic pipe when a full column of water remains in the pipe. This damage occurs most often near fittings. Laying pipe eighteen to twenty-four inches deep will protect the pipe from exposure to sunshine as well as cold temperatures. In areas where rodents are prevalent, gopher control practices may be required to reduce damage to the plastic pipe. Using a booster pump to move the water from a well to a high point for storage and subsequent delivery by gravity flow is a common practice. Siphons or hydraulic rams operated by water power are other means of raising water over terrain obstructions.

Storage and Watering Facilities

Storage facilities. Unless natural flow is enough to promptly and continuously supply the maximum amount of water needed, stockwater developments must be provided with storage facilities. The

FIG. 118. *Plastic pipe being laid in a trench in Box Elder County, Utah, to carry water to areas with no natural water sources.* (Soil Conservation Service photo)

amount of storage capacity needed depends upon the type of water development, uniformity of water yield, kind of livestock using the facility, and number of stock watering at one time. When sheep water as a band rather than as individuals, larger storage facilities are needed than when livestock or game animals water singly or in small groups.

Water sources which provide irregular or erratic flows of water such as reservoirs or catchment basins, which depend upon storms of moderate intensity, require larger storage facilities. It has been suggested that central storage facilities from which water is piped out to outlying areas should provide a seven-day capacity with engine driven pump or fourteen days when windmills are used [59].

Storage facilities may consist of metal, fiberglass, cement, or masonry tanks or cisterns, or even surplus railroad tank cars. Polyethylene plastic or butyl rubber can be used to line cinder block structures, masonry structures, or grain bins for water storage [16, 44, 98]. Procedures for developing plastered concrete water storage tanks have also been described [95]. These procedures basically include making a shell of welded wire reinforcing fabric covered with chicken wire, plastering both sides of the walls with concrete, running a concrete floor and built-up area around the base of the walls, and painting the inside of the tank with an emulsified asphalt paint.

Bottomless structures lined with plastic or butyl rubber are preferably partially buried to eliminate rodent burrowing underneath and reduce ice formation [44]. Permanent tops or floating covers should be provided on storage tanks wherever evaporation is apt to be serious. Collapsible butyl rubber bags can also be purchased for water storage, and they have been particularly recommended for use in conjunction with rain traps [46]. Deep pits can be lined with plastic or butyl rubber and covered with the same material to prevent both leakage and evaporation [43, 45].

Watering facilities. Many types and shapes of watering troughs and tanks are used successfully on the range. Most are constructed from cement, masonry, or heavy metal. To be effective and require minimum maintenance, watering facilities should be constructed of strong materials, be well reinforced, and have adequate supports. They should be located on level, solid ground, be firmly anchored to the ground, and sealed against leakage. Drilling basins into rock can provide permanent watering places, or cement can be used to enlarge or create new rock pools [92]. The latter practices have been particularly important in developing small quantities of water for game animals in arid areas.

Since cattle water individually or a few at a time, a ten-foot elongated trough provides adequate head room (Figure 119). When sheep water as a band, much greater trough space must be provided. Troughs used by sheep bands should not be less than seventy-five feet long [77]. Trough or tank tops should not be higher than twenty to twenty-four inches above ground for cattle or twelve to sixteen inches above ground for sheep. Guard rails are often desirable to prevent livestock from falling into the tanks.

Stock tanks should be located on well-drained, nonerosive sites of easy access to animals. Hauling in broken rock or gravel around the tank may be necessary to drain off overflow or run-in water. Ramps are often needed to allow livestock, game animals, and birds safe access to water [92]. Even abandoned dug wells and flooded mine shafts can sometimes be made safely accessible to animals when their use for drinking places is necessary. A tank located along a division fence allows stock on both sides to make use of it. Corrals or holding pastures should be supplied with stockwater, if possible, particularly where livestock may be left overnight.

Stock tanks or troughs supplied water from storage tanks should be equipped with a float valve to prevent waste while maintaining a full water level. If the amount of water coming into a tank is not controlled, provide an overflow to carry away excess water. This will

FIG. 119. *Semicylindrical water tank for cattle.*

500

prevent formation of mud holes and bogs around the water tank, and the overflow water can be piped to a supplementary tank or to an emergency reservoir or both. Clogging of the overflow can be prevented by providing a U-pipe on top of the standpipe with a hole bored in the top to prevent air lock.

Keeping ice-free water available to livestock during winter is made possible by use of coal, fuel oil, or electric water warmers. The use of solar radiation has been made with reasonable effectiveness in keeping stock tanks ice free. Livestock may not drink enough water if compelled to drink it ice cold or may be prevented from drinking at all if tanks freeze over and the ice is not broken by hand [18, 33]. Water heating up to temperatures of 50° F is apt to increase production while reducing labor costs. Drawing upon heat reserves of the body to warm cold ingested water failed to influence the body temperatures of sheep in one Canadian study [2]. But in very cold weather the ingestion of cold water may require additional conversion of productive energy to heat energy to maintain body warmth.

Range and Pasture Fences

Uses and benefits. Intensive and effective management of grazing lands is dependent upon adequate fencing. The term *range* originally implied expansive, unfenced grazing lands. However, as the native grazing lands of the West began to be recognized as an important but limited resource, it became apparent that a fundamental requirement in the management of grazing lands was to provide provisions for control of grazing. Although stockwater developments, salting practices, placement of other supplements, and occasional drifting of livestock are useful indirect methods of controlling livestock, fencing and herding are the only direct means of regulating seasons, numbers, kinds, and distribution of grazing animals.

The uses and benefits that can be expected from fencing are many. Although seldom will all of these benefits be realized with any specific fence, fencing pastures and rangeland can provide the following advantages [30, 50, 64]:

1. Permanently establish boundary lines.
2. Permit rotation, deferment, and resting of grazing lands.
3. Regulate season of grazing use.
4. Regulate stocking rates.
5. Provide better distribution of grazing.
6. Protect soil and related resources on erosive sites.

7. Separate animals by kind, sex, age, or stage of production.
8. Permit the grazing of cattle and sheep in the same pasture.
9. Control breeding, improve breeding efficiency, and provide lambing and calving areas.
10. Reduce transmissable animal diseases.
11. Control straying, trespassing, and injury to livestock.
12. Reduce labor required in handling livestock.
13. Eliminate need for herding sheep in many areas.
14. Permit closer observation and care of livestock.
15. Exclude livestock from hazardous areas such as bogs, poisonous plants, and highways.
16. Protect new seedlings and other range improvements.
17. Increase grazing capacity of grazing land as much as 25 percent through better management.
18. Obtain heavier livestock sale weights, greater number of saleable offspring, and cleaner wool in some cases.
19. Maintain friendly relationships with neighbors.

Increasing labor costs, lack of capable herders, and narrow profit margins are forcing more sheepmen in the West to look to herderless management of sheep under fence [8]. Running range sheep loose in fenced pastures began in eastern Wyoming and the Southern Great Plains, and from there the idea and practice has spread to other range areas. Major savings in labor costs must be compared with increased fence investments, and management must assume part of the sheep supervision previously provided by herders [69]. Also, failure to adequately control predators in the area will prevent loose running of sheep from being a practical alternative to herding. Both barbed-wire and woven-wire fences have been successfully used as range sheep fences.

Ranchers having converted from herding to management of sheep under fence generally indicate they would not return to herding because of the benefits of fenced management [6, 69]:

1. Increased grazing capacity and improved range condition.
2. Improved distribution by spreading out grazing, less trampling damage, sheep reach difficult spots, and sheep water singly whenever they desire.
3. Improved condition of ewes and better mothering of twins because of less disturbance.
4. Similar to slightly improved lamb weights.
5. Cleaner wool.
6. More breed choices with sheep since the herding instinct is of reduced importance.
7. Easier to run sheep and cattle in the same pasture.

However, new problems that may result from changing to herderless management under fence include (1) some trailing, (2) more winter and storm losses, (3) more predator losses and theft, (4) local concentration of grazing, and (5) interference with big game migration. Sheep reportedly can be herded part time and fenced part time, but an adjustment period is involved each time the change is made.

Planning and location. A fencing plan should be a significant part of the total ranch or range plan. The complete fencing plan of the ranch or range unit including gates, corrals, autogates, and water gaps should be plotted on the planning map and carefully correlated with projected management and improvement plans. Recommendations for fence construction will vary according to use, class of livestock, and type of range.

Classification of fences as boundary, drift, or division fences implies the major function attributed to specific fences [33]. *Boundary fences* prevent animals from trespassing on neighboring properties and commonly follow ownership lines. *Drift fences* keep livestock from getting off or on a range or confine herds to specific elevations. *Division fences* are used to divide range into smaller units for greater management or provide enclosure or exclosure for special uses. In practice the major function of a fence is seldom this distinct, and a fence may provide most or all of these functions.

Boundary fencing has been more common on rangeland than cross fencing. Although fencing around the perimeter of a ranch or range administrative unit warrants first priority consideration, subsequent interior fencing is generally required to maximize production and production efficiency from the range resource. Legal boundary fences are specified by state laws in most states, and the required type and construction details vary from state to state [20, 61]. Livestock owners in most states are liable for damages caused by their animals trespassing on other people's property if the livestock owners do not habitually build and maintain good boundary fences. Although boundary fences should meet acceptable standards of effectiveness and comply with state law, proper design and construction of interior fences are also important.

The following considerations should be considered in locating the fence on the ground:

1. Provide best possible distribution of grazing around existing water.
2. Insure that adequate water currently exists or can be provided in each pasture.

3. Separate range sites where feasible, particularly where different management will be required on each.
4. Separate seeded areas from native ranges.
5. Provide enough pastures to carry out appropriate grazing systems.
6. Exclude livestock from eroded areas, poisonous plant areas, springs, ponds, dams, and spillways.
7. Size of pastures in relation to grazing capacity and planned use.
8. Complement existing fences.
9. Convenience of access of stock and vehicles and effect on big game movements.
10. Effective use of natural barriers such as rocky ridges and ledges, gullies, and deep water.
11. Prevent interference with fire breaks, roads, trails, and general operation.
12. Prevent livestock trailing along fences through erosive areas (Figure 120).
13. Place division fences along crests of ridges where possible for better distribution of grazing.

Types of Fences

Conventional wire fences. Barbed wire fences are the most common type used for range and pasture fencing. Reasons for the popularity of barbed wire fences include the reduced bulk required for materials handling, reduced labor required in construction, acceptable

FIG. 120. *Improper location of water tank and fence resulting in livestock concentration and wind erosion in the Nebraska Sandhills.*

504

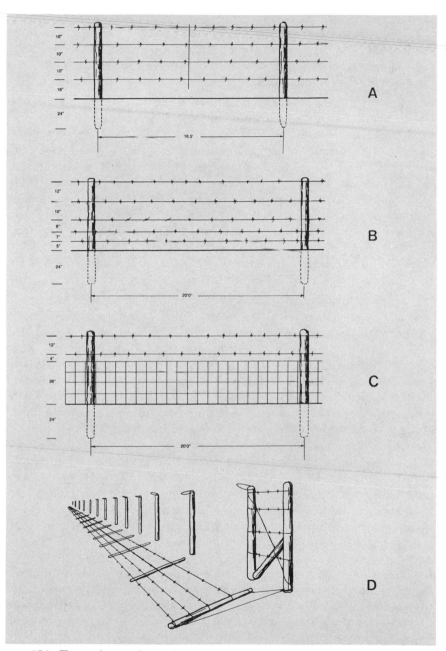

FIG. 121. *Types of range fences.* A, *conventional fence with four strands of barbed wire, posts 16.5 feet apart, and one stay between posts;* B, *sheep-tight, five-wire, barbed wire fence;* C, *combination woven wire and barbed wire fence;* D, *let-down fence for snowpack areas.*

maintenance and longevity, and adaptability to various types of land. Probably the most "conventional" fence used in range fencing is the four-strand barbed wire fence with posts one rod apart and one stake or stay between posts (Figure 121). This type of fence, generally with specified alternative types, is commonly designated by state laws as a legal boundary fence [61].

Although a three-strand barbed wire fence may suffice for division fences where cattle pressures are not too heavy, it should not normally be considered for use as a boundary fence. Recommended wire heights from ground level for four-strand cattle fences are 16-26-36-46 inches [50] or 16-24-32-42 inches [86]. Recommended wire heights for three-strand interior fences are 18-30-42 inches [50] or 18-28-40 inches [86]. Since sheep are more apt to crawl under than through or over a fence like cattle, wires must be placed closer to the ground to turn sheep. Although sheep accustomed to fences can be held successfully with four-wire fences, five strands are commonly used (Figure 121B). Recommended wire heights for a five-strand fence for sheep are 5-11-18-26-38 inches [69]. However, for fencing sheep from cattle, recommended wire heights are 5-12-20-30-42 inches [50].

Effective sheep fences can also be made from 26-inch or 32-inch woven wire with one or more strands of barbed wire on top, the barbed wires primarily for controlling cattle and damage by cattle [69]. An effective fence for dividing sheep and cattle is a 26-inch woven wire below with two strands of barbed wire at 30 inches and 42 inches above ground level (Figure 121C) [50]. In 1961 the latter fence cost $910 per mile in Wyoming (including 25 percent labor charge) compared to $760 for a five-strand barbed wire fence [69]. In Hawaii in 1961 four-strand barbed wire fences cost $1,100 to $1,250 per mile with 60 percent of the cost being materials while woven wire fences ranged from $1,300 to $1,600 per mile [64]. Fence costs today are considerably higher but vary depending upon kind of terrain and availability of local labor and supplies. Fencing costs per mile on steep, rough, brushy range often run 50 percent more than on level, brush-free range.

Let-down fence. The let-down fence (Figure 121D), also commonly called the lay-down or snow fence, is now commonly used in mountainous areas where snowpacks of four feet or more commonly loosen or break wires and pull over posts [81]. The let-down fence is basically a four-strand fence that can be laid down as a unit but remains under tension at all times. Brace posts and line posts are used as in an ordinary fence, and the wires are stretched between the brace posts. However, the wires are fastened to stub posts set on the ground next to permanent posts. Wire loops are stapled at the bottom and top of each wooden line post or placed through holes bored in metal line posts. The fence is raised in the spring by placing the stub posts in the loop and is laid flat on the ground in reverse steps in the fall.

Although the let-down fence greatly reduces fence maintenance to repair damage from snowpack weight and slippage, it also has the following disadvantages: (1) cannot protect against strays or trespass after lowered, (2) must be manually raised and lowered annually, (3) the over-winter contact with the ground rusts wires more quickly, and (4) initial cost is somewhat higher [81]. This suggests that let-

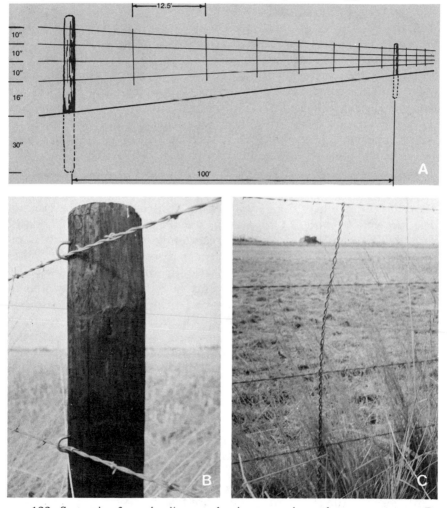

FIG. 122. *Suspension fence.* A, *diagram showing post, wire, and stay arrangement;* B, *stapling method;* C, *wire spacing maintained by wire stay* (University of Nebraska photos). *Note: stays should be clipped below the bottom wire to prevent dragging on ground or vegetation.*

down fence should be limited to sites where snow build-up is actually a problem. An alternative type of let-down fence utilizes no stub posts, but attaches the wires to the line posts by means of a removable staple key [86].

Suspension fence. The suspension fence differs from conventional four-strand barbed wire fences in spacings of 80 to 120 feet between line posts (Figure 122). Line posts serve the sole functions of maintaining spacings between wires and the desired height above the ground level. The second major difference is that the wires are not attached solid to the line posts but are allowed to slip when placed under sudden extra tension from either direction. Twisted wire stays are placed about every 15 feet between line posts to prevent cattle from forcing the wires apart. In contrast to conventional fences which use the barricade principle to stop livestock, the suspension fence functions as a series of panels with resilience and whiplike action to turn back livestock. The top and bottom of each suspended section also react in opposite directions when pressure is applied. A semisuspension fence with posts spaced 50 feet apart has been used but does not have the live action of a true suspension fence [50].

The quality and setting of corner and brace posts are particularly important with the suspension fence. Corner and brace posts should have top diameters of six to eight inches, be set firmly in the ground, and be well anchored. Only at the wooden brace or stretch posts, provided at one-quarter-mile intervals or at shorter intervals if required by slope changes, are the wires tied solidly to the post. Wires are fastened to wooden line posts with two-inch-long staples or by metal clips placed over the wires and held by a six-penny nail at each end. The staples or strips must be left loose so the wire is free to move back and forth past the posts. Where steel line posts are used, the wires can be attached with standard wire clips. Vertical wire spacings should be similar to standard barbed wire fences. It is important that the bottom wire and the ends of the wire stays do not drag the ground or brush or the effectiveness of the suspension fence will be reduced. Wires should be tightly stretched so there is no more than three inches of sag between posts.

Suspension fences were originally developed in Texas as interior or cross fences but appear to work equally well as boundary fences. Based on experimental evidence and rancher experience [25, 31, 55, 90], it is concluded that the suspension fence has the following advantages over the standard barbed wire fence:

1. Requires approximately 200 fewer line posts per mile (53 versus 264 posts) when placed at 100-foot intervals rather than at one-rod intervals.

2. Cost of materials has been from one-half to two-thirds the cost of the standard fence.
3. A saving of about fifty man-hours per mile when wooden posts are used or twelve man-hours when steel posts are used.
4. Requires less maintenance (by about 50 percent) for breakage and wires being pulled loose from posts.
5. Because of its resilience, animals are not as likely to run through or break a strand. (However, calves may roll under the fence when cattle are crowded against the suspension fence.)
6. Because of its whipping action, bulls are discouraged from fighting through the fence.
7. Turns cattle equally as well or better than standard barbed wire fences.
8. Lasts at least as long as standard barbed wire fences when properly constructed and maintained.
9. Tilts to allow tumbleweeds to go under or over the fence.

Although suspension fences have been used successfully in the Northern Great Plains, their adaptation to mountain snowpack areas is unproven. Although suspension fences are well adapted on level to rolling sites and uniform slopes, they appear unadapted to rough, uneven terrain.

Twenty-seven modifications of the suspension fence have been installed at the Pasture Research Center near Norbeck, South Dakota [55]. Based on 1968 prices, cost of suspension fencing with four strands, posts 100 feet apart, and stays every 15 feet was $407 per mile compared to $685 per mile for conventional fencing. The $407 was broken down into $304 for materials and $103 for labor. Suspension fencing costs varied from $326 per mile for three strands, posts 150 feet apart, and stays every 20 feet to $497 per mile for five strands, posts 100 feet apart, and stays 10 feet apart. Preliminary evidence has indicated that the three-wire types were damaged more by livestock putting their heads through the wires than four-wire types.

Wooden fences. Rail and pole fences are less commonly built now than formerly because of the high cost of labor and scarcity of local sources of poles and rails. However, they quite often have a longer maintenance-free life expectancy. Except in isolated areas where poles are readily available, pole fences are now mostly limited to holding pastures and corrals. Because of aesthetic appearance, they are preferred for many recreational and special-use areas [86]. Types of pole fences include the zig-zag fence (also called the worm or snake fence), the log-and-block fence, the buck pole fence, the straight rail fence, post-and-pole fence, and the post-pole-and-wire fence [33, 82, 86].

Board, plank, or slab fences may be substituted for pole fences when available and the cost justifies. Boards or planks are often used for fencing around ranchsteads because they are attractive and safe when used for paddocks, corrals, or horse pastures.

Miscellaneous fences. Temporary fences are less frequently used in range fencing than on temporary and irrigated pastures. Wooden panels, snowfence, or woven wire can be used in making temporary corrals or holding areas for livestock, fencing around haystacks, protecting livestock from predators, or for breeding paddocks. The electric fence is a popular type of temporary fence. It has the advantage of low cost in building and maintaining, in using simple fencing materials, and being easily built or removed. But it has disadvantages in that livestock require training to respond, it will not stop wild or frightened animals, it requires a steady source of current, and it must be kept free of plants and other obstructions that can cause shorting out.

The addition of a second wire below but not electrified generally provides additional effectiveness (Figure 76). Electric fences can be used temporarily on range to protect new seedings or a low recovery of local areas sprayed with herbicides. However, most range fencing should be permanent since the original need will generally continue indefinitely.

Coyote-proof sheep fences are commonly made of net or woven wire, snow fence, or panels. However, an electric anti-coyote fence was found effective in preventing depredation of sheep in Idaho [23]. This 5-foot fence consisted of alternating ground and charged wires, with the wires ranging from four inches apart at the bottom to six inches apart at the top. Materials cost only about $1600 per mile compared to $2400 for a conventional woven wire fence.

Construction. Clearing the proposed fence line of brush, trees, and other obstructions before construction begins makes construction easier, quicker, and cheaper [50, 64]. Also, a cleared fence line is easier and cheaper to maintain. Clearing a fence right-of-way with a blade is generally best, but care must be taken on erosive sites that erosion is not accelerated. Clearing to allow a maintenance road on at least one side of the fence not only makes future fence inspection easier but also provides access roads and makes driving of livestock easier.

On steep grades and erosive soils, site preparation should avoid removal of the herbaceous ground cover, or accelerated erosion may result. In heavily timbered or very rough and rocky terrain, clearing must generally be reduced to a minimum. On sites where scalping in preparation for fence construction is practiced, seeding to grass should follow. However, revegetation of fence lines cleared of vegetation in desert areas is particularly difficult, and such areas are particularly susceptible to infestation of noxious plants such as halogeton and bitterweed.

Fence posts used in range and pasture fencing are generally wooden or metal and rarely concrete or rock. Wooden rather than metal posts should always be used for corner posts and brace posts and for line posts in snow country [86]. However, steel posts are equally satisfactory for line posts where post strength is not a major factor and are preferred in rocky country where drilling is necessary to set posts. Steel line posts have the advantage of acting as grounds for wires, are fireproof, are relatively light, compact, and easy to handle, and are easily driven into most soils [82, 86].

The tree species, the amount of heartwood, and the dryness of the climate all affect the life expectancy of untreated wooden posts [7]. Black locust and Osage orange or hedge posts are highly resistant to decay and should last fifteen to twenty-five years without treatment in humid areas or up to thirty to forty years in arid areas. Posts from (1) cedar and juniper with at least two-thirds of the diameter of heartwood, (2) mulberry, (3) catalpa, (4) baldcypress, (5) white oak, (6) redwood, (7) sassafras, and (8) Gambel oak, if cut from live plants in the spring, will normally last up to fifteen years without treatment or twenty to twenty-five years when treated [7, 56]. Posts from other tree species including the pines, the oaks, and cedar and juniper with large amounts of sapwood should be treated with preservatives such as coaltar creosote, pentachlorophenol, or zinc chloride. However, when properly treated, posts from the latter tree species group should last twenty years or more.

Where fences must be built across areas such as rock ledges or outcrops without sufficient soil depth to hold a wooden or metal post, special practices or structures must be used. If an air compressor is available and steel posts can be used, holes can be drilled for driving in the posts. Under certain conditions, dynamite can be used in digging holes for wooden posts. In some cases spans of a suspension fence can be used to traverse a rock ledge; or a rock jack can be made by making a cylindrical crib and filling with loose rocks, and a stub post can then be fastened directly to the rock jack; or line posts can be set on sills or braced or can be made in the shape of a figure "4" to give it a broader base support. A fence type alternating one rock jack and three figure "4" posts has been appropriately called the rock-jack and figure-four fence [33, 86].

Water flow and abrupt changes in topography require special fence adaptations. Placing a stretch post where an abrupt change oc-

511

curs, whether in the bottom of a depression or top of a ridge, solves many fencing problems. Wire fences can be conformed to the curvature of many depressions not subject to flooding or active gullying by tying down with deadman rocks or posts or well braced upright posts. When fences must cross deep channels or arroyos, the fence may best be stretched across the top and a short fence segment placed in the bottom to prevent livestock crossing.

When crossing perennial streams, interrupted streams, and channels subject to flash flooding, water gaps or floodgates should be provided. Floodgates across ditches, gullies, and narrow waterways prevent stock from swimming or crawling under the fence but shed debris that might form a dam causing the fence to wash out. Top-hinged, self-cleaning floodgates made of wire or wooden panels to fit the channel contour are particularly effective and require minimum maintenance (Figure 123). To cross wide depressions a break-away fence constructed to break in the center or at one end and swing to the side to allow flood waters to pass prevents permanent damage to the fence. However, following each major water flow the water gap should be checked and repaired at the break if broken [50].

In areas where lightning is a hazard, barbed-wire and woven-wire fences should be grounded to prevent or reduce livestock loss and human injury. Thorough grounding is also essential to prevent deterioration of the wire itself by lightning-rusting and loss of temper and thereby reducing longevity of use [49]. Twisted-wire stays installed every 16 to 20 feet and making firm contact with the soil reduce lightning damage but cannot be used in this fashion with suspension

FIG. 123. *Diagram of top-hinged, self-cleaning floodgate for use in crossing flood channels.* (Adapted from McNamee and Kinne [50])

512

fences. Using steel line posts or alternating steel and wooden posts provide an effective grounding except on very dry sites [30]. The most effective ground is obtained by driving eight-foot, one-half-inch pipes into the ground every 150 feet and leaving sufficient pipe above the ground to make direct connection with each wire strand.

Since corner and end posts are the foundation of any fence, their design and construction are very important (Figure 124). Failure at

FIG. 124. *Strong, dependable corner assembly used with a suspension fence at the University of Nebraska North Platte Station.* (Nebraska Farmer photo)

FIG. 125. *Strategically located autogates such as these at the Southern Plains Experimental Range, Woodward, Oklahoma, are not only time saving but reduce maintenance.*

513

FIG. 126. *Post drivers are commercially available for either front or rear tractor mounting.* (Photos by Danuser Machine Co. and Shaver Manufacturing Co.)

FIG. 127. *The Fury Fencer drives posts, dispenses wire, and stretches the wire in a single operation.* (U.S. Steel photo)

these points destroys the effectiveness of the fence and requires reconstruction and restretching of wires. One of several types of effective brace assemblies must be used if the fence is to have lasting effectiveness [50, 82]. Other construction considerations in range and pasture fencing include proper gauge and splicing of wire, adequate but not excessive stretching of wires, proper attachment of wires to the post, and setting posts at proper depths and spans. Providing fences with many strategically located gates and autogates across roads frequently used is not only convenient but saves much time and reduces maintenance costs (Figure 125).

The use of power equipment has greatly benefited fence building. Many or nearly all of the steps formerly accomplished with hand tools can now be done with power equipment. Power diggers save time and reduce the work of digging, but tamping is still required. Power post drivers can be used for setting metal line posts or wooden posts if relatively straight and bluntly pointed on the large end (Figure 126). Post driving is feasible in most soils where holes can be power dug and has the advantage of rapidly and firmly setting posts without additional tamping required. One man-hour of labor has been shown sufficient to set four posts with hand tools, six posts with power digger and hand tamping, or fifteen posts using a power driver [57]. Post drivers are available for direct mounting on the side or rear of a tractor or mounting on a two-wheel trailer. A pavement

FIG. 128. *Deer fence at the Kerr Wildlife Area, Kerrville, Texas, constructed with two widths of woven wire and two strands of smooth wire on the top. Note the fence line was cleared of brush prior to fence construction.*

breaker attachment powered by compressed air has also been adapted to driving steel posts [29].

A commercially available unit developed by United States Steel, referred to as the *Fury Fencer* (Figure 127), drives posts, dispenses either woven wire or barbed wire or any combination of the two, and brings the wire up to the desired tension in a one-pass operation [50]. This fencer requires two or three men to operate when pulled by a farm tractor, can build more than a mile of fence daily, can handle either wood or steel posts, and can be used on level or rolling land wherever a tractor can go. However, the greatest increase in efficiency is on straight fences over gentle terrain [65].

Big game fences. It is sometimes desirable to fence big game either in or out of certain areas. To exclude or confine deer requires special fencing (Figure 128). Upright deer fences recommended for rolling terrain and open timber in Texas have been 8.5 feet high, constructed of two widths of woven wire and one strand of smooth wire on the top, and with 12-foot line posts spaced 15 feet apart and set 2.5 feet into the ground [27]. Although effective with deer, this type of fence is costly, with labor and materials running $2,460 per mile in 1965. In California a 7-foot fence on level ground or 10-foot fence on slopes was suggested using woven wire [47]. It was pointed out that the fence must be set clear to the ground since deer will crawl under if space remains. However, it appears unnecessary to bury the net wire since deer uncommonly dig under a fence set to ground level [5].

Various modifications of overhanging or slanting deer fences have been used in attempting to reduce construction costs. Such fences are based on the observations that deer must be rather close to the base of a fence before attempting to jump it [5, 38]. Blaisdell and Hubbard [5] suggested attaching an outrigger to the top of standard 4.5-foot fences and extending in a sloping decline to a point on the ground 8 feet from the base of the fence and in the direction of the deer pressure. A 4-foot horizontal segment extending from the top of a standard fence in the direction of deer pressure has also been tried [38]. These have been found only partially effective.

A single-segment fence was considered equally effective to a more expensive vertical-overhanging fence in California comparisons [38, 47]. The slanting fence consisted of a 6-foot segment of mesh wire fastened to the ground with stakes on one edge and extending upward at an angle of approximately 45°. In cross section the 6-foot mesh wire served as the hypotenuse to a right triangle formed by the ground side and the upright (unfenced) side. The vertical-overhanging fence differed in that the lower end of the 6-foot mesh wire was not attached directly to the ground but to the top of a 2-

516

foot wire-covered vertical segment. The costs of the slanting fence and the vertical-overhanging fences were $1,000 and $1,700 per mile compared to $2,500 per mile (1958 prices) for a conventional 8-foot upright deer fence. The high side of both overhanging styles was the only effective side against deer pressures; deer first tried to crawl under but were subsequently discouraged from jumping because of the net wire extending above them. The principal disadvantage of these style fences was that livestock easily damaged them. Increasing the height of the vertical segment on the vertical-overhanging fence kept sheep off the top and turned sheep from both directions, but turned deer from one direction only.

Big game considerations in livestock fencing. It is often desirable to design fences that will regulate domestic livestock but still allow free passage of all big game. Such a design must consider how big game get by a fence. Antelope commonly run up to a fence and go under the bottom wire or through the fence near the bottom [78]. Even though capable of jumping eight feet high, antelope seldom jump over two feet high unless under stress induced by man [76].

Sheep-tight fences do not allow antelope to crawl under or go between the lower wires. It was concluded from studies in Wyoming

FIG. 129. *Antelope pass structure in Wyoming patterned after a standard cattle guard being jumped by antelope.* (Bureau of Land Management photo by Raymond D. Mapston)

that a thirty-two-inch net wire fence or a twenty-six-inch net wire with one strand of barbed wire four inches above the net wire were the best styles to turn sheep but allow antelope to jump over [76]. However, few fawns were found to cross. Cattle fences with the bottom wire no closer than sixteen inches to the ground permit antelope to go under while readily turning cattle [78].

Designing fences to turn both cattle and sheep while allowing antelope to pass freely is more difficlt. Drop gaps in such fences are suggested for use when livestock are not present, particularly when placed where antelope cross regularly [76]. Dirt ramps constructed to the height of the fence appear effective for antelope crossing but allow sheep to cross also. However, standard cattleguards 6 feet long generally prevent cattle and sheep from passing but are readily jumped by antelope. Antelope pass structures constructed similar to cattleguards are readily used by antelope when placed near fence corners or along well used trails (Figure 129) [51]. It has been suggested that the passes be limited to 5.5 feet wide if vehicular crossing is undesirable.

If livestock fences are not constructed over forty inches high, mature deer, elk, and moose in good condition readily jump over [78]. However, young fawns or elk or moose calves may have some diffi-

FIG. 130. *Working corrals and associated livestock handling facilities are required for effective ranch operation.*

culty in negotiating the fence. Where elk and moose frequently cross, it is suggested that a pole be used in place of the top wire to reduce damage to the fence. Elk and moose tend to drag their hind legs when jumping over a fence, and the pole is more readily seen or less easily damaged during the crossing. Also, deer are less afraid of a pole than wire and are less apt to get their legs trapped in crossing.

Corrals and Related Facilities

A well-planned corral system and effective livestock handling equipment on the ranch make handling of stock easier, save labor, cut livestock damage and shrinkage to a minimum, and return their investment many times over (Figure 130) [33]. Although the central corral system is generally located at the ranch headquarters, secondary corral facilities located at strategic, outlying sites are generally required, particularly where major divisions of the ranch property are not contiguous.

Corral systems should be designed to meet all projected needs but not be overly elaborate. Construction should be simple, strong, durable, and dependable. The use of materials locally available will help to keep the costs reasonable. The essential features for a practical corral system should include the following considerations [18, 33]:

1. Place on a convenient, accessible, well-drained site.
2. Arrange so animals can be quickly and easily sorted.
3. Include sufficient pens to accommodate all groups of livestock.
4. Provide small holding pastures near the corrals.
5. Provide water in the corrals or at least in the adjacent holding pastures.
6. Use smooth corral sides and protect corners to prevent bruising and crippling of livestock.
7. Make strong enough to hold animals.
8. Avoid pockets and square turns; rounded sides are ideal.
9. Design so maintenance and upkeep will be low.

Adequate facilities should be included in the corral system for constraining, castrating, branding, dehorning, loading and unloading, weighing, sorting, holding, spraying and dipping, and sheep shearing [33]. A squeeze chute capable of completely restraining livestock and of being operated by one man should be included. Feed bunks, racks, and self-feeders should be provided to the extent needed and preferably built on skids for easy moving. Detailed plans and blueprints for the construction of corrals and related facilities are available from agricultural experiment stations and extension services or in printed materials [1, 18, 20, 33, 34, 53, 54, 80, 84, 91].

Sheds and shelters are less important for mature breeding cattle than for feeder cattle or fattening cattle [33]. Protection for mature range livestock is generally required only in cold, humid climates where winter rains are frequent [18]. Elsewhere, natural shelter in rough or wooded areas or around windbreaks is usually sufficient. European breeds of livestock are well adapted to cold environments. When adequate feed is available, thermogenesis is usually sufficient to preserve the normal temperature of cattle even with environmental temperature as low as –18°C [2]. But under range conditions, cattle and sheep may be unable to eat sufficient forage to maintain body weight without converting additional productive energy to heat energy. Consumption of cold water or snow further increases the heat requirement.

Trails and Walkways

To provide access ways and obtain more uniform utilization of range forage, the construction of trails and bridges is often required

FIG. 131. *Constructing an access trail with a hula dozer to obtain better range utilization.* (Arcadia Equipment Co.)

520

in steep, rough country (Figure 131). Providing access into watering places is often as important as providing access to ungrazed forage. Also, graded stock trails when properly located protect watersheds by lessening grazing and trailing over erosive sites. Trail construction may be required though heavy brush and timber as well as across difficult terrain. Trails must be located where they will not accelerate erosion as a result of trailing and be provided with drain bars so that runoff can be diverted to prevent excessive maintenance requirements. Constructing livestock access ways wide enough to serve as vehicular access roads has the additional advantages of permitting the hauling in of salt and other supplements, fencing materials, horses and equipment, and livestock and aid the control of fires.

A system of cattle walkways and earthen windbreaks on salt marsh rangelands of the Gulf Coast provides not only better grazing distribution but also protection from cold wind and sleet or rain during freezing weather [74, 96]. With some construction alterations, the system can also provide the means of keeping out sea water during high water periods. The walkways and earthen windbreaks can be constructed with a dragline. The walkways should be built up at least two feet above normal high-water levels while the settled height of the windbreak portions should be no less than six feet. By staggering the borrow pits on both sides of the built-up areas, ready grazing is allowed in either direction. The borrow pits can provide drinking water and also act as firebreaks. The built-up areas also serve as calving areas, bedgrounds, refuge from high water, driveways, and lines for fence construction.

Literature Cited

1. Albaugh, Reuben, C. F. Kelly, and H. L. Belton. 1952 (Revised). Beef handling and feeding equipment. California Agric. Ext. Serv. Cir. 414.
2. Bailey, C. B., R. Hironaka, and S. B. Slen. 1962. Effects of the temperature of the environment and the drinking water on the body temperature and water consumption of sheep. Canadian J. Anim. Sci. 42(1):1–8.
3. Beale, Donald M., and Arthur D. Smith. 1970. Forage use, water consumption, and productivity of pronghorn antelope in western Utah. J. Wildlife Mgt. 34(3):570–582.
4. Biswell, H. H., and A. M. Schultz. 1958. Effects of vegetation removal on spring flow. California Fish & Game 44(3):211–230.
5. Blaisdell, James A., and Richard L. Hubbard. 1956. An outrigger type deer fence. California Forest & Range Expt. Sta. Res. Note 108.
6. Blankenship, James O. 1969. Herderless sheep management on mountain ranges. ASRM, Abstract of Papers, 22nd Annual Meeting, pp. 19–20.
7. Blew, J. Oscar, Jr., and Francis J. Champion. 1967 (Rev.). Preservative treatment of fence posts and farm timbers. USDA Farmers' Bul. 2049.
8. Brown, Lloyd N. 1958. Small earth dams. California Agric. Expt. Sta. Cir. 467.
9. Burman, R. D., M. A. McNamee, and R. L. Lang. 1958. Reservoirs for range stockwater development. Wyoming Agric. Expt. Sta. Cir. 67.

10. Butcher, John E. 1966. Snow as the only source of water for sheep. Amer. Soc. Anim. Sci., Western Sec., Proc. of Annual Meeting 15:205–209.
11. Butcher, John E., Lorin E. Harris, and Robert L. Raleigh. 1959. Water requirements for beef cattle. Utah Farm & Home Science 20(3):72–73.
12. Cardon, P. B., E. B. Stanley, W. J. Pistor, and J. C. Nesbitt. 1951. The use of salt as a regulator of supplemental feed intake and its effect on the health of range livestock. Arizona Agric. Expt. Sta. Bul. 239.
13. Chiarella, Joseph U. 1962. The use of paved water catchments in the Southwest. ASRM, Abstract of Papers, 15th Annual Meeting, pp. 63–65.
14. Choi, Sung S., and J. E. Butcher. 1961. The influence of alternate day watering on feed and water consumption of sheep maintained under two temperatures. J. Anim. Sci. 20(3):678.
15. Costello, David F., and Richard S. Driscoll. 1957. Hauling water for range cattle. USDA Leaflet 419.
16. Currier, W. F. 1973. Water harvesting by trick tanks, rain traps, and guzzlers. In Water-Animal Relations Symposium Proceedings, Kimberly, Idaho, pp. 169–181.
17. Davis, Delmer I. 1974. The importance of evaluating livestock water. Texas Agric. Ext. Misc. Pub. 1157.
18. Davis, R. L., and W. F. Edgerly. 1963 (Rev.). Feedlot and ranch equipment for beef cattle. USDA Farmers' Bul. 1584.
19. Embry, L. B., M. A. Hoelscher, R. C. Wahlstrom, et al. 1959. Salinity and livestock water quality. South Dakota Agric. Expt. Sta. Bul. 481.
20. Ensminger, M. E. 1962 (Third Ed.). The stockman's handbook. The Interstate Printers and Publishers, Danville, Ill.
21. Frasier, Gary W. 1975. Water harvesting: A source of livestock water. J. Range Mgt. 28(6):429–434.
22. Gastler, C. F., and E. O. Olsen. 1957. Dugout water quality. South Dakota Farm & Home Research 8(2):20–23.
23. Gates, N. L., J. E. Rich, D. D. Godtel, and C. V. Hulet. 1978. Development and evaluation of anti-coyote electric fencing. J. Range Mgt. 31(2):151–153.
24. Goebel, Carl J. 1956. Water development on the Starkey Experimental Forest and Range. J. Range Mgt. 9(5):232–234.
25. Halff, Thomas. 1957. Fence it cheaper and easier with the suspension fence. J. Range Mgt. 10(6):257–258.
26. Halloran, Arthur F., and Oscar V. Deming. 1956. Water development for desert bighorn sheep. USDI, Fish & Wildl. Serv., Wildl. Mgt. Series Leaflet 14.
27. Halls, L. K., C. E. Boyd, D. W. Lay, and P. D. Goodrum. 1965. Deer fence construction and costs. J. Wildl. Mgt. 29(4):885–888.
28. Hamilton, C. L., and Hans G. Jepson. 1940. Stock-water developments: wells, springs, and ponds. USDA Farmers' Bul. 1859.
29. Heede, Burchard H. 1964. A pavement breaker attachment to drive steel fence posts. J. Soil & Water Cons. 19(5):182–183.
30. Henderson, G. E. 1954. Planning farm fences. Univ. of Georgia, Agric. Engr. Dept., Athens.
31. Hoffman, G. O., B. J. Ragsdale, and W. J. Waldrip. 1964. Save with suspension fences. Texas Agric. Ext. Serv. Leaflet 637.
32. Howe, E. Crosby, and Deon D. Axthelm. 1970. Control of algae and moss in water tanks. Nebraska Agric. Ext. Campaign Cir. 220.
33. Hubbard, William A. 1969. Farm and ranch equipment for beef cattle. Canada Dept. of Agric. Pub. 1390.
34. Huber, M. G., John Landers, and Dean Frischknecht. 1961. Beef cattle equipment. Oregon Agric. Ext. Serv. Bul. 751.
35. Humphrey, R. R., and R. J. Shaw. 1957. Paved drainage basins as a source of water for livestock or game. J. Range Mgt. 10(2):59–61.
36. Hutchings, Selar S. 1958. Hauling water to sheep on western ranges. USDA Leaflet 423.
37. James, Lynn F., John E. Butcher, and Kent R. Van Kampen. 1970. Relationship between Halogeton glomeratus consumption and water intake by sheep. J.

Range Mgt. 23(2):123–127.

38. Jones, Milton B. and Wm. M. Longhurst. 1958. Overhanging deer fences. J. Wildl. Mgt. 22(3):325–326.

39. Julander, Odell. 1966. How mule deer use mountain rangeland in Utah. 1966. Utah Academy of Sci. Proc. 43(2):22–28.

40. Lamb, Samuel H., and Rex Pieper. 1971. Game range improvement in New Mexico. New Mexico Inter-Agency Range Comm. Rep. 9.

41. Lauritzen, C. W. 1960. Ground covers for collecting precipitation. Utah Farm & Home Sci. 21(3):66–67, 87.

42. Lauritzen, C. W. 1966. Farm ponds and plastic liners. Utah Farm & Home Sci. 27(3):90–92.

43. Lauritzen, C. W. 1967. Butyl for the collection, storage, and conveyance of water. Utah Agric. Expt. Sta. Bul. 465.

44. Lauritzen, C. W. 1967. Rain traps of steel. Utah Farm & Home Sci. 28(3):79–81.

45. Lauritzen, C. W., Frank W. Haws, and Allan S. Humpherys. 1956. Plastic film for controlling seepage losses in farm reservoirs. Utah Agric. Expt. Sta. Bul. 391.

46. Lauritzen, C. W., and Arnold A. Thayer. 1966. Rain traps for intercepting and storing water for livestock. USDA Agric. Info. Bul. 307.

47. Longhurst, Wm. M., Milton B. Jones, Ralph R. Parks, et al. 1962. Fences for controlling deer damage. California Agric. Ext. Serv. Cir. 514.

48. Mackie, Richard J. 1970. Range ecology and relations of mule deer, elk, and cattle in the Missouri River Breaks, Montana. The Wildl. Soc. Wildl. Mono. 20.

49. McIlvain, E. H., and M. C. Shoop. 1965. Deterioration of barbwire by lightning. J. Range Mgt. 18(3):153–154.

50. McNamee, Michael A., and Edwin Kinne. 1965. Pasture and range fences. Rocky Mtn. Regional Publication 2. (University of Wyoming, Laramie.)

51. Mapston, Raymond D., Rex S. Zobell, Kenneth B. Winter, and William D. Dooley. 1970. A pass for antelope in sheep-tight fences. J. Range Mgt. 23(6):457–459.

52. Martin, S. Clark, and Donald E. Ward. 1970. Rotating access to water to improve semidesert cattle range near water. J. Range Mgt. 23(1):22–26.

53. Midwest Plan Service. 1960. Sheep equipment plans. Nebraska Agric. Ext. Serv. Cir. 60-712.

54. Midwest Plan Service. 1963. Beef equipment plans. Nebraska Agric. Ext. Serv. Cir. 63-716.

55. Moore, R. A., H. G. Young, M. E. Larson, and G. B. Haiwick. 1968. Long span fences. South Dakota Agric. Expt. Sta. Bul. 546.

56. Moore, Raymond R. 1952. Treating fence post for long life. Utah Agric. Ext. Serv. Fact Sheet 14.

57. Neetzel, John R. 1966. Building better farm fences. Minnesota Agric. Ext. Serv. Bul. 272.

58. Nichol, A. A. 1938. Experimental feeding of deer. Arizona Agric. Expt. Sta. Tech. Bul. 75.

59. Patterson, Theodore C. 1967. Design considerations for small pipelines for distribution of livestock water on rangelands. J. Range Mgt. 20(2):104–107.

60. Pearson, H. A., D. C. Morrison, and W. K. Wolke. 1969. Trick tanks: water developments for range livestock. J. Range Mgt. 22(5):359–360.

61. Perlman, Harvey. 1966. Nebraska fence laws. Nebraska Agric. Ext. Cir. 65-829.

62. Peterson, H. V. 1961. Reconnaissance methods of locating stock wells in arid regions of the United States. International Association of Science and Hydrology Pub. 57, pp. 628–640.

63. Peterson, Harold V., and Virgil T. Heath. 1963. Stock water facilities for the Pacific Southwest. J. Soil & Water Cons. 18(3):103–108.

64. Philipp, Perry F. 1961. The economics of ranch fencing in Hawaii. Hawaii Agric. Expt. Sta. Agric. Econ. Bul. 20.

65. Range Seeding Equipment Comm. 1965 (Rev.). Handbook of range seeding

equipment. USDA and USDI.

66. Reginato, Robert J., Francis S. Nakayama, and J. Bennett Miller. 1973. Reducing seepage from stock tanks with uncompacted, sodium-treated soils. J. Soil and Water Cons. 28(5):214-215.

67. Renfro, George, Jr. 1968. Sealing leaking ponds and reservoirs. USDA, Soil Cons. Serv. TP-150.

68. Roberts, N. Keith, and E. Boyd Wennergren. 1965. Economic evaluation of stockwater developments. J. Range Mgt. 18(3):118-123.

69. Roberts, W. P., Jr. 1961. Fencing vs. herding of range sheep. Wyoming Agric. Expt. Sta. Mimeo. Cir. 156.

70. Rollins, Myron B. 1969. Controlling seepage with playa sediments. Nevada Ranch & Home Rev. 4(5):14-15.

71. Rollins, M. B., A. S. Dylla, and G. A. Myles. 1963. Experimental bentonite sealing. Nevada Agric. Expt. Sta. Bul. 229.

72. Rollins, M. B., and B. L. McNeal. 1976. Nevada playa sediments as pond and canal sealers. Nev. Agric. Expt. Sta. Bul. B41.

73. Shen, R. T. 1959. Sealing farm ponds and reservoirs with bentonite. Wyoming Agric. Ext. Serv. Cir. 162.

74. Shiflet, Thomas N. 1963. Earthen windbreaks, a new management device for salt marsh rangelands. J. Range Mgt. 16(6):332-333.

75. Sneva, Forrest A., L. R. Rittenhouse, and V. E. Hunter. 1977. Stockwater's effect on cattle performance on the high desert. Ore. Agric. Expt. Sta. Bul. 625.

76. Spillett, J. Juan, Jessop B. Low, and David Sill. 1967. Livestock fences—how they influence pronghorn antelope movements. Utah Agric. Expt. Sta. Bul. 470.

77. Stoddart, L. A., and A. D. Smith. 1955 (second edition). Range improvements, chapt. 16. In Range management. McGraw-Hill Co., New York City.

78. Sundstrom, Charles. 1966. Fence designs for livestock and big game. Range Improvement Notes 11(2):3-11. (U.S. Forest Service, Intermountain Region.)

79. Sykes, Glenton G. 1937. Desert water tanks. USDA, Southwestern Forest & Range Expt. Sta., Tucson, Arizona. Mimeo.

80. Trenary, O. J., and Paul S. Pattengale. 1955. Equipment for handling beef cattle. Colorado Agric. Ext. Serv. Bul. 441.

81. Turner, George T. 1960. A lay-down fence for snow country. J. Range Mgt. 13(1):43-44.

82. USDA, Agric. Res. Serv. 1961. Farm fences. USDA Farmers' Bul. 2173.

83. USDA, Agric. Res. Serv. 1967. Rain traps for wildlife and recreation. Agric. Res. 16(4):13.

84. USDA, Agric. Res. Serv. 1969. Housing and equipment for sheep. USDA Farmers' Bul. 2242.

85. USDA, Agric. Res. Serv. 1969. Suggested guide for weed control, 1969. USDA Agric. Handbook 332.

86. USDA, U.S. Forest Serv. 1969. Structural range improvement handbook. U.S. Forest Serv., Intermountain Region, Ogden, Utah.

87. USDA, Soil Cons. Serv. 1967. National handbook for range and relating grazing lands. Soil Conservation Service, Washington, D.C.

88. Vallentine, John F. 1963. Water for range livestock. Nebraska Agric. Ext. Serv. Cir. 63-156.

89. Waldrip, Wm. J. 1960. Evaluation of chemical films for retarding evaporation under field conditions. Texas Agric. Expt. Sta. Prog. Rpt. 2158.

90. Waldrip, William J. 1962. Fence costs—six kinds compared. Amer. Cattle Producer 44(4):8.

91. Watson, Ivan, and Charley Taylor. 1952 (Rev.). Drawings of ranch equipment. New Mexico Agric. Ext. Ser. Cir. 210.

92. Weaver, Richard A., Floyd Vernoy, and Bert Craig. 1959. Game water development on the desert. California Fish & Game 45(4):333-342.

93. Welchert, W. T., and Barry N. Freemen. 1969. The horizontal well as a new method of range water development. Progressive Agriculture in Arizona 21(6):8-11.

94. Welchert, W. T., and Barry N. Freeman. 1973. 'Horizontal' wells. J. Range Mgt. 26(4):253–256.
95. Welchert, W. T., and J. N. McDougal, Jr. 1965. How to make a plastered concrete water-storage tank. Arizona Agric. Ext. Serv. Bul. A-41.
96. Williams, Robert E. 1959. Cattle walkways—an aid to coastal marsh range conservation. USDA Leaflet 459.
97. Winchester, C. F., and M. J. Morris. 1956. Water intake rates of cattle. J. Anim. Sci. 15(3):722–740.
98. Wunderlich, R. Eugene. 1968. Water storage developments to benefit livestock and wildlife. ASRM, Abstract of Papers, 21st Annual Meeting, pp. 21–22.

Appendix

Table of weights, measures, and equivalents.

1. Length measure

1 mile = 5,280 feet = 1,760 yards = 320 rods = 80 chains = 1,609.34 meters = 1.609 kilometers

1 chain = 66 feet = 22 yards

1 rod = 16.5 feet = 5.5 yards

1 yard = 3 feet = 36 inches = 0.914 meter = 91.44 centimeters

1 foot = 12 inches = 0.305 meter = 30.48 centimeters

1 inch = 2.54 centimeters = 25.4 millimeters

1 kilometer = 1,000 meters = 0.621 mile

1 meter = 10 decimeters = 39.37 inches = 3.281 feet = 1.094 yards

1 decimeter = 10 centimeters = 3.94 inches

1 centimeter = 10 millimeters = 0.394 inch

1 millimeter = 0.04 inch

2. Area measure

1 township = 36 sections = 36 square miles = 23,040 acres

1 section = 1 square mile = 640 acres = 2.59 square kilometers = 259.0 hectares

1 acre = 43,560 square feet = 160 square rods = 0.405 hectares = 4047 square meters

1 square rod = 0.006 acres = 272.25 square feet = 25.293 square meters

1 square foot = 0.093 square meter

1 square inch = 6.451 square centimeters

1 square kilometer = 1,000,000 square meters = 100 hectares = 0.3861 square mile = 247.1 acres

1 hectare = 10,000 square meters = 2.471 acres

1 square centimeter = 0.155 square inch

3. Capacity measure

1 gallon = 4 quarts = 8 pints = 16 cups = 256 tablespoons = 768 teaspoons = 231 cubic inches = 128 fluid ounces = 3.785 liters

1 quart = 2 pints = 4 cups = 64 tablespoons = 192 teaspoons = 0.946 liter

1 pint = 2 cups = 32 tablespoons = 0.473 liter = 473 milliliters

1 cup = 16 tablespoons

1 tablespoon = 3 teaspoons = 0.5 fluid ounce

1 liter = 1.057 quarts = 0.264 gallon = 61.02 cubic inches

4. Weight measure

1 ton = 2,000 pounds = 907.18 kilograms
1 hundredweight (cwt.) = 100 pounds = 45.359 kilograms
1 pound = 16 ounces = 453.59 grams = 0.454 kilograms
1 ounce = 28.35 grams
1 kilogram = 2.205 pounds = 1,000 grams = 1,000,000 milligrams
1 gram = 0.0022 pound = 0.0352 ounce = 1,000 milligrams
1 gallon water = 8.346 pounds = 3785.655 grams
1 cubic foot of water = 62.43 pounds
1 cubic centimeter of water at 39.2°F = 1 gram

5. Rate equivalents

1 lb./A = 1.04 g./100 sq. feet = 2.83 g./sq. rod = 0.0023 lb./100 sq. feet
1 lb./A = 1.121 kg./hectare = 11.21 g./square meter
1 kg./hectare = 0.891 lb./A
1 p.p.m. = 1 mg./kg.
1% = 10,000 p.p.m. = 8.34 lb. a.e./100 gal. water
2 lb. a.e.h.g. = 0.2% a.e. solution (when carrier and commercial product weigh the same per gallon)
1 cup/sq. rod = 10 gal./acre
1 mi./hr. = 1.61 km./hr. (1 km./hr. = .62 mi./hr.)
1 liter/hectare = 0.383 gal./A
1 gal./A = 2.612 liters/hectare

6. Miscellaneous

Degrees Fahrenheit = (9/5 × degrees Celsius) plus 32
Degrees Celsius = 5/9 × (degrees Fahrenheit minus 32)
Area of circle = 3.1416 × radius^2
Circumference of circle = 3.1416 × diameter
Volume of cylinder = 3.1416 × radius^2 × height
Grams per 96 square feet = pounds per acre

Indexes

COMMON NAME–SCIENTIFIC NAME
INDEX OF PLANTS

529

Falsehellebore (*Veratum californicum*), 134, 172, 487
Falsemesquite (*Calliandra eriophylla*), 233
Fescue (*Festuca* spp.), 92
Fescue, hard (*Festuca ovina* var. *duriuscula*), 294
Fescue, Idaho (*Festuca idahoensis*), 223, 229, 294, 417
Fescue, meadow (*Festuca elatior*), 457
Fescue, sheep (*Festuca ovina*), 469
Fescue, sixweeks (*Vulpia octoflora,* formerly *Festuca octoflora*), 453
Fescue, tall (*Festuca arundinacea*), 262, 267, 294
Filaree (*Erodium* spp.), 83, 225, 430, 459
Fir (*Abies* spp.), 63, 229
Fir, Douglas (*Pseudotsuga menziesii*), 217, 219, 229
Fleabane, aspen (*Erigeron speciosus*), 455
Foxtail, creeping (*Alopecurus arundinaceous*), 295
Foxtail, meadow (*Alopecurus pratensis*), 267, 294
Gallberry, common (*Ilex glabra*), 172, 242
Galleta (*Hilaria jamesii*), 225, 226, 396, 469
Geranium (*Geranium* spp.), 75, 455
Gilia, slenderleaf (*Collomia linearis*), 455
Globemallow, scarlet (*Spaeralcea coccinea*), 456
Goldaster, hairy (*Heterotheca villosa,* formerly *Chrysopsis villosa*), 455
Goldenrod (*Solidago* spp.), 116
Goldenrod, larchleaf (*Haplopappus larcifolius*), 233
Goldenrod, rayless (*Haplopappus heterophyllus*), 172
Grama (*Bouteloua* spp.), 425, 439
Grama, black (*Bouteloua eriopoda*), 37, 225, 226, 235, 267, 294, 375, 450
Grama, blue (*Bouteloua gracilis*), 39–40, 44–45, 123, 227, 234, 235, 236, 255, 265, 267, 294, 375, 396, 414, 415, 416, 421, 423, 438, 440, 458, 467
Grama, Rothrock (*Bouteloua rothrockii*), 396
Grama, sideoats (*Bouteloua curtipendula*), 45, 123, 227, 236, 252, 255, 268, 294, 317, 328, 381, 382, 396, 416, 429, 438
Grama, sprucetop (*Bouteloua chondrosiodes*), 441
Greasewood (*Sarcobatus vermiculatus*), 95, 108, 172, 263
Grounsel, lambstongue (*Senecio integerrimus*), 223
Grounsel, threadleaf (*Senecio longilobus*), 172
Gumweed (*Grindelia squarrosa*), 116, 172
Hackberry (*Celtis* spp.), 77
Halogeton (*Halogeton glomeratus*), 32, 46–47, 54, 86, 134, 172, 311, 470, 482
Hardinggrass (*Phalaris tuberosa* var. *stenoptera*), 232, 294, 323–324, 417
Hardwoods (many species)
Hawksbeard, tapertip (*Crepis acuminata*), 223
Hopsage (*Grayia spinosa*), 453

SUBJECT INDEX

Boldface page entries denote topics covered in depth, including chapter and division headings. Livestock and game species are not indexed individually; refer to appropriate practices and treatments.

SUBJECT INDEX

Boldface page entries denote topics covered in depth, including chapter and division headings. Livestock and game species are not indexed individually; refer to appropriate practices and treatments.

542